INTEGRATED CHEMISTRY

A TWO-YEAR GENERAL AND ORGANIC CHEMISTRY SEQUENCE

VOLUME ONE, FIRST YEAR

PRELIMINARY EDITION

Timothy R. Rettich
David N. Bailey
Jeffrey A. Frick
Forrest J. Frank

Illinois Wesleyan University

Houghton Mifflin Company Boston New York

DEDICATIONS

TIMOTHY R. RETTICH:

To Barbara, and to all my students, especially Tony, for having taught me so much.

DAVID N. BAILEY:

To Sally and my students who make life worthwhile – even in the most difficult times.

FORREST J. FRANK:

To Dottie, Sharon, Nathan, and Brenden who are still teaching me about life.

JEFFREY A. FRICK:

To Phyllis and Abbey, for making each day brighter.

ACKNOWLEDGMENTS

The following students made significant contributions to this text, including artistic work, drawing of chemical structures, the production of spectra, and the layout of the text: Sarah Beth Anderson, Michael Davis, Sarah Grigsby, Laura Hornbeck, David Martel, Dustin Mergott, Kristel Monroe, Angie Nettleton, Scott Robowski, Ethan Schrum, and Sarah Studnicki.

Thanks to Mark Featherly for the photographs in this book.

The authors express their appreciation to the many chemistry teachers who have offered their insights and encouragement to this project. We are especially grateful to the members of our advisory committee: William Bordeaux, Mauri Ditzler, Graham Ellis, Pat Holt, Doris Kolb, Ken Kolb, and John Moore.

We are grateful to the following colleges and universities who adopted a new chemistry curriculum and served as trial sites as part of the development of this text: Bellarmine College, Huntington College, Illinois College, and Illinois Wesleyan University.

This project was supported, in part, by the National Science Foundation. Opinions expressed are those of the authors and not necessarily those of the Foundation. (Undergraduate Course and Curriculum Development Grant, DUE 9455718)

Printed in the U.S.A.

Library of Congress Catalog Card Number: 99-71921

ISBN: 0-395-98092-5

123456789–HS–03 02 01 00 99

What is "Integrated Chemistry"?

This text is designed as an introduction to college-level chemistry. It differs from other so called "general chemistry" texts in that it does not narrowly focus on inorganic and physical chemistry, but it does include those disciplines, as well as organic, analytical, and biochemistry. If this course is your only exposure to college chemistry, it is our belief that you will have a broader, more realistic appreciation of what chemistry is and what chemists do because of this approach.

Who Should Take Integrated Chemistry?

If you are interested in the biological sciences or allied health fields, it is our belief that you will find this text's early exposure to organic chemistry (not typically found in most first-year curricula) to benefit your study in those areas. The four-semester integrated sequence will prepare you for standardized tests by the American Chemical Society, and/or for the chemically related topics found in both the life science and physical science sections of the Medical Colleges Admissions Test.

If you are a chemistry major or minor, the four-semester sequence of integrated chemistry provides the same topic coverage as seen in older curricula, but with a different order. The connection between topics and over-arching themes are stressed more in this approach than in conventional curricula. Four semesters of integrated chemistry will prepare you for additional course work in chemistry. Since you will be exposed to all the branches of chemistry sooner using this integrated approach, it is our hope that you will find one or more areas of chemistry to be especially exciting. This will make your transition from a chemistry student to a chemical professional easier and more rewarding.

If you are interested in physics or engineering, and plan to have just a one-year chemistry course, this text will provide you with a basic understanding of a wide range of chemical materials, concepts, and terminology. This broader presentation of chemical topics should make for more and better connections between chemistry and engineering, applied physics, or material science.

Similarly, non-science majors with an interest in chemistry will find that integrated chemistry provides a solid introduction to a wider range of chemical topics than presented in most first-year college chemistry courses.

Some Suggestions:

If you have not had high school chemistry, or if you just wish to be better prepared for material upcoming in this course, an extra effort in studying the material in Chapter One will be especially valuable.

Note that no single textbook can ever contain all the material that is relevant to the study of the subject. Like other introductory texts, this textbook provides an overview;

it presents a limited number of specific examples, and develops theory from one (or occasionally two or three) perspectives. Additional worked-out examples, different approaches to (or levels of) theoretical concepts, more problems to test one's understanding, can be found in other chemistry texts. You should make use of your college library, in addition to this text, as you study.

Each chapter contains some "concept check" problems, that are worked out for you. It is best if, when you read the book, you first attempt your own solution to the problem before reading the solution.

There are problems at the end of each chapter. The answers to some of these problems are given in Appendix H. Again, it is advisable to try to work through to an answer on your own before checking the index. Learning to solve problems backwards (i.e., starting with the answer) may not translate into good problem solving skills on exams where, in general, the answers are not presented on the last page. There is a range of difficulty in these end of chapter problems, from simply following the paradigm laid out in a prior concept check to some problems designed to challenge the most capable students.

Most chapters also have "overview problems". These are typically multiple concept problems that incorporate some aspect of the material presented in the current chapter with material or concepts treated earlier in the course. These are not necessarily the hardest problems in the chapter, but they do require that you maintain a command of the important concepts from each chapter.

The study of chemistry is a cumulative effort. A strong foundation is required for effective future learning. It is our hope that this first year course in integrated chemistry will provide you with that basis.

Background for the Project

Collectively, the authors of this text have spent nearly 90 years teaching in the typical first year (general chemistry) and second year (organic chemistry) sequence. Over that time, our experience showed that capable and motivated students often found chemistry too abstract, too theoretical, and uninspiring. Students would likely finish a year of the typical curriculum with little or no idea of what most chemists actually do. Even those students who were sufficiently inspired to continue their studies in science beyond organic chemistry apparently made few of the vital connections between the traditional first and second year courses. Those reinforcing links, the parallels that were obvious to the teachers, were completely hidden from the students actually going through the process the first time. Our curriculum was tightly bound by disciplinary divisions: intermediate level inorganic and physical chemistry concepts had to be covered the first year (whether students were prepared to learn it or not); hybridization and molecular structures that were taught the first year were repeated the second year because the compounds contained carbon. We had a curriculum that, although commonly taught across most college campuses, satisfied neither the instructors nor the students.

In place of the traditional curriculum, we dreamed of a course that would bring all the major branches of chemistry into view starting the very first semester; a course that allowed a student to appreciate much of what practicing chemists actually do; a course that led students from observations to theories, not the reverse. We realized, however, that without the appropriate texts, our dream course would never materialize. Thanks to the support of the National Science Foundation and Houghton Mifflin Company, to the chemistry departments at Bellarmine College, Huntington College, Illinois College, Illinois Wesleyan University, and San Jose State University (who adopted working versions of these texts), to the encouragement of numerous other colleagues, and most of all, to our students, we have developed texts suited to an integrated two-year chemistry sequence.

Why Adopt an Integrated Curriculum

History is full of pedagogical vogues and fashions. The only ones that remain for long are those that are demonstrably better than their predecessors. After offering this curriculum at several trial sites for several years, we believe that this integrated curriculum has staying power. This is based upon the new curriculum surpassing the old curriculum as determined by a variety of measurements: student retention, student scores on standardized exams, student course evaluations, and feedback from faculty participating in trial studies. Faculty who would like to see more specific information should visit the Houghton Mifflin website, or contact the authors directly.

Using the Text:

As even a quick survey of the table of contents will reveal, the order and scope of topics in this text differs markedly from the usual first year "general chemistry" text. Much of the material in this text is cumulative, drawing extensively from earlier chapters. Altering the order of chapters, or eliminating an early chapter, will likely change the effectiveness of the presentation in the chapters that are covered.

Since some college chemistry students did not take (or remember nothing from) a high school chemistry course, the book begins in chapter one with an introduction to (or review of) some fundamental terminology used throughout the course. Depending upon the strength of the academic background of the average student, the instructor may wish to spend considerable time or little time in covering this material. This chapter stresses chemical terminology as a preparation towards understanding upcoming chemical concepts. It does not emphasize metric units, significant figures, or unit conversions. These are introduced on an "as needed" basis, with reference to material in the appendix.

Chapter Two uses the analytical chemistry of separation as an introduction to mixtures, compounds, and solutions. Chapters Three and Four introduce IR and NMR spectrometry as a means of understanding chemical structure and nomenclature. Some instructors may be used to thinking of spectroscopy as a topic suitable only for more advanced students. In fact, this introduction to the means by which "real world" chemistry is done has proven to be readily understood by first semester college freshmen of widely varying academic ability. If instrumentation is available for student use during laboratory, that is also a definite plus. Otherwise, appropriate computer software and web resources are available to make this material even more engaging to the student.

Chapters Five and Six introduce a wide variety of chemical reactions to the students. Again, as was the case with spectroscopy, the approach we take is first to let the students see what is readily observable (what reactants lead to what products) as a basis for building an understanding for later theoretical interpretation. These two chapters also incorporate some fundamentals of stoichiometric relationships and mole concepts.

Chapters Seven and Eight examine molecular structure in greater depth, and introduces VSEPR and isomers including enantiomers. Chapters Nine through Twelve are fairly conventional approaches to the topics of atomic structure, the periodic table, bonding theories, and acids and bases. The student's prior exposure to organic structure and nomenclature, however, allow for a greater range and depth of topic. For example, since students now readily recognize carboxylic acids and amines, they can far better understand their acidic and basic properties

Chapters Thirteen and Fourteen introduce resonance, aromaticity, and aromatic reactions. Both electrophilic and nucleophilic aromatic substitution mechanisms are covered in some detail. These are used as examples in the discussions in Book Two of the thermodynamics and kinetics of reactions, and should be covered in the first year if at all possible.

Chapters Fifteen and Sixteen are traditional studies of the gas, liquid, and solid phases. Again, student familiarity with both inorganic and organic materials facilitates a broader discussions of the intermolecular forces that explain the various distinctive physical properties examined. These two chapters are not crucial to the material developed in the second year course, and coverage here may be minimized at the instructor's discretion.

Volume One, First Year

Volume Two, Second Year

INTEGRATED CHEMISTRY

A Two-Year General and Organic Chemistry Sequence
Volume One: First Year

CHAPTER FOURTEEN: AROMATIC REACTIONS

CHAPTER FIFTEEN: COMPLETE CHAOS AND IDEAL BEHAVIOR 403

CHAPTER SIXTEEN: LIQUIDS AND SOLIDS 427

APPENDICES

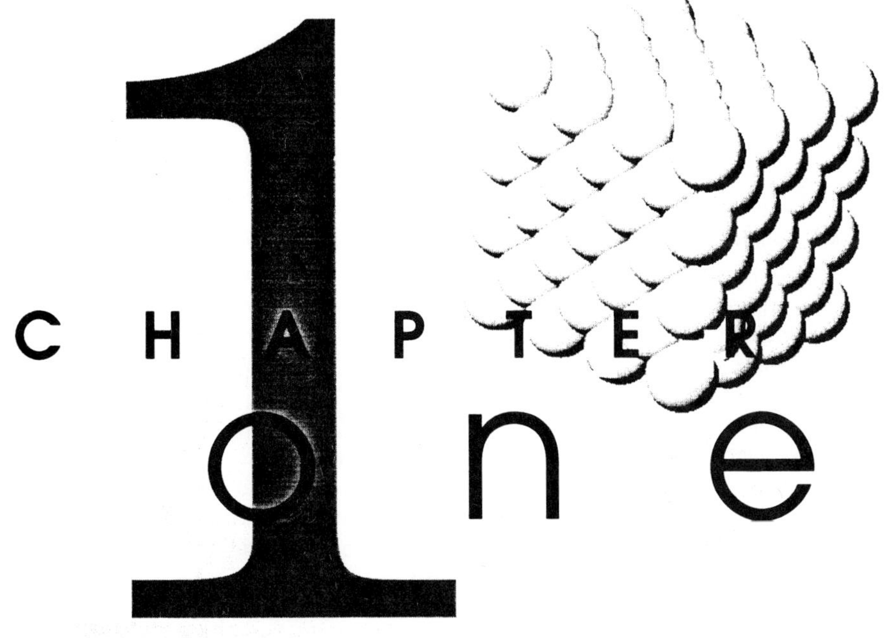

SCIENCE AND Structure

SCIENCE AND Structure

Paper or Plastic?

Questions. Choices. Life is full of both. A trip to the mall, convenience store, or supermarket quickly proves the point. Are you paying with cash (paper) or credit (plastic)? Alcohol in your gasoline or not? Your beverage with or without caffeine, naturally or artificially sweetened, and in a metal can, glass or plastic bottle? And finally, after making all these choices, another decision about the bag to contain your purchases: paper or plastic?

On what basis do we answer such questions and make such choices? Frequently cost and convenience are the prime concerns. Sometimes, however, the true cost of a choice does not appear when its bar code is run across the scanner. Our everyday choices may involve using a type of credit, with the balance charged to the environment. Such bills often appear years or decades overdue, with considerable interest charges: acid rain, overflowing landfills, ozone holes, etc. How does the cost to fix the problem once it becomes intolerable compare to the costs of avoiding the problem in the first place? For example, would it make a significant difference if the product we purchased was labeled as "environmentally friendly, chlorofluorocarbon free?" And do decisions made by an individual on a planet of over 5 billion decision makers really matter?

Questions that combine issues of politics, economics, philosophy, and natural science are going to be complex. Even seemingly simple questions posed by the public, such as: "Is this water safe to drink?" have complex answers. What distinguishes science from other intellectual endeavors is how science goes about posing such questions, determining and interpreting answers based upon experimentation, and rephrasing the questions. This occurs in a cyclic process known as the scientific method. The best scientific answers to such complex questions are seldom the "yes" or "no" that society in general wants. Often the only valid scientific answer is a blur of averages and statistics placed inside a maze of assumptions and qualifying statements. No wonder that a senator, after sitting through the testimony of numerous expert witnesses testifying to a complex issue, supposedly wished aloud for "a one-handed scientist"; that is, someone who would not follow every statement with the provision "On the other hand..."

It is human nature to be both curious and impatient; to want definite and final answers to questions. It is the nature of the scientific method, however, to continually check for falsifiability, to remain skeptical of one's own theories via continually refined experiments. The word "science" is etymologically linked to "knowledge," but is also related to words meaning "to separate, to cut." Our word "scissors" also comes from this root. It may be useful to relate the two blades of the scissors to the theory and experiment of the scientific method. The correct functioning of the whole is possible only when the separate parts act in opposition. Experimentation without theory is simply a pile of useless data. An hypothesis untested by observation and experiment can never grow into a theory; it is not capable of responding to new information. It is the counterbalanced, self-correcting nature of the opposing forces of theory and experiment that keep science zig-zagging towards knowledge. Science is not a closed book of established facts. It is an ongoing, dynamic process where yesterday's theory must wrestle today's experiment to see what will be considered true tomorrow. Welcome to the fray.

1.2 CHEMISTRY: THE SCIENCE OF STRUCTURE

The forerunners of modern science, natural philosophers, long questioned the structure of matter. One question that obsessed them was: "If you keep subdividing matter, do you ever reach an end? If so, what is it?" The original hypothesis of an ultimate particle of matter is traditionally assigned to Leucippus and his student Democritus (460-370 BC). The atomism of Democritus held that the universe was created of two things: the **atoms** and the void. The void, or empty space, was infinite in extent. The atoms were the indivisible, smallest particle of a substance. The Greek word "atomos" literally means "not cuttable."

FIGURE 1.1: Representation of atoms in solid lead.

Later in this book we will trace the development of the atomic theory from its philosophical origin to current theory. The history of how scientists have thought about the atom is an example of the zig-zag approach to truth that science often takes. Initially a scientist devises a model to describe what is unknown by reference to that which is readily observed and already known. As more information is obtained, the scientist must refine or discard the initial model. For now, we will focus on a model of matter derived from a simple interpretation of atoms as marbles or BBs, only much, much smaller.

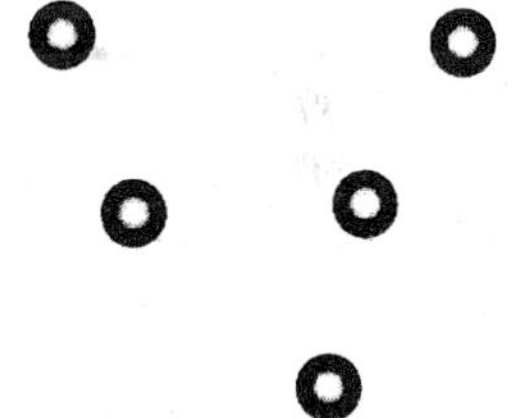

FIGURE 1.2: Representation of atoms in gaseous helium.

Figure 1.1 shows a schematic structure of atoms of the metal lead. Figure 1.2 shows a schematic structure of atoms in the gas helium. These models show the atoms and the void proposed by Democritus, and they match our mental image of BBs in a box. The solid lead, however, has the atoms packed tightly together in an orderly fashion. The gas helium has the atoms spread out and randomly distributed. The question remains: "Do atoms really look like this?" The answer is that we do not know. No one has ever directly seen the atomic level. Scientists have, however, built instruments including the scanning tunneling microscope and the atomic force microscope which can produce images that seem to agree with this basic model of matter.

We can explain some of the properties of matter based upon even these simple models. **Density** is a physical property. It is defined as follows:

$$\textbf{Density} = \frac{\text{mass}}{\text{volume}}$$

Atoms have mass, and the void does not. As the ratio of mass to volume, density is related to the number of atoms per unit volume (e.g., per cubic centimeter). For solid lead at typical room conditions, this number is about 3.3×10^{22} atoms per cubic centimeter. For the gas helium at typical room conditions, this number is about 2.5×10^{19} atoms per cubic centimeter. This means that a sample of solid lead has over one thousand times the number of atoms per cubic centimeter than a sample of gaseous helium. Helium, with far fewer atoms per volume, is thus much less dense than lead. One consequence of this is that we are not surprised to see a helium balloon float in the air, as opposed to the behavior of the proverbial lead balloon.

The large amount of void in a sample of helium, and the small amount of void in lead, also results in differences in other properties. For example, atoms transfer heat by bumping into one another. A sample of lead, with atoms very close together, conducts heat much more quickly than does a sample of gas, where atoms bump into each other far less frequently.

Concept Check: Density

If one knows any two of the terms (mass, volume, density) the third is calculated from the other two. What is the mass of 1.00 cubic foot (2.83×10^4 cm^3) of gold, if gold has a density of 18.88 g cm^{-3}?

Solution: mass = (density) (volume) = (18.88 g cm^{-3}) (2.83×10^4 cm^3) = 5.34×10^5 g (nearly 1,200 pounds!)

1.3 A QUICK TOUR INSIDE THE ATOM

Although atoms originally received that name as a result of the theory that held that they were the indivisible basis of matter, we currently know that a wide variety of sub-atomic particles exists within the atom. The **nucleus** of the atom contains positively charged **protons**. Each proton has a +1 charge and a mass of approximately 1 atomic mass unit (amu). Also in the nucleus are **neutrons**. Each neutron has a mass of approximately 1 amu, but no charge. Outside the nucleus are **electrons**, each with a -1 charge and very little mass. For an atom, which has no net charge, this means an equal number of protons and electrons.

All atoms of an **element** may not be exactly identical. Hydrogen, for example, always has 1 proton in its nucleus. The number of protons in the nucleus is called the **atomic number**, and this value determines the identity of the element. Atomic number 1 means hydrogen (H), atomic number 2 means helium (He), and so on. The sum of the number of protons and neutrons in the nucleus is known as the **mass number**. A helium nucleus with 2 protons and one neutron has a mass number of 3. This is symbolized as $^{3}_{2}He$. In general, the **isotope** or nuclide (the specific atomic nucleus) is designated as $^{A}_{Z}X$, where X is the symbol for the element, Z is the atomic number, and A is the mass number. Sometimes the atomic number is omitted, since it is redundant given the elemental symbol. Thus, instead of being written $^{14}_{6}C$ (indicating 6 protons and 8 neutrons), it is occasionally written as ^{14}C, since carbon, by definition, must have 6 protons in its nucleus. $^{14}_{6}C$ is pronounced "C-fourteen," and indicates the particular isotope of carbon being discussed. When an element is symbolized without mass number superscript, the specific isotope used is unimportant, and a statistical mixture of isotopes that occur naturally is assumed. We will investigate some consequences of this in more detail later.

Concept Check: Nuclide Symbols

Meitnerium (Mt) is the element most recently officially named by IUPAC (the international body with jurisdiction over such things.) Meitnerium also has the most protons of any of the named elements, with 109. Write the nuclide symbol of Meitnerium if it has 142 neutrons.

Solution: *Since this isotope has 109 protons and 142 neutrons, the mass number is computed as the sum: 109 + 142 = 251, which is used as a superscript. The number of protons is the atomic number, which is used as a subscript. The answer is*

$$^{251}_{109}Mt$$

When the number of protons in the nucleus of the atom equals the number of electrons around the nucleus, the species is a neutral atom. If the number of electrons differs from the number of protons, the species has a net charge and is called an ion. If there are more electrons than protons, there is a net negative charge and the species is called an **anion**. If there are fewer electrons than protons, there is a net positive charge and the species is called a **cation**. The charge on a species is shown in the upper right hand corner. A hydrogen atom has one proton and one electron. Loss of that one electron results in a positive species indicated H^+, commonly known as the hydrogen ion. A different ion of hydrogen could form if the neutral hydrogen atom gains an electron, forming the anion H^-, known as hydride. A chlorine atom has seventeen protons and seventeen electrons. If a chlorine atom gains one extra electron, the species is an anion symbolized Cl^-, and it is named chloride. Note that multiple charges are possible. A magnesium atom that has lost two electrons is symbolized by Mg^{2+}, named the magnesium ion. Note that the number precedes the positive sign. An oxide anion is formed when an oxygen atom gains two electrons. This is symbolized by O^{2-}.

1.4 FORCES OF ATTRACTION BETWEEN ATOMS, MOLECULES

As we shall discuss later in some detail, there are different kinds of forces between atoms. Some strong forces, called bonds, hold atoms together to form **molecules**. Molecules are stable combinations of atoms. For example, an atom of hydrogen is symbolized H; an atom of oxygen is symbolized O; a molecule of water is symbolized H_2O. This formula indicates that two atoms of hydrogen and one atom of oxygen stick together in an entity called a water molecule.

Other forces between separate atoms or molecules are weaker, but still important. These forces depend upon the distance separating the atoms or molecules. Unless molecules are literally smashed together (at which point repulsive forces predominate,) the closer molecules are together, the stronger the force of attraction between them. As the distance of separation increases, the force of attraction decreases. Atoms or molecules in the gas phase, which already have relatively large distances of separation, will move more or less independently of one another. For example a sample of helium gas introduced into an empty chamber will expand to occupy the entire chamber. It does this because the random motion of the molecules can not be held in check by the weak forces of attraction between the widely separated molecules. A sample of a liquid or solid will not expand in this fashion, but keeps its own volume. The molecules in a liquid or a solid are still moving, but their proximity to other molecules (and the resulting forces of attraction) keeps them from expanding like a gas. The liquid and solid phases are referred to as being "condensed phases" for this reason. The liquid phase and gas phase are both referred to as being "fluid"; that is, liquids and gases both assume the shape of the container (although only the gas will assume the volume of the container as well).

When the forces between the separate atoms or molecules are strong, a solid results, which has its own definite shape as well as volume. Very strong forces between atoms of carbon, for example, can result in diamond, one of the hardest materials known. Because of these forces, it is difficult to move an atom of carbon in diamond relative to the other carbon atoms. This makes the material hard.

1.5 ELEMENTS, MOLECULES, COMPOUNDS AND THE PERIODIC TABLE

Hydrogen, oxygen, carbon, lead and helium are examples of elements. Elements can exist as individual atoms like those seen in the previous models of helium and lead, or in combinations of atoms called molecules. Hydrogen and oxygen exist as diatomic molecules, symbolized by H_2 and O_2. Molecules of phosphorus and sulfur are symbolized as P_4 and S_8. These molecules are shown in Figure 1.3. Elemental molecules have only one type of atom, but depending on the element in question those molecules may contain one, several, or many atoms of that element.

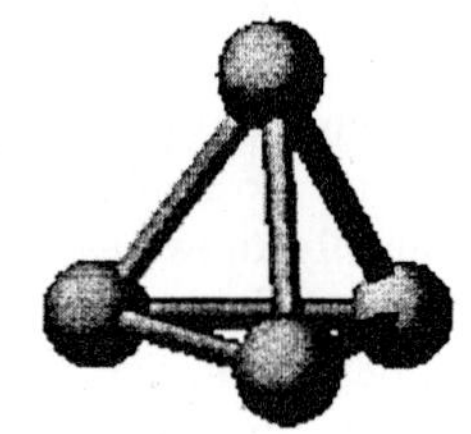

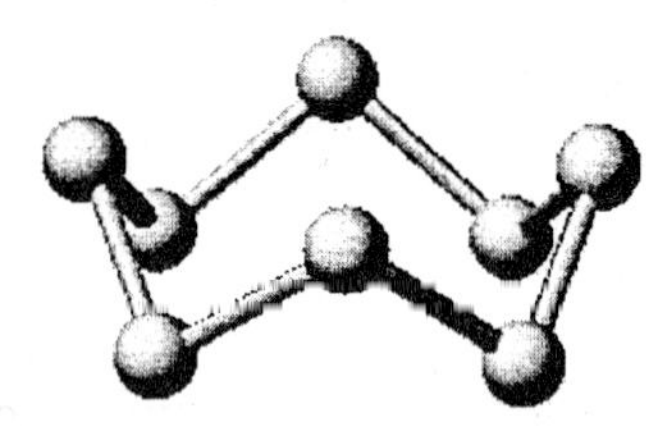

FIGURE 1.3: Models of P_4 and S_8.

Elements are distinguished from species like water, H_2O, and table salt, NaCl, which are called **compounds**. There are just over one hundred chemical elements, but the numerous ways of combining these elements result in literally millions of compounds. Later in the text we will examine the meaning of "compounds" in greater detail. Now we simply note that compounds, like elements, have specific properties that allow them to be classified. We have already seen one way of categorizing chemical species by physical properties, calling them solid, liquid or gas. Elements and compounds can be classified on this basis: at room temperature helium and methane (CH_4) are both gases; bromine and water are both liquids; carbon and table salt (NaCl) are both solids.

Elements have an additional, unique classification scheme, illustrated by the periodic table. As shown in Figure 1.4, the table is organized in rows (called periods) and columns (called groups or families). Later in the book we will examine the organization of the table in greater detail. For now, we simply note that position in the table is related to physical and chemical properties. That columns in the table have been called "families" is due to elements in that family sharing distinguishing characteristics. The periodic table shown in Figure 1.4 is divided diagonally. Elements below and to the left of the diagonal are described as **met-**

als, elements above and to the right are described as **non-metals**, and certain elements adjacent to the diagonal are described as **metalloids**.

The Periodic Table

IA	IIA											IIIA	IVA	VA	VIA	VIIA	VIIIA
1 H																	2 He
3 Li	4 Be											5 B	6 C	7 N	8 O	9 F	10 Ne
11 Na	12 Mg											13 Al	14 Si	15 P	16 S	17 Cl	18 Ar
19 K	20 Ca	21 Sc	22 Ti	23 V	24 Cr	25 Mn	26 Fe	27 Co	28 Ni	29 Cu	30 Zn	31 Ga	32 Ge	33 As	34 Se	35 Br	36 Kr
37 Rb	38 Sr	39 Y	40 Zr	41 Nb	42 Mo	43 Tc	44 Ru	45 Rh	46 Pd	47 Ag	48 Cd	49 In	50 Sn	51 Sb	52 Te	53 I	54 Xe
55 Cs	56 Ba	57 La	72 Hf	73 Ta	74 W	75 Re	76 Os	77 Ir	78 Pt	79 Au	80 Hg	81 Tl	82 Pb	83 Bi	84 Po	85 At	86 Rn
87 Fr	88 Ra	89 Ac	104 Rf	105 Ha	106 Sg	107 Ns	108 Hs	109 Mt									

58 Ce	59 Pr	60 Nd	61 Pm	62 Sm	63 Eu	64 Gd	65 Tb	66 Dy	67 Ho	68 Er	69 Tm	70 Yb	71 Lu
90 Th	91 Pa	92 U	93 Np	94 Pu	95 Am	96 Cm	97 Bk	98 Cf	99 Es	100 Fm	101 Md	102 No	103 Lr

FIGURE 1.4:
The Periodic Table

1.6 METALS, NON-METALS, AND METALLOIDS

Nearly 70% of the known elements are termed metals. Nearly all the metals and metalloids are solids; the few exceptions are liquids. Over half of the non-metallic elements, however, are gases. The distinctions between the properties of metals and non-metals are numerous, and will provide important chemical insights.

Metals are good conductors of heat and electricity. They are lustrous (shiny), malleable (can be hammered into sheets), and ductile (can be drawn into wires). Metallic character increases as one moves down a group or to the left in a period. Non-metals are generally insulators, as opposed to conductors. Non-metals vary widely in terms of appearance, but as one goes down a group, the characteristics more closely approximate those of metals. The halogens (Group VII) for example, range from yellow-green gases for fluorine and chlorine, to a dark orange liquid for bromine, to a shiny silver-grey solid for iodine. Non-metallic solids tend to be neither malleable nor ductile, and frequently display brittleness.

The chemical properties of metals and non-metals are likewise distinctive. Oxygen, a non-metal, forms compounds with nearly every element, both metals and non-metals. But the properties of these oxygen-containing compounds, called oxides, are very different. One key feature has to do with acid/base properties. We will investigate acids and bases in more detail later. For now, we simply note that chemical indicators can have different colors when in acidic or basic solution. For example, litmus is pink in acid and blue in base. Sodium oxide (Na_2O) is a metal oxide that dissolves in water to form an aqueous solution that turns litmus blue; sodium oxide in particular and metal oxides in general are bases. Sulfur dioxide (SO_2) is a non-metal oxide that dissolves in water to form an aqueous solution that turns litmus pink; sulfur dioxide in particular and non-metal oxides in general are acids. Sulfur dioxide in the atmosphere plays an important role in the production of acid rain.

The metalloid elements are boron, silicon, germanium, arsenic, antimony and tellurium. Their physical properties are intermediate between metals and non-metals. For example, instead of being conductors like metals, or insulators like non-metals, these elements can act as semi-conductors. Silicon, and to a lesser extent germanium and arsenic have had their semi-conducting properties exploited in the development of advanced electrical components. Likewise the chemical properties of these elements are intermediate between metals and non-metals. For example, the oxides of some of the metalloids are amphoteric. This means they can function as either a weak acid or as a weak base. We will examine more periodic trends of the elements later in the text.

1.7 ELECTRONS, IONS AND OXIDATION NUMBERS

Perhaps the most significant difference in terms of the chemical properties of metals and non-metals relates to the electrons around the atoms of those elements. Electrons are negatively charged sub-atomic particles. Metals tend to lose their electrons easily, non-metals tend to hold on to their electrons. The amount of energy required to remove an electron from an atom is called its **ionization potential**. Metals have lower ionization potentials than non-metals, and thus electrons may be easily removed from metals. When a neutral metal atom (having no charge) loses an electron, the atom becomes a cation (a positive ion.)

A related concept is **electron affinity**, which is measured by the energy released when a neutral atom captures an electron. Non-metals have much higher electron affinities than metals. Non-metal atoms (except those in Group VIII, the inert gases) love to get an extra electron; neutral metal atoms are not eager to add to their collection of electrons. Since non-metals have both high ionization potentials (they do not like to lose electrons) and high electron affinities (they like to gain electrons,) neutral non-metal atoms frequently end up with extra electrons, and thus become anions.

When an atom loses full or partial control over one or more of its electrons, it is said to have a positive **oxidation number**, and thus resemble a cation. When an atom gains full or partial control over one or more electrons in addition to those of its neutral atom, it is said to have a negative oxidation number, and thus resemble an anion. When combined with other elements, metals almost without exception have positive oxidation numbers. When a compound is made from a metal and a non-metal, the metal will always have a positive oxidation number and the non-metal will always have a negative oxidation state. Sodium chloride (NaCl) for example has metallic sodium with a +1 oxidation number and the non-metal chlorine with a -1 oxidation number. In Figure 1.5, the total transfer of an electron from a neutral sodium atom to a neutral chlorine atom is pictured. The result is a positive sodium ion, Na^+, and a negative chloride ion, Cl^-. In some compounds, however, it is more appropriate to imagine a partial transfer of control of electrons, as opposed to forming formal cationic and anionic species.

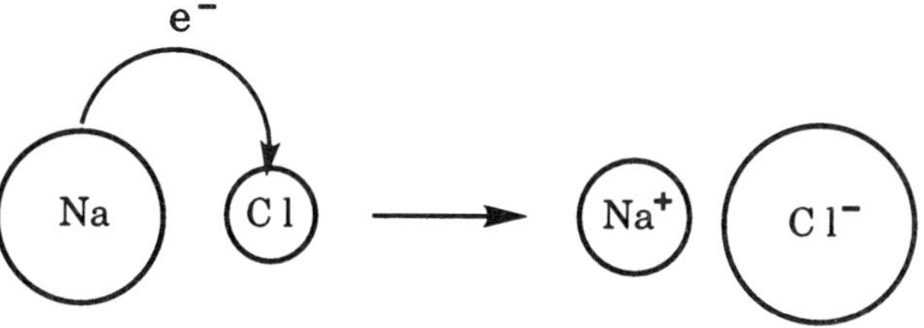

FIGURE 1.5: Electron transfer in making an ionic compoind.

Carbon monoxide (CO), for example, is described as having carbon with a formal positive oxidation number of +2 and oxygen with a formal oxidation number of -2. We will discuss ways of determining oxidation numbers later in the text. For our current purposes, we need only to understand that CO is not a combination of carbon cations and oxygen anions, but instead is a molecule made by sharing (albeit unequally) the electrons between the carbon and oxygen atoms. The degree to which an atom tries to control the electrons shared with another atom is termed **electronegativity**. In Figure 1.6, we see a diagonal trend, as electronegativity increases going up and to the right. Note that this parallels the trend of metal versus non-metal character. The most metallic elements are in the lower left corner of the periodic table; these metals have low electronegativity. The least metallic elements are in the upper right corner. These non-metals have a high electronegativity. In our example compound of carbon monoxide, oxygen and carbon are both non-metals, but oxygen is farther to the right in the period. Oxygen has a greater electronegativity than carbon, has greater control over the shared electrons, and thus is said to have a negative oxidation number. As we see in the next section differing values of electronegativity are crucial in explaining how atoms bond.

increasing electronegativity →

s^1	s^2	d^1	d^2	d^3	d^4	d^5	d^6	d^7	d^8	d^9	d^{10}	s^2p^1	s^2p^2	s^2p^3	s^2p^4	s^2p^5	s^2p^6
1 H	2 He																
3 Li	4 Be											5 B	6 C	7 N	8 O	9 F	10 Ne
11 Na	12 Mg											13 Al	14 Si	15 P	16 S	17 Cl	18 Ar
19 K	20 Ca	21 Sc	22 Ti	23 V	24 Cr	25 Mn	26 Fe	27 Co	28 Ni	29 Cu	30 Zn	31 Ga	32 Ge	33 As	34 Se	35 Br	36 Kr
37 Rb	38 Sr	39 Y	40 Zr	41 Nb	42 Mo	43 Tc	44 Ru	45 Rh	46 Pd	47 Ag	48 Cd	49 In	50 Sn	51 Sb	52 Te	53 I	54 Xe
55 Cs	56 Ba	57 La	72 Hf	73 Ta	74 W	75 Re	76 Os	77 Ir	78 Pt	79 Au	80 Hg	81 Tl	82 Pb	83 Bi	84 Po	85 At	86 Rn
87 Fr	88 Ra	89 Ac	104 Rf	105 Ha	106 Sg	107 Ns	108 Hs	109 Mt									

58 Ce	59 Pr	60 Nd	61 Pm	62 Sm	63 Eu	64 Gd	65 Tb	66 Dy	67 Ho	68 Er	69 Tm	70 Yb	71 Lu
90 Th	91 Pa	92 U	93 Np	94 Pu	95 Am	96 Cm	97 Bk	98 Cf	99 Es	100 Fm	101 Md	102 No	103 Lr

FIGURE 1.6: Periodic trends in electronegativity.

1.8 IONIC AND COVALENT BONDING

Previously we have seen that metals have low ionization potentials, low electronegativities, and low electron affinities. Conversely, non-metals have high ionization potentials, electronegativities and electron affinities. We use this information as the basis for a brief examination of two extreme types of bonding. One is covalent bonding, the other is ionic bonding.

Covalent bonds form between elements of similar electronegativity. Since carbon and oxygen, both non-metals, have similar electronegativities, neither is strong enough to wrest complete control of the electrons from the other. Instead, they are forced into an uneasy truce, a working arrangement. They share the electronic wealth, but oxygen (being slightly stronger) gets a slightly bigger share. The result is discrete molecules of carbon monoxide, each containing one atom of carbon and one atom of oxygen. At room conditions, the attractive forces between molecules of carbon monoxide are weak. As a result, carbon monoxide is a gas.

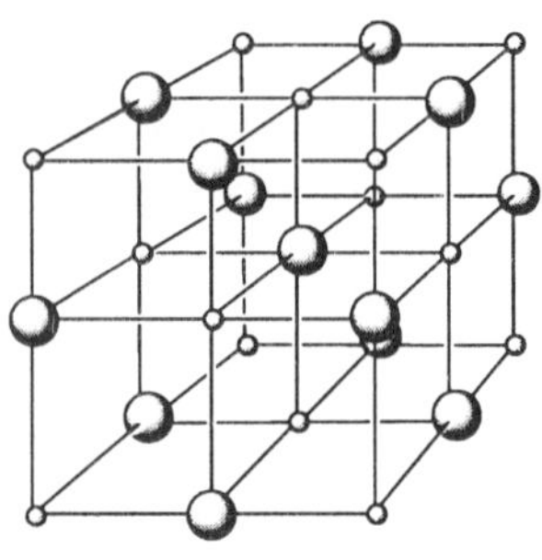

FIGURE 1.7: Model of NaCl.

The distribution of the electronic wealth in the compound sodium chloride (NaCl) and the resulting physical properties are quite different from carbon monoxide. As seen in Figure 1.6, the metal sodium has a much lower electronegativity than the non-metal chlorine. In a case of wrestling for control of the electrons, it is no contest: chlorine wins, and winner takes all (i.e., complete control of the contested electron.) The low ionization potential of sodium means that it is ready to give up an electron. The high electron affinity of chlorine means that it is ready to take an electron. The bonding in sodium chloride is described as **ionic bonding**, since the compound is composed of sodium cations and chloride anions.

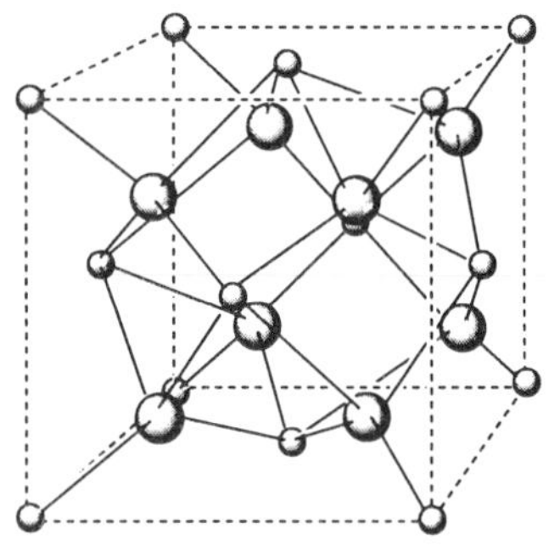

FIGURE 1.8: Model of CaF_2.

From earlier physics courses, you may have learned that electrical opposites attract. The force responsible for this is known as **coulombic attraction**. That is the major force of interaction between ions of sodium and chlorine. A model of solid sodium chloride is shown in Figure 1.7. Note that each sodium cation has 6 chloride ions surrounding it, and that each chloride ion has 6 sodium ions surrounding it. This regular arrangement, known as a lattice, extends in all 3 dimensions. Within a single visible crystal of salt, there are enormous numbers of sodium ions and chloride ions. There is no set number of sodium or chloride ions in a crystal; that varies with the size of the crystal. The sodium and chloride ions are simply present in a one to one ratio, hence the formula NaCl. Calcium fluoride (CaF_2) shown in Figure 1.8, similarly has an enormous number of ions in a crystal, with the exact number depending on the size of the crystal. However, for every calcium ion (Ca^{2+}), there are two fluoride ions (F^-) in order to maintain electrical neutrality.

CONCEPT CHECK: CHEMICAL PROPERTIES AND PERIODIC TRENDS

Calcium oxide (CaO) is known as lime or calx. It is commercially important in the production of bricks, mortar, plaster, and steel. Based on the periodic table and the formula, do you think calcium oxide is ionic or covalent? Does CaO exists as individual molecules or as an extended lattice? Is an aqueous solution of CaO acidic or basic?

Solution: CaO is a compound of calcium, a Group II (alkaline earth) metal, and oxygen, a non-metal. Compounds of metals and non-metals are expected to be ionic. Thus solid calcium oxide exists in an extended lattice of Ca^{2+} cations and O^{2-} anions. Metal oxides are expected to form basic solutions when dissolved in water. Note that calcium oxide reacts with water to form $Ca(OH)_2$, known as slaked lime, a strong base.

1.9 ELECTRON CONFIGURATION

Electrons play a crucial role in chemistry. Chemists have devised a classification scheme for electrons in an atom. Electrons are often referred to in terms of the energy level they occupy. These energy levels are designated by a code consisting of a number followed by a letter. The numbers are integers, beginning with one. The letters are typically s, p, d and f. The numbers represent the period (also called the shell) in which the electron is found. The first period is assigned the number 1, the second period 2, and so on. It is sometimes convenient (though an oversimplification) to imagine these as relating to circular orbits around the nucleus (Figure 1.9.) The farther the orbit from the nucleus, the higher the energy level and the larger the number designating the orbital.

The letters s, p, d and f refer to the shape of the electron distribution in space. Figure 1.10 shows the spherically symmetric s orbitals. The image on the left is imagined as an instantaneous photographic exposure repeated over time, showing the relative probability of finding the individual electron at various locations around the nucleus. The image on the right is a contour surface, containing the electron with a specified probability, e.g., 95 or 99% of the time. It is sometimes thought of as an electron cloud, with the electron spread throughout the indicated region. Neither verbal explanation is probably correct, but the picture of the electron distribution in space is commonly accepted. A 2s orbital is spherically symmetric like the 1s orbital, but the 2s orbital is bigger. An electron in this energy level on average is farther out from the nucleus than an electron in the 1s energy level.

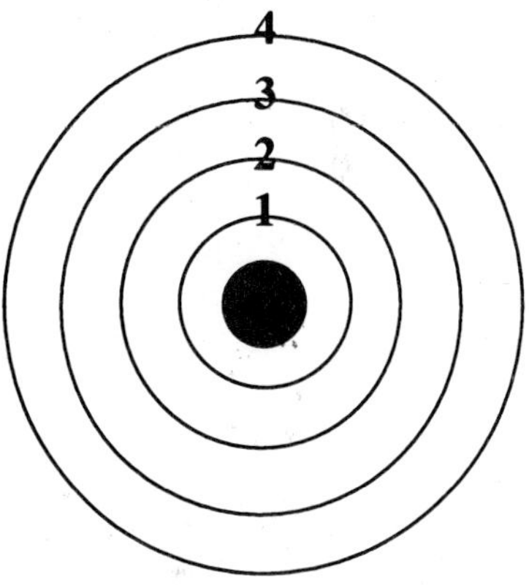

FIGURE 1.9: Model of electron orbits/energy levels in an atom.

The shapes of the p and d orbitals are shown in Figures 1.11 and 1.12. Note that while any s energy level is a single orbital, any p energy level has three distinct orbitals. These dumbbell-shaped orbitals are oriented along the x, y and z axes, and thus are designated as p_x, p_y and p_z. The p_x, p_y and p_z orbitals are of equal energy. A d energy level has five distinct orbitals of equal energy, designated d_{xy}, d_{xz}, d_{yz}, $d_{x^2-y^2}$ and d_{z^2}. The f energy level has seven distinct orbitals of equal energy, but these are not of particular interest in this course.

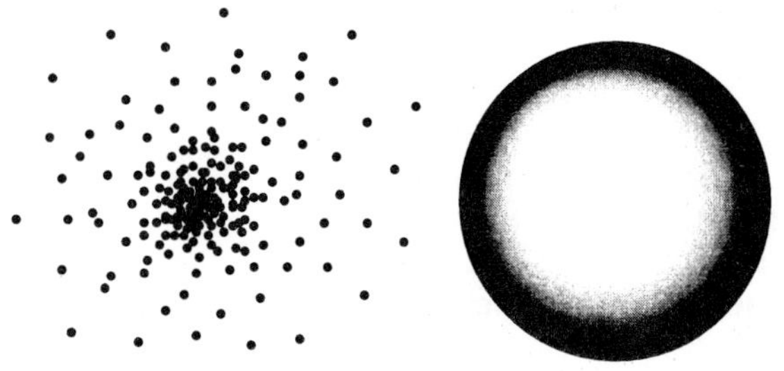

FIGURE 1.10: Spatial distribution of electrons in a 1s orbital.

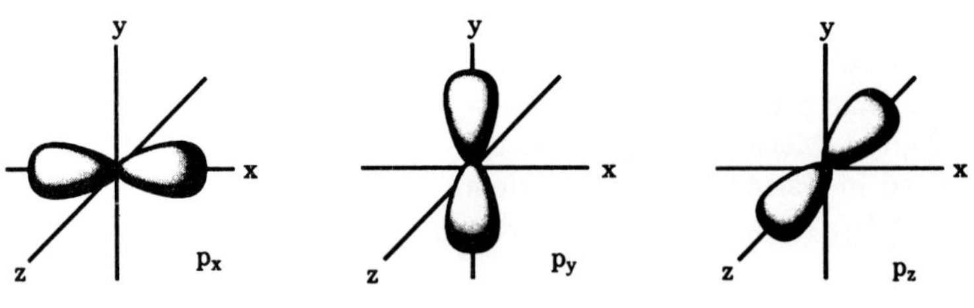

FIGURE 1.11:
Spatial distribution of electrons in p orbitals.

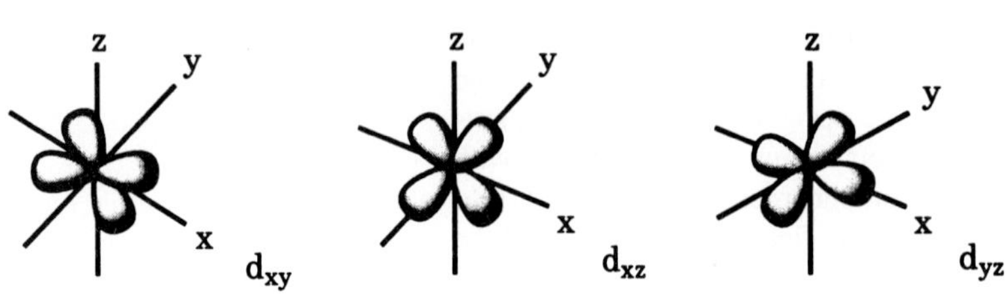

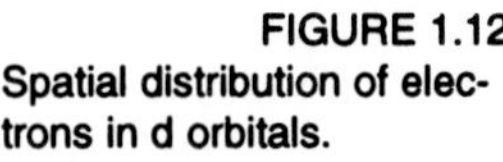

FIGURE 1.12:
Spatial distribution of electrons in d orbitals.

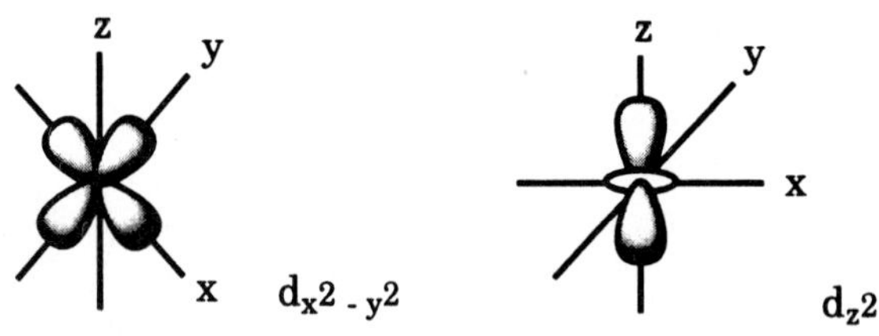

An s energy level has one orbital, a p energy level three orbitals, a d energy level five orbitals, and an f energy level seven orbitals. This is sometimes pictured as shown in Figure 1.13. Note that each of these dashes represents an orbital. Each of these orbitals can contain a maximum of two electrons.

The first period contains only an s orbital, which is designated 1s. There can be zero, one or two electrons in this energy level, designated as $1s^0$, $1s^1$ or $1s^2$, respectively. The second period contains s and p orbitals. These would be designated as 2s and 2p. If this level were completely occupied by electrons, we would designate this by showing $2s^2\ 2p_x2\ 2p_y2\ 2p_z2$, or more commonly, simply $2s^2\ 2p^6$. The third period contains s, p and d orbitals. The fourth period contains s, p, d and f orbitals. The fifth period would add a g energy level, and so on. In common chemical practice, we are concerned with the first seven periods, and with orbitals s, p, d and f. The lowest energy level is the 1s, followed by 2s, 2p, 3s, 3p, 4s, 3d, 4p and so on. The ordering can be remembered diagramatically as shown in Figure 1.14.

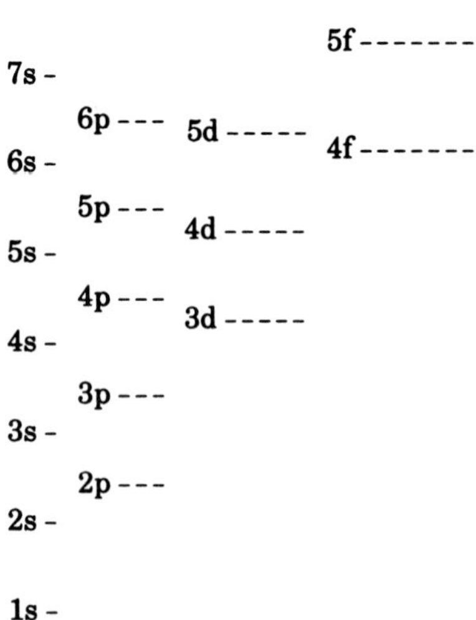

FIGURE 1.13:
Dash representation of electron energy levels.

An alternate to the mnemonic represented by the diagram in Figure 1.14 is to understand the structure of the periodic table as represented in Figure 1.15. Note that the first two columns on the left represent the "s block" electrons, which begin at the 1s level. The six columns on the right of the periodic table represent the "p block" electrons, which begin at the 2p level. Between these extremes we see the ten column wide "d block" of transition metals, which begin at the 3 d level. Finally, the fourteen column wide "f block" of the inner transition elements are tucked in at the bottom of the periodic table. These begin at the 4 f level. If you are already comfortable with both the period and orbital designation as a function of position on the periodic table, the table itself can be used as a reminder of orbital energy levels; that is, 1s is before 2s is before 2p is before 3s is before 3p is before 4s is before 3d, and so on. If you are not yet comfortable with period and orbital designation on the periodic table, Figure 1.14 may prove easier for now.

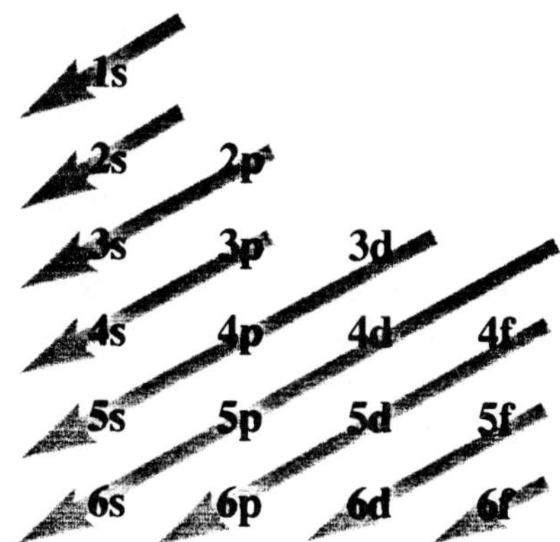

FIGURE 1.14: Mnemonic for order of electron orbitals.

An atom will have a number of electrons equal to its atomic number. Argon, for example, has atomic number 18. The 18 protons require 18 electrons to

s^1	s^2	d^1	d^2	d^3	d^4	d^5	d^6	d^7	d^8	d^9	d^{10}	s^2p^1	s^2p^2	s^2p^3	s^2p^4	s^2p^5	s^2p^6
1 H	2 He																
3 Li	4 Be											5 B	6 C	7 N	8 O	9 F	10 Ne
11 Na	12 Mg											13 Al	14 Si	15 P	16 S	17 Cl	18 Ar
19 K	20 Ca	21 Sc	22 Ti	23 V	24 Cr	25 Mn	26 Fe	27 Co	28 Ni	29 Cu	30 Zn	31 Ga	32 Ge	33 As	34 Se	35 Br	36 Kr
37 Rb	38 Sr	39 Y	40 Zr	41 Nb	42 Mo	43 Tc	44 Ru	45 Rh	46 Pd	47 Ag	48 Cd	49 In	50 Sn	51 Sb	52 Te	53 I	54 Xe
55 Cs	56 Ba	57 La	72 Hf	73 Ta	74 W	75 Re	76 Os	77 Ir	78 Pt	79 Au	80 Hg	81 Tl	82 Pb	83 Bi	84 Po	85 At	86 Rn
87 Fr	88 Ra	89 Ac	104 Rf	105 Ha	106 Sg	107 Ns	108 Hs	109 Mt									

58 Ce	59 Pr	60 Nd	61 Pm	62 Sm	63 Eu	64 Gd	65 Tb	66 Dy	67 Ho	68 Er	69 Tm	70 Yb	71 Lu
90 Th	91 Pa	92 U	93 Np	94 Pu	95 Am	96 Cm	97 Bk	98 Cf	99 Es	100 Fm	101 Md	102 No	103 Lr

FIGURE 1.15: The Periodic Table with electron configurations.

make a neutral argon atom. The **electron configuration** for argon is: $1s^2\ 2s^2\ 2p^6\ 3s^2\ 3p^6$. Note that the total of the superscripts is 18, the number of electrons. Krypton, element 36, has 36 electrons in the neutral atom. The electron configuration for krypton is $1s^2\ 2s^2\ 2p^6\ 3s^2\ 3p^6\ 4s^2\ 3d^{10}\ 4p^6$. The electron configurations for all the elements are listed in the appendix.

The carbon atom has 6 total electrons. Two of these electrons are in the first shell ($1s^2$ and four of the electrons are in the second shell ($2s^2\ 2p^2$.) The electrons in the outermost shell are called the valence electrons. Those in shells closer to the nucleus are called core electrons. The element chlorine, atomic number 17, has 2 electrons in the first shell, 8 electrons in the second shell, and 7 electrons in the third shell. The seven electrons in the outermost shell (the third period) are the **valence electrons**. The other 10 electrons are **core electrons**. The element bromine, atomic number 35, has an electron configuration of $1s^2\ 2s^2\ 2p^6\ 3s^2\ 3p^6\ 4s^2\ 3d^{10}\ 4p^5$. The 1s, 2s, 2p, 3s, 3p and 3d electrons are considered core electrons, and the 4s and 4p electrons are considered valence electrons. The Group numbers listed near the top of the periodic table (Figure 1.4) indicate the number of valence electrons that atoms in that group have. Sodium, a Group I element, has 1 valence electron; magnesium, a Group II element, has 2 valence electrons; aluminum has 3 valence electrons, and so on.

CONCEPT CHECK: ELECTRON CONFIGURATION AND THE PERIODIC TABLE

The symbol [inert gas] is used to designate the electron configuration of the specified inert gas. For example, the electron configuration of lithium is $1s^2\ 2s^1$, or [He] $2s^1$. An atom has an electron configuration of [Ne] $3s^2\ 3p^4$. How many total electrons does this atom possess? How many are core electrons and how many are valence electrons? To what Group does this element belong?

SOLUTION: Neon has 10 protons and 10 electrons. The unknown element has 6 additional electrons, for a total of 16 electrons. The first 10 of these electrons (those in common with neon) are core electrons. The next 6 are valence electrons. With 6 valence electrons, the element is in Group VI. (Note that the neutral atom with 16 total electrons is sulfur.)

1.10 LEWIS DOT DIAGRAMS OF ATOMS AND SIMPLE IONS

One common way of indicating the electron environment around an atom or group of atoms is to surround the elemental symbol with a number of dots equal to the number of valence electrons. Examples of elements from the third period are shown in Figure 1.16. By convention, there are normally four positions in which one places dots around the elemental symbol: above, below, left, and right. By convention, one places dots singly into each of the four positions until forced to start pairing with the fifth electron. These symbolisms and conventions should not be taken too literally. Silicon, for example, does not have 4 unpaired valence electrons, nor are the electrons located in a plane, 90° apart.

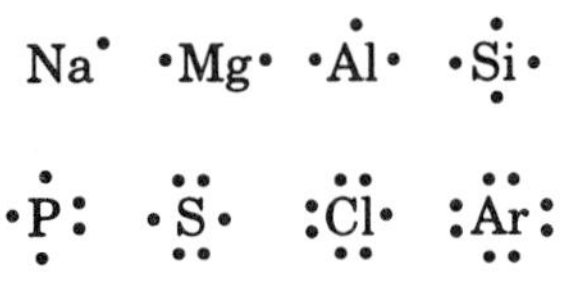

Na• •Mg• •Al• •Si•

•P: •S• :Cl• :Ar:

FIGURE 1.16: Lewis dot diagram of third period of atoms.

Simple cations and anions may also be represented with **Lewis dot diagrams**. The only difference is in the number of valence electrons indicated. As we have seen previously, the atom chlorine, in Group VII, has seven valence electrons. By gaining an extra electron, chlorine becomes the anion chloride. The Lewis dot diagram for chloride is symbolized by $[:\ddot{\underset{\cdot\cdot}{Cl}}:]^-$ The ion Cl^- and the atom Ar both have 8 valence electrons and 18 total electrons. They are said to be isoelectronic. The calcium atom has 20 total electrons. The loss of two of these electrons results in the calcium ion, Ca^{2+}. This has 18 total electrons, and is also isoelectronic with argon atom. Frequently the charges on simple monatomic ions are directly related to the number of electrons that must be gained or lost to become isoelectronic with the nearest inert gas (Group VIII) element.

CONCEPT CHECK: LEWIS DOTS AND ELECTRON CONFIGURATIONS OF IONS

Write the Lewis dot diagram and the electron configuration for As^{3-}.

SOLUTION: Arsenic atom is a Group V element. As the neutral atom, it would have 5 valence electrons. The 3- charge on the arsenide ion indicates that it has three electrons in addition to the 5 valence electrons in the neutral species. This is a total of 8 electrons, or 4 pairs of electrons, shown in the Lewis dot diagram:

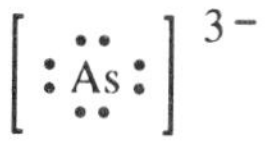

$$\left[:\ddot{\underset{\cdot\cdot}{As}}: \right]^{3-}$$

The electron configuration of arsenide ion would have 3 electrons more than that for the neutral arsenic atom. Arsenic atom has an electron configuration:

As: $1s^2\ 2s^2\ 2p^6\ 3s^2\ 3p^6\ 4s^2\ 3d^{10}\ 4p^3$

The arsenide ion would have three more electrons, which enter the lowest energy level possible, the 4p

As^{3-}: $1s^2\ 2s^2\ 2p^6\ 3s^2\ 3p^6\ 4s^2\ 3d^{10}\ 4p^6$

1.11 LEWIS DOT DIAGRAMS OF SIMPLE MOLECULES

It is common that atoms beyond the first period will find some way of getting eight valence electrons surrounding them. One way, as we have seen, is by the atom gaining or losing sufficient electrons to become isoelectronic with an inert gas. For non-metals, another option is to covalently share electrons in a way that provides the optimum number of electrons for all atoms involved in a molecule. Diatomic molecular elements are common examples of this. The Lewis dot structures for two fluorine atoms and for F_2, the fluorine molecule, are shown below:

$$:\ddot{\underset{\cdot\cdot}{F}}\cdot + \cdot\ddot{\underset{\cdot\cdot}{F}}: \longrightarrow :\ddot{\underset{\cdot\cdot}{F}}\cdot\cdot\ddot{\underset{\cdot\cdot}{F}}:$$

By sharing a pair of electrons, both of the two fluorine atoms achieve an octet of electrons. The pair of electrons shared between the two fluorine atoms is called a bonding pair of electrons. A single bond is composed of a pair of electrons. When illustrating chemical bonds, it is common to use a dash for a pair of electrons instead of the complete Lewis dot diagram. This is shown as:

$$F—F$$

The number of electrons that must be shared in any molecule in which all atoms assume their inert gas configurations can be calculated by the "NAS rule." "N" stands for the number of electrons needed by each atom to achieve an inert gas status. For first period elements (e.g., hydrogen) this means 2 electrons for each hydrogen atom. For elements in other periods, it is assumed that 8 electrons are needed to fill the valence shell. For fluorine, the N value for each atom is 8, for a total N value of 16. "A" stands for the number of electrons available; i.e., the valence electrons each atom brings to the molecule. Since each fluorine atom has

7 valence electrons, the two atoms combined have an A value of 14. The value of S is computed by the equation:

$$N - A = S$$

Substituting the values of N and A yields a value of 2 for S. There must be two electrons shared in the Lewis dot diagram for F_2. We start the Lewis dot diagram for F_2 by putting in the shared electrons first, and then completing the octet around each atom, distributing the total number of available electrons.

Another example is O_2. The "N" value for each oxygen is 8, so the total N value is 16. The "A" value for oxygen, a Group VI element, is 6. For two oxygen atoms, the combined A value is 12. Solving N-A = S yields a value of 4 electrons to be shared in the oxygen molecule. This is shown in Figure 1.17. The 4 electrons shared between the two oxygen atoms represent two bonds. The oxygen molecule is said to have a double bond.

:O::O:

Lewis dot diagrams can be drawn for larger molecules as well. Water, H_2O, has an "N" value of 8 for oxygen, and two for each of the hydrogen atoms. Remember, as a first period element, 2 electrons complete the inert gas configuration for a hydrogen atom. The total N value for H_2O is calculated by:

N = (1 O) (8 e- needed per O) + (2 H) (2 e- needed per H) = 12 e- total needed

A = (1 O) (6 e- available per O) + (2 H) (1 e- available per H) = 8 e- total available

The number of shared electrons, calculated as N - A, is 12 - 8 = 4. We are now prepared to draw the Lewis dot diagram for water. In many cases, a central atom may be identified, and other atoms are arranged symmetrically around that atom. When there is some doubt as to what element is the central atom in the molecule, remember the following rules:

1. Hydrogen will never be the central atom in a molecule;
2. Otherwise, the more electropositive (less electronegative) element will likely be the central atom.

We thus construct the skeletal arrangements of the molecule by putting the oxygen in the center, and arranging the hydrogen atoms around it symmetrically:

H O H

Next, the four shared electrons are added. All atoms must be connected by a minimum of one pair of electrons.

```
H:O:H
```

Finally, the remaining available electrons are distributed. Note that the shared pair between hydrogen and oxygen already satisfies the inert gas configuration for hydrogen, a first period element. No further electrons can be put around hydrogen. The remaining four available electrons are located on the oxygen atom, completing its octet.

```
  ..
H:O:H
  ..
```

Remember that Lewis dot diagrams simply indicate bonding and unshared electrons. They do not display geometry. For instance, despite the Lewis dot diagram for water, the three atoms are not in a straight line.

Next, consider formaldehyde, CH_2O. Calculations show that both carbon and oxygen need eight electrons, the two hydrogens each need two electrons, so the total number of electrons needed is 20. The number of electrons available are 4 for carbon, 6 for oxygen, and 1 each for the two hydrogen atoms, for a total of 12. $S = N - A = 20 - 12 = 8$. A total of 8 electrons, or 4 pairs of electrons must be shared. We construct the skeleton of the molecule by putting the most electropositive element (except hydrogen) in the center. Since oxygen is to the right of carbon in the periodic table, oxygen is more electronegative (less electropositive.) We therefore put carbon in the center, and arrange the two hydrogen atoms and the oxygen atom around it.

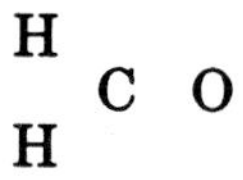

```
H
   C  O
H
```

Next, bonding pairs of electrons are used to connect each atom one time.

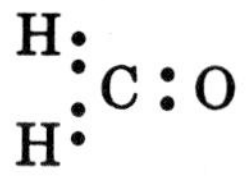

```
H:
  C:O
H:
```

One additional pair of electrons must be shared. Since the hydrogen atoms already have the maximum number of electrons allowed for first period elements, the second pair of shared electrons must be between the carbon and the oxygen.

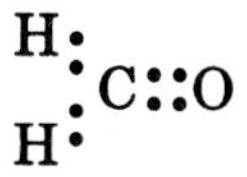

```
H:
  C::O
H:
```

The Lewis dot diagram is completed by using the 4 remaining available electrons to satisfy the electron needs of the individual atoms. Since the hydrogen atoms and carbon are already satisfied, the remaining electrons go around oxygen, completing the diagram.

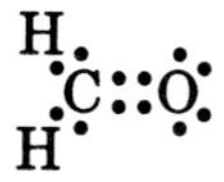

```
H:    ..
  C::O:
H:    ..
```

Again, when called upon to illustrate the single bonds between the hydrogen atoms and carbon, and the double bond between carbon and oxygen, the formaldehyde molecule may be represented as:

$$\begin{matrix} H \diagdown \\ \quad C=O \\ H \diagup \end{matrix}$$

We note that some compounds have Lewis dot diagrams for which NAS calculations fail. Nitric oxide (NO) for example, has an odd number of electrons, and nitrogen and oxygen can not both have their octets satisfied. Some compounds of elements in Group III form in a fashion in which 6 electrons (not 8) are typically around the atom. And some compounds of elements in the third (or subsequent) periods can fit more than an octet of electrons around the atom. We will address such exceptions more thoroughly in Chapter 7.

Before then, our focus the next few chapters will be on the chemical and physical properties of compounds, and their underlying chemical structures. For the compounds observed in these chapters, NAS will suffice. It will also be very helpful to remember that the following are typically true for neutral molecules: a carbon atom forms four bonds, a nitrogen atom forms three bonds, an oxygen atom forms two bonds, and a hydrogen atom forms one bond.

Anion: negatively charged atom or group of atoms.

Atom: the smallest particle of an element that retains its chemical identity.

Atomic number: the number of protons in the nucleus.

Cation: positively charged atom or group of atoms.

Compound: pure substance containing two or more elements chemically combined in a fixed ratio.

Coulombic force: force of repulsion (between like charges) or force of attraction (between opposite charges) which is directly proportional to the product of the charges and inversely proportional to the distance of separation.

Core electrons: electrons in filled or closed shells, closer to the nucleus and at a lower energy than valence electrons.

Covalent compound: compound formed from only non-metallic elements.

Density: ratio of mass to volume

Electron: fundamental subatomic particle with negative charge and little mass. It is found in the atom, but outside the nucleus.

Electron affinity: a measure of the ability of a neutral atom to attract an electron.

Electron configuration: the designation of the number of electrons in each occupied orbital (energy level) of an atom or ion.

Electronegativity: a measure of the ability of an atom in a molecule to attract a shared pair of electrons.

Element: pure substance composed of only one kind of atom.

Ionic compound: compound including elements that differ greatly in electronegativity (i.e., a metal and a nonmetal).

Ionization potential: the energy required to remove an electron from a chemical species.

Isotopes: atoms of the same element that differ in the number of neutrons in the nucleus.

Lewis dot diagram: representation in which the valence electrons are represented by dots around the elemental symbol.

Mass number: the sum of the protons and neutrons in the nucleus

Metal: an element with characteristic low ionization potential, low electronegativity, high conductivity, luster, malleability, ductility, etc. Metallic character increases going down a group and to the left in a period.

Metalloid: an element with properties intermediate between those of metals and nonmetals.

Molecule: smallest particle of matter that retains the chemical identity of the element or compound.

Neutron: a neutral subatomic particle found in the nucleus.

Nonmetal: an element with characteristic high ionization potential, high electronegativity, low conductivity, etc. Nonmetal character increases going up a group and to the right in a period.

Nucleus: the dense, positively charged portion of the atom. It contains protons and neutrons.

Oxidation number: an assignment of imaginary electrical charge to an atom in a compound as a means of keeping track of the electrons in certain types of reactions.

Proton: a positive subatomic particle found in the nucleus.

Valence electrons: those outermost, highest energy electrons in an atom.

HOMEWORK Problems

ATOMIC SYMBOLS

1. For each of the following species, indicate the number of protons, neutrons and electrons:
   (a) ^{18}O
   (b) ^{40}Ar
   (c) $^{29}Al^{3+}$
   (d) $^{32}S^{2-}$

2. For each of the following species, indicate the number of protons, neutrons and electrons:
   (a) ^{16}N
   (b) ^{30}Si
   (c) $^{25}Mg^{2+}$
   (d) $^{19}F^{-}$

3. For each of the following species, indicate the number of protons, neutrons and electrons:
   (a) ^{22}Ne
   (b) ^{32}P
   (c) $^{56}Fe^{3+}$
   (d) $^{128}Te^{2-}$

4. For each of the following species, indicate the number of protons, neutrons and electrons:
   (a) ^{132}Xe
   (b) ^{52}Cr
   (c) $^{114}Cd^{2+}$
   (d) $^{127}I^{-}$

5. Write the correct nuclide symbol for each of the following:
   (a) 15 protons, 16 neutrons, 15 electrons
   (b) 53 protons, 74 neutrons, 54 electrons
   (c) 26 protons, 30 neutrons, 23 electrons
   (d) 92 protons, 148 neutrons, 90 electrons

6. Write the correct nuclide symbol for each of the following:
   (a) 11 protons, 12 neutrons, 10 electrons
   (b) 34 protons, 46 neutrons, 36 electrons
   (c) 30 protons, 35 neutrons, 28 electrons
   (d) 94 protons, 150 neutrons, 90 electrons

7. Write the correct nuclide symbol for each of the following:
   (a) 29 protons, 32 neutrons, 29 electrons
   (b) 39 protons, 50 neutrons, 39 electrons
   (c) 88 protons, 140 neutrons, 86 electrons
   (d) 85 protons, 125 neutrons, 86 electrons

8. Write the correct nuclide symbol for each of the following:
   (a) 18 protons, 22 neutrons, 18 electrons
   (b) 37 protons, 48 neutrons, 37 electrons
   (c) 55 protons, 77 neutrons, 54 electrons
   (d) 16 protons, 16 neutrons, 18 electrons

Ionic vs. Covalent Compounds

9. Classify each of the following compounds as ionic or covalent:
   (a) $MgCl_2$
   (b) P_4O_{10}
   (c) SF_6
   (d) $CsNO_3$

10. Classify each of the following compounds as ionic or covalent:
   (a) N_2O_5
   (b) $NiCl_2$
   (c) XeF_4
   (d) $Sr(IO_3)_2$

11. Which of the following substances exist as a collection of discrete molecules? Which exist as ions in an extended lattice?
   (a) $BaSO_4$
   (b) SO_2
   (c) Cl_2O_7
   (d) NaBr

12. Which of the following substances exist as a collection of discrete molecules? Which exist as ions in an extended lattice?
   (a) N_2H_4
   (b) $FeCl_3$
   (c) CF_4
   (d) $Cu(CN)_2$

Periodic Trends

13. Referring only to a periodic table, rank the following in terms of increasing ionization potential:
   (a) carbon, cesium, copper
   (b) fluorine, francium, phosphorus
   (c) neon, nickel, nitrogen

14. Referring only to a periodic table, rank the following in terms of increasing electron affinity:
   (a) bismuth, bromine, barium
   (b) chromium, chlorine, calcium
   (c) scandium, sulfur, silicon

15. Referring only to a periodic table, rank the following in terms of increasing electronegativity:

(a) oxygen, beryllium, carbon
(b) gold, lead, barium
(c) selenium, polonium, tellurium

Electron Configurations

16. Referring only to a periodic table, write the electron configuration of each of the following:
e.g., krypton: $1s^2\ 2s^2\ 2p^6\ 3s^2\ 3p^6\ 4s^2\ 3d^{10}\ 4p^6$

(a) oxygen
(b) aluminum
(c) germanium

17. Specify the number of valence electrons for each of the species in problem 16.

18. Referring only to a periodic table, write the electron configuration of each of the following:

(a) sulfur
(b) gallium
(c) magnesium

19. Specify the number of valence electrons for each of the species in problem 18.

20. Referring only to a periodic table, write the electron configuration of each of the following:

(a) chlorine
(b) calcium
(c) antimony

21. Specify the number of valence electrons for each of the species in problem 20.

22. Referring only to a periodic table, write the electron configuration of each of the following:

(a) bromine
(b) strontium
(c) tin

23. Specify the number of valence electrons for each of the species in problem 22.

24. Referring only to a periodic table, write the electron configuration of each of the following:

(a) fluorine
(b) potassium
(c) arsenic

25. Specify the number of valence electrons for each of the species in problem 24.

LEWIS DOT DIAGRAMS

26. Write the Lewis dot diagrams of the following atoms/ions:
    (a) N
    (b) F^-
    (c) Si
    (d) S^{2-}

27. Write the Lewis dot diagrams of the following atoms/ions:
    (a) B
    (b) Br^-
    (c) S
    (d) N^{3-}

28. Write the Lewis dot diagrams of the following atoms/ions:
    (a) O
    (b) Cl^-
    (c) Si
    (d) Se^{2-}

29. Write the Lewis dot diagrams of the following molecules. (Note the structure given indicates which atoms are bonded together.)
    (a) carbon dioxide O C O
    (b) hydrogen sulfide H S H
    (c) silane SiH_4 (silicon is central atom)
    (d) acetylene H C C H

30. Write the Lewis dot diagrams of the following molecules. (Note the structure given indicates which atoms are bonded together.)
    (a) sulfur dioxide O S O
    (b) arsine AsH_3 (arsenic is central atom)
    (c) sulfur difluoride F S F
    (d) ethylene CH_2CH_2 (two hydrogens are singly bonded to each carbon)

31. Write the Lewis dot diagrams of the following molecules. (Note the structure given indicates which atoms are bonded together.)
    (a) iodine molecule I_2
    (b) phosphine PH_3 (phosphorus is central atom)
    (c) hydrogen cyanide H C N
    (d) allene H_2 C C C H_2

32. Lewis dot structures may also be drawn for polyatomic ions. The charge on the ion influences the number of available electrons, "A". A charge of -n on an anion indicates that the value for A calculated for the neutral species should be increased by n. A charge of +n on a cation indicates that the value for A calculated for the neutral species should be decreased by n. Write the Lewis dot diagram for the following compund ions:
    (a) nitrite ion (NO_2^-)
    (b) ammonium ion (NH_4^+)
    (c) hypochlorite ion (OCl^-)
    (d) carbonate (CO_3^{2-})

MIXTURES AND Separations

MIXTURES AND Separations

Nature has produced an incredible variety of both organic and inorganic chemicals. Practical use or theoretical understanding of any individual chemical substance is usually possible only when that substance is reasonably pure (i.e., free from contamination by other substances that would significantly alter its observable properties.) Chemistry thus often involves the separation of one component of a mixture.

The separation of complex mixtures to find a particularly useful substance is a long-standing problem, but one with great potential rewards. Nowhere is this more clearly demonstrated than in the isolation of medicinal drugs from nature's chief chemical factory: the plant world. Most folkloric treatments of diseases in man and animal utilize concoctions made by processing all or part of plants in various ways. Often the concoction was a veritable "witches' brew," with numerous components present, some of which could be noxious or even toxic. History shows that the discovery of effective concoctions was widespread, and presumably resulted from trial and error approaches.

Some examples of medicinally useful chemicals obtained from plants are: quinine, an anti-malarial agent, found in the bark of the cinchona tree; salicylates, a family including aspirin, found in the bark of the willow tree; opium, a group of narcotic analgesics including morphine, found in the seed pod of the opium poppy; cocaine, a topical anesthetic, found in the leaves of the coca plant; taxol, an anti-cancer agent found in the bark of the Pacific yew tree; and caffeine, a stimulant present in leaves of tea and in coffee beans.

Once a natural product reaches a critical level of commercial interest, concerns about the reliability of the supply of raw material, quality control and cost effectiveness may result in an effort to synthesize the material, as opposed to harvesting the product from nature. Drugs originally discovered in nature are then often produced solely in the laboratory. Researchers interested in improving on nature may even take certain clues from the structure of natural products and from results of computer models to design and synthesize a variety of new compounds with related structures (e.g., synthetic drugs with greater potency or fewer side effects.) The chemist doing synthesis in the lab will face the same question as the chemist working with the natural products: "How do we isolate the pure substance?" In this chapter we will look at some of the physical and chemical means by which mixtures, both natural and artificial, are separated and by which pure substances are isolated.

2.2 SEPARATION BASED UPON PHYSICAL PROPERTIES

Density

Sometimes a mixture can be resolved simply by utilizing the physical properties of the different components of the mixture. Gold was recognized by the ancients as a very useful metal: its luster, malleability, resistance to corrosion, and relative scarcity combined to make gold highly treasured. While there are many ways of isolating pure gold from other material, one used by prospectors is "panning for gold." The miner's wide, shallow pans in Figure 2.1(a) are dug into a creek bed and swirled with water. Lower density material such as dirt and sand is spun out of the pan and small pieces of gold (a high density material) remain in the pan. The laboratory counterpart of this physical process is **centrifugation**. Figure 2.1(b) shows a centrifuge and a test tube of solid suspended in a liquid. By centrifugation, the dense solid is pushed below the less dense liquid. The liquid, known as a decantate, is carefully poured off without disturbing the sediment, known as a precipitate, in a process known as decanting. Separation by density is further refined by the use of the ultracentrifuge. This spins samples at a very high speed, generating a very high force, that allows separations of certain complex biological materials.

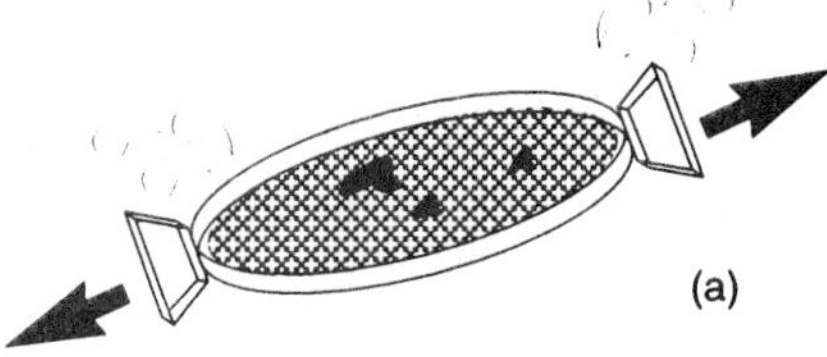

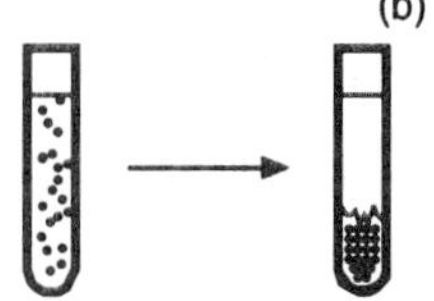

FIGURE 2.1: (a) Gold panning and (b) centrifugation.

Sidebar: Gold (a dense solid and dispersed solution)

A particularly vexing example of isolating a desired component from a natural resource is that of gold dissolved in sea water. With 9 pounds of gold dissolved in every cubic mile of ocean, and 320 million cubic miles of ocean, there are, at current prices, approximately $17 trillion dollars worth of gold waiting to be gathered.

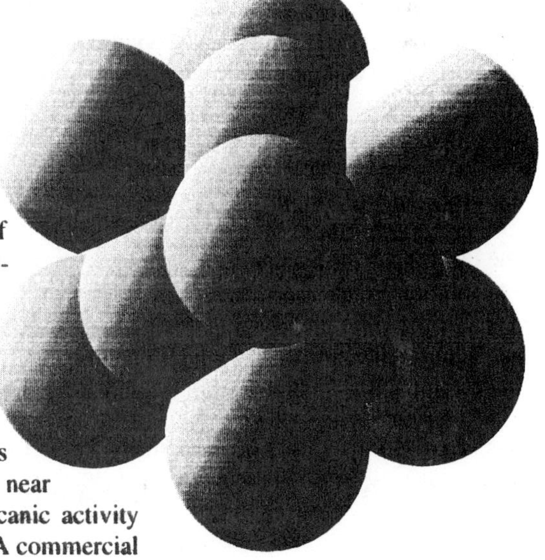

FIGURE 2.2: Unit cell of gold.

Melting Point

The melting point of a solid is another physical property that can be exploited in order to isolate a substance from a mixture. Sulfur is a yellow solid and historically was sometimes found in pure form near volcanoes; hence the biblical name for sulfur: "brimstone." Volcanic activity underground can concentrate sulfur and deposit it at the surface. A commercial means of isolating sulfur from geologic deposits that operates in a fashion similar to that of the volcano is the Frasch process.

As seen in the diagram, superheated steam is piped underground into an area rich in sulfur. The low-melting sulfur liquefies in the extreme heat, and the liquid sulfur-steam-water mixture can then be pumped to the surface, leaving behind the other, much higher melting solids. Evaporation of the water from the pumped liquid mixture leaves behind reasonably pure sulfur. Sulfur is isolated in huge amounts and is eventually

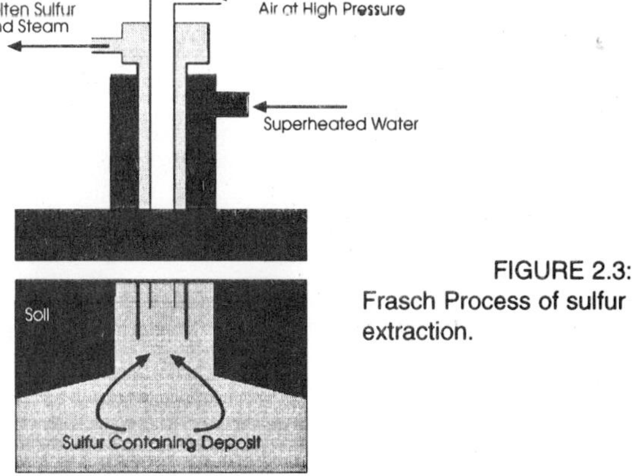

FIGURE 2.3: Frasch Process of sulfur extraction.

transformed into sulfuric **acid**, the number one chemical produced worldwide.

Purification by melting has a high-technology application in the semiconductor industry, where zone refining is used in the ultra purification of silicon. As seen in Figure 2.4, a "zone" of solid impure silicon is melted by a movable heater. When the heater moves, so does the zone of melted silicon. The impurities in the silicon are more soluble in liquid silicon than in solid silicon. Thus the impurities follow along in the moving liquid zone, and are eventually swept to one end of the silicon sample. This end may be cut off, leaving behind a sample of "refined" (i.e., purified) silicon.

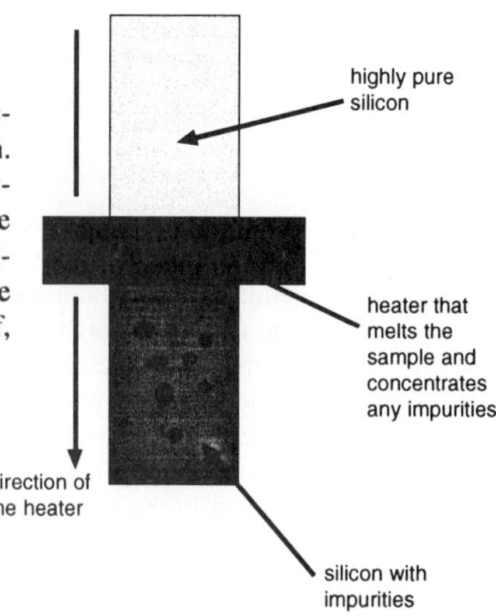

FIGURE 2.4: Zone refining of silicon.

Boiling Point

A volatile liquid (one with a relatively low boiling point) can frequently be separated from less volatile material via the process of simple distillation. As shown in Figure 2.5, a mixture of a volatile liquid (water) and a non-volatile dissolved solute (salt) can be heated until the water boils.

As the water vapor rises, it enters the condenser. The lower temperature of the condenser reverses the process, and the steam is liquefied back into water. The "distillate," collected from the condenser, contains only the volatile component, in this case water.

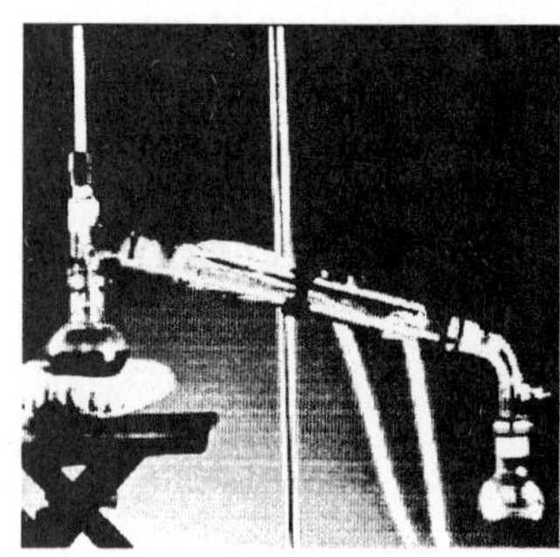

FIGURE 2.5: Simple distillaton of an aqueous salt solution.

Suppose that a mixture contains two volatile liquids, A and B. Simple distillation would result in a distillate containing proportionately more A (the component of higher volatility, lower boiling point) but would contain some B as well. One possible solution to this would be fractional distillation, shown in Figure 2.6.

The long heated column separating the boiling mixture from the condenser permits repeated equilibration between the rising vapor and the descending liquid. The usual result is a separation of volatile components. An important industrial separation process using this method is the refining of petroleum. Crude petroleum is a mixture of a vast number of hydrocarbons, substances containing only carbon and hydrogen. In general, as the number of carbon atoms in the hydrocarbon increases, so does the boiling point of the substance:

It is not important for the petroleum refineries to separate each and every com-

Table 2.1 B.P. *vs.* Carbon Number in a Hydrocarbon

Molecule Name	Number of Carbons	Boiling Point (•c)
methane	1	-164
ethane	2	-88.6
propane	3	-42.1
butane	4	-0.6
pentane	5	36.1
hexane	6	69.0
heptane	7	98.4
octane	8	125.7
nonane	9	150.8
decane	10	174.1
eicosane	20	343.0

TABLE 2.1 Boiling point vs. number of carbon atoms in a hydrocarbon.

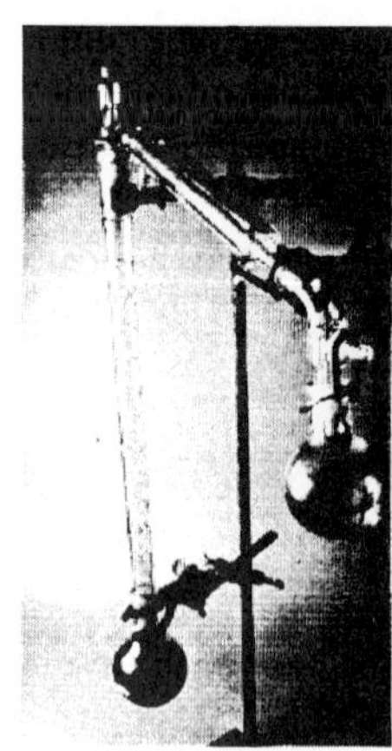

FIGURE 2.6: Fractional distillation apparatus.

ponent in the mixture. Instead, fractions within a specified boiling point range are collected, since this is sufficient for most commercial applications. Each fraction collected is an example of a **solution**. A variety of methods, including better fractional distillation, could be used to separate each solution into its various components.

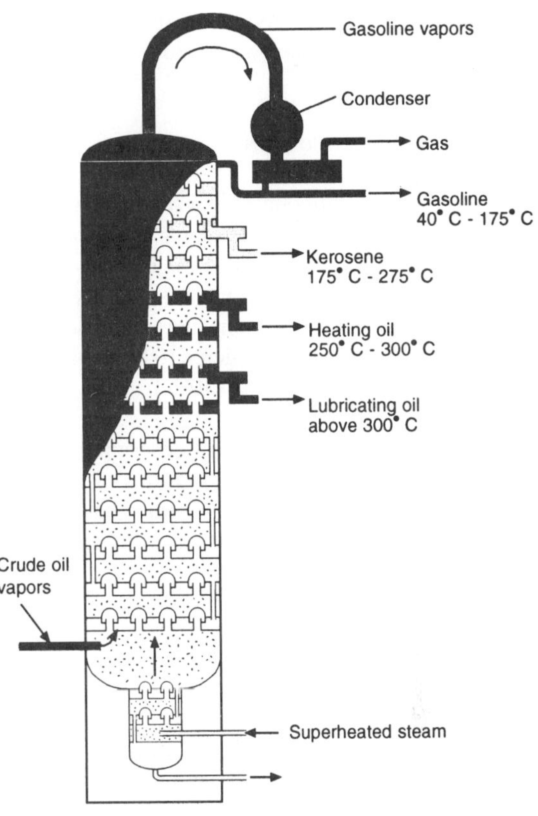

FIGURE 2.7: Petroleum refining: an example of fractional distillation.

2.3 MIXTURES AND SOLUTIONS

TYPES OF MIXTURE

Mixtures can be classified as homogeneous or heterogeneous. A **heterogeneous mixture** has more than one phase present. A mixture of solid sand and liquid water, Figure 2.8 (a) shows a heterogeneous mixture, i.e., it contains two phases.

A phase is a region of space with uniform characteristics, including composition. A sampling of this mixture from near the bottom of the beaker in Figure 2.8 (a) would analyze as mostly sand, whereas a sampling of the same mixture from near the top of the beaker would be analyzed as essentially pure water. Ice water, Figure 2.8 (b), is similarly a heterogeneous mixture. Although both ice and water are H_2O, the solid phase (ice) has distinctly different characteristics than the liquid phase (water.) Figure 2.8 (c) shows a mixture of two immiscible (not mutually soluble) liquids such as oil and vinegar. Although both are liquids, there are two phases, each with distinguishing characteristics. The presence of more than one phase makes these mixtures heterogeneous.

Homogeneous mixtures are also known as solutions. A solution can be gas phase, solid phase, or liquid phase. The air in the atmosphere is a gas phase solution consisting of nitrogen, oxygen, argon, carbon dioxide, water vapor, etc. Natural gas is a solution made of methane, ethane, propane, nitrogen, etc. An example of a solid solution is stainless steel. One type of stainless steel contains iron, chromium, nickel and carbon. Another example of a solid solution is brass, a mixture of copper and zinc. The most prevalent type of solution in chemistry is a liquid phase solution. The fraction called gasoline in Figure 2.7 is an example of a liquid homogeneous solution. It is composed of numerous liquid components. But it is not necessary that all components in a liquid solution be liquid. A dilute aqueous solution of solid sodium chloride (i.e., a small amount of salt

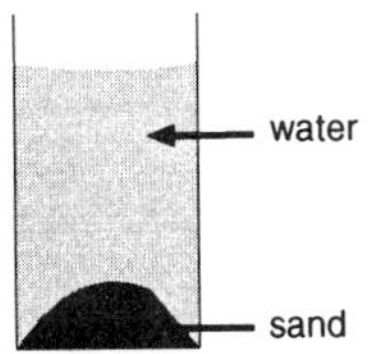

a) Heterogeneous mixture of snd and water

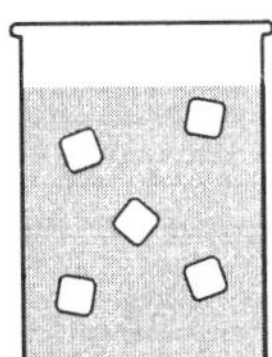

b) Heterogeneous mixture of ice water

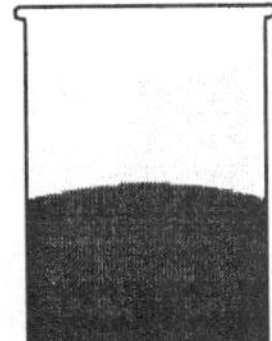

c) Heterogeneous mixture of two immiscible liquids

FIGURE 2.8: Heterogeneous mixtures.

added to excess water) is described as being a liquid solution. It is termed a liquid since it meets the criteria for that phase set out in Chapter One, and it is termed a solution since the mixture is homogeneous (uniform in properties throughout.)

SOLVENT, SOLUTE AND CONCENTRATION

In describing the components of a solution, two common terms are used: solvent and solute. The **solvent** is the major component present. The remaining components of the solution are termed the solutes. In the example of a dilute aqueous salt solution, water was the solvent and salt was the solute. In air, which is about 78% nitrogen, nitrogen is considered the solvent, with the other gases being solutes. Yellow brass is considered to be a solution composed of the solvent copper (70%) and the solute zinc (30%.)

As we have just seen, one way of describing a solution is by percentage composition. This percentage may be in terms of volume or in terms of mass. Saying air is 78% (by volume) nitrogen means that 100 volume units (e.g., liters) of air are composed of 78 liters of nitrogen, 21 liters of oxygen, 0.9 liters of argon, etc. Percentage is defined as:

$$\% = \frac{\text{part}}{\text{whole}} \times 100$$

When both the part and the whole are measured in terms of volume, this is known as a volume percent. Brass is a solid solution of copper and zinc. Brass typically is 70% copper by mass. That means 100 grams of brass contains 70 grams of copper and 30 grams of zinc.

Percentage is one means of expressing the **concentration** of a specific solute in a solution. Another term used to express the concentration of a solute is molarity. This refers to the amount of chemical substance (moles) of solute per liter of solution. The concept of moles is addressed later in this text. For now, we can assume that each compound has a unique mass equivalent to one mole. A mole of sodium chloride corresponds to about 58.44 grams. When this mass of sodium chloride is weighed, dissolved in water, and diluted in a volumetric flask to a final volume of 1.000 liter, a 1.000 molar aqueous sodium chloride solution is made. (Figure 2.9) This is commonly abbreviated 1.000 M NaCl(aq). The use of more than this amount of salt, or less than this amount of water, results in a higher concentration of salt in solution, and a larger value of the molarity.

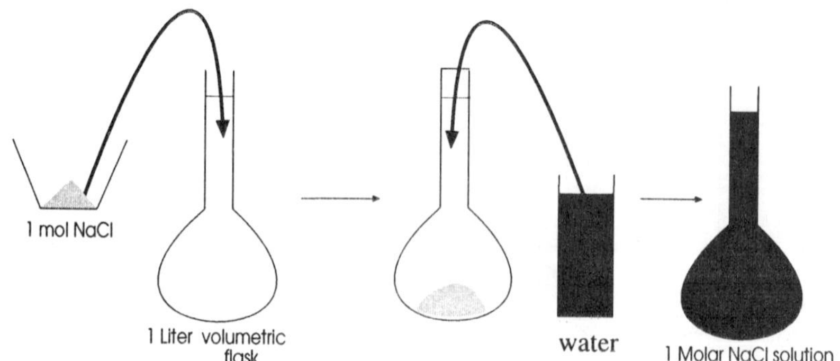

FIGURE 2.9: Steps in preparing 1.0 molar aqueous salt solution.

Concept Check: % by Mass and % by Volume

When 50.0 cm^3 of pure sulfuric acid (H_2SO_4, density = 1.84 g cm^{-3}) is carefully added to 200.0 cm^3 of pure water (H_2O, density = 0.9982 g cm^{-3}) the resulting solution (once cooled back to room temperature) has a density of 1.1366 g cm^{-3}. Describe the final solution in terms of % by mass of sulfuric acid and % by volume of sulfuric acid.

Solution: since the volume and density of the pure components are known, the masses of each are readily determined. Those lead readily to the percent by mass:

$$50.0\ \text{cm}^3\ H_2SO_4\left(\frac{1.84\ \text{g}}{1\ \text{cm}^3}\right) = 92.0\ \text{g}\ H_2SO_4$$

$$200.0\ \text{cm}^3 H_2O\left(\frac{0.9982\ \text{g}}{1\ \text{cm}^3}\right) = 199.6\ \text{g}\ H_2O$$

$$92.0\ \text{g} + 199.6\ \text{g} = 291.6\ \text{g solution}$$

$$\%\ \text{by mass}\ H_2SO_4 = \left(\frac{92.0\ \text{g}\ H_2SO_4}{291.6\ \text{g solution}}\right) \text{x}\ 100 = 31.5\%$$

Since we know the initial volume of pure sulfuric acid, all we need to determine the % by volume is the final total volume. This is obtained from the total mass and the density:

$$291.6\ \text{g solution}\left(\frac{1\ \text{cm}^3}{1.1366\ \text{g}}\right) = 256.6\ \text{cm}^3\text{solution}$$

$$\%\ \text{by volume}\ H_2SO_4 = \left(\frac{50.0\ \text{cm}^3 H_2SO_4}{256.6\ \text{cm}^3\text{solution}}\right) \text{x}\ 100 = 19.5\%$$

Note the significant difference when describing the solution by mass (31.5%) or by volume (19.5%.)

2.4 SOLUBILITY AND SEPARATION

Mixtures can sometimes be easily separated by the difference in solubility of their various components. A mixture of two solids, for example sand and salt, might be separated by their differing solubility in water. Sand is nearly insoluble in water, but salt dissolves in water. Sufficient water could be added to the mixture of sand and salt to dissolve all the salt into solution. The solid sand could then be separated from the aqueous salt solution by gravity **filtration** as seen in Figure 2.10.

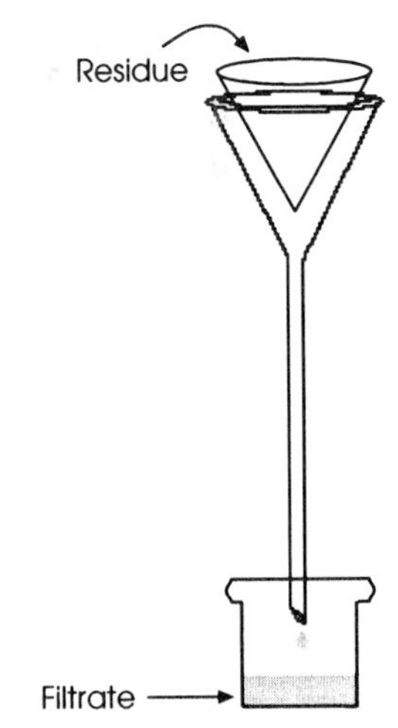

FIGURE 2.10: Gravity filtration.

The solid sand is retained on the filter paper, and is termed the residue. The salt solution passes through the filter and is collected. This solution is called the filtrate. The salt solution (filtrate) can then be heated. The water will evaporate, but the salt will remain. Both the sand and salt have thus been successfully isolated from the mixture.

The solubility of chemical species in various solvents is of great practical importance. A minor change in physiological fluids can result in an important change in solubility, causing effects ranging from "the bends" experienced by deep sea divers to kidney stones and gout. Commercially, solubility is critical in the formulation of the many synthetic liquid mixtures. Solubility is also the key to much of the processing of natural products mentioned earlier. A common laboratory method of purifying a solid sample is **recrystallization**. This is a purification process based upon solubility. When a solid rapidly forms in a solution, the solid often traps impurities. To isolate a solid sample of higher purity, a chemist will select a solvent in which the impurities are more soluble than the substance being purified. The solid sample is added to a minimal amount of the solvent and the solution is heated until the entire sample dissolves. When the solution cools, the desired substance recrystallizes but the impurities remain in solution.

Solubility is important, but the nearly unlimited combinations of specific solutes with specific solvents mean that individual solubility data can not be effectively memorized. Instead, by classifying compounds into one of a few categories and by learning several solubility rules (or guidelines) a chemist is able to predict and utilize solubility based upon the underlying chemical principles. We begin our study of these principles in the following sections.

2.5 STRONG AND WEAK ACIDS AND BASES

One definition of an acid is a compound that, when added to water, increases the hydrogen ion $[H^+]$ concentration. There is a wide variety of organic and inorganic acids. Organic acids are normally formed from the carboxyl group, -COOH. Inorganic acids are typically either binary hydrogen compounds (HF, HCl, H_2S, H_2Se) or ternary compounds of hydrogen and oxygen (HNO_3, HNO_2, H_3PO_4, H_2SO_4, H_2SO_3, etc.)

SIDEBAR: INORGANIC NOMENCLATURE

Inorganic compounds can often be named by following a few simple rules. Binary salts (compounds containing one metallic element and one nonmetallic element) are named as two words: the first is the name of the metal; the second word takes the stem of the name of the nonmetal and adds the suffix "ide". Examples include NaCl (sodium chloride), Na_2S (sodium sufide), and MgF_2 (magnesium fluoride.) These are simple examples because sodium in a compound is always +1, and magnesium in a compound is always +2.

Some metals have variable oxidation states. For example iron is often found in a +2 or a +3 oxidation state. Copper is sometimes found in a +1 or a +2 oxidation state. Again, we name binary salts by first stating the name of the metal. When the metal may be found in more than one oxidation state, this is followed parenthetically by the oxidation state of the metal in Roman numerals. The nonmetallic element is named as before: the stem name of the element with an "ide" suffix. Examples include CuCl (copper (I) chloride), $CuCl_2$ (copper (II) chloride), FeO (iron (II) oxide), and Fe_2O_3 (iron (III) oxide). Note that there are also common names for these compounds that use another system wherein the lower oxidation state of the metal is designated by an "ous" suffix, while the higher oxidation state is designated by an "ic" suffix. In this system, CuCl would be called cuprous chloride, and $CuCl_2$ would be termed cupric chloride.

One can also have binary inorganic compounds wherein both elements are nonmetals. These compounds are named in a fashion similar to the binary salts. The nonmetal element that is more metallic is named first, followed by the stem of the more nonmetallic element with an "ide" suf fix. For example, HF is named hydrogen fluoride. Hydrogen is listed first in the formula and named first because hydrogen is more metallic (i.e., it has a lower electronegativity) than fluorine. When several bin–ary compounds exist for the same two nonmetals, Greek prefixes are used to designate the number of atoms of the element present in the molecule. For example: COis carbon monoxide; CO_2 is carbon dioxide; SO_3 is sulfur trioxide; CCl_4 is carbon tetrachloride; PCl_5 is phosphorous pen tachloride; and SF_6 is sulfur hexafluoride.

Most ternary and higher inorganic compounds (3 or more elements) involve polyatomic ions that have specific names. See Appendix E for a listing of many of the more common polyatomic ions. For example, we see that the PO_4^{3-} ion is called phosphate. Thus the compound Na_3PO_4 is named as before, metal first, followed by the nonmetal anion: sodium phosphate. Note that the "ide" suffix used for an elemental anion is changed to "ate", signifying a polyatomic anion including two or more elements. Note also that Appendix E lists similar anions that differ only in the suffix, either "ate" or "ite". For example SO_4^{2-} is sulfate; SO_3^{2-} is sulfite.

Binary hydrogen compounds, like HCl, are named by the rules above as hydrogen chloride. When compounds of hydrogen and a non-metal are placed in aqueous solution, they are named as acids. The name hydrogen is shortened to a prefix "hydro-", and while the "ide" suffix of the second word is replaced by "ic acid". Thus an aqueous solution of HCl is called hydrochloric acid. H_2Se is hydrogen selenide. An aqueous solution of H_2Se is hydroselenic acid.

Ternary acids may be formed by adding hydrogen to an oxo anion listed in Appendix E. The name of the acid is determined from the stem name of the anion. If the anion's suffix was "ate", that suffix is replaced by "ic acid". If the anion's suffix was "ite", that suffix is replaced by "ous acid".

An acid is considered strong if most of it exists in ionized form when added to water. Consider HCl as our first example. Hydrogen chloride is a covalent compound. It is a gas at typical room temperature and pressure. An aqueous solution of hydrogen chloride is called hydrochloric acid. When bought as a commercial product to clean cement or treat a pool, it is sometimes sold as muriatic acid. These concentrated solutions of HCl are about 35% HCl by mass, or about 12 molar HCl. But there are few, if any, molecules of HCl in solution! It is essentially completely ionized into H^+ and Cl^-. Consequently HCl is considered a strong acid. Of the binary acids, only HCl, HBr and HI are considered strong acids in aqueous solution. Inorganic ternary acids with the general formula H_mXO_n are strong when $n-m \geq 2$. For example, perchloric acid, $HClO_4$, has $m=1$ and $n = 4$; thus $n-m = 3$, and $HClO_4$ is a strong acid. H_2SO_4, sulfuric acid, has n-m ≥ 2 and is also a strong acid. Phosphoric acid, H_3PO_4, has n-m = 1, and is considered weak. Tables 2.2 and 2.3 list the common inorganic acids, and categorizes them as strong or weak.

TABLE 2.2: COMMON STRONG INORGANIC ACIDS TABLE 2.2

FORMULA	NAME
HCl	Hydrochloric Acid
HBr	Hydrobromic Acid
HI	Hydroiodic Acid
$HClO_4$	Perchloric Acid
HNO_3	Nitric Acid
H_2SO_4	Sulfuric Acid (first proton only)

TABLE 2.3: COMMON WEAK INORGANIC ACIDS TABLE 2.3

FORMULA	NAME
HF	Hydrofluoric Acid
H_3PO_4	Phosphoric Acid
HNO_2	Nitrous Acid
H_2SO_3	Sulfurous Acid
H_3BO_3	Boric Acid

CONCEPT CHECK: INORGANIC ACIDS

The ion AsO_4^{3-} is called arsenate. What is the name of an aqueous solution of H_3AsO_4, and is that acid weak or strong?

The name of the acid is given by dropping the "ate" of arsenate and adding "ic acid". Thus a solution of H_3AsO_4 is called arsenic acid. The value of n-m for this acid is 4-3 = 1; thus H_3AsO_4 is a weak acid.

Organic acids have the general form R-COOH, where R stands for a variable group. All organic acids are weak acids. We can compare the relative strengths of acetic acid and hydrochloric acid in the following fashion. We stated earlier that a 35% (12M) HCl solution was nearly 100% ionized. A 35% by mass acetic acid solution is only 0.2% ionized. A 12M acetic acid solution is about 0.1% ionized. The ionization reactions of hydrochloric acid and acetic acid are shown below. Note that the arrows going in both directions mean some reactants and some products are both present. The large arrow pointing to the right for hydrochloric acid indicates that the ionized form predominates; the large arrow pointing to the left for acetic acid indicates that the molecular form predominates.

$$HCl \rightleftharpoons H^+ + Cl^- \quad \text{EQ. 2.1}$$

$$CH_3COOH \rightleftharpoons H^+ + CH_3COO^- \quad \text{EQ. 2.2}$$

TABLE 2.4: COMMON ORGANIC ACIDS TABLE 2.4

FORMULA	NAME
HCOOH	Formic Acid
CH_3COOH	Acetic Acid
CH_3CH_2COOH	Propionic Acid
C_6H_5COOH	Benzoic Acid
$C_6H_4(COOH)_2$	Phthalic Acid
$C_8H_7O_2COOH$	Acetylsalicylic Acid (aspirin)

Acids tend to have a characteristic sharp and sour taste. For example, the key component of vinegar is acetic acid. In the earliest published account of aspirin (by E. Stone in Philosophical Transactions of the Royal Society of London, 1763) the author notes that his decision to test for medicinal properties was based upon the unusual taste of the bark of the willow tree, which contained salicylic acid.

One definition of a **base** is a compound that, when added to water, increases the hydroxide ion [OH^-] concentration. Strong bases are typically soluble metal hydroxides. These are listed in Table 2.5. By dissolving in water to form the metal cation and hydroxide anion, these salts increase the concentration of hydroxide ions and thus are bases. Metal hydroxides that do not dissolve do not increase the amount of OH^- present in solution.

TABLE 2.5

TABLE 2.5: COMMON STRONG BASES

FORMULA	NAME
LiOH	Lithium Hydroxide
NaOH	Sodium Hydroxide
KOH	Potassium Hydroxide
RbOH	Rubidium Hydroxide
$Sr(OH)_2$	Strontium Hydroxide
$Ba(OH)_2$	Barium Hydroxide
TlOH	Thallium (I) Hydroxide

Solutions of strong bases are generally strong cleaning agents (e.g., lye, drain cleaners, etc.) When spilled on the skin, these solutions rapidly react with the tissue, and cause a slippery feeling. Prudent practices in the lab, of course, exclude tests involving tasting or touching of chemicals.

Ammonia, NH_3, is a common example of a weak inorganic base. NH_3 is a covalent compound and a gas at room temperature and pressure. When dissolved in water, it partially ionizes to form additional OH^- in solution. Aqueous ammonia is sometimes called ammonium hydroxide. Only a small fraction of ammonia in water undergoes the ionization reaction shown below.

$$NH_3 + H_2O \rightleftharpoons NH_4^+ + OH^- \qquad \text{EQ. 2.3}$$

Organic bases, like organic acids, are weak. Organic amines, for example methyl amine, CH_3NH_2, undergo a small amount of ionization in a reaction similar to that for ammonia, except that it forms $CH_3NH_3^+$ along with the OH^-.

One key feature of acids and bases is that they react with each other. Aqueous acids and bases react to form a salt and water. A strong acid like HCl reacts with a strong base like NaOH to form water and the salt NaCl. In a similar fashion, a weak acid, like benzoic acid, C_6H_5COOH, reacts with a strong base, like NaOH, to form water and the salt sodium benzoate, C_6H_5COONa. A strong acid, like HCl, can react with a weak base, like methyl amine (CH_3NH_2) to form a salt called methylammonium chloride (CH_3NH_3Cl). Later in this text we will take a more in-depth look at acids, bases, and their reactions.

WEAK VERSUS STRONG

Do not confuse the strength of an acid (whether or not it is largely dissociated in water) with concentration (how much of the total species is present in solution.) One can have either concentrated or dilute solutions of a strong acid. One can have either concentrated or dilute solutions of a weak acid.

2.6 STRONG, WEAK AND NON-ELECTROLYTES

Ionic compounds that dissolve in water also readily form ions in aqueous solution. For example, sodium chloride dissolves in water to form Na^+ and Cl^-. We have already seen that some covalent compounds, like hydrogen chloride, also readily dissolve in water to form ions. Species that dissolve in water to form ions readily conduct electricity. Figure 2.11 shows a means for testing a solution for its ability to conduct electricity. When pure water is used as the liquid, essentially no current passes between the electrodes, and the light bulb remains unlit. When gaseous HCl or solid NaCl is added to the water, ions form which can carry charge. This establishes an electric current, which causes the bulb to light. Species that dissolve in water to form significant amounts of ions are called **strong electrolytes**. Note that strong acids (which form H^+) and strong bases (which form OH^-) by definition are strong electrolytes.

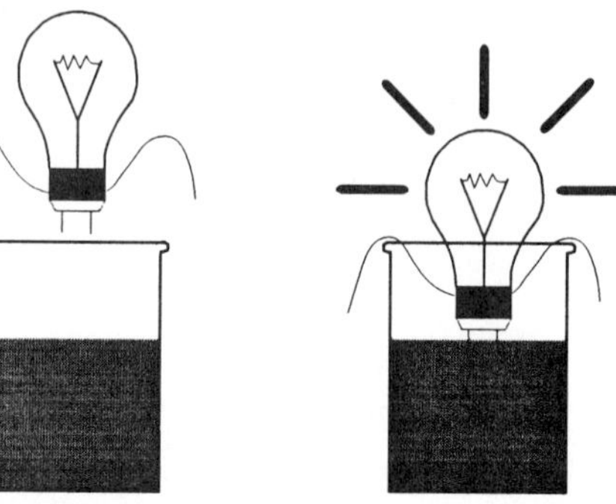

FIGURE 2.11
Conductivity apparatus.

Other species that dissolve in water cause a dim lighting of the bulb. This is also due to ionization, but the ionization occurs to a much smaller extent than for a strong electrolyte. A typical example of this would be a weak acid. As we have previously noted, when a weak acid or base is put into water, only a few percent of these molecules are ionized. Species that undergo a small amount of ionization are referred to as **weak electrolytes**. Their aqueous solutions still carry an electric current, but far less than a strong electrolyte.

A third category of solutes includes those which when added to water cause no change in the number of ions present in solution. Such solutes fail to cause any lighting of the bulb, because no current carrying ions are added. These solutes are termed **non-electrolytes**. Sucrose, regular table sugar, $C_{12}H_{22}O_{11}$, dissolves in water but fails to conduct any current. Covalent compounds that dissolve in water but do not produce ions are examples of non-electrolytes.

2.7 POLARITY AND SOLUBILITY

Classifying compounds in terms of their relative acid/base or electrolytic strengths will facilitate later discussion, but such classifications do not fully explain solubility. For example, while strong electrolytes and strong acids will always be soluble in water, weak acids and non-electrolytes may also readily dissolve in water. An important general guideline of solubility that helps resolve this issue is that "like dissolves like." Oil and water do not mix, for example, because one is polar and the other non-polar. This generalization means that polar solutes tend to dissolve in a polar solvent like water, while non-polar solutes tend to dissolve in a non-polar solvent like hexane. Determining whether a particular solute or solvent is or is not polar is thus crucial to determining solubility.

A complete understanding of polarity will only be possible later in the course when molecular geometry is explored. For example, that H-O-H is bent (non-linear) is decisive in water being polar as opposed to non-polar. For now, Table 2.6 lists common liquid solvents in terms of their measured dipole moment (a quantitative measure of charge separation within the molecule.) The larger the dipole moment, the greater the polarity of the molecule. Those molecules with a zero dipole moment are completely non-polar. As you can see, there are gradations in polarity, not clearly differentiated polar versus non-polar.

TABLE 2.6: POLARITY OF SOME COMMON SOLVENTS

TABLE 2.6

NAME	DIPOLE MOMENT (Debye units)
carbon tetrachloride	0
benzene	0
toluene	0.36
chloroform	1.01
diethyl ether	1.15
dichloromethane	1.60
ethanol	1.69
methanol	1.70
ethyl acetate	1.78
water	1.85
acetaldehyde	2.69
acetone	2.88
acetonitrile	3.92

According to our rule of "like dissolves like," whether or not a particular solid, liquid or gaseous solute dissolves in one of the solvents listed depends on the solute polarity as well. Solid iodine, I_2, happens to be a non-polar molecule. Since the two iodine atoms are equally matched in electronegativity, the bond is perfectly covalent. Since the two atoms define a line, the molecule is also linear. As a consequence, I_2 is a linear, non-polar molecule with a zero dipole moment. Solid iodine is not very soluble in pure water, a polar solvent, but is readily soluble in carbon tetrachloride, a non-polar substance.

Strong acids, strong bases and strong electrolytes in general are all considered polar, and will dissolve most readily in polar solvents. Weak electrolytes and non-electrolytes may still have polar properties that do not translate into ion formation. Methanol, CH_3OH, is shown as a fairly polar molecule in Table 2.6 . Methanol and water are termed miscible, meaning mixable in all proportions; e.g., 1 g of methanol will mix with 99 grams of water, 99 grams of methanol will mix with 1 gram of water, as well as any other combination. Methanol is a non-electrolyte, and exists in aqueous solution as molecules. This high solubility in water is due to several factors, including methanol's polarity. As the length of the alcohol is increased, the polarity of the molecule decreases. The solubility of the alcohols in water similarly decreases, as shown in Table 2.7.

TABLE 2.7: SOLUBILITY OF SOME ALCOHOLS IN WATER

TABLE 2.7

ALCOHOL FORMULA	SOLUBILITYY (g alcohol/ 100 g H_2O)
CH_3OH	infinite
CH_3CH_2OH	infinite
$CH_3CH_2CH_2OH$	infinite
$CH_3(CH_2)_3OH$	7.9
$CH_3(CH_2)_4OH$	2.3
$CH_3(CH_2)_5OH$	0.6
$CH_3(CH_2)_6OH$	0.2

The alcohols in Table 2.7 all have the general formula R-OH. The solubility of alcohols in water is attributed to the -OH group of the alcohol. This portion of the molecule is called "hydrophilic", literally meaning "water-loving". The R portion of the alcohol is called "hydrophobic", literally meaning "water-fearing". As the length of the R group increases in

2.8 SOLUBILITY OF IONIC COMPOUNDS IN WATER

Water is the most common solvent. We have seen that whether a covalent compound will or will not dissolve in water is related to the polarity of the compound, or possibly related to whether or not the covalent compound will ionize in water (e.g., a strong acid.) Although covalent compounds that readily form ions in water are water soluble, not all ionic compounds are water soluble. The reasons that ionic solids may not dissolve in water to form separate ions are numerous, but let us examine just one.

The force that holds cations and ions together in an ionic solid is the coulombic force of attraction, seen in Chapter One. This force of attraction F is proportional to the charges of the cation and the anion; the greater the charges, the greater the attraction. Other things being equal, the attraction between cation M^+ and anion A^- is expected to be much less than the attraction between cation M^{2+} and anion A^{2-}. An example of a M^+A^- salt is sodium chloride. About 357 g of NaCl will dissolve in a liter of water at room temperature. An example of a $M^{2+}A^{2-}$ salt is barium sulfate. Only about 0.0023 g of $BaSO_4$ will dissolve in a liter of water at room temperature. While forces other than the charge on the ions are important, such widely disparate solubility data show that ionic compounds have vastly different solubilities. Again, since the number of combinations of cations and anions is so large, we need to follow a few rules as opposed to trying to memorize individual solubility data for each salt. These guidelines are given in the following list.

1. All common alkali metal and ammonium salts are soluble. (i.e., salts with Li^+, Na^+, K^+, Rb^+, Cs^+, and NH_4^+ cations)

2. All nitrates, perchlorates, chlorates, and acetates are soluble. (i.e., salts containing NO_3^-, ClO_4^-, ClO_3^-, and CH_3COO^- anions)

3. All common chlorides (Cl^-), bromides (Br^-) and iodides (I^-) are soluble except those of Ag^+, Hg_2^{2+}, and Pb^{2+}.

4. All sulfates (SO_4^{2-}) are soluble except those of Sr^{2+}, Ba^{2+}, Hg_2^{2+}, and Pb^{2+}. The sulfates of Ag^+ and Ca^{2+} are sparingly soluble.

5. All carbonates (CO_3^{2-}), chromates (CrO_4^{2-}) and phosphates (PO_4^{3-}) are insoluble (except those covered by rule #1).

6. All hydroxides (OH^-) are insoluble except those of the alkali met als and Tl^+. Hydroxides of Ca^{2+}, Sr^{2+} and Ba^{2+} are sparingly soluble.

7. All sulfides (S^{2-}) are insoluble except those of the alkali and alkaline earth metals and the ammonium ion.

8. As a general rule, salts formed from a multiply charged cation and multiply charged anion will be insoluble.

The order of the rules is important. When a contradiction arises between rules, the lower number rule has priority. For example, consider whether potassium chromate is soluble. Rule #5 says all chromates are insoluble, rule # 1 says all potassium salts are soluble. Rule #1 has priority and potassium chromate is in fact soluble. Some salts will have neither their cation nor their anion listed explicitly in the rules. $AlAsO_4$ is such a salt. It is composed of Al^{3+} cations and AsO_4^{3-} anions. Rule 8 can be invoked in this case to state that this aluminum salt will be insoluble in water. In certain cases, a corollary may be attempted, noting chemical similarities to ions listed in the rules. Arsenate AsO_4^{3-}, is closely related to phosphate PO_4^{3-}, and rule #5 would, by analogy, have correctly predicted aluminum arsenate to be insoluble. Care must be taken when reasoning by these analogies. Magnesium chloride ($MgCl_2$) for example, is soluble in water (over 1600 g in 1 liter of water); although fluoride is generally similar to chloride, magnesium fluoride (MgF_2) is not water soluble (0.087 g in 1 liter of water.) Also, these solubility rules are for pure water. As we will see later, adjusting the level of acidity of the water can profoundly influence the solubility of some species. Finally, note that Appendix E lists some common ions.

2.9 SOLUBILITY AND EXTRACTION

Extraction via solubility is a common occurrence. A drip coffee pot separates the desired components of coffee from the coffee grounds by slowly adding hot water. The water soluble components pass through the filter and form the aqueous solution called "coffee." The solid grounds remaining on the filter are the less water soluble components of the coffee beans. One way of identifying the components of coffee would be to isolate and characterize each of the components. Since there are numerous components in coffee, we would have to devise a complex scheme of separation.

For the present, let us focus on isolating a single component from a mixture. For example, a synthesis designed to make one product almost always results in a mixture. This mixture may include initial reactants as well as unintended side-products of the reaction. How does one isolate the desired product? Another example is the isolation of the desired component from a complex mixture of natural origin.

An example of extraction of a solute from a solution is to mix the solution with an immiscible solvent in which the desired solute is preferentially dissolved. In the qualitative analysis of aqueous iodide ion, the procedure typically calls for chemically treating the aqueous solution of iodide ion (I^-) to change it into iodine (I_2). The iodine-containing aqueous solution is then mixed in a separatory funnel (Figure 2.12) with a non-polar solvent like carbon tetrachloride. The non-polar iodine is roughly one hundred times more soluble in the non-polar carbon tetrachloride layer than in the water layer. After shaking the iodine containing aqueous phase with the carbon tetrachloride phase, the organic layer is much more concentrated in iodine. The bottom organic layer in the funnel is separated from the aqueous layer. Evaporating the organic solvent will then yield the isolated iodine solid.

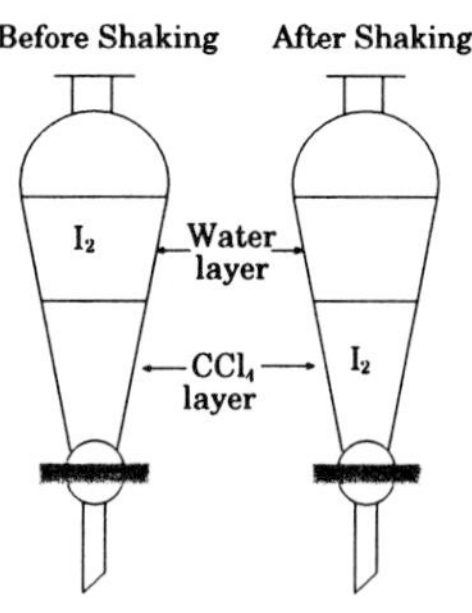

FIGURE 2.12: Separatory funnels.

Note that the more dense layer will be on the bottom of the separatory funnel, and the less dense on top. The densities of some common solvents are shown in the following table.

TABLE 2.8

TABLE 2.8: APPROXIMATE DENSITIES OF COMMON SOLVENTS

NAME	DENSITY (g/mL)
petroleum ether	0.66
ethyl ether	0.74
ethyl alcohol	0.79
acetone	0.79
methyl alcohol	0.81
benzene	0.90
water	1.00
dichloromethane	1.32
chloroform	1.49
carbon tetrachloride	1.59

Note that densities can change with temperature, and that as solutes are concentrated into the various solvents, the solution's density may change from that of the pure solvent. In some cases, Solvent A may be less dense than solvent B. The layers would be A on top, B on bottom. But if a significant amount of solute was extracted into solvent A, it is possible that its density could exceed that of solvent B, and the layers could be reversed in the funnel after the extraction.

The process of simple liquid-liquid extraction can be summarized as follows. Let X be the component one wishes to extract from a solution that has A as its solvent. The individual solute X would have a certain solubility in solvent A and another solubility in solvent B. In order to extract and isolate X, solvent B must be chosen so that:

1. **B and A are immiscible (not mutually soluble; two layers form)**
2. **X is more readily soluble in B than in A.**
3. **Solvent B and X can be readily separated**

In practice, the two solvents are almost always water and some organic solvent. The choice of solvent and how the extraction is performed depends on the specific solubilities of X and the origin of the mixture.

SIDEBAR: COCAINE STRUCTURE, PROPERTIES, AND ISOLATION

Cocaine, pictured at right, is a weak base with the formula $C_{17}H_{21}O_4N$. This is the so-called "free-base" form of the drug. Cocaine is only slightly soluble in water (1.67 g/liter) but it does act as a base, giving a characteristic color change with acid base indicators (e.g., turning red litmus blue.) The basic properties are due to the reaction:

$$C_{17}H_{21}O_4N + H_2O \longrightarrow C_{17}H_{21}O_4NH^+ + OH^- \quad \text{EQ. 2.4}$$

The most commonly available form of cocaine is cocaine hydrochloride, a salt formed by the reaction of cocaine with hydrochloric acid.

$$C_{17}H_{21}O_4N + HCl \longrightarrow C_{17}H_{21}O_4NH^+ Cl^- \quad \text{EQ. 2.5}$$

This chloride salt is quite water soluble (2500 g/liter) in accordance with our solubility rules. The compound's high water solubility allows it to be readily absorbed through the mucous membranes when sniffed up the nose. (Nasal decongestants use this same approach: the antihistamine is a sparingly soluble weak base made water soluble as the hydrochloride salt.)

N—CH_3, O, C—O—CH_3, O—C—C_6H_5, H, O

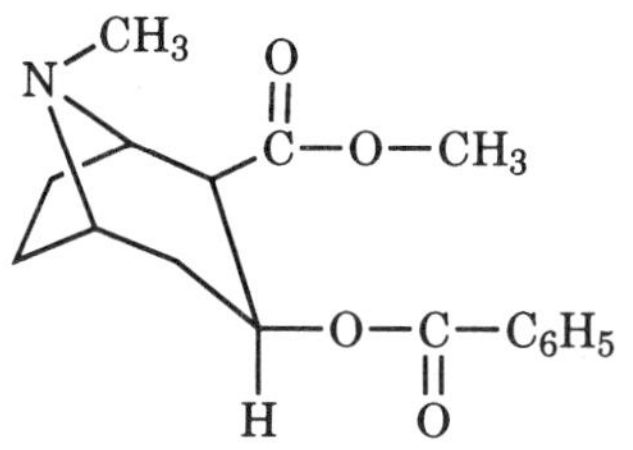

FIGURE 2.13: Structure of cocaine.

Cocaine is extracted from coca plant leaves by grinding and mixing with water. The filtered aqueous solution contains a low concentration of the free base. The aqueous solution is then mixed with a small amount of a non-polar organic solvent, into which the low-polarity base is extracted. Hydrochloric acid is then added to the organic solution. The highly polar cocaine hydrochloride salt precipitates out of the non-polar organic solution as a white solid.

Pure cocaine may be recovered from the hydrochloride of cocaine in a process known as "free-basing." First the solid hydrochloride salt is reacted with a basic aqueous solution, then the free base is extracted in a low-polarity solvent like ethyl ether. The hydrochloride salt of cocaine decomposes as it approaches its melting point. The free base cocaine, as a covalent compound, has a lower melting point and is more volatile than the hydrochloride salt. Thus it is the free base form of cocaine that is used when smoked. Cocaine's low polarity allows it to be absorbed by the fatty cell membranes of the lungs when smoked.

The solubility of cocaine in non-polar material facilitates cocaine's influence on the fat-like tissue of the brain. Cocaine, after passing across the blood/brain barrier, interferes with the neurotransmitter norepinephrine. This causes both the brief period of stimulation and the longer period of depression, anxiety, and craving for more cocaine. This may cause a psychological addiction to cocaine, which frustrates many who wish to free themselves of its influence. Scientists have recently

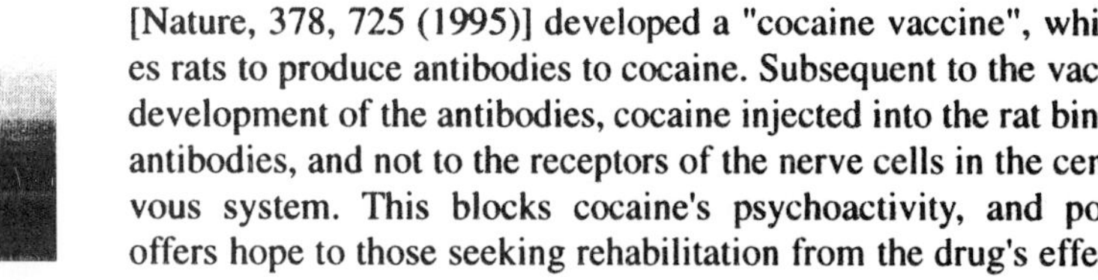

[Nature, 378, 725 (1995)] developed a "cocaine vaccine", which causes rats to produce antibodies to cocaine. Subsequent to the vaccine and development of the antibodies, cocaine injected into the rat binds to the antibodies, and not to the receptors of the nerve cells in the central nervous system. This blocks cocaine's psychoactivity, and potentially offers hope to those seeking rehabilitation from the drug's effects.

PROBLEM

Consider the example of the isolation of an organic acid synthesized in a non-polar organic solvent. In addition to the solvent (ethyl ether, or henceforth simply ether) and the acid, benzoic acid (C_6H_5COOH), there could be other products as well as excess starting materials for the reaction, such as bromobenzene, C_6H_5Br. The acid formed is a weak acid. Let us assume that the other materials present are highly soluble in ether and are not acidic.

An initial attempt might be to extract benzoic acid from the ether solvent using pure water. The reaction mixture in ether and pure water could both be added to a separatory funnel. Water and ether are not miscible, so two layers will form and the first criterion is met. Unfortunately benzoic acid is roughly one hundred times more soluble in ether than in pure water. Consequently relatively little of the benzoic acid will leave the organic phase and enter the aqueous phase, and such an extraction does not work.

A second attempt would be more involved, but much more effective in isolating benzoic acid. Again, the synthesis mixture, including benzoic acid and other solutes, is present in an ether solution. This synthesis mixture is reacted with an aqueous solution of NaOH in a separatory funnel. Benzoic acid and aqueous sodium hydroxide react to form the salt sodium benzoate (NaC_6H_5COO) in the reaction shown below:

$$C_6H_5COOH + NaOH \longrightarrow Na^+ + C_6H_5COO^- + H_2O \qquad \text{EQ. 2.6}$$

After mixing, the lower aqueous layer is removed. This layer contains sodium benzoate. (Remember that polar species generally do not dissolve in non polar solvents like ether, but that sodium salts are quite soluble in water.) This aqueous solution is then treated with a strong acid, for example HCl. This reaction reforms the benzoic acid.

$$C_6H_5COO^- + H^+ \longrightarrow C_6H_5COOH \qquad \text{EQ. 2.7}$$

Remember, since benzoic acid is a weak acid, it exists largely in its molecular, unionized form. We now have an aqueous solution of molecular benzoic acid (and excess H^+, Na^+ and Cl^-.) We now can take advantage of benzoic acid's high solubility in ether by repeating the extraction process: the aqueous benzoic acid solution is added to the separatory funnel along with pure ether. After mixing, the lower aqueous layer is removed and discarded. The ether layer now contains benzoic acid. This may seem as if we have gone in a circle: we started and ended with a solution of benzoic acid in ethyl ether solution. The difference is that at the end of the process described, only benzoic acid remains in the ether layer; the other original contaminants have been removed. Once benzoic acid is isolated in the ether layer, we can simply evaporate the ether, and pure benzoic acid remains. The extraction and isolation of benzoic acid is complete. This process is shown diagramatically in Figure 2.14.

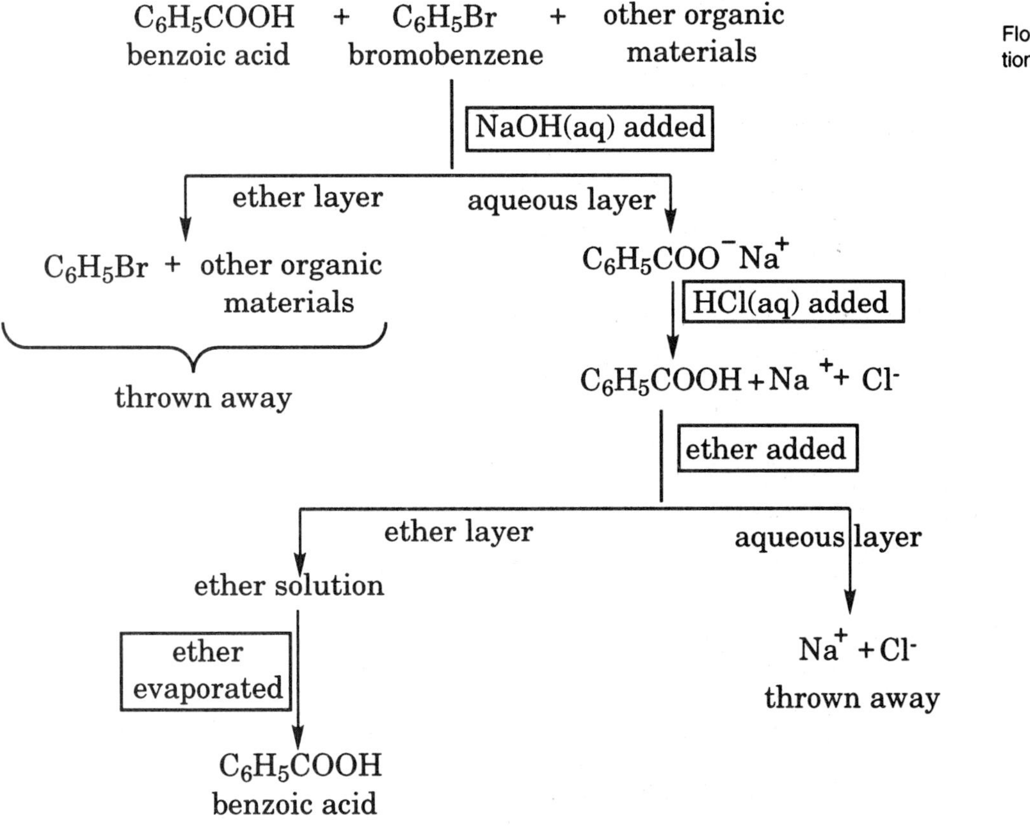

FIGURE 2.14: Flow diagram for extraction of pure benzoic acid.

2.10 CONTINUOUS EXTRACTION

In the previous section, we examined the extraction and removal of a solute from liquid solution using another solvent and a separatory funnel. If the solute was not particularly soluble in the solvent used for extraction, little solute is isolated in a one step process (e.g., the attempt to extract benzoic acid from ether using pure water.) One could do this type of extraction over and over: adding the extracting solvent to the solution in the funnel, shaking, removing the extracting solvent layer, add more extracting solvent to the funnel, etc. If the solubility of the solute in the extracting solvent is small in comparison to the solute's solubility in the original solvent, such multiple extractions are very inefficient. Instead, an apparatus like those pictured in Figure 2.15 may be used.

In Figure 2.15 (a), an organic solvent is heated in a boiling flask. The vapors rise into a cooler region called the condenser. Here the organic vapor converts back into a liquid that falls down the inner tube to the bottom of the second flask, containing an aqueous solution of the solute. As the organic liquid emerges from the fritted glass filter at the bottom of the tube, it rises through the aqueous layer, dissolving some solute along the way. The organic layer, now containing some solute, floats to the top of the flask and eventually flows down into the boiling flask. The organic solvent is again evaporated, and the process continues. Eventually (sometimes over a period of days) the solute is concentrated in the boiling flask. The process described works assuming that the organic phase is less dense than water. This is most often the case. If the organic phase is more dense than water, then the apparatus in Figure 2.15 (b) may be used.

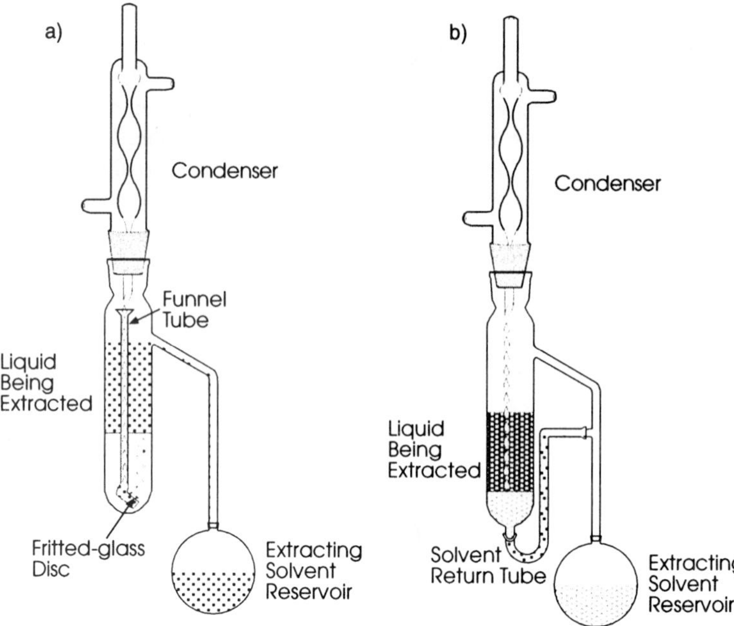

FIGURE 2.15: Continuous extraction apparatus a) and b).

Figure 2.15 showed two means of continuous liquid-liquid extraction. Often one wishes to extract one or more solutes from a solid solution (e.g., quinine from cinchona bark) in a continuous fashion. Figure 2.16 shows a Soxhlet extractor that facilitates continuous extraction from a solid mixture into a liquid solution. The solid mixture is placed in a porous thimble. The extracting solvent is boiled, the vapors rise and then return to the liquid phase in the condenser. The hot liquid then saturates the solid in the thimble until the liquid level rises to the top of the siphon tube. The liquid, and the solute it has dissolved, are then siphoned back into the boiler. The pure solvent then boils, and the process continues. Eventually, the extracted solute(s) are concentrated in the boiler.

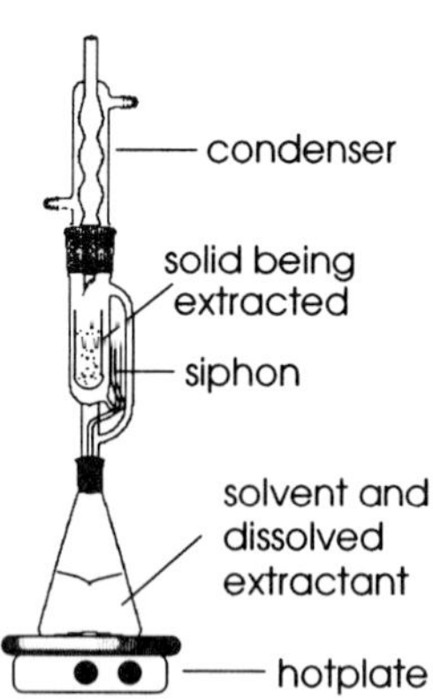

FIGURE 2.16: Soxhlet extractor.

SIDEBAR: EXTRACTION AND ISOLATION OF FIREFLY LUCIFERIN

One of the main chemicals that enables the light emission of fireflies is luciferin. It can be extracted from fireflies in the following abbreviated procedure [from Advances in Natural Products Chemistry, Edited by S. Natori, N. Ikekawa and M. Suzuki.] The abdomens of frozen fireflies were removed and ground in cold acetone (to remove the fat.) After filtering, the residue was washed with hot water (to extract the desired component; other unwanted water soluble components were also extracted.) The aqueous extract was acidified, and extracted several times with ethyl acetate (extracting the desired components and a few other components.) The ethyl acetate solution was evaporated, and the residue was further purified by column chromatography (discussed below) using a cellulose

column and a solvent mixture of ethyl acetate, ethanol and water. A yellow-green fluorescent band was collected, and further purified by additional chromatography, extraction and crystallization.

2.11 CHROMATOGRAPHY

In the previous two sections, we have examined means by which one or more solutes can be extracted from a mixture, discarding the remainder like coffee grounds after having brewed the coffee. Often in chemical practice it is advantageous to isolate every component of a mixture. The most common means of achieving that end is **chromatography**. The process of chromatography dates back to the early 1900s when Mikhail Tswett, a Russian biochemist educated in Italy and teaching in Poland, conducted an interesting experiment. He placed a mixture of plant pigments at the top of a column containing solid calcium carbonate. He then added a liquid solvent, petroleum ether, and allowed it to flow down the column. The various plant pigments were separated into a series of colorful bands as they progressed down the column. Tswett chose chromatos, the Greek word for color, as the basis for naming the process chromatography. It is perhaps more than coincidence that Tswett was focused on "color," since that is the meaning of "Tswett" in Russian.

PREPARATIVE COLUMN CHROMATOGRAPHY:

Tswett's experiment can be shown in a simplified form in Figure 2.17. Consider a mixture, with components A, B, and C. Also consider the column as being composed of a stationary solid phase and a mobile liquid phase (also known as the eluting solvent.) Components A, B, and C have some degree of attraction to the solid stationary phase, and some degree of affinity for the liquid mobile phase. A sample containing all three components is put at the top of the column (#1). Then the stopcock at the bottom of the column is opened and additional eluting solvent is added to maintain a flow through the column. The forces of attraction between A and the stationary and mobile phases differ from those same forces for B and C. Suppose that A is weakly attracted, B is moderately attracted and C is strongly attracted to the stationary phase. On average, A spends more time in the mobile phase than B, which spends more time in the mobile phase than C. As the mobile phase continues down the column (#2-6), A, B, and C are gradually separated. The correct choice of stationary phase, mobile phase and length of column should ultimately result in a total separation of A, B, and C. The

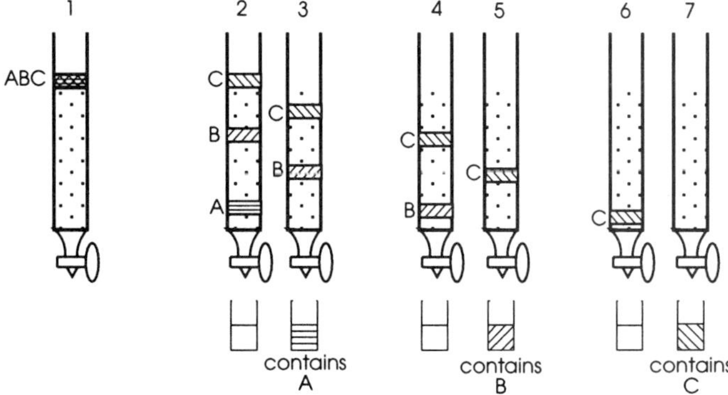

FIGURE 2.17:
Column chromatography.

eluting bands can be collected in individual beakers, the solvent evaporated, and the isolation of A, B, and C is complete.

Thin Layer/Paper Chromatography

Sometimes one does not need to isolate significant quantities of every component. A simpler and generally faster approach is to use thin layer chromatography (TLC) illustrated below.

As shown in Figure 2.18, small thin strip of absorbent solid (stationary phase) has spots of solutions containing various components placed near one end. That end of the strip is placed into the eluting solvent, with the solvent level below the spots. As the eluting solvent migrates up the adsorbing strip, the components of the mixture in the original spot will distribute themselves between the stationary and mobile phases. Components more strongly attracted to the mobile phase and less strongly attracted to the stationary phase will advance more rapidly than other components in the mixture. In Figure 2.18, a three-component mixture (on the right-hand edge) is resolved into three separate spots in this fashion.

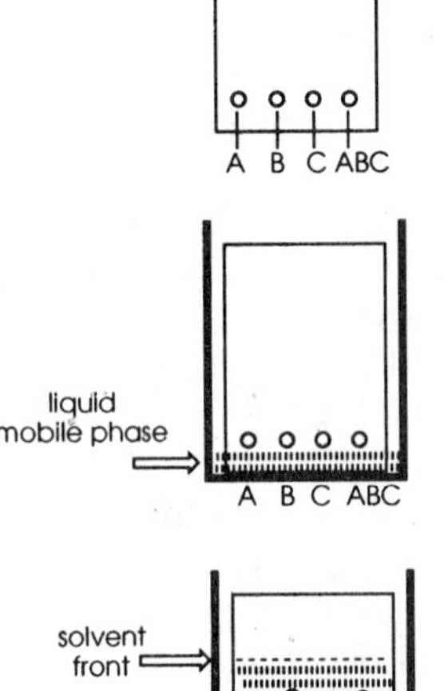

FIGURE 2.18: TLC chamber and plate.

Three pure compounds A, B, & C, along with a mixture of A, B, & C, are placed on a plate close to the bottom edge

The plate is placed into a jar containing the liquid mobile phase.

The liquid mobile phase rises through the four spots until the mobile phase is close to the top of the plate. The height to which the solvent rises is called the solvent front.

As the components move up the TLC, their relative progress can be measured in terms of the length migrated by a single component and the length migrated by the eluting solvent, as shown in the figure. The retention factor (abbreviated R_f) can be determined by the equation:

$$R_f = \frac{\textbf{length traveled by solvent}}{\textbf{length traveled by eluting solvent}}$$

TLC and paper chromatography can be used to identify mixtures. This is a good qualitative method of determining whether or not a specific component is present. This can be done by running one or more knowns along with the unknown. In Figure 2.18, known A, B and C were spotted alongside a mixture of the three in order to identify each spot in the mixture. If the R_f values between the spot in the unknown and that for the known match, that is evidence in favor of the compound being present in the unknown. TLC is commonly used as a rapid means of determining the success of a synthesis, or the subsequent purification by extraction. Appearance of many spots indicates an impure mixture. Appearance of a single spot with the correct R_f value indicates a successful synthesis and/or purification.

Concept Check: TLC and Rf Values

In a TLC experiment, known compounds A, B, and C were spotted onto strips coated with silica gel. When eluted with ethyl ether, compounds A, B, and C had R_f values of 0.31, 0.85, and 0.52. An unknown solid thought to contain A, B, C, or possibly some combination, was analyzed in the same fashion. Two spots were observed: one had moved 3.7 cm, the other had moved 6.1 cm from the start. The solvent front had eluted 7.3 cm when the strip was removed from the jar. Which components were likely present in this unknown? Which components were likely absent?

SOLUTION: The R_f values are ratios of the distance traveled by the solute to the distance traveled by the solvent front. Spot one and spot two have R_f values calculated below:

$$R_f\,(\text{spot one}) = \frac{3.7\text{ cm}}{7.3\text{ cm}} = 0.51$$

$$R_f\,(\text{spot two}) = \frac{6.1\text{ cm}}{7.3\text{ cm}} = 0.84$$

Spot one has an R_f value sclose to that observed for C. Spot two has an R_f value close to that observed for B. The unknown likely contains B and C, but likely does not contain A.

High Performance Liquid Chromatography: This approach is similar to preparative column chromatography, but more advanced. Instead of relying on gravity to pull the mobile phase along, a high performance liquid chromatograph (HPLC) generally uses high pressure to push the mobile phase through the stationary phase. See Figure 2.19.

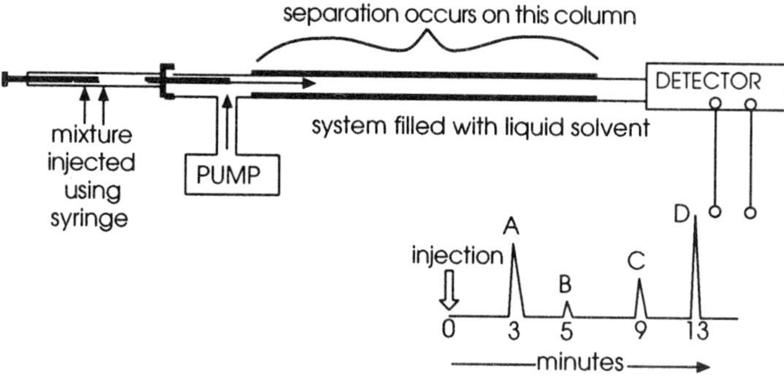

FIGURE 2.19: HPLC diagram.

HPLC can use a variety of detectors to monitor the eluent coming off of the column. Simple preparative column chromatography relies on the human eye to see a compound's visible color in a band eluting from the column. If the component should be colorless, one is generally out of luck. Using a refractive index detector or an ultraviolet light absorbance detector as part of the HPLC allows detection of components that would be invisible to the human eye. The type of columns used in HPLC allow a much greater resolution of components in a mixture. When separation of similar compounds might prove extremely difficult on a preparative column, they may be readily separated using HPLC.

GAS CHROMATOGRAPHY

Gas chromatography (GC) is similar to liquid chromatography, but differs in that the mobile phase is gas, not liquid. As shown in Figure 2.20, the arrangement is similar to that for an HPLC.

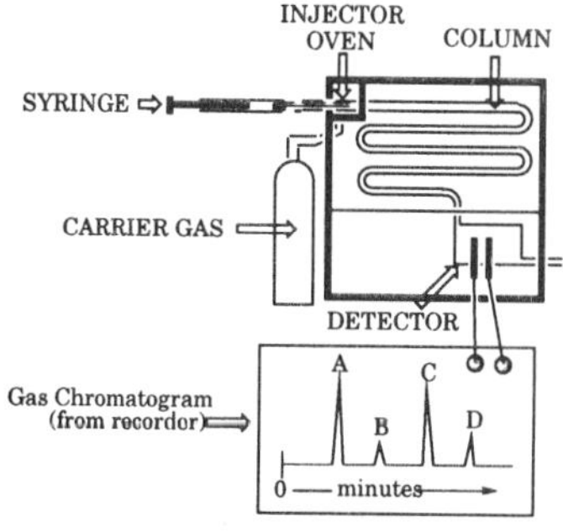

FIGURE 2.20: GC diagram.

The use of gas as the mobile phase allows different types of columns to be used, including exceptionally long columns to improve the separation between components that elute near one another. Gas Chromatographs also commonly use variable temperature to facilitate analysis. Components analyzed by GC must be volatile enough that they can enter the gas phase within the allowed temperature range. Non-volatile materials, or compounds that readily decompose with heating can not be effectively analyzed by GC, but might be analyzed by HPLC. GC detection by thermal conductivity, flame ionization or electron capture can be exceptionally sensitive. Analysis of trace components (those present at a very low concentration, even parts per billion) is most often done by GC.

Chromatography in Combination

When non-destructive means of detecting the individual components in a mixture are used, the eluent from a chromatograph can be analyzed further. Components separated by chromatography can thus be analyzed and identified by other analytical techniques. These so-called hyphenated instruments essentially put another instrument on the end of a chromatograph to analyze the individual components.

One specific technique we mention here is mass spectroscopy. The combination of gas chromatography and mass spectroscopy (GC-MS) is both common and powerful. After separation by GC, each peak in the chromatogram is separately analyzed by mass spectroscopy. This allows unique characterization of each component of the mixture separated by the GC.

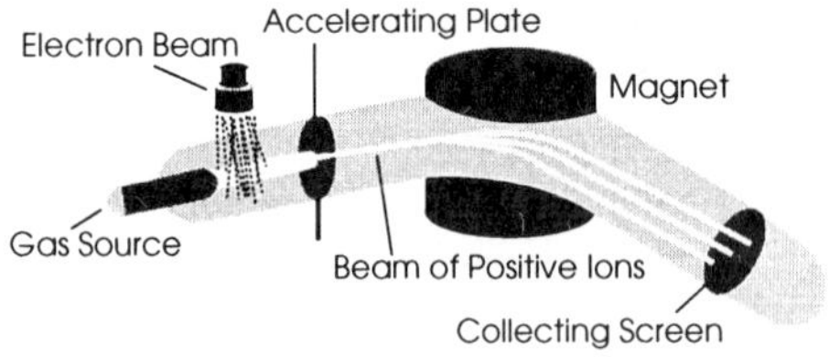

FIGURE 2.21: Block diagram of neon isotopes separated by MS.

There are various means by which mass spectroscopy can function. The one pictured in Figure 2.21 is a magnetic sector mass spectrometer. A stream of carrier gas from the GC contains the isolated sample (in this case, neon). The sample is swept into an evacuated chamber, and a stream of high energy electrons bombard the sample. The strength of the electron beam is adjustable. This knocks off one (or rarely, more) electron(s), forming positively charged neon cations. These neon cations are attracted towards a negative plate, and pass thru a small collimating opening. This beam of cations then enters a magnetic field. A moving particle in a magnetic field is bent a certain amount, depending on the mass to charge ratio (known as m/z). As shown in this figure, the sample of neon is split into three separate trajectories, with varying intensities. Each are collected by a detector and plotted out in a mass spectrum. The mass to charge ratio of about 20 is most predominant, followed by a much smaller peak at 22, and a very small peak at 21. These represent the three naturally occuring isotopes of neon.

When the sample introduced into the mass spectrometer is polyatomic, the mass spectrum becomes more complicated, and sometimes strongly depends upon the nature and strength of the electron bombardment. If the bombardment is not too strong, frequently a significant fraction of the molecules impacted will become molecular cations:

$$\mathbf{M = M^+ + e\text{-}} \qquad \text{EQ. 2.8}$$

The M^+ product of the molecule M is refered to as the parent peak in the mass spectrum. It may or may not be the largest peak in the mass spectrum (which are all normed to the largest peak present.)

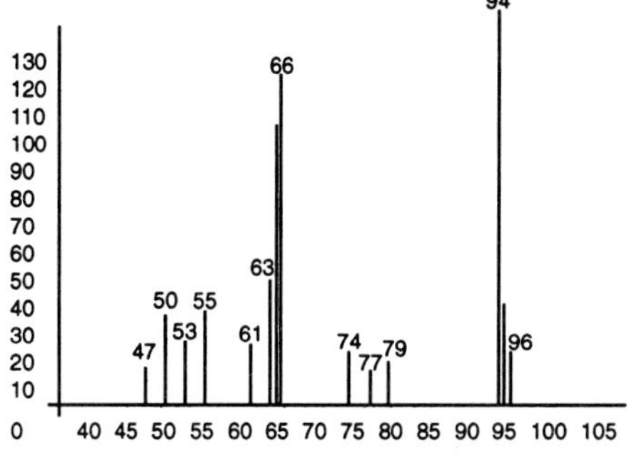

FIGURE 2.22: Mass spectrum of phenol.

In Figure 2.22, we see the largest peak, and nearly the highest mass peak, at m/z = 94. This value alone (even assuming only +1 charge present) can not identify the species; a mass of about 94 can be obtained many ways: C_7H_{10}; C_6H_8N, C_6H_6O, etc. Information in addition to the rough mass of the parent peak is needed. This information may be obtained in several ways. High resolution mass spectrometers, for example, can determine small but significant differences in masses. For example, C_7H_{10}, C_6H_8N, and C_6H_6O weigh 94.07825, 94.06567, and 94.04186 amu respectively. They are distinguished by the exact values of the isotopic masses involved.

The natural isotopic distribution gives another clue to deciphering the mass spectrum. Note that in Figure 2.22, the parent peak at 94 is followed by a much smaller peak at 95. The ratio of peak intensities at 94 and 95 show that these peaks are almost certainly due to ^{12}C and ^{13}C isotopes (^{12}C = 98.892%; ^{13}C = 1.108%.) When the molecule contains chlorine or bromine, both of which have significant fractions of two naturally occuring isotopes, the ratios of peaks two mass units apart are a clear cut atomic signature indicating the presence of the specific halogen. ^{35}Cl = 75.53%; ^{37}Cl = 24.47%; ^{79}Br = 50.52%; ^{81}Br = 49.48%

Another characteristic allowing one to identify a substance from its mass spectrum is given by the mass and relative abundance of peaks in addition to the parent peaks. These form the so-called the "fragmentation pattern". The molecular cation formed in the ionization chamber may break into several pieces, each of which could, under some circumstances, acquire a positive charge and have its mass determined. The specific fragmentation pattern is characteristic of a molecule in the way that fingerprints can identify individuals. If the pattern has been previously recorded for a known sample, it can usually be identified later when observed as an unknown.

Other techniques are used besides mass spectroscopy to analyze the separated components of a mixture. Chapters 3 and 4 look at two powerful analytical tools: infrared spectroscopy and nuclear magnetic resonance.

Acid: any species that, when dissolved in water, increases the H^+ concentration.

Base: any species that, when dissolved in water, increases the OH^- concentration.

Centrifugation: the process of separating substances of differing densities by the action of centrifugal force.

Chromatography: a wide variety of experimental techniques whereby components of a mixture are separated by means of their difference in attraction to stationary and mobile phases.

Concentration: term expressing the amount or fraction of a particular substance in a solution.

Decantation: the pouring off of a liquid (decantate) in a fashion that avoids disturbing the sediment (precipitate.)

Distillation: a means of separating components of a mixture by means of differing boiling points.

Extraction: a wide variety of experimental techniques whereby a particular component is removed from a mixture or solution.

Filtration: means of separating a heterogeneous mixture into the solid residue and the liquid filtrate.

Heterogeneous mixture: a mixture of components with more than one phase present. Physical and chemical properties differ among the separate phases in the mixture.

Homogeneous mixture: also known as a solution; a single phase wherein properties are uniform throughout the mixture.

Non-electrolyte: covalent species that, when dissolved in water, does not produce ions.

Solute: one of the minor components in a homogeneous mixture.

Solution: (see homogenous mixture.)

Solvent: the major component in a homogeneous mixture.

Strong acid/base: an acid/base that is significantly ionized in aqueous solution.

HOMEWORK Problems

SOLUTIONS AND CONCENTRATION

1. A certain type of magnet is made from the following amounts of each ingredient: 38.4 g iridium, 64.0 g nickel, 16.0 g cobalt, 201.6 g iron. What is the % by mass of each of the components of the mixture?

2. Estimate the approximate % by volume of each component in a mixture of 20.0 ml ethanol, 30.0 ml of acetone, and 25.0 ml of water.

3. A low melting point solder is prepared by combining 75.0 g of lead, 56.2 g of tin, and 18.8 g of bismuth. What is the % by mass of each of the components of the mixture?

4. Specify which of the components are solutes and which are solvents in the previous problem.

5. The density of methanol at room temperature is 0.791 g cm^{-3}. The density of ethyl bromide at room temperature is 1.461 g cm^{-3}. What is the % by mass of each component when 50.0 ml of methanol and 50.0 ml of ethyl bromide are mixed?

6. What is the % by volume of methanol in an aqueous solution that is 40.0% methanol (by mass). The 40.0% by mass methanol solution has a density of 0.9347 g cm^{-3}. The density of pure methanol at room temperature is 0.791 g cm^{-3}.

7. A certain ore has a density of 3.50 g cm^{-3}. This ore is 6.25% copper (by mass.) What volume of ore (in cubic meters) must be processed in order to obtain 1000.0 kg of pure copper? What mass of waste from the ore (that part not containing copper) must be disposed of in order to produce this amount of copper?

ACIDS AND BASES

8. Your body produces both lactic acid and hydrochloric acid. Explain how these two acids differ in terms of their strength (i.e., degree to which they ionize to form H^+.)

9. Metal oxides are basic anhydrides; that is when dissolved in water, they form basic solutions (increased OH^- concentration.) What bases are formed when the following are dissolved in water? Specify if the solution is a weak or strong base.
   (a) Na_2O
   (b) CaO
   (c) NH_3

10. Nonmetal oxides are acidic anhydrides; that is when dissolved in water, they form acidic solutions (increased H^+ concentrations.) What acids are formed when the following are dissolved in water? Specify if the solution is a weak or strong acid.
    (a) SO_3
    (b) SO_2
    (c) Cl_2O_7
    (d) P_4O_{10}

11. The strength of commercial vinegar samples (aqueous solutions of acetic acid) can be analyzed by determining the number of H^+ ions in a liter of the vinegar sample. Is this number of H^+ ions equal to, less than, or greater than the number of acetic acid molecules present per liter. Why?

12. The strength of commercial samples of muriatic acid (aqueous solutions of hydrogen chloride) can be analyzed by determining the number of H^+ ions in a liter of muriatic acid. Is this number of H^+ ions equal to, less than, or greater than the number of hydrogen chloride molecules present per liter. Why?

Properties of Ions and Ionic Compounds; Electrolytes; Polarity

13. Write the Lewis dot diagrams of the following cations. Note that the charge on the cation represents the number of electrons lost; this reduces the calculated value of A in the NAS calculation.
    (a) H_3O^+
    (b) NO^+
    (c) SF_3^+
    (d) PCl_4^+

14. Write the Lewis dot diagrams of the following anions. Note that the charge on the anion represents the number of extra electrons; this increases the calculated value of A in the NAS calculation.
    (a) NO_3^-
    (b) SO_3^{2-}
    (c) ClO_4^-
    (d) PO_4^{3-}

15. Name each of the following:
    (a) $K_2Cr_2O_7$
    (b) $Al(NO_3)_3$
    (c) $Ni_3(PO_4)_2$
    (d) $CaBr_2$
    (e) Hg_2Cl_2

16. Specify whether each of the compounds in problem #15 is/is not water soluble.

17. Name each of the following:
    (a) $Pb(C_2H_3O_2)_2$

(b) $CoCO_3$
(c) $(NH4)_3PO_4$
(d) $Fe(OH)_3$
(e) $NiCl_2$

18. Specify whether each of the compounds in problem #17 is/is not water soluble.

19. Write the formula for each of the following
    (a) hydrofluoric acid
    (b) nitric acid
    (c) sodium hydroxide
    (d) ethanol
    (e) iron (III) bromide

20. Classify each of the compounds in problem #19 as ionic or covalent

21. Classify each of the compounds in problem #19 as strong, weak or non-electrolyte.

22. Classify each of the following as ionic or covalent.
    (a) $HClO_2$
    (b) NH_3
    (c) Li_2CO_3
    (d) CH_3OH
    (e) SF_6

23. Classify each of the compounds in problem #22 as strong, weak or non-electrolyte.

24. Would you expect potassium carbonate to be more soluble in ethanol or toluene? Explain.

25. Would you expect nitrogen gas (N_2) to be more soluble in water or in liquid benzene? Explain.

EXTRACTION

26. Suppose you have synthesized a weak organic base in an organic solvent such as chloroform. After disappointing results from extracting the compound with pure water, it was determined that making the water acidic before using it to extract the weak base greatly improved the extraction. Offer an explanation of these results.

27. A weak acid was synthesized in aqueous solution. At this dilute aqueous concentration, the species is over 50% ionized. While attempting to extract the acid into an organic solvent, it was observed that the extraction was far more successful when the water was first treated with a strong acid prior to extraction. Offer an explanation of these results.

28. You are given a mixture of 3 solid organic compounds, all of which are sol-

uble in ether, but none of which readily dissolve in water. You wish to isolate the three organic compounds (A, the weak acid; B, the weak base; and N, the neutral compound) as the pure solids. You are also given the following solutions: ether, water, aqueous HCl, and aqueous NaOH. Describe an extraction/evaporation process by which pure A, pure B and pure N can be isolated. HINT: a flow chart would be a big help!

CHROMATOGRAPHY

29. What kind of chromatography could be used to achieve the following objectives? Briefly explain your choice.
    (a) to determine if a prescription medicine (pill) contains 2, 3 or 4 separate ingredients.
    (b) to obtain one gram of a pure sample of a compound found in fairly large concentration in a plant or tree.
    (c) to find the number and relative amounts of components in a complex mixture like gasoline.

30. What kind of chromatography could be used to achieve the following objectives? Briefly explain your choice.
    (a) to collect samples of 3 dyes combined in a commercial mixture.
    (b) to determine if a white powder obtained during a police raid is heroin or aspirin.
    (c) to determine what drug was used in an overdose by patient in the emergency room (given the pumped stomach contents.)

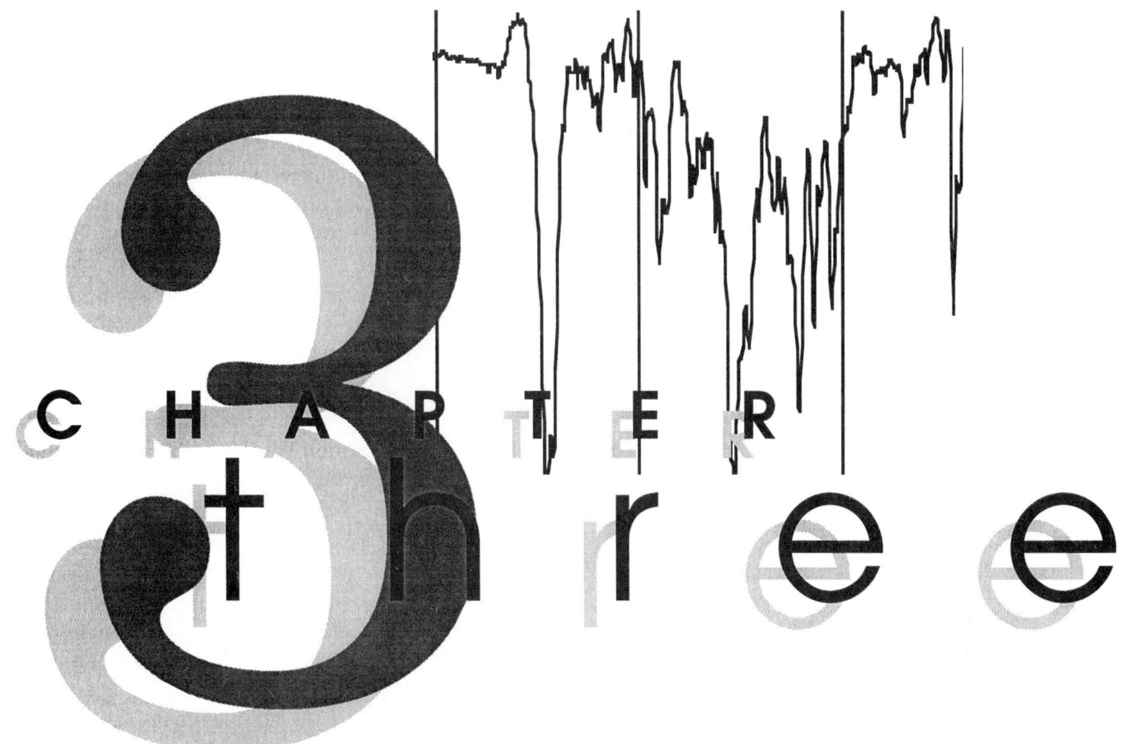

Spectroscopy and Structure

SPECTROSCOPY AND Structure

In the previous chapter, we briefly examined the problem of separating complex mixtures and isolating pure compounds. Two examples of compounds obtained from natural sources were quinine (the anti-malarial agent) isolated from the bark of the cinchona tree and morphine (the important pain killer) isolated from the opium poppy.

After separating and purifying these compounds, chemists attempted to determine their chemical structures. This was not a simple task, as evidenced by the complex structures of the two compounds in question.

Chemists used many methods to determine these and numerous other complex structures, and new methods are being developed all the time. In this chapter, we consider one of the powerful methods for structure determination: infrared (IR) spectroscopy. By determining what **frequency** of **infrared radiation** a substance absorbs, we can gather clues about the unknown structure. In the next chapter we introduce a method whose recent development has relied on high speed computers and superconducting magnets. This method is termed nuclear magnetic resonance spectroscopy (NMR). By the way, NMR has nothing to do with the radiation associated with atomic bombs or nuclear power plants. As we will see, the radiation in NMR does not involve radioactive decay, and it is not harmful. In fact, an exciting development of this method is the technique called MRI (magnetic resonance imaging) which is used for medical diagnosis.

a) H_3CO H–C H N H C=C H H N

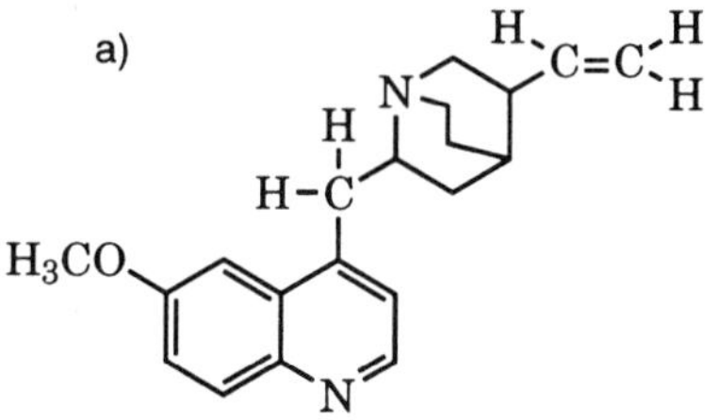

b) CH_3 N HO O OH

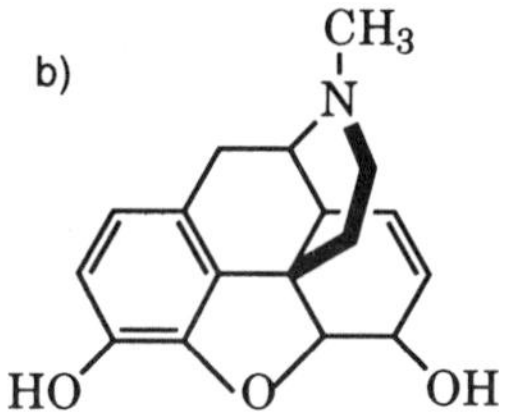

FIGURE 3.1: Structure of a) quinine and b) morphine.

3.2 THE FUNCTIONAL GROUP

Chemical structure and chemical properties are closely related. One way of determining how a group of compounds with similar properties is structurally related is to determine a distinctive group of 2-5 atoms that occur in all those chemically similar compounds. This collection of atoms shared by similar compounds is known as a **functional group**. For example, when a carbon atom is attached to oxygen which in turn is attached to a hydrogen atom (such as in ethyl alcohol, seen below) the COH is a functional group called **alcohol.**

Alcohols are often symbolized as R-OH, where R represents the **backbone** to which the functional group is attached. As we saw earlier, in Table 2.6, varying the number of carbons in the backbone connected to the functional group produces a family of related compounds, all of which are alcohols.

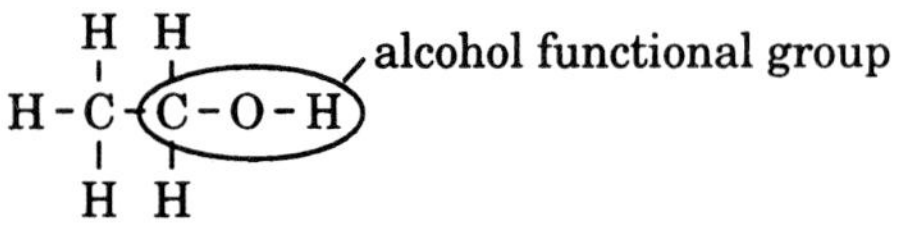

H-C-C-O-H (alcohol functional group)

FIGURE 3.2: Ethanol, an example of the alchohol functional group.

Functional groups are not limited to organic compounds, and the backbone need not be a chain of carbon atoms. Silanols, for example, are a collection of inorganic compounds with distinctive properties. Though quite different from organic alcohols, silanols do share some properties with alcohols. Silanols have the common structural element shown below.

For our current purposes, the discussion of functional groups will be limited to organic examples. Table 3.1 lists most of the important organic functional groups.

H_3C—Si—OH (OH, CH_3) (silanol functional group)

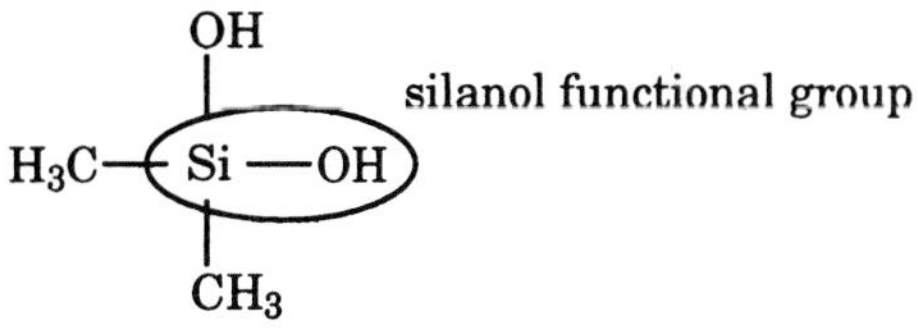

FIGURE 3.3: The silanol functional group.

Several different functional groups may contain similar elements. For example the **carboxylic acid, ketone** and **aldehyde** groups all contain a carbon oxygen double bond. Both the carboxylic acid and alcohol functional groups have hydroxyl (OH) bonded to a carbon. But each of these listed groups

TABLE 3.1

TABLE 3.1: ORGANIC FUNCTIONAL GROUPS

single bonds only		with C=O bond		with double/triple bonds excluding C=O	
alcohol	C—O—H	**ketone**	C—C(=O)—C	**nitrile**	C≡N
amines	C—N(H)—H	**aldehyde**	C(=O)-H	**alkene**	C=C
	C—N(H)—C	**carboxylic acid**	C(=O)—OH	**alkyne**	C≡C
	C—N(C)—C	**amides**	C(=O)-N(H)-H	**aromatic**	e.g. benzene C_6H_6
ether	C-O-C		C(=O)-N(H)-C		e.g. phenyl C_6H_5
halides	C—X (X = F, Cl, Br, I)		C(=O)-N(C)-C		
		ester	C(=O)—O—C		

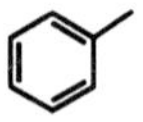

has a distinctive set of chemical properties. Thus a carboxylic acid reacts in an entirely different manner than an alcohol, a ketone, or an aldehyde. Similarly, although the **amine** and **amide** groups both have carbon nitrogen single bonds, and their group names differ in only one letter, they are very different functional groups!

At the right side of the functional group table are three groups that contain double or triple carbon to carbon bonds. They are distinctive functional groups, but often they are treated as simply part of the backbone. The backbone, however, is usually made of carbon to carbon single bonds. Several functional groups can occur in a single compound. Two examples are shown with the functional groups labeled.

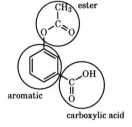

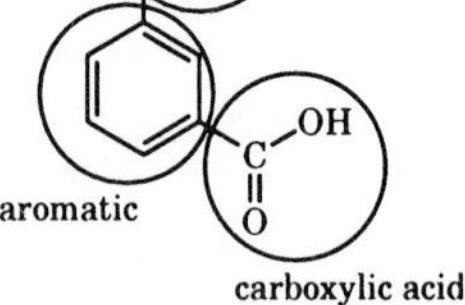

FIGURE 3.4: Compounds with mulitple functional groups.

3.3 THE USE OF IR SPECTROSCOPY TO DETERMINE FUNCTIONAL GROUPS

Infrared spectroscopy is widely used as a quick method of determining what characteristic functional groups may be present in a compound whose structural formula is not already known. To illustrate how the infrared spectrum identifies functional groups, let us turn to a drug example that often confronts the police. During a drug raid, the police found a white powder that they suspected might be cocaine. The suspect claimed that it was powdered sugar. The police sent the powder to the crime lab, where the chemist determined the infrared (IR) spectrum of the white powder. This spectrum was compared with literature spectra for samples of cocaine and sugar (sucrose.) The spectra are shown below.

Note the presence of the large absorption peak around 1700 cm^{-1} in the known cocaine spectrum; this absorption peak is absent in the powdered sugar spectrum, and is also absent in the spectrum of the unknown white powder. This absorbance difference indicates that cocaine has a functional group that is not present in sugar. Note the structural differences between molecules of cocaine and sugar.

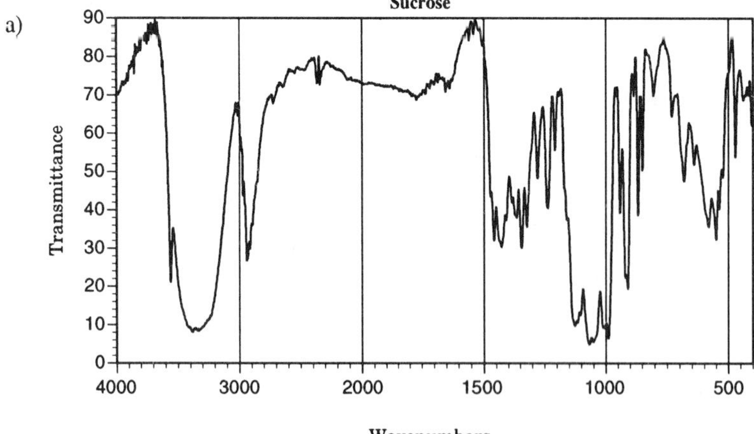

FIGURE 3.5: IR Spectra of a) known powdered sugar (sucrose)..(cont'd on next page).

b)

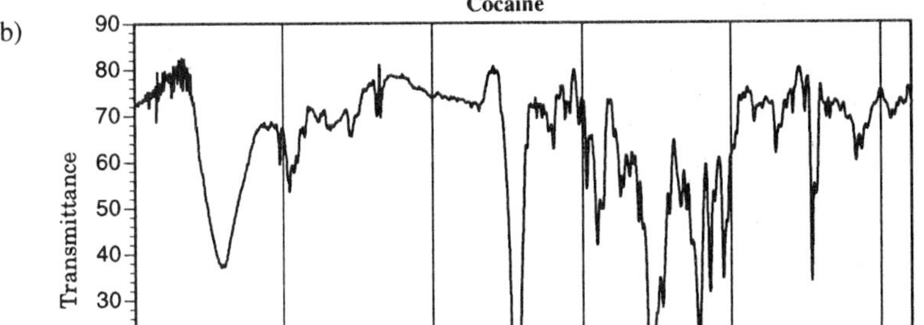

c)

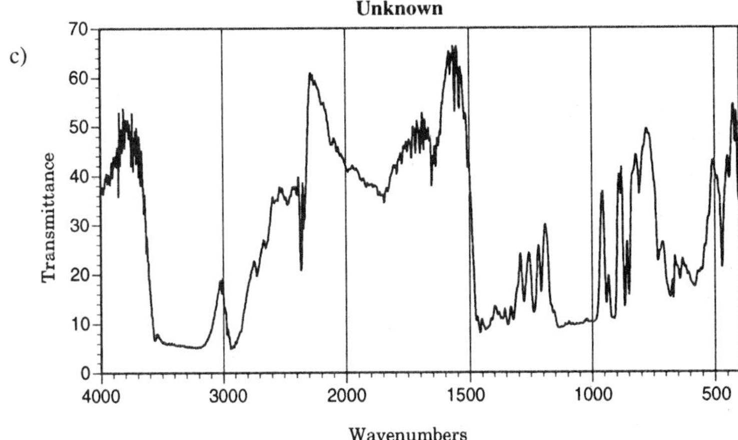

FIGURE 3.5 (cont'd): IR Spectra of b) known cocaine and c) unknown white powder.

The unknown white powder has an IR spectrum quite unlike that for cocaine but nearly identical to that for sugar, known as sucrose. Thus the chemist can confidently tell the police that the white powder was not cocaine. If the IR analysis resulted in a spectrum similar to that of cocaine, additional confirmatory tests would have been conducted.

a) [structure of cocaine: $N-CH_3$; $C(=O)-O-CH_3$; $O-C(=O)-C_6H_5$; H]

b) [structure of sucrose: CH_2OH, H, O, H, OH, H, OH, H, OH, O, CH_2OH, O, H, OH, H, CH_2OH, OH, H]

FIGURE 3.6: Structures of a) cocaine and b) sucrose.

Every compound has a distinctive IR spectrum that indicates its unique structure. IR spectra are produced by shining infrared radiation onto a prepared sample, and determining which frequencies are absorbed, and with what intensity. To better understand this, we need to examine the **electromagnetic spectrum** and the properties of light.

3.4 ELECTROMAGNETIC RADIATION

Electromagnetic radiation, or simply light, can be thought of as an oscillating electric field giving rise to a magnetic field (as in an electromagnet) and an oscillating magnetic field that gives rise to an electric field (as in an electric dynamo).

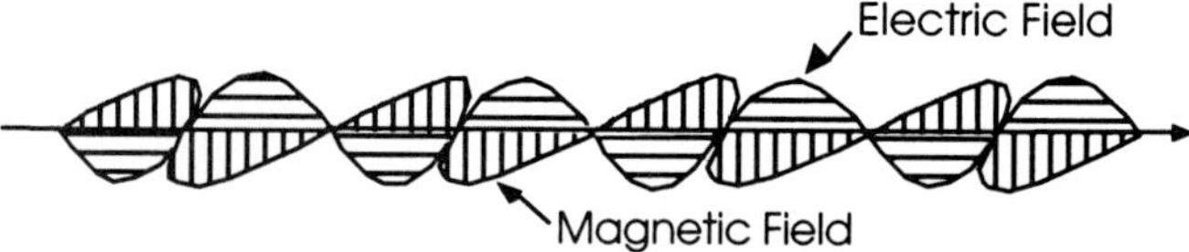

FIGURE 3.7: Electromagnetic radiation.

Light considered as a wave has an amount of energy inversely proportional to **wavelength**, λ, (pronounced "lambda") and directly proportional to frequency, ν, (pronounced "nu"). Wavelength is the distance between identical positions on the wave.

Frequency is the number of wave cycles that pass a point per second and is expressed as **Hertz (Hz)** or **cycles per second (cps)**. For example, in Figure 3.8, if 3 wave crests, moving in the direction represented by the arrow, pass point X in one second, the frequency would be 3 Hertz (3 cycles per second). You use frequency each time you tune your radio to FM or AM; a typical radio scale is provided in Figure 3.9.

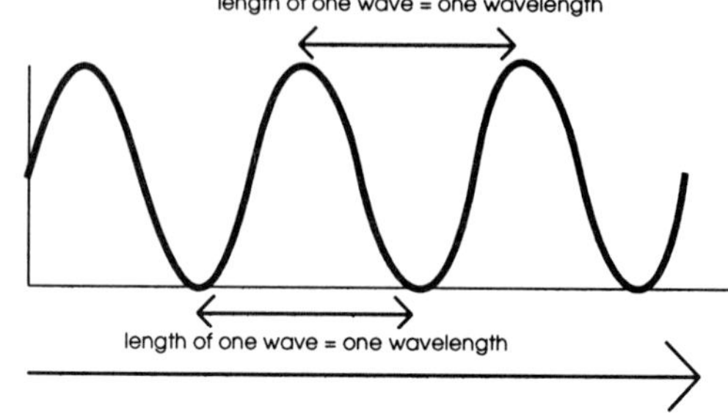

FIGURE 3.8: Wavelength of light.

The wavelengths of the radio frequencies given above range from 3-600 meters. Electromagnetic radiation encompasses several distinct regions, as shown below, and varies widely in both wavelength and frequency.

Frequency and wavelength are related by the equation:

$$c = \lambda\nu \quad \text{EQ. 3.1}$$

where "c", the speed of light, equals the product of λ, the wavelength, times ν, the frequency. Since the speed of light is constant, wavelength and frequency are always inversely related; i.e., as one increases, the other decreases. The energy of light is directly proportional to the frequency, and thus energy is inversely proportional to the wavelength, as shown in the equation

$$E = h\nu = \frac{hc}{\lambda} \quad \text{EQ. 3.2}$$

where h is Planck's constant, with a specific value given in the appendix. Thus in Figure 3.10, low energy, long wavelength, low frequency radiation is found on the right hand side, while high energy, short wavelength and high frequency radiation is found on the left hand side.

Radio Scale		
FM	88 - 108	MHz
AM	530	1700 kHz

FIGURE 3.9: The range of common radio frequencies.

The infrared region of the spectrum is most commonly observed in the form of heat lamps, used for keeping convenience food warm. While to our eyes the radiation from these lamps looks red, most of the energy is given off at wavelengths longer than that associated with the color red. These are wavelengths to which the human eye does not respond. Living objects, whether human, animal or plant, radiate heat. This radiated heat is observed in the IR region of the electromagnetic spectrum. There are special photographic films and cameras able to take pictures using IR radiation instead of visible light. Thus satel-

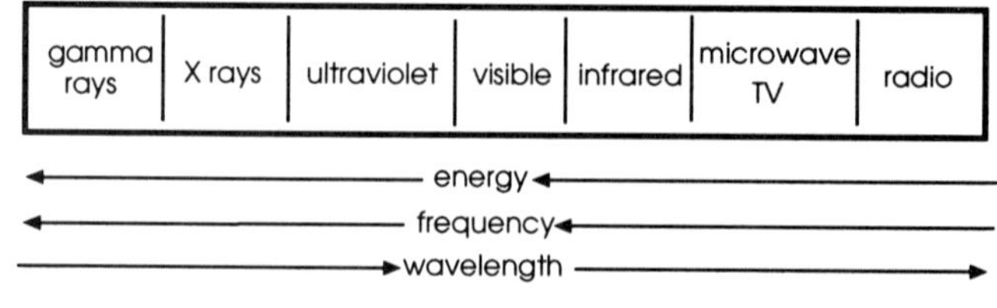

FIGURE 3.10: Electromagnetic spectrum.

lites in the sky with these cameras can detect which trees in a forest are alive and which are dead. Also special night vision goggles have been developed that allow people to see in the dark by converting the IR radiation given off by living beings into visible radiation.

Concept Check: Wavelength, Frequency, Wavenumber and Energy

It is standard procedure when using an FTIR spectrometer to correct for "background" absorbance. These are absorbances due to the air itself. The diatomic molecules of nitrogen and oxygen do not absorb in this region, but water vapor and carbon dioxide do absorb. A typical background spectrum is shown. The absorbance at 2349 cm^{-1} is due to CO_2. What wavelength and frequency is this? What energy is this radiation?

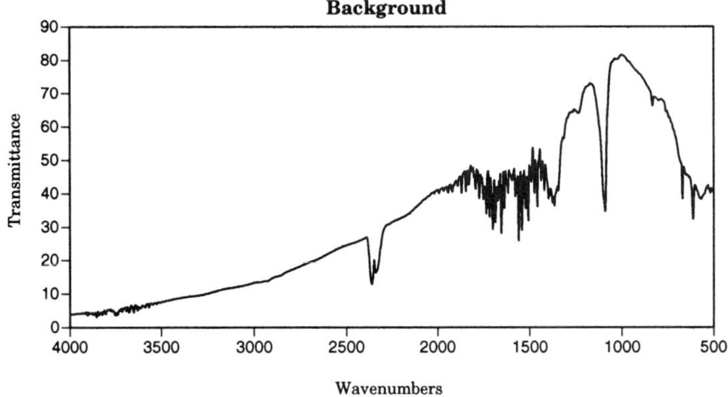

Solution: The value of 2349 cm^{-1} is the wavenumber, which is the inverse of the wavelength. The value of the wavelength is thus the reciprocal of the wavenumber:

$$\lambda = \frac{1}{2349\ cm^{-1}} = 4.257 \times 10^{-4}\ cm$$

The frequency can then be computed using the wavelength and c, the speed of light.

$$\nu = \frac{c}{\lambda} = \frac{2.99792 \times 10^{8}\ m\ s^{-1}}{\left(4.257 \times 10^{-4} cm\right)\left(\frac{1\ m}{100\ cm}\right)} = 7.042 \times 10^{13} s^{-1}$$

From the frequency and h, Planck's constant, the energy is determined:

$$E = h\nu = \left(6.62617 \times 10^{-34} Js\right)\left(7.042 \times 10^{13} s^{-1}\right) = 4.666 \times 10^{-20} J$$

3.5 THE INFRARED SPECTROPHOTOMETER AND IR SPECTRA

Infrared radiation is used for functional group determination because it is just the right energy to cause the bonds of molecules to vibrate. How much light is absorbed and the exact frequency of light absorbed depend upon the functional group and the structure of the molecule. The instrument used to determine how much of which specific IR frequencies are absorbed by a sample is called a spectrophotometer or, more simply, a spectrometer.

FIGURE 3.11: Block diagram of an infrared spectrophotometer.

The source in the diagram above continuously emits all the frequencies of the infrared region of the spectrum. The monochromator separates these different frequencies of radiation in much the same fashion that a prism can separate white light into the colors of the rainbow. These separated frequencies are then focused by the monochromator onto an exit slit, beginning with frequencies at the high end of the IR region and proceeding to lower frequencies. These different frequencies of IR radiation then hit the sample at different times. The detector determines how much light is transmitted by the sample, and the final result is an IR spectrum. This is typically a plot of percent transmittance (y-axis) versus **wavenumber** (x-axis), as shown in the figure below. The wavenumber (in cm^{-1}) is the inverse of the wavelength. The wavenumber is proportional to frequency, and is the most commonly used unit for the x-axis in IR spectra.

When the line is at the top of the graph, 100% of the light is being transmitted, and energy is not being absorbed by the compound. When the compound does absorb radiation, the percentage of light reaching the detector is decreased, and the result is a decreased signal, or an inverse peak. Thus at a, c, e, and g in Figure 3.12, the compound does not absorb radiation, but at b, d, and f radiation is absorbed.

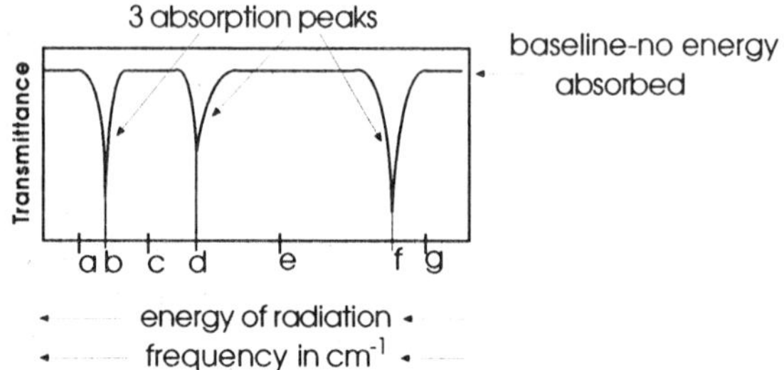

FIGURE 3.12: A simple IR absorption spectrum.

The entire infrared region extends from 14,000 cm^{-1} to 200 cm^{-1}. Most IR spectrophotometers measure only from 4000 cm^{-1} down to 600 cm^{-1}, although some are adapted to measure down to 200 cm^{-1}. The spectra seen in this course will be measured in the 4000 to 600 cm^{-1} range. Recent advances have produced the Fourier Transform Infrared (FTIR) spectrophotometer, which is more advanced than the spectrometer diagramed in Fig 3.11. The same spectral information, however, is still provided. One advantage of modern IR spectrometers is their ability to compare and contrast spectral peaks in different samples, extending far beyond the single peak example shown in Figure 3.5. Nearly instantaneous comparisons can be made using all the absorbances present in an unknown and a host of known spectra. A suspected illicit drug would thus not only be compared to cocaine, but to heroin, morphine, and many others as well.

When IR radiation is absorbed, it has just the correct energy to cause bonds to

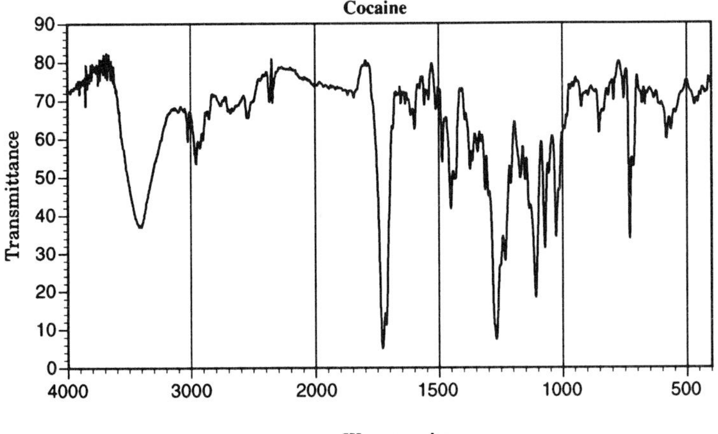

FIGURE 3.13:
IR spectrum of cocaine.

vibrate, bend, twist, and stretch without breaking. For example, a C=O bond will stretch like a metal spring when it is bombarded with energy of about 3.5 x 10^{-23} Joules. This corresponds to a wave number of 1760 cm^{-1}. Cocaine, which has two such C=O bonds, absorbs strongly in this region, as seen in Figure 3.13.

FIGURE 3.14:
Structure of cocaine.
Note C=O bonds.

The stretching for the C=O bond is shown below.

Energy is absorbed, which causes an increased amplitude of vibration in the carbonyl bond. This causes the decreased transmittance, or increased absorbance, observed in the spectrum. The C=O absorption appears somewhere within the

C=O
C=O
C=O
C=O
C=O

energy of about 1760 cm^{-1} causes the C=O to stretch in this manner

peak results from C=O absorbing energy

1760 cm^{-1}

FIGURE 3.15:
Strectching of the C=O bond and IR absorbance.

range of 1760-1650 cm^{-1}. The exact range is influenced by the structure around the C=O. The table below shows how the maximum absorbance for the C=O stretch varies according to the specific functional group containing that bond.

So far, our discussion has focused only on the stretching of the C=O bond. The stretching of bonds in molecules generally occurs in one of two fashions: either symmetrically, or antisymmetrically. Other vibrations involve a change in bond angle as opposed to bond length. These are called bending, scissoring or twist-

TABLE 3.2: IR ABSORBANCE RANGES OF CARBONYL STRETCH IN VARYING FUNCTIONAL GROUPS

TABLE 3.2

1690 - 1650 cm^{-1}	amides
1725 - 1700 cm^{-1}	carboxylic acids
1725 - 1705 cm^{-1}	ketones
1740 - 1720 cm^{-1}	aldehydes
1750 - 1735 cm^{-1}	esters

ing. Some specific types of vibrations are illustrated below. The figure below shows how water can stretch symmetrically, stretch asymmetrically, or bend.

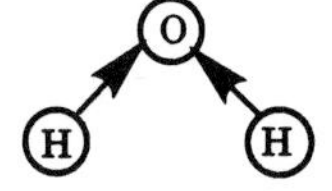

Symmetric Stretching

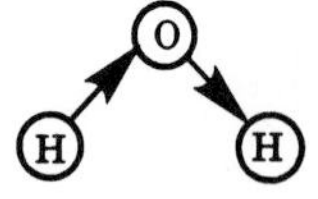

Asymmetric Stretching

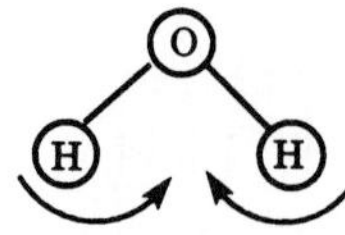

Bending or Scissoring

FIGURE 3.16: Vibrations of the water molecule.

In general it takes more energy (higher wavenumber) to cause a bond to stretch or bend when the bond is stronger, as seen in Table 3.3.

TABLE 3.3

TABLE 3.3: BOND STRENGTH VERSUS WAVENUMBER AT MAXIMUM ABSORBANCE

BOND TYPE	BOND STRENGTH (kJ/mole)	WAVE NUMBER (cm^{-1})
C—C	348	1200
C=C	612	1660
C≡C	838	2200

The following figure gives a thumbnail sketch of where in the IR spectrum certain functional groups are likely to absorb. A more complete listing is given in the Table 3.17.

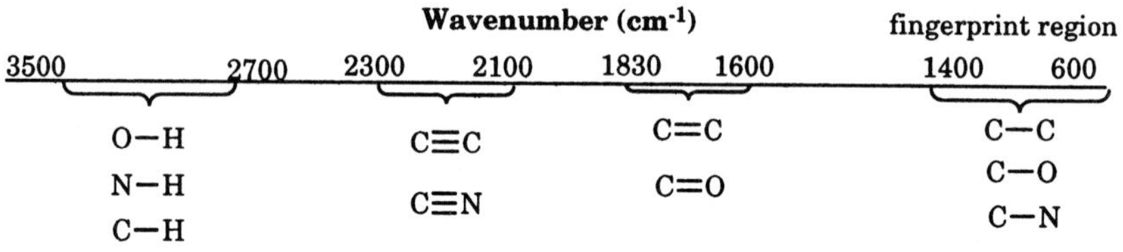

Wavenumber (cm^{-1})

3500 – 2700	2300 – 2100	1830 – 1600	fingerprint region 1400 – 600
O—H	C≡C	C=C	C—C
N—H	C≡N	C=O	C—O
C—H			C—N

FIGURE 3.17: Infrared absorbance maxima for various bonds.

Compounds are considered related if they display similar properties. In figures 3.18, we show the IR spectra of three unknown compounds, all with the same number of carbon, hydrogen and oxygen atoms in their molecules: $C_4H_{10}O$. A brief comparison of these spectra shows a similarity between (a) and (c) that is not shared by (b).

Compounds (a) and (c), with strong absorbances centered around 3300 cm^{-1}, indicate the presence of the C-O-H group. Compounds (a) and (c) are in fact alcohols. Compound (b) lacks the absorbance characteristic of alcohols. Compound (b) has the absorbance characteristic of the C-O-C structure, and is called an **ether**. IR absorbance characteristics give us a strong hint as to what family of compounds a substance belongs. That is a great aid in determining the compound's molecular identity (i.e., name). As we will see in the following section, and in the entire next chapter, the name of a specific chemical compound is directly related to its chemical structure.

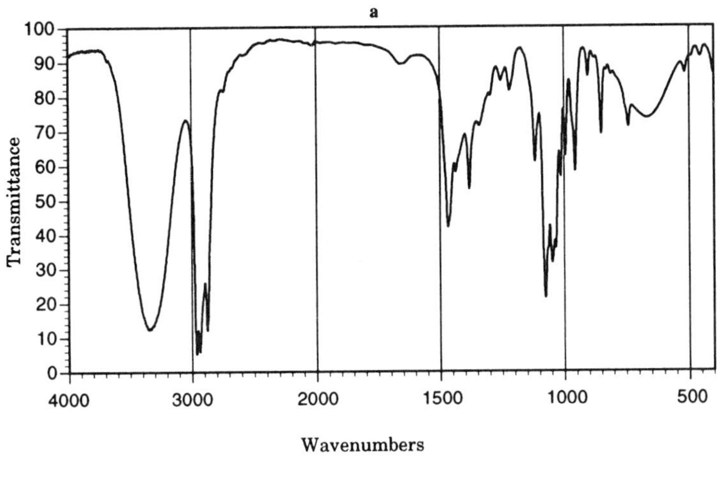

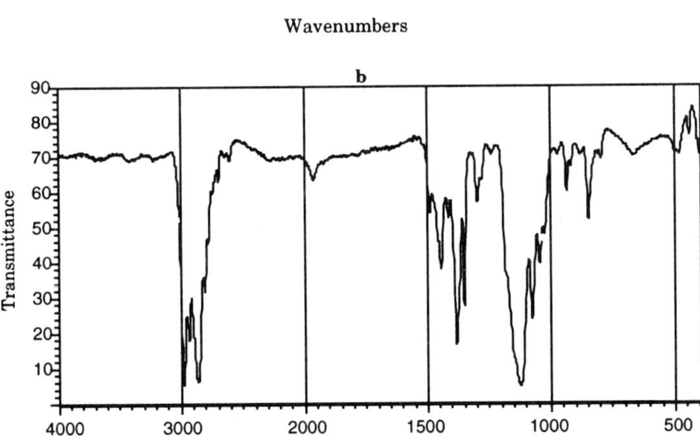

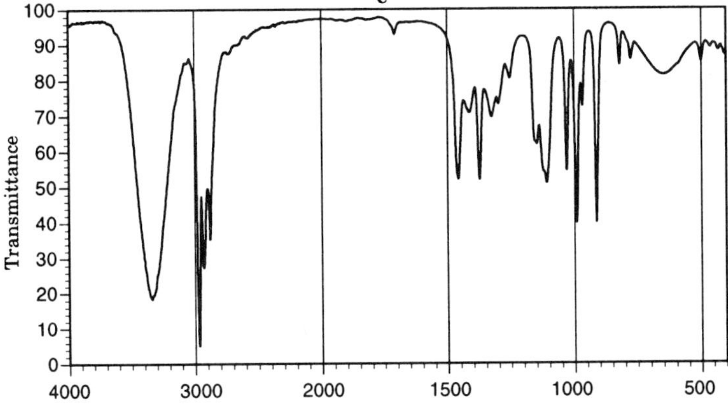

FIGURE 3.18: IR spectra of three compounds with the formula $C_4H_{10}O$.

3.6 INTRODUCTION TO NAMING COMPOUNDS

When chemistry was a younger science and fewer chemicals were known, they were named in no systematic way. The person who discovered the chemical might name it after a person. For example, barbituric acid (and subsequently all the barbituates) was named after Barbara, the girlfriend of the discoverer. A compound might be named from the circumstances of its discovery. Formic acid was named for the Latin word for ants, which is where formic acid was first found. Such names are called common names and are still sometimes used. As more chemicals were found and named, it became apparent that compounds named in such a fashion caused a great deal of confusion and frustration; no one could hope to remember the names of the thousands of chemicals a practicing scientist might be called upon to use. In order to systematize the names of compounds, especially the myriad of organic compounds, rules of nomenclature based on structure were developed. While this system allows one to produce the molecular structure from the name, and the name from the structure, it is not always simple to use. Certainly, even a pharmacist is likely to have trouble deciphering the name [2S-(2a,5a,6b)]-3,3-dimethyl-7-oxo-6-[(phenylacetyl)amino]-4-thia-1-azabicyclo[3.2.0]heptane-2-carboxylic acid monopotassium salt. But if asked for penicillin G potassium, the prescription is easily filled. These are the systematic and common names for the compound pictured at right.

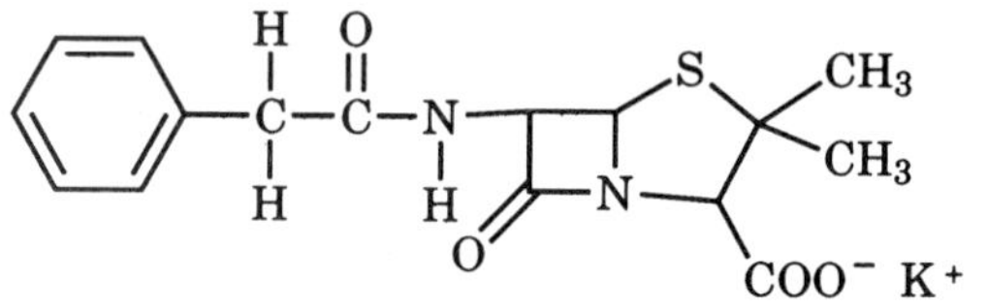

FIGURE 3.19: What is in a name?.

For the last few decades, every known chemical has been given one official name by the Chemical Abstracts Service of the American Chemical Society according to rules set up by the **International Union of Pure and Applied Chemistry (IUPAC)**. Yet the systematic name is sometimes so cumbersome (as in the previous example) that the common name is still used. The nomenclature for chemicals is similar to that for individuals. The official name (on the birth certificate) frequently is shortened, or an entirely different name (nickname) is substituted for the original. To avoid such confusion, the federal government deals with individuals via social security numbers. These numerical "names" are also more easily handled by computers. Thus each chemical is now also given its very own Chemical Registry Number (CRN), a sort of social security number. This facilitates the search and retrieval of information about the compound available in computer data bases. The CRN for penicillin G potassium is [113-98-4].

3.7 NAMING SIMPLE ORGANIC COMPOUNDS

In this section, we will examine the naming of small organic compounds (one, two or three carbons in the backbone) with one functional group. Longer carbon chains, backbones that are not chains of carbon to carbon single bonds, and molecules with more than one functional group are more complicated, and are dealt with later.

The name of these organic compounds is based upon the name of the hydrocarbon forming the carbon backbone. The names of the three hydrocarbons representing chains one, two and three carbons in length are shown in Table 3.4.

The common name of a compound is often formed by the name of the appropri-

TABLE 3.4

TABLE 3.4: THE NAMES OF SIMPLE CARBON-HYDROGEN SPECIES (ALKANES)

HYDROCARBAN FORMULA	HYDROCARBON NAME	FRAGMENT FORMULA	FRAGMENT NAME	SYSTEMATIC STEM
CH_4	methane	CH_3 -	methyl	methan-
CH_3CH_3	ethane	CH_3CH_2 -	ethyl	ethan-
$CH_3CH_2CH_3$	propane	$CH_3CH_2CH_2$ -	propyl	propan-

ate backbone fragment (known as a radical), followed by the name of the functional group. The systematic name of a compound takes the stem from the hydrocarbon backbone, and changes the suffix to indicate the functional group added to the backbone. These points are illustrated with numerous examples, classified according to their functional groups.

ALCOHOLS

The simplest alcohol contains one carbon atom with the structural formula CH_3OH. This alcohol is sometimes known as wood alcohol (based upon the fact that it was first obtained by heating wood in limited air.) The common name, methyl alcohol, is made from the name for the backbone fragment CH_3 (sometimes known as a methyl group, or methyl radical) plus the name of the functional group. The IUPAC systematic name is methanol. This is based upon the stem "methan". The terminal "e" is dropped from methane, and the suffix "ol" is added, indicating an alcohol.

The next simplest alcohol has the structural formula CH_3CH_2OH. This is known as grain alcohol, or drinking alcohol. The common name, derived from the radical CH_3CH_2, is ethyl alcohol. The systematic name, based on the stem "ethan" is ethanol.

When we get to the hydrocarbon propane, $CH_3CH_2CH_3$, different molecules are made when the alcohol group replaces a hydrogen on either the first or the second carbon. This results in two different alcohols, with different physical and chemical properties. These molecules are made from the same number of the same type of atoms, and are called "isomers", meaning equal parts. Note that putting an alcohol group on the third carbon is equivalent to putting an alcohol group on the first carbon. When the alcohol group is on a terminal carbon, the prefix "normal" is added to the common name. Thus $CH_3CH_2CH_2OH$ is known as normal-propyl alcohol, or n-propyl alcohol. The systematic name recognizes the variable position of the OH group by numbering the carbons. $CH_3CH_2CH_2OH$ is known as 1-propanol. (Note: 3-propanol would be the same molecule, so the convention is that the lowest possible number is assigned.) When the OH group is on the middle carbon (next to the terminal carbon) the common name is modified with the prefix "iso". The common name of $CH_3CHOHCH_3$ is isopropyl alcohol. The systematic name indicates that the OH is attached to the second carbon from the end by identifying the compound as 2-propanol.

The spectrum for 1-propanol is shown in figure 3.20. The most characteristic IR absorption peak for alcohols is the OH stretch that is usually broad and centered about 3300 cm^{-1}. An absorbance due to a C-O stretch also appears in the region 1300-1000 cm^{-1}. Again, note the features this spectrum shares with the spectra of other alcohols, in Figure 3.18 (a) and (c).

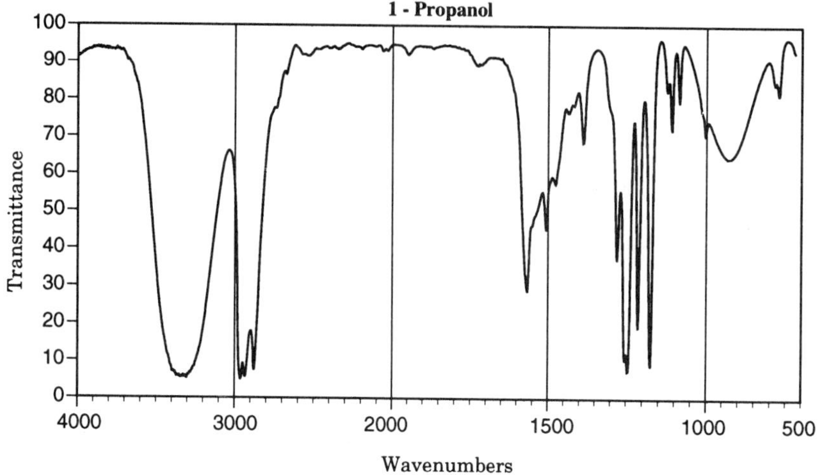

FIGURE 3.20: IR spectrum of 1-propanol.

There is another type of alcohol formed when the OH group is attached directly to a **benzene ring**. This is a special functional group called a phenol. Phenols have IR bands similar to alcohols, due to the same structural element, but phenols differ from common alcohols in many other chemical properties. Both phenol and benzene are examined later this chapter.

CONCEPT CHECK: NOMENCLATURE AND IR SPECTRA OF ALCOHOLS

The four-carbon hydrocarbon butane is $CH_3CH_2CH_2CH_3$. What would be the common name and systematic name of the corresponding alcohol: $CH_3CH_2CH_2CH_2OH$? In what IR region would you expect the functional group to absorb?

SOLUTION: The common fragment name formed from butane would be "butyl". Since the alcohol is on the end carbon, the common name would be "butyl alcohol". The systematic stem of butane would be "butan-". The suffix "ol" would be added to indicate an alcohol, and the numerical prefix "1-" would be used to indicate that the -OH is on the end carbon: "1-butanol". The OH stretch in alcohols will be a broad, strong band roughly between 3650 and 3200 cm^{-1}. (Taken from the table of typical IR absorbances at the end of this chapter.)

ETHERS

Ethers have the general formula R-O-R', where R and R' represent the hydrocarbon fragments bonded to oxygen. The ether most frequently encountered is $CH_3CH_2OCH_2CH_3$, known by the common names of diethyl ether, ethyl ether, or just ether. When two different groups are attached to the oxygen, the common name is determined by naming the R and R' fragments in alphabetical order, followed by the word "ether". For example, $CH_3OCH_2CH_2CH_3$ has the common name methyl propyl ether. IUPAC convention names ethers by selecting the R group that is longest, then naming the -OR' group as an alkoxy substituent. Thus diethyl ether is also named ethoxyethane, while methyl propyl ether becomes methoxypropane by IUPAC convention. All ethers contain a carbon-oxygen single bond. The stretching of this bond results in an infrared absorption between 1050 and 1200 cm^{-1}. Unfortunately, other types of absorption occur in this range as well, and ethers are generally not characterized by their IR spectra.

Amines

The common name for the one carbon amine, CH_3NH_2, is formed by combining the radical stem methyl with the name of the functional group: methylamine. Ethylamine is the common name for $CH_3CH_2NH_2$; n-propyl amine for $CH_3CH_2CH_2NH_2$; iso-propylamine for $CH_3CH(NH_2)CH_3$.

The hydrogen atoms on the nitrogen can be replaced with carbon atoms to give the compounds below. When the amine has two hydrogens and one carbon it is termed a primary amine; if the nitrogen is bonded to one hydrogen and two carbons, the compound is a secondary amine; if the nitrogen is bonded to three carbons and no hydrogens, the compound is a tertiary amine; a compound with nitrogen bonded to four groups, any of which can be hydrogen and carbon, is termed a quaternary amine, and must be positively charged.

IUPAC names for amines are similar to alcohols. The stem has "amine" added as a suffix. The prefix "N" is used to indicate additional substituents on the nitrogen atom. Table 3.5 summarizes information about several common amines.

Table 3.5: Amine Nomenclature

TABLE 3.5

Formula	Common Name	Systemic Name	Type Amine
$CH_3-N(H)-H$	methylamine	methanamine	primary
$CH_3-N(H)-CH_3$	dimethylamine	N-methylmethanamine	secondary
$CH_3-N(CH_3)-CH_3$	trimethylamine	N,N-dimethylmethanamine	tertiary
$CH_3-N^+(CH_3)_2-CH_3$	tetramethyl ammonium ion	N,N,N-trimethyl-methanammonium ion	quaternary

1 - Propylamine

Transmittance

100 90 80 70 60 50 40 30 20 10 0

4000 3000 2000 1500 1000 500

Wavenumbers

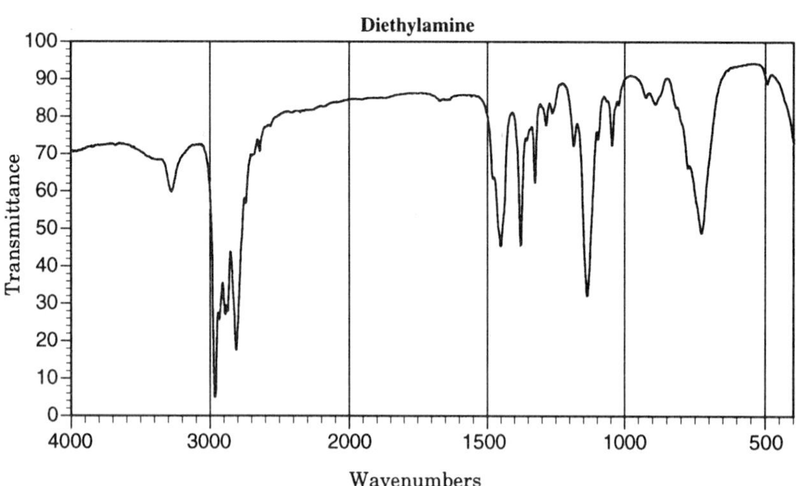

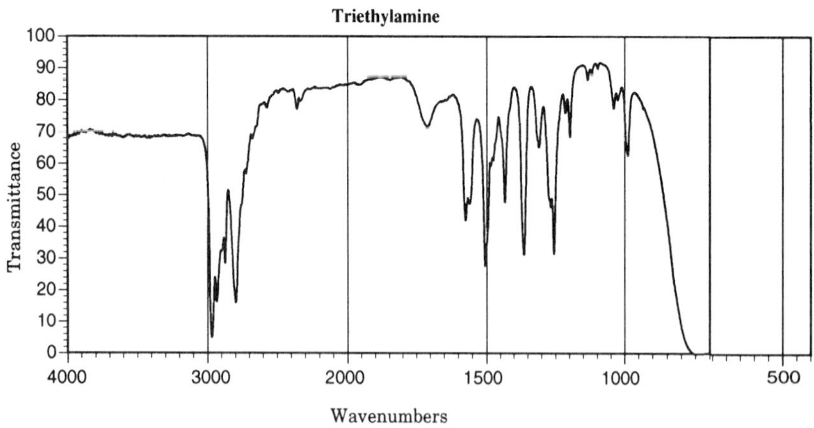

FIGURE 3.21: IR spectra of a) propylamine, b)diethylamine, c)triethylamine.

The first two spectra in the figure above show an N-H stretching absorption in the region of 3200-3500 cm^{-1} for each N-H bond. This absorbance occurs about the same wavenumber as the absorbance due to the OH bond in alcohols, although the N-H band is usually sharper than the O-H band. In a primary amine, e.g. , propylamine, there are two stretching absorptions corresponding to two ways of stretching for the two N-H bonds. The two N-H bonds can stretch together (symmetrical stretch) or they can stretch opposite (asymmetrical stretch). A secondary amine, e.g. , diethylamine, has only one N-H bond and thus only one N-H stretching peak. Triethylamine lacks any N-H bond, and thus does not absorb in this region. A C-N absorption appears in all three spectra around 1000-1200 cm^{-1}. This band is of limited use because C-C and C-O stretches absorb in the same region.

CONCEPT CHECK: NOMENCLATURE OF AMINES

Give the systematic name of the following compound, and indicate if it is a primary, secondary, tertiary, or quaternary amine:

$$CH_3CH_2CH_2N\ (CH_2CH_3)_2$$

SOLUTION: The nitrogen atom is bonded to two two-carbon chains, and to one three-carbon chain. We name the compound as a derivative of the simple amine formed from the longest carbon chain; thus it is a derivative of propanamine. That two ethyl groups are attached to the nitrogen is signified by the prefix: "N,N-". Thus the IUPAC name of the compound is N,N-diethylpropanamine. Since the nitrogen is directly bonded to three carbons, the amine is described as "tertiary".

HALIDES

Organic halides contain chlorine, bromine, iodine or fluorine atoms attached to the carbon backbone. The common name is patterned after inorganic halides such as sodium chloride: the halide is named last, preceded by the organic component, the more positive part. For example, CH_3Cl is methyl chloride; CH_3CH_2Br is ethyl bromide. The IUPAC names use the prefixes chloro, bromo, iodo and fluoro. For example, CH_3Cl has the systematic name of chloromethane; CH_3CH_2Br is bromoethane. Numbers are used in the systematic names when necessary to distinguish the carbon atom to which the halogen is bonded. For example, ICH_2-CH_2-CH_3 is called 1-iodopropane; CH_3-CHI-CH_3 is called 2-iodopropane. The IR spectra of organic halides do not contain peaks that easily distinguish the functional group, so these IR absorptions will not be introduced now.

KETONES

The ketone functional group contains a carbon oxygen double bond. Ketones are distinguished from other functional groups that contain a carbon oxygen double bond by having the oxygen attached to a carbon that is in a continuing chain of carbons. Ketones have the formula $R_1(C{=}O)R_2$, where R_1 and R_2 are bonded to the carbonyl carbon via carbon-carbon single bonds. The smallest ketone, shown in Figure 3.22, has three carbons. This compound has common names of

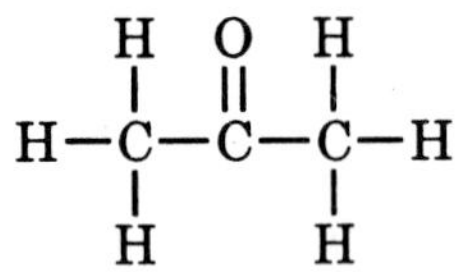

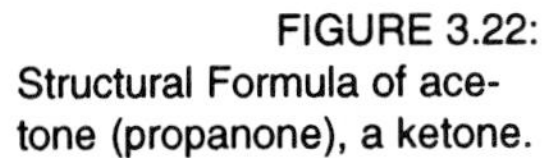

FIGURE 3.22: Structural Formula of acetone (propanone), a ketone.

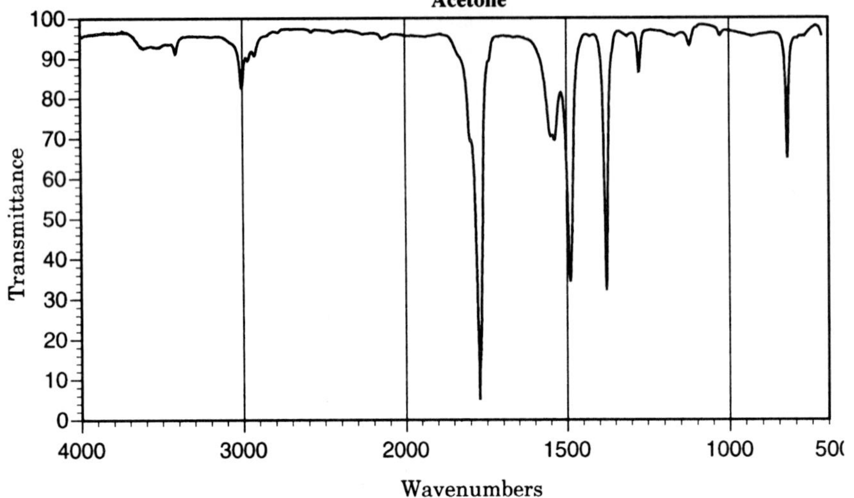

FIGURE 3.23: IR spectrum of acetone (propanone).

dimethyl ketone and acetone. IUPAC naming takes the name for the backbone with three carbon atoms, propane, and replaces the "e" with "one", indicative of a ketone functional group. Thus the systematic name of the compound in Figure 3.22 is propanone.

The carbonyl functional group is readily observed in the spectrum for propanone, since the C=O stretch is usually the strongest absorption in the IR. This band appears around 1715 cm^{-1}.

Carboxylic Acids

The carboxylic acid functional group is very common, and it also serves as the basis for naming several other functional groups: aldehydes, **esters**, and amides. The two simplest and most well known carboxylic acids are usually known by their common names, formic and acetic acids. The IUPAC naming scheme used in most other carboxylic acids requires adding "oic acid" to the systematic name for the hydrocarbon backbone, as shown below.

Table 3.6: Carboxylic Acid Nomenclature

TABLE 3.6

Formula	Common Name	Systemic Name
O ‖ H—C—OH	formic acid	methanoic acid
H O │ ‖ H—C—C—OH │ H	acetic acid	ethanoic acid
H H O │ │ ‖ H—C—C—C—OH │ │ H H	propionic acid	propanoic acid

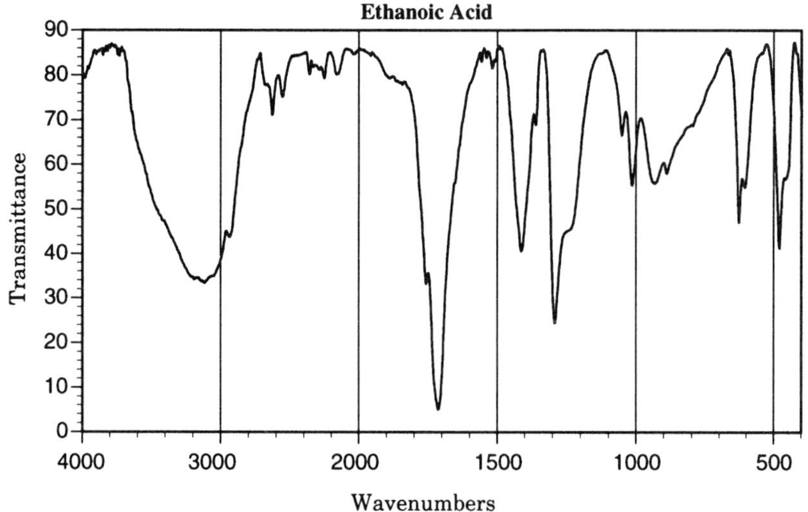

FIGURE 3.24: IR spectrum of acetic acid (ethanoic acid).

The IR spectra of carboxylic acids are characterized by C=O absorptions about 1710 cm^{-1} and O-H absorptions which are very broad and centered about 3000 cm^{-1}, the same place as C-H bonds.

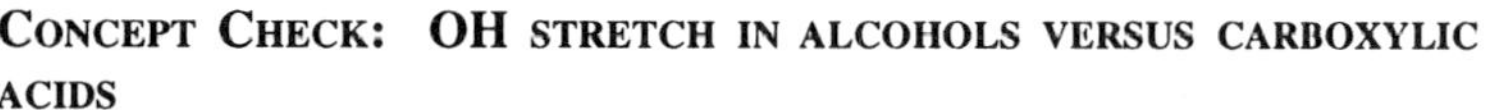

CONCEPT CHECK: OH STRETCH IN ALCOHOLS VERSUS CARBOXYLIC ACIDS

Alcohols and carboxylic acids both contain the OH group. The stretching of the OH bond is quite distinct in the IR spectrum. How do the IR spectra of alcohols and acids differ in this respect?

SOLUTION: If we examine the OH absorbance of an alcohol (like that of 1-propanol in Figure 3.20) and compare it to the OH absorbance of a carboxylic acid (like that of acetic acid in Figure 3.24) we see that the absorbance is broader and at a lower wavenumber for an acid (typically 3500-2500 cm^{-1}) while the absorbance is narrower and at a higher wavenumber for an alcohol (typically 3500-3200 cm^{-1}.)

ALDEHYDES

The **aldehyde** functional group contains a C=O with a general formula R(C=O)H. This means the **carbonyl group** is at the end of a carbon chain; only one carbon is bonded to the carbonyl carbon in the case of an aldehyde, as compared to two carbons bonded to the carbonyl carbon in the case of a ketone. The common name is derived using the name used for acids, dropping the "-ic acid" and adding "aldehyde". The systematic names take the backbone name and replace the -e with -al.

The carbonyl absorption is about 1710 cm^{-1} with the two characteristic aldehyde C-H stretching peaks appearing in the 2720-2820 cm^{-1} region.

TABLE 3.7: ALDEHYDE NOMENCLATURE

TABLE 3.7

FORMULA	COMMON NAME	SYSTEMIC NAME
H—C(=O)—H	formaldehyde	methanal
H—CH₂—C(=O)—H	acetaldehyde	ethanal
H—CH₂—CH₂—C(=O)—H	propionaldehyde	propanal

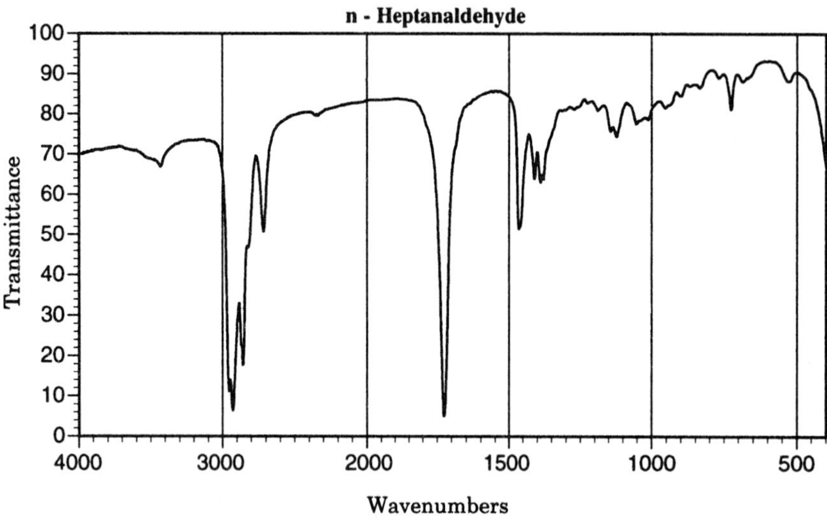

FIGURE 3.25: IR spectrum of n-Heptanaldehyde.

CARBOXYLIC ACID SALTS

Organic salts are named in both the common and the systematic manner in a fashion similar to inorganic salts. The cation is named first. The anion, formed by the loss of H^+ from the organic acid, is named by replacing the carboxylic acid ending "-ic acid" with "-ate".

The IR spectra of these salts will not be discussed.

TABLE 3.8: NAMES OF SALTS OF SIMPLE ORGANIC ACIDS

TABLE 3.8

FORMULA	COMMON NAME	SYSTEMIC NAME
H—C(=O)—O^- Na^+	sodium formate	sodium methanoate
H—CH₂—C(=O)—O^- K^+	potassium acetate	potassium ethanoate
H—CH₂—CH₂—C(=O)—O^- Rb^+	rubidium propionate	rubidium propanoate

Carboxylate Esters

Esters have the general formula R_1COOR_2, where R_1 and R_2 represent carbon radicals, such as CH_3-, CH_3CH_2-, etc. The R_1COO part is thought of as being derived from the organic acid R_1COOH. The -R_2 radical is thought of as replacing the H in the previous carboxylic acid. Thus esters are named like the salts of organic acids. The R_2 radical is named first, followed by the organic anion formed by the ionized carboxylic acid. As seen before, the anion is named by replacing the ending -ic acid with -ate. Several examples are shown below.

TABLE 3.9

Table 3.9: Names of simple organic esters

Formula	Common Name	Systemic Name
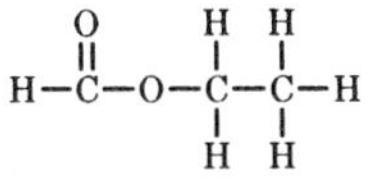	ethyl formate	ethyl methanoate
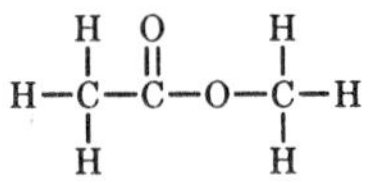	methyl acetate	methyl ethanoate
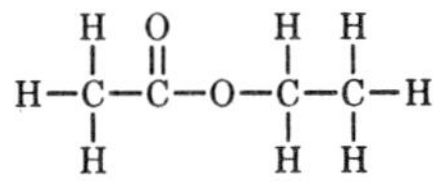	ethyl acetate	ethyl ethanoate

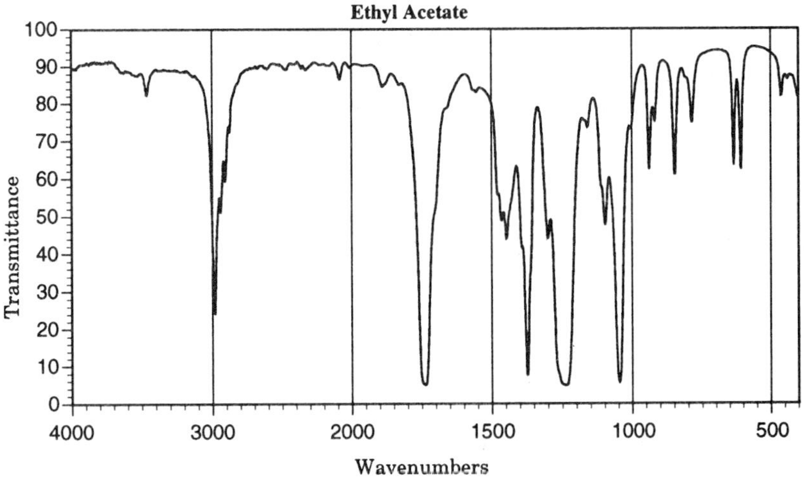

FIGURE 3.26: IR spectrum of ethyl acetate (ethyl athanoate).

Note the strong carbonyl absorption band in the IR spectrum of ethyl acetate. The C=O stretching occurs at the relatively high value of 1735 cm^{-1} characteristic of esters (as compared to aldehydes and ketones, which absorb at lower frequencies.) The ester group is also characterized by C-O stretches in the 1050-1250 cm^{-1} range.

Concept Check: Nomenclature of esters

Give the common name of the following ester. (Hint: a four-carbon hydrocarbon is known as butane.)

```
             O
             ‖
             C
   CH3CH2 ⁄    ⁀ O—CH2CH2CH2CH3
```

Solution: The ester above has a three-carbon chain and a four-carbon chain. Note that the three-carbon chain contains the carbonyl carbon (i.e., one doubly bonded to oxygen), while the four-carbon chain carbons have all single bonds. Since the three-carbon chain contains the COO- , the ester is named as a propionate (the anion of propionic acid.) As indicated in the text, the name of the fragment replacing the H in the acid is used as a prefix. The common name of the four carbon fragment from butane is "butyl". Thus, the compound is named butyl propionate.

Note that propyl butanoate is not the same as butyl propionate. Propyl butanoate, shown below, is a different molecule. It is derived from butanoic acid:

```
                O
                ‖
                C
   CH3CH2CH2 ⁄    ⁀ O—CH2CH2CH3
```

Carboxylic Acid Amides

Primary amides have the general formula $R(CO)NH_2$ with the possiblility that one or two of the hydrogens on the nitrogen are replaced by R groups. Amides are named from the carboxylic acids by replacing the ending -ic or -oic acid with -amide as below.

TABLE 3.10

Table 3.10: Names of simple organic primary amides

Formula	Common Name	Systemic Name
H–C(=O)–N(H)–H	formamide	methanamide
H–C(H)(H)–C(=O)–N(H)–H	acetamide	ethanamide
H–C(H)(H)–C(H)(H)–C(=O)–N(H)–H	propionamide	propanamide

There are two variations in amide structure in which the hydrogens are replaced with carbon groups on the nitrogen such as $CH_3CONHCH_3$ and $CH_3CON(CH_3)_2$. The names of the carbon groups attached directly to the nitrogen are named first, preceded by N- to show they belong on the nitrogen.

TABLE 3.11: NAMES OF SECONDARY AND TERTIARY ORGANIC AMIDES

FORMULA	COMMON NAME	SYSTEMIC NAME
$CH_3-C(=O)-NH-CH_3$	N-methylacetamide	N-methylethanamide
$CH_3-C(=O)-N(CH_3)-CH_3$	N,N-dimethylacetamide	N,N-dimethylethanamide

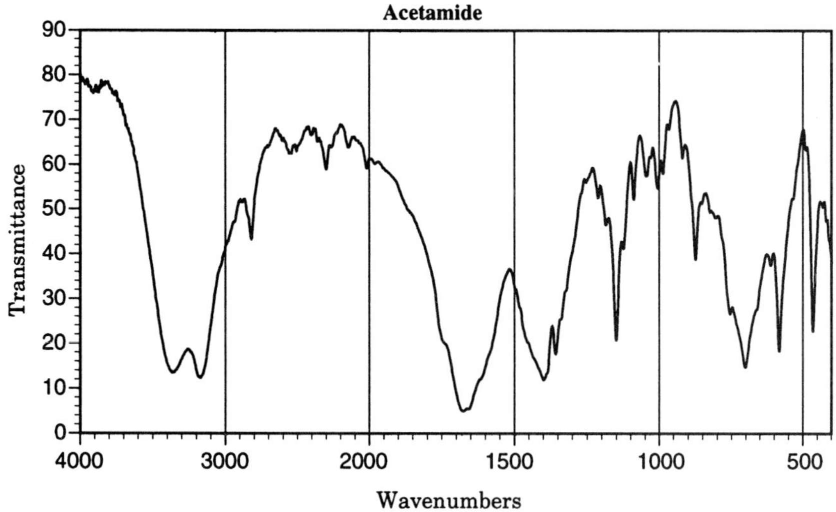

FIGURE 3.27: IR spectrum of a)acetamide, b) N-methylacetamide, (continued on next page).

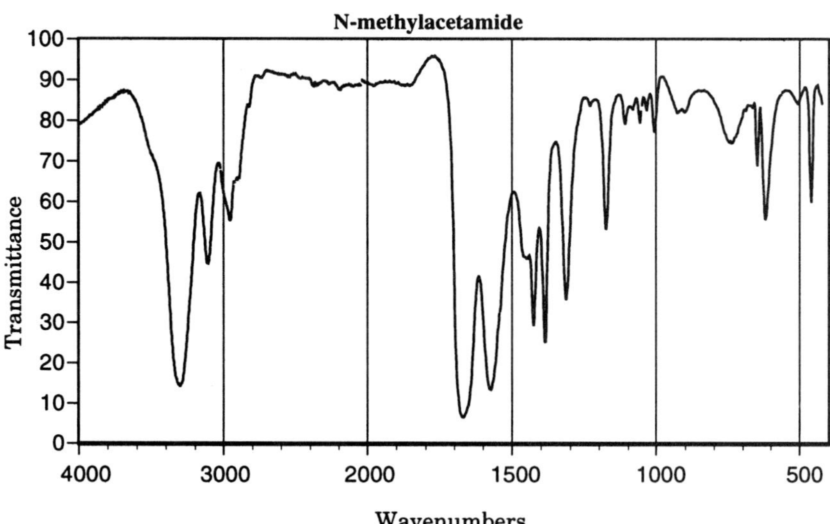

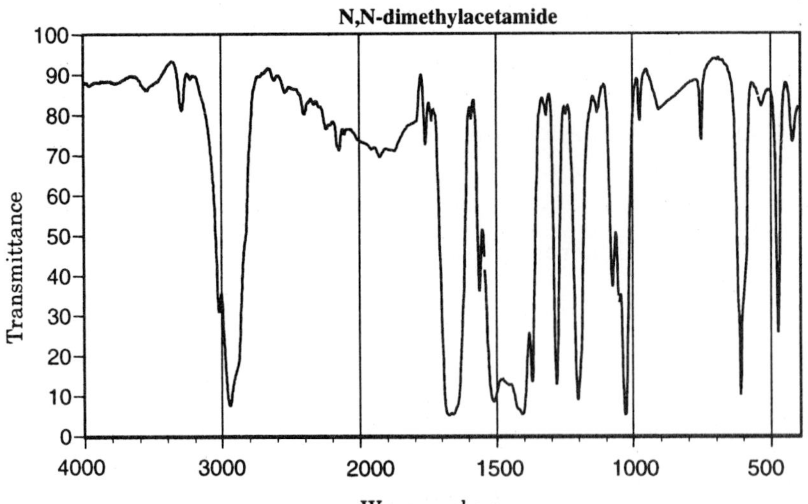

FIGURE 3.27(cont'd): c)N,N-dimethylacetamide.

The carbonyl absorption in the IR spectra of amides appears at lower wavenumbers than that observed for other carboxylic acid derivatives: approximately 1640-1680 cm^{-1}. In the N-H stretch region note that two absorptions appear for the NH_2 group in (a), whereas one appears for the $NHCH_3$ group in (b), and no absorptions appear for the $N(CH_3)_2$ group in (c). This is the same situation found earlier for the amines, and the absorptions again correspond to symmetric and asymmetric vibrations of the N-H bond.

Nitriles

The **nitrile** functional group features a carbon-nitrogen triple bond. One common name is derived from the carboxylic acid name. The nitrile is named by dropping the "ic acid" from the carboxylic acid and adding "onitrile". Another common name derives from the inorganic nomenclature analogous to cyanide salts. The IUPAC name is taken from the hydrocarbon chain and adding nitrile.

TABLE 3.12

TABLE 3.12: NAMES OF ORGANIC NITRILES

FORMULA	COMMON NAME	SYSTEMIC NAME
$H_3C-C\equiv N$	acetonitrile (methyl cyanide)	ethanenitrile
$H_3C-CH_2-C\equiv N$	propionitrile (ethyl cyanide)	propanenitrile

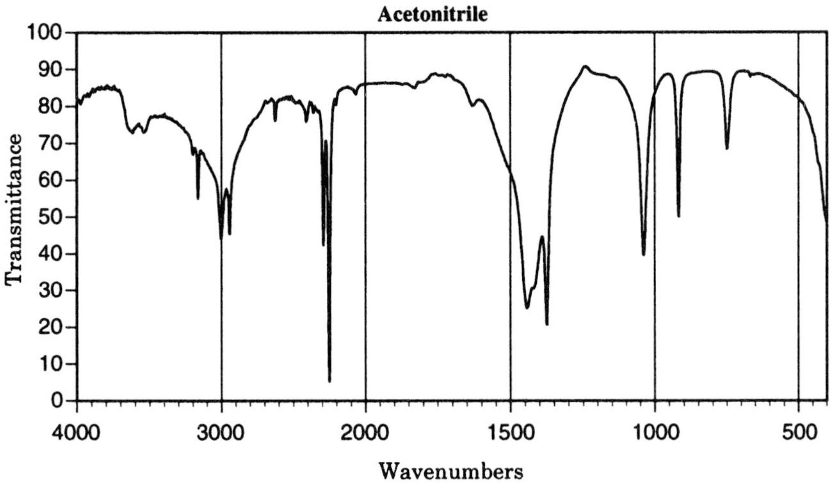

FIGURE 3.28: IR spectra of acetonitrile (ethanenitrile).

Note that the nitrile group has a particularly distinctive peak that appears in a region not usually occupied by other absorptions: 2250cm^{-1}. The only other IR absorption that is close to that for nitrile is the absorption for the stretching of the carbon-carbon triple bond. That occurs at slightly lower wavenumbers, about 2200 cm^{-1}.

3.8 MULTIPLE CARBON CARBON BONDS AS FUNCTIONAL GROUPS

There are several groupings of carbon to carbon bonds that can be considered as either backbone structure or as functional groups. We introduce some of the terminology in this section, and we will explore the topic more fully in the next chapter.

ALKENES, ALKYNES

Hydrocarbons have sequences of carbon to carbon bonds. In all the previous examples, the carbon to carbon backbone bonds have been single bonds. Yet carbon can also form double and triple bonds with another carbon atom. The naming of compounds with double or triple carbon to carbon bonds is similar, and we consider these two groups together. The systematic IUPAC name of hydrocarbons with a double bond replaces the "ane" suffix with "ene". For hydrocarbons with triple bonds, the "ane" ending is replaced with "yne". As seen with many small molecules, there are widely used common names as well as the systematic names. Note that the common names can easily cause confusion. For example, the common name "acetylene", although ending with the "ene" suffix of **alkenes**, is actually an **alkyne**.

TABLE 3.13: NAMES OF ALKENES AND ALKYNES

TABLE 3.13

FORMULA	COMMON NAME	SYSTEMIC NAME
$H_2C{=}CH_2$ (H—C=C—H with H on each C)	ethylene	ethene
$CH_3{-}CH{=}CH_2$ (H—C(H)(H)—C(H)=C(H)—H)	propylene	propene
H—C≡C—H	acetylene	ethyne
$CH_3{-}C{\equiv}C{-}H$ (H—C(H)(H)—C≡C—H)	methyl acetylene	propyne

Remember, as we saw in Table 3.3, the bond strength increases as the bond order increases. The characteristic stretching frequency for carbon carbon single, double and triple bonds consequently shifts from lower energies (1200 cm^{-1}) for alkanes, to higher energies (1660 cm^{-1}) for alkenes, and finally to yet higher energies (2200 cm^{-1}) for alkynes.

Earlier, we saw that the absorbance due to the stretching of a C=O group depends on its local environment. This is also true for the absorbance due to stretching of the C=C or C≡C group. The strength of an individual absorbance due to these multiple bonds may vary depending on the position of the multiple bond within the carbon chain. The frequency of IR absorbance due to the carbon carbon multiple bond is also affected by the individual features of the molecule. Most notable among these possible influences is conjugation. An isolated double bond absorbs around 1640-1680 cm^{-1}. When one double bond is exactly one single bond removed from another double bond, the two double bonds are said to be conjugated. Conjugated double bonds absorb at a lower frequency, as shown in the following table:

TABLE 3.14: ABSORBANCES OF ISOLATED AND CONJUGATED ALKENES

TABLE 3.14

STRUCTURE	DESCRIPTION	MAXIMUM ABSORBANCE
C=C—C—C=C	Isolated	1640-1680 cm^{-1}
C=C—C=C—C	Conjugated	1620-1640 cm^{-1}

Conjugated double bonds can be thought of as spreading the multiple bond character over the one intervening single bond, as illustrated in the figure below:

C=C—C=C—C ⟶ C⋯C⋯C⋯C⋯C

FIGURE 3.29: Conjugated double bonds as an extended bonding system.

The bonds in the extended bonding system shown in the figure above are intermediate in strength between single and double bonds. Conjugated double bonds are consequently somewhat weaker than isolated double bonds (which are too far removed from one another to form such an extended system). Thus conjugated double bonds are easier to stretch, and their absorbance is shifted to a lower frequency, around 1620-1640 cm^{-1}.

A multiply bonded carbon atom that is also bonded to a hydrogen atom will have a somewhat stronger C-H bond than the typical alkane carbon hydrogen bond. Thus in the Figures below, we see that the normal alkane C-H stretch occurs at a lower frequency (2900-3000 cm^{-1}) than observed for =C-H stretches (3000-3300 cm^{-1}).

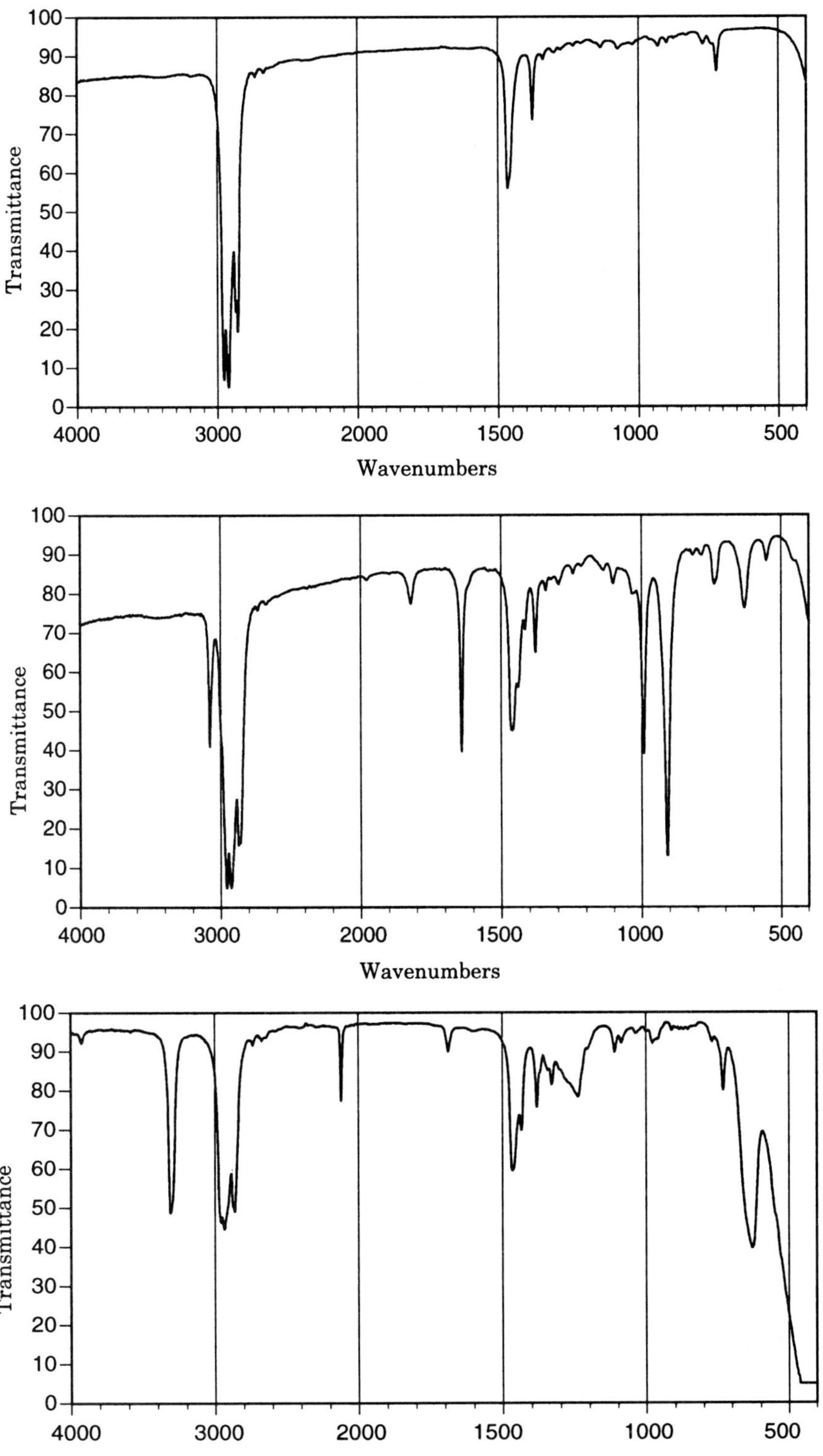

FIGURE 3.30: IR spectra of a)alkane, b)alkene, c)alkyne.

BENZENE

Benzene, with the formula C_6H_6, is a molecule with the six carbons in a ring. Various representations of benzene are shown in the figure below.

Note that when viewed in the Lewis dot form, benzene contains a completely conjugated alkene system throughout the ring. Other means of writing the benzene structure emphasize this extended bonding structure, which clearly reduces the carbon carbon bond strength below that of isolated double bonds.

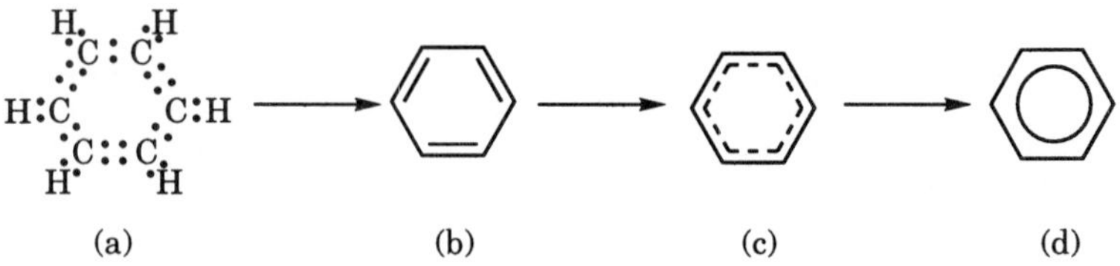

(a) (b) (c) (d)

FIGURE 3.31: Structures of benzene: a)Lewis Dot strutures, b)fixed double bonds, c)conjugation (by dash symbol), d)conjugation (by circle symbol).

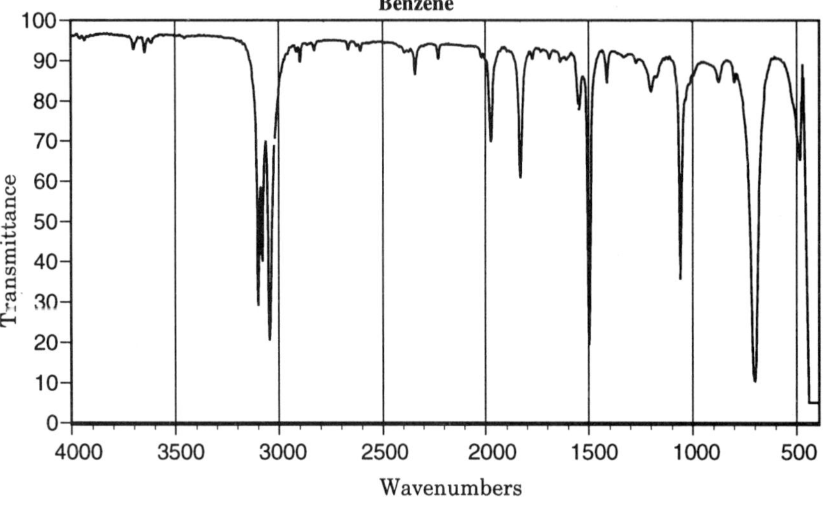

FIGURE 3.32: IR spectrum of benzene.

The C=C stretch in benzene is seen to occur near 1500 cm^{-1}. This value, when compared to those in Table 3.14, shows that a unique kind of extended bond system (and energy lowering) occurs in benzene. This is termed aromaticity, a topic that will be dealt with later in the text. Also note that the C-H stretch of hydrogens bonded to the benzene ring carbons absorbs just above 3030 cm^{-1}. This is also characteristic of aromatic compounds, several of which are shown in the next figure.

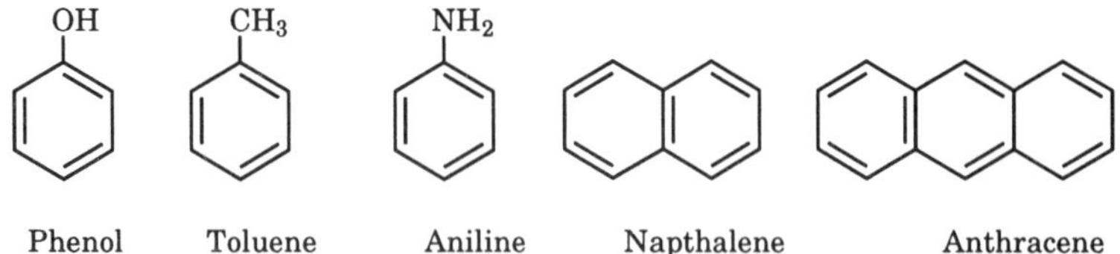

FIGURE 3.33: Structure of various aromatic compounds.

3.9 INDEX OF HYDROGEN DEFICIENCY

The infrared spectrophotometer is only one of many sophisticated analytical instruments. Information produced using one analysis technique frequently compliments that obtained by another analytical method. As the final point in this chapter, we consider the structural information that may be obtained from simply knowing the molecular formula. Using only the molecular formula, the number of double bonds and/or rings (known as the **index of hydrogen deficiency, IHD**) can be determined.

Let us consider once again the simple hydrocarbons listed in Table 3.4. These are alkanes, with all single carbon carbon bonds. The formulas of these simple chain-like alkanes were CH_4, C_2H_6, and C_3H_8. These hydrocarbons are termed "saturated": that is, the maximum number of hydrogen atoms are bonded, since there are no carbon carbon multiple bonds (or rings) that would diminish the possible number of hydrogens attached. Saturated chain-like alkanes have the general formula of C_nH_{2n+2}. A four carbon long straight chain alkane would have n=4, and thus a formula of $C_4H_{2(4)+2} = C_4H_{10}$.

Suppose, for example, that an unknown compound X has a molecular formula of C_4H_8. Since a saturated four carbon long alkane would have ten hydrogen atoms, X is missing two hydrogens. Four of the structures possible are given in Figure 3.34.

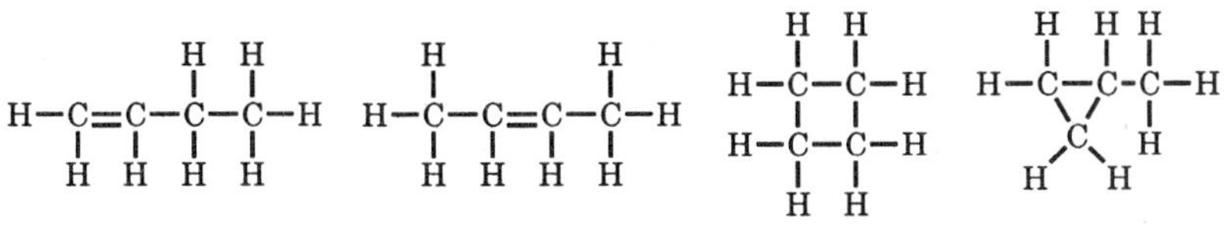

```
                                           H  H         H   H H
    H  H        H         H                |  |         |   | |
    |  |        |         |              H—C——C—H    H—C——C-C—H
H—C=C—C—C—H  H—C—C=C—C—H                 |  |          \ /  |
  | | |  |      | | |  |                 H—C——C—H        C   H
  H H H  H      H H H  H                   |  |         / \
                                           H  H        H   H
```

FIGURE 3.34: Possible structutres of C_4H_8.

Note that the two missing hydrogen atoms result in structures with either one double bond, or one ring in the molecular structure. The actual value of the index of hydrogen deficiency in a compound with n carbon atoms is given by the equation:

$$IHD = \frac{(2n + 2) - EHN}{2} \qquad \text{EQ. 3.3}$$

EHN symbolizes the effective hydrogen number. The value of EHN is calculated according to the following equation, where X is the number of halogen atoms and Y is the number of nitrogen atoms in the molecule.

$$EHN = \#\ \text{hydrogen atoms} + X - Y \qquad \text{EQ. 3.4}$$

Note that the number of oxygen atoms present in a molecule does not influence the value of the effective hydrogen number, and thus does not influence the index of hydrogen deficiency.

Six molecular formulas are listed below. The calculations of the effective hydrogen number and the index of hydrogen deficiency are shown, as well as one possible structure for each formula.

TABLE 3.15: EXAMPLE CALCULATIONS OF EHN AND IHD.

TABLE 3.15

FORMULA	EHN	IHD
A. C_5H_8	8	$\frac{(2(5)+2)-8}{2}=2$
B. C_5H_7Cl	7 + 1 = 8	$\frac{(2(5)+2)-8}{2}=2$
C. C_5H_9N	9 - 1 = 8	$\frac{(2(5)+2)-8}{2}=2$
D. C_5H_6O	6	$\frac{(2(5)+2)-6}{2}=3$
E. C_5H_8O	8	$\frac{(2(5)+2)-8}{2}=2$
F. C_6H_6	6	$\frac{(2(6)+2)-6}{2}=4$

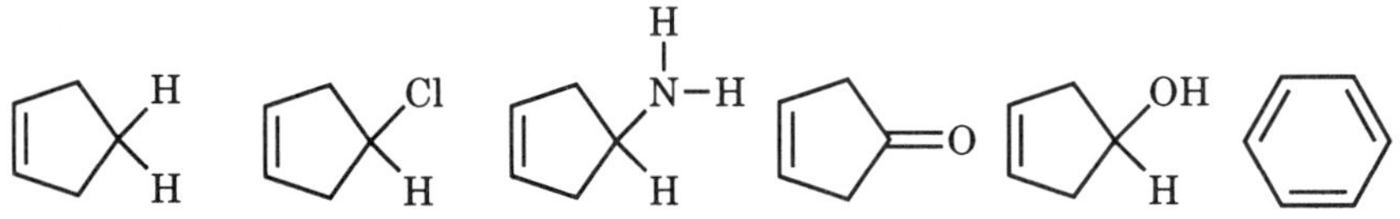

A C_5H_8 B C_5H_7Cl C C_5H_9N D C_5H_6O E C_5H_8O F C_6H_6

FIGURE 3.35
Possible structures for compounds in Table 3.15.

Note that for A, B, C, and E, IHD values of 2 were calculated. This means that the possible molecular structures for these compounds all have either two double bonds, one ring and one double bond, or two rings. Compound D has an IHD value of three; one possible structure shows one ring and two double bonds. Compound F has an IHD value of four; one possible structure has one ring and three double bonds. Note that each of these samples has several other possible structures as well; but the total number of rings/multiple bonds in each possibility is the same.

As you work through spectral problems, remember that the molecular formula (and the index of hydrogen deficiency calculated therefrom) will help determine if certain structural elements or functional groups are possible.

CONCEPT CHECK: IHD CALCULATIONS AND MOLECULAR STRUCTURE

Phenacetin, an aspirin substitute, has a chemical formula of $C_{10}H_{13}NO_2$. Calculate the IHD value for phenacetin.

SOLUTION: Phenacetin, with 13 hydrogen atoms and 1 nitrogen atom, has an effective hydrogen number of 13-1 = 12. The 10 carbon atoms indicate that n = 10. Substituting into the equation yields:

$$IHD = \frac{(2(10)+2)-12}{2} = 5$$

The actual structure of phenacetin is shown below. From inspection of this structure determine the IHD value expected (not calculated), and the identity of the various functional groups in phenacetin.

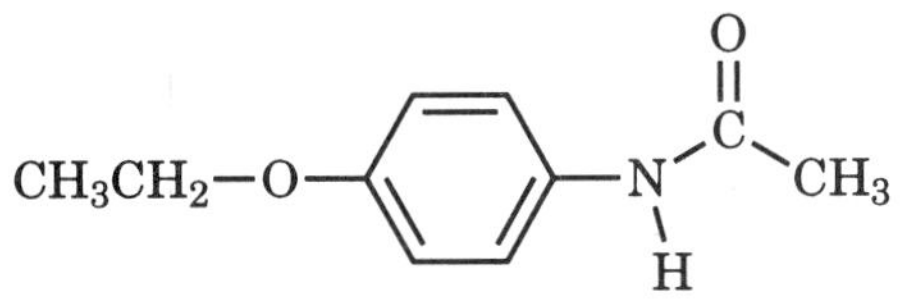

O
C
CH_3CH_2—O— —N CH_3
H

Solution: One ring, three carbon-carbon double bonds, and one carbon-oxygen double bond sums to an expected IHD of 5. In addition to the aromatic group, both an ether and an amide group are present.

3.10 A FINAL NOTE ABOUT IR SPECTRA

Most of the frequencies used to identify functional groups lie above 1400 cm^{-1}. Absorbances that occur at lower frequencies result from complex interactions of all the bonds in the molecule. This region of the spectrum, from about 1400 to 400 cm^{-1}, is called the **fingerprint region**. Typically there are numerous and complex absorbances in this region that help positively identify individual compounds.

Our previous analysis of IR spectra focused on identifying possible families of compounds to which an unknown might belong. Spectral libraries can be searched, looking at spectra for compounds in that chemical family. Data bases can be searched by computer to provide lists of the best overall matches between an unknown spectrum and all the other spectra stored in its memory. Such searches are limited by the contents of the spectral library, and possibly by the purity of the compounds examined. Known compounds can also be directly analyzed to obtain comparison spectra. If the unknown is shown to have an exact match with a known compound's IR spectrum, including the fingerprint region, there is near certainty in the identification of the unknown.

3.11 INTERPRETING IR SPECTRA—TWO EXAMPLES

When attempting to identify a compound by analysis of the IR spectrum, a formula of the compound would be available. In general it is advisable to start by determining the IHD from the molecular formula. The combination of a molecular formula and the IHD value will determine which functional groups may be present or could not be present. After this, the spectrum should be examined for confirmation of any possible functional groups. Once a tentative identification has been made, reference to the literature allows a direct comparison between the spectra of the unknown and the known compounds.

Example One: A compound with the molecular formula $C_3H_6O_2$ has the IR spectrum shown on the next page.

FIGURE 3.36:
IR spectrum of $C_3H_6O_2$.

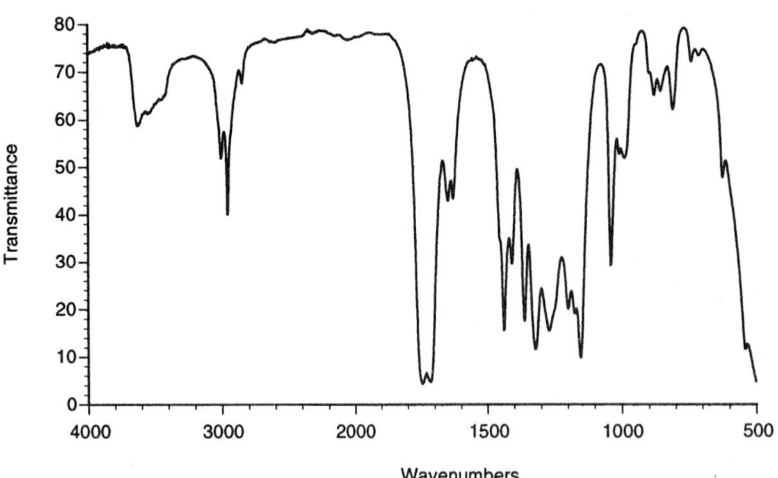

A compound with the molecular formula $C_3H_6O_2$ has an IHD of 1. Based on the formula, this could be the result of a carbon-carbon double bond, a ring, or a carbon-oxygen double bond. The presence of oxygen in the formula indicates ethers and alcohols are possibilities. That there is an oxygen atom present, and that the IHD is greater than zero indicate that aldehydes and ketones are possible. The absence of nitrogen in the formula eliminates amides. Since there are two oxygen atoms present, carboxylic acids and esters are also possible.

The IR spectrum has several notable features present, while others are absent. The most important characteristic in the spectrum is the absorbance at 1745 cm^{-1}. The carbonyl functional group is clearly present. Carboxylic acids, ketones, and aldehydes are still possible, though they are less likely, since these have carbonyl absorbances that usually are lower than 1745 cm^{-1}. Two strong absorbances, one at 1050 cm^{-1}, and another at 1250 cm^{-1}, are characteristic of carbon-oxygen single bonds. The OH absorbances typical of alcohols or carboxylic acids are missing from the spectrum, and these groups may be eliminated as possibilities. The presence of an oxygen singly bonded to a carbon atom, another oxygen doubly bonded to a carbon atom, the high wavenumber of the carbonyl, and the absence of an oxygen hydrogen stretch all strongly correlate to an ester. Absolute identification of the ester based only on this information is not possible. For example, the spectral evidence would be consistent with either ethyl formate or methyl acetate. Specific identification is possible only by comparison with known spectra.

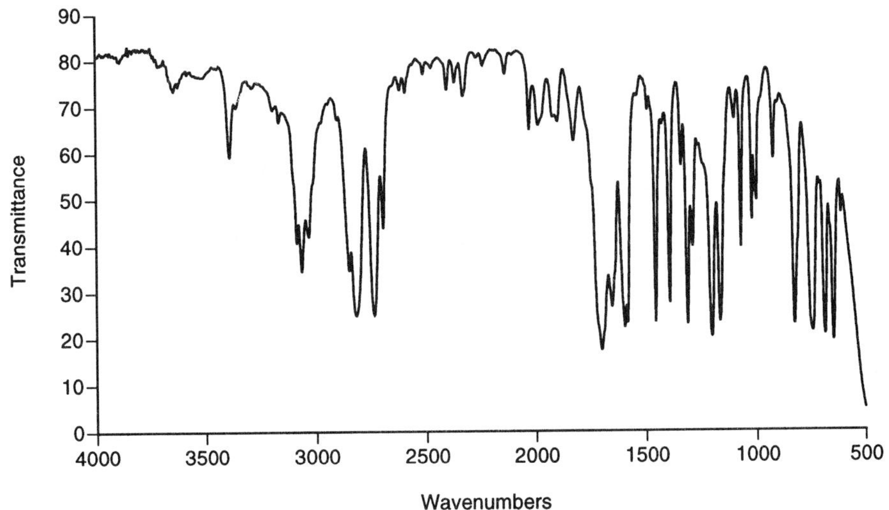

FIGURE 3.37:
IR spectrum of C_7H_6O.

EXAMPLE TWO: A compound with the molecular formula C_7H_6O has the IR spectrum shown below.
The IHD calculated from C_7H_6O is 5. When faced with such a large value for the IHD, it is often practical to consider the presence of a benzene ring, C_6H_6. The three double bonds and the ring itself account for an IHD of 4. The presence of just one oxygen means that alcohols, ethers, aldehydes and ketones are possible, but that acids and esters are not.

Examination of the IR spectrum reveals several key features. Carbon hydrogen stretching is observed between 3100 and 3000 cm^{-1}, indicating the presence of an aromatic ring. A carbonyl stretch is observed at 1700 cm^{-1}. Two fairly strong absorbances are noted between 2700 and 2900 cm^{-1}. Those two bands, in conjunction with a carbonyl, are strongly indicative of an aldehydic hydrogen carbon stretch. Attaching the phenyl group (C_6H_5) as the R in the aldehyde RCHO, agrees with the overall formula of C_7H_6O. The compound would be named benzaldehyde, and comparison with the spectrum of a known sample would confirm its identification.

TABLE 3.16: SUMMARY OF FUNCTIONAL GROUPS AND NOMENCLATURES

GROUP/STRUCTURE		SUFFIX	EXAMPLE	COMMON / IUPAC NAME
alcohol	$C-O-H$	-ol	CH_3CH_2OH	ethyl alcohol, ethanol
amines	$C-\underset{H}{\underset{\vert}{N}}-H$	-amine	$CH_3CH_2NH_2$	ethylamine, ethanamine
	$C-\underset{H}{\underset{\vert}{N}}-C$		$(CH_3CH_2)_2NH$	diethylamine, N-ethylethanamine
	$C-\underset{C}{\underset{\vert}{N}}-C$		$(CH_3CH_2)_3N$	triethylamine, N,N-diethylethanamine
ether	$C-O-C$	ether	$CH_3CH_2OCH_2CH_3$	diethyl ether, ethoxyethane
halides	$C-X$	halide	CH_3CH_2Cl	ethyl chloride, chloroethane
ketone	$C-\overset{O}{\overset{\Vert}{C}}-C$	-one	$CH_3-\overset{O}{\overset{\Vert}{C}}-CH_3$	dimethyl ketone, propanone, acetone
aldehyde	$C-\overset{O}{\overset{\Vert}{C}}-H$	-al	$CH_3-\overset{O}{\overset{\Vert}{C}}-H$	acetaldehyde, ethanal
carboxylic acid	$C-\overset{O}{\overset{\Vert}{C}}-OH$	-oic acid	$CH_3-\overset{O}{\overset{\Vert}{C}}-OH$	acetic acid, ethanoic acid
ester	$C-\overset{O}{\overset{\Vert}{C}}-O-C$	-oate	$CH_3-\overset{O}{\overset{\Vert}{C}}-OCH_3$	methyl acetate, methyl ethanoate
amides	$C-\overset{O}{\overset{\Vert}{C}}-\underset{H}{\underset{\vert}{N}}-H$	-amide	$CH_3-\overset{O}{\overset{\Vert}{C}}-NH_2$	acetamide, ethanamide
	$C-\overset{O}{\overset{\Vert}{C}}-\underset{H}{\underset{\vert}{N}}-C$		$CH_3-\overset{O}{\overset{\Vert}{C}}-\underset{H}{\underset{\vert}{N}}-CH_3$	N-methylethanamide
	$C-\overset{O}{\overset{\Vert}{C}}-\underset{C}{\underset{\vert}{N}}-C$		$CH_3-\overset{O}{\overset{\Vert}{C}}-\underset{CH_3}{\underset{\vert}{N}}-CH_3$	N,N-dimethylethanamide
nitrile	$C\equiv N$	-nitrile	$CH_3C\equiv N$	acetonitrile, ethanenitrile
alkene	$C=C$	-ene	$H_3C-\overset{H}{\overset{\vert}{C}}=CH_2$	propylene, propene
alkyne	$C\equiv C$	-yne	$H_3CC\equiv CH$	methyl acetylene, propyne

TABLE 3.16

KEY WORDS & Concepts

Alcohol: functional group containing $-\overset{|}{\underset{|}{C}}-O-H$

Aldehyde: functional group containing $-\overset{O}{\overset{\|}{C}}-H$

Alkene: functional group containing C=C

Alkyne: functional group containing C≡C

Amide: functional group containing $-\overset{O}{\overset{\|}{C}}-\underset{|}{N}-$

Amine: functional group containing $-\overset{|}{\underset{|}{C}}-\underset{|}{N}-$

Aromatic: typically a ring shaped compound with alternating single and multiple bonds.

Aryl: an aromatic structural component; phenyl, shown at right, is a common example

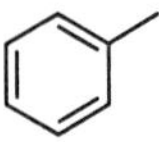

Backbone: part of structure containing chiefly C to C single bonds and to which functional groups are attached

Benzene: an aromatic compound with the formula C_6H_6

Carbonyl group: C=O grouping belonging to several functional groups

Carboxylic acid: functional group containing $-\overset{O}{\overset{\|}{C}}-O-H$

Electromagnetic spectrum: range of all possible frequencies of energy

Ester: functional group containing $-\overset{O}{\overset{\|}{C}}-O-\overset{|}{\underset{|}{C}}-$

Ether: functional group containing $-\overset{|}{\underset{|}{C}}-O-\overset{|}{\underset{|}{C}}-$

Fingerprint region: region of IR spectrum from 1400-600 cm^{-1}

Frequency (ν)**:** wave characteristic; the number of wavefronts (cycles) passing a point per second; units of Hertz or cps

Functional group: a grouping of about 2-5 atoms that include a carbon and usually at least one atom of an element other than carbon (commonly hydrogen, oxygen, nitrogen, halogen).

Hertz (Hz, cps): measurement of frequency

Index of hydrogen deficiency (IHD): method by which the number of double bonds and or rings can be found using only the molecular formula of a compound.

Infrared radiation: region of the electromagnetic spectrum in which molecules vibrate and stretch.

IUPAC: systematic method for naming chemicals from the International Union of Pure and Applied Chemistry

Ketone: functional group containing $-\underset{|}{\overset{|}{C}}-\overset{\overset{O}{\|}}{C}-\underset{|}{\overset{|}{C}}-$

Nitrile: functional group containing $-C\equiv N$

Wavelength (λ): characteristic length between equivalent positions along a wave.

Wavenumber: the inverse of the wavelength, usually in units of cm^{-1}.

Infrared Stretching Absorption Ranges

Bond	Group	Stretching (cm^{-1})
C-H	alkanes	2960-2850(s)
C-H	alkenes	3080-3020 (m)
C-H	aromatic	3100-3000 (v)
C-H	aldehyde	2900, 2700 (m, 2 bands)
C-H	alkyne	3300(s)
C≡C	alkyne	2260-2100 (v)
C≡N	nitrile	2260-2220 (v)
C=C	alkene	1680-1620 (v)
C=C	aromatic	1600-1450 (v)
C=O	ketone	1725-1705 (s)
C–O	aldehyde	1740-1720 (s)
C=O	ester	1750-1735 (s)
C=O	carboxylic acids	1725-1700 (s)
C=O	amide	1690-1650 (s)
O-H	alcohols	3650-3200 (s,broad)
O-H	carboxylic acids	3500-2500 (s,broad)
N-H	amines	3500-3300 (m)
N-H	amides	3500-3350 (m)
C-O	alcohols, ethers, esters	1300-1000 (s)

s means strong absorption
m means medium absorption
w means weak absorption
v means variable absorption

HOMEWORK Problems

Identification of functional groups in compounds

Circle and label all functional groups in the following chemicals.

1. 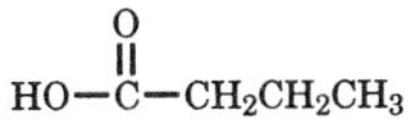 $HO-\overset{O}{\overset{\|}{C}}-CH_2CH_2CH_3$

2. (benzene ring)$-\overset{O}{\overset{\|}{C}}-H$

3. $CH_3CH_2-\overset{O}{\overset{\|}{C}}-O-H$

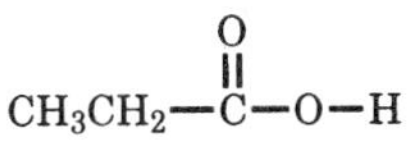

4. 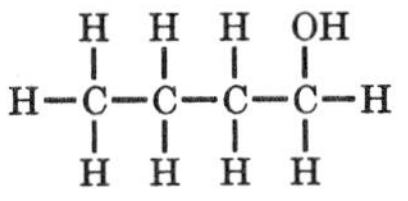 $H-\underset{H}{\overset{H}{C}}-\underset{H}{\overset{H}{C}}-\underset{H}{\overset{H}{C}}-\underset{H}{\overset{OH}{C}}-H$

5. 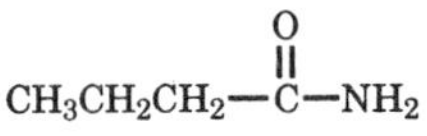 $CH_3CH_2CH_2-\overset{O}{\overset{\|}{C}}-NH_2$

6. 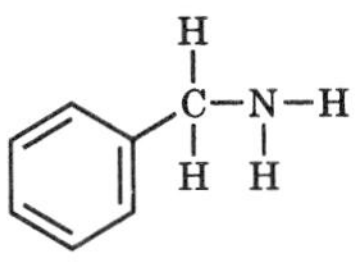 (benzene ring)$-\underset{H}{\overset{H}{C}}-\underset{H}{N}-H$

7. 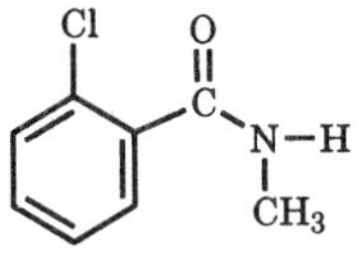 Cl; (benzene ring)$-\overset{O}{\overset{\|}{C}}-\underset{CH_3}{N}-H$

8. 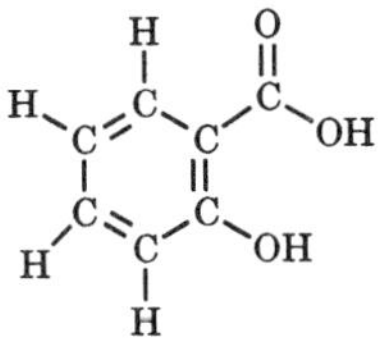 H, C, C, C, OH, O, OH, H, H, H

9. 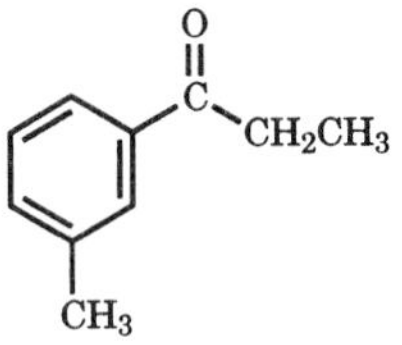 O, C, CH_2CH_3, CH_3

10. morphine

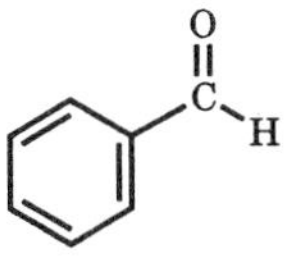

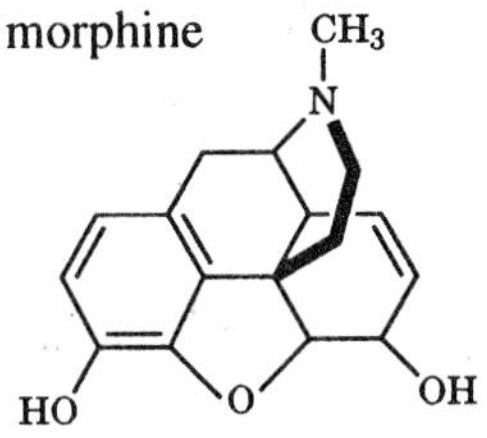

CH_3, N, HO, O, OH

11. quinine

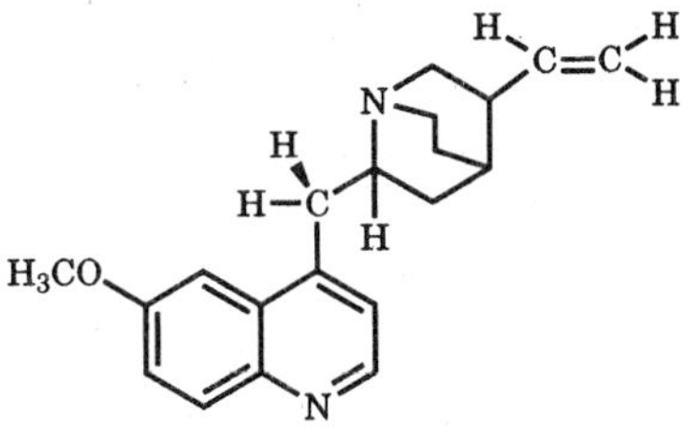

H, C=C, H, H, N, H, H—C, H, H_3CO, N

12. heroin

CH_3, N, $H_3C-\overset{}{C}-O$, O, O, $O-C-CH_3$, O

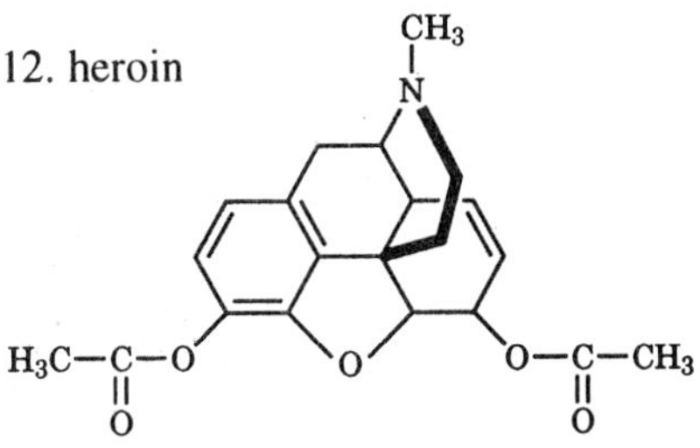

13. cocaine

N, CH_3, O, $C-O-CH_3$, H, O—C, H, O

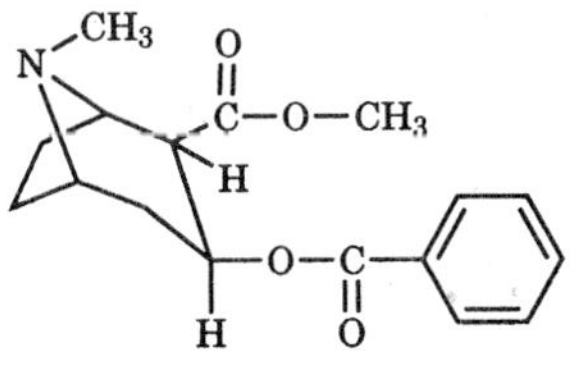

INDEX OF HYDROGEN DEFICIENCY

14. Determine the value of the IHD for each of these formulas and draw one of the possible structures.

(a) C_4H_8
(b) C_4H_8O
(c) $C_5H_{11}Cl$
(d) C_7H_8

15. Determine the value of the IHD for each of these formulas and draw one of the possible structures.

(a) C_3H_6
(b) $C_5H_{10}O$
(c) C_2H_3Cl
(d) $C_4H_{11}N$

16. Determine the value of the IHD for each of these formulas and draw one of the possible structures.

(a) C_3H_5N
(b) $C_3H_6O_2$
(c) C_2H_3OCl
(d) $C_2H_5O_2N$

IR SPECTRA

17. What functional group is present in each of the following compounds whose spectrum is given below? In most cases there is only one functional group. Note that the empirical formula is also given; use it to find the index of hydrogen deficiency. Hint: find the best cm^{-1} value from the spectrum of the important peaks before you proceed very far.

(a) $C_6H_{12}O$

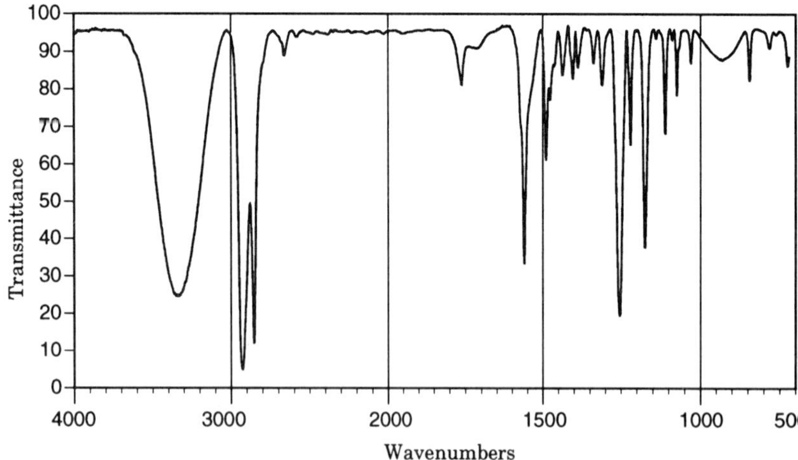

(b) $C_4H_{10}O$

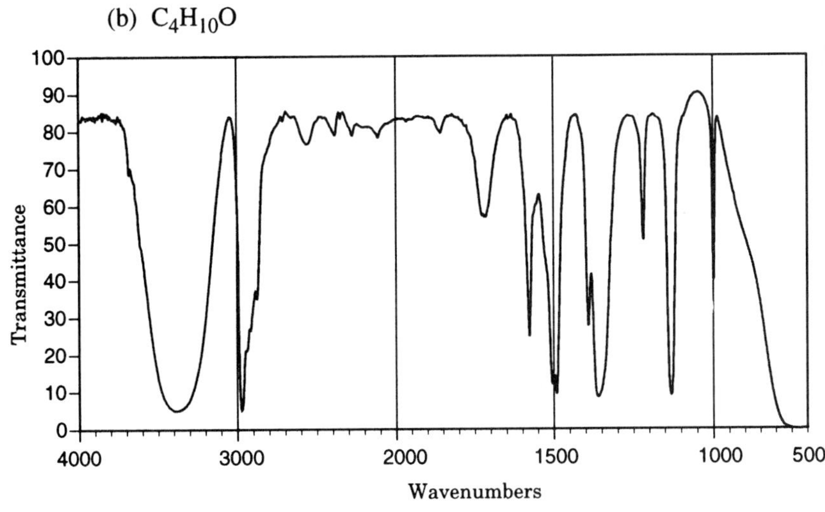

(c) $C_8H_8O_3$

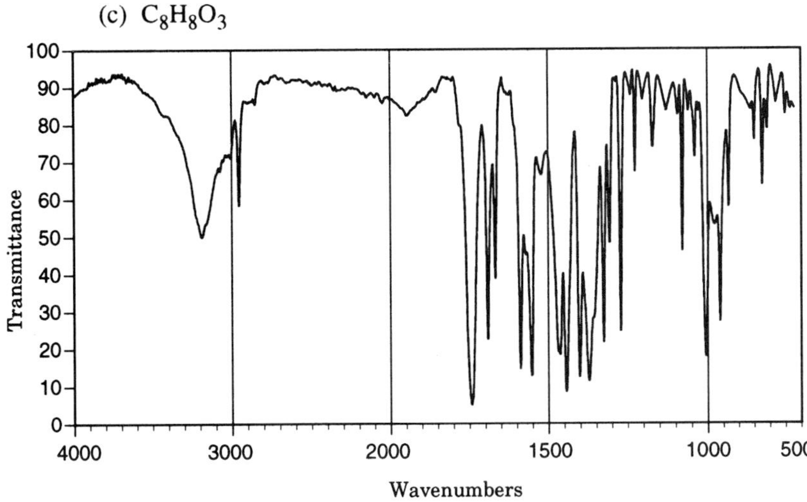

(d) $C_7H_6O_2$

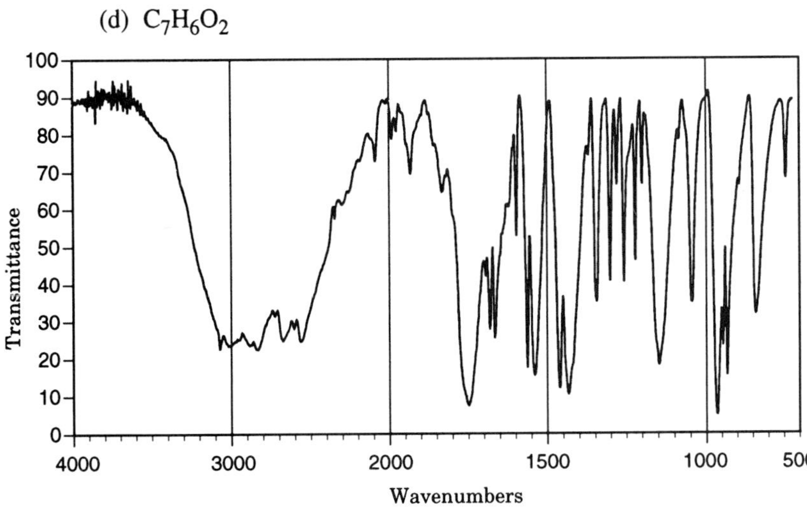

(e) C_7H_7NO

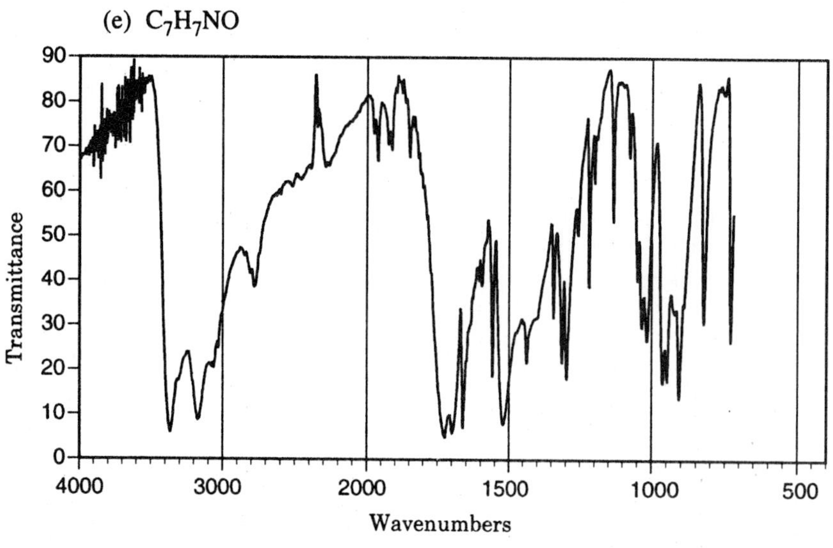

(f) $C_3H_6O_2$

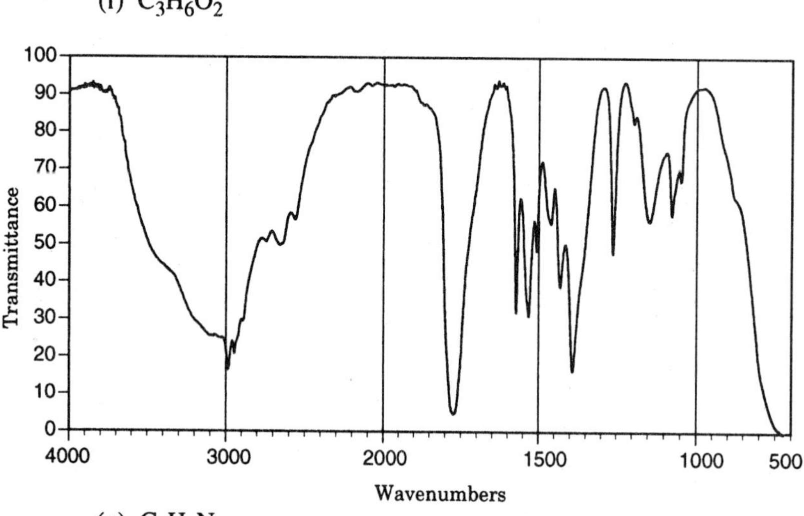

(g) C_2H_3N

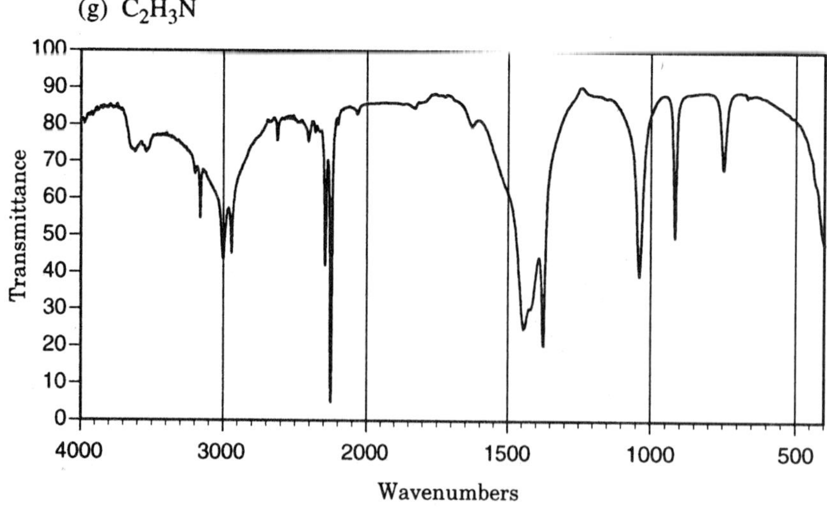

18. A compound containing only carbon, hydrogen and oxygen has an IHD of 0. The IR spectrum shows a broad absorbance from 3600 to 3200 cm^{-1}.

(a) Could the compound be a carboxylic acid? Why or why not?
(b) Could the compound be an ether? Why or why not?
(c) Could the compound be an alcohol? Why or why not?
(d) Could the compound be an aldehyde? Why or why not?

19. A compound containing only carbon, hydrogen and oxygen has an IHD of 1. The IR spectrum shows a strong absorbance around 1700 to 1750 cm^{-1}.

(a) Could the compound be a carboxylic acid? Why or why not?
(b) Could the compound be an ester? Why or why not?
(c) Could the compound be an amine? Why or why not?
(d) Could the compound be an alkene? Why or why not?

20. A compound containing only carbon, hydrogen and nitrogen has an IHD of 1. The IR spectrum shows a broad absorbance between 3500 and 3300 cm^{-1}.

(a) Could the compound be a carboxylic acid? Why or why not?
(b) Could the compound be a nitrile? Why or why not?
(c) Could the compound be an amine? Why or why not?

21. A compound contains carbon, hydrogen, nitrogen and oxygen. The compound is thought to be either an amino acid (carboxylic acid with an amine group attached to the carbon backbone) or an amide. How would the two possibilities compare in terms of:

(a) possible IHD values?
(b) characteristic IR absorbance ranges?

CHAPTER THREE OVERVIEW Problems

22. One compound with the formula of $C_3H_6O_2$ is propanoic acid (propionic acid).

(a) Draw the Lewis dot diagram of this compound.
(b) Is propanoic acid ionic or covalent?
(c) Is it a strong, weak or non-electrolyte?

23. Would you expect propanoic acid to be more soluble in carbon tetrachloride or in water, and why?

24. One compound with the formula of $C_4H_{10}O$ is diethyl ether. Draw the Lewis dot structure of this compound.

25. At 25 degrees C, water saturated with diethyl ether is 6.05% (by mass) ethyl ether. This ether-saturated aqueous solution has a density of 0.997 grams per cm^3.

(a) How many grams of ethyl ether are present in 10.00 liters of ether-saturated aqueous solution?

(b) How many grams of water are present in 10.00 liters of ether-saturated aqueous solution?

26. Suggest a solvent suitable for extracting ether from an ether-saturated aqueous solution. Would the non-aqueous layer into which the ether is extracted be the top or the bottom layer in the separatory funnel?

27. One compound with the formula of $C_3H_6O_2$ is ethyl formate. Another is methyl acetate.

(a) Draw the Lewis dot structure of ethyl formate

(b) Draw the Lewis dot structure of methyl acetate.

28. At room temperature, ethyl formate has a density of 0.917, and methyl acetate has a density of 0.934 grams per cm^3.

(a) Which is heavier, and by how much; 1.000 liter of ethyl formate, or 1.000 liter of methyl acetate?

(b) Which contains more molecules; 1.000 liter of ethyl formate, or 1.000 liter of methyl acetate?

29. One compound with the formula C_2H_3N is acetonitrile. Draw the Lewis dot structure of acetonitrile.

30. Acetonitrile is used as a non-aqueous solvent for inorganic salts. To what physical property of acetonitrile might this be attributed?

BACKBONE AND NMR

BACKBONE AND NMR

CHAPTER FOUR

Last chapter we learned how the absorbance of infrared radiation could characterize groups of atoms that have a distinctive bonding pattern. IR spectroscopy was seen as an important means of determining functional groups within a molecule. In this chapter we continue our introduction to molecular structure, with a focus on those atoms that have a non-zero magnetic moment, μ. Such atoms are said to have a nuclear spin (symbol I), and behave like tiny bar magnets. **Nuclear magnetic resonance** spectroscopy (NMR) is a technique that studies the properties of nuclei with non-zero nuclear magnetic spin. NMR will help us to learn more about molecular structure and, in particular, the structure of the backbone to which the functional groups are attached.

The nucleus of an atom is composed of protons and neutrons. When the nucleus has an odd number of protons (the atomic number Z) or an odd number as the sum of protons and neutrons (the mass number A), the nucleus has a non-zero nuclear spin number (I), and can be analyzed by NMR spectroscopy. Such individual nuclei (or isotopes) are shown in Table 4.1.

TABLE 4.1: SOME PROPERTIES OF NMR ACTIVE NUCLEI

TABLE 4.1

NUCLEUS	Z	A	I	% NATURAL ABUNDANCE	MAGNETIC MOMENT
^{1}H	1	1	0.5	99.985	2.7927
^{2}H	1	2	1.0	0.015	0.8574
^{3}H	1	3	0.5	0	2.9788
^{13}C	6	13	0.5	1.108	0.7022
^{13}N	7	13	0.5	0	-0.322
^{14}N	7	14	1	99.63	0.4035
^{15}N	7	15	0.5	0.37	0.2830
^{15}O	8	15	0.5	0	0.719
^{17}O	8	17	2.5	.037	-1.8930
^{17}F	9	17	2.5	0	4.720
^{19}F	9	19	0.5	100.0	2.6273
^{31}P	15	31	0.5	100.0	1.1305

Note that some common nuclei, such as ^{12}C and ^{16}O, are not included in this table. Nuclei with an even number of both protons and neutrons have an I value of 0. Only those nuclei with I > 0 will have a nuclear magnetic moment. For such nuclei to be easily measured, the I value must be 0.5. For ease of analysis, it helps if the specific isotope is naturally present in compounds to a large extent, and if the moment of the individual magnet is fairly strong (i.e., a large value of μ). The isotope ^{1}H is ideal in this respect: it has I = 0.5; it is by far the most common hydrogen isotope; and it acts as a relatively strong nuclear magnet. The isotopes ^{19}F and ^{31}P also meet these criteria. ^{13}C, while not as abundant as the magnetically inactive ^{12}C isotope, is reasonably strong, and sufficiently represented in ordinary organic samples to be detected. As we will see, however, a stronger magnetic field will be required for ^{13}C analysis than for ^{1}H analysis.

The nuclei ^{1}H, ^{13}C, ^{19}F, and ^{31}P are commonly used in NMR spectroscopy. These nuclei will align themselves in an external magnetic field in one of two ways, as pictured in the figure below. In the absence of an applied magnetic field, the nuclear spins (represented by arrows) are randomly oriented (A). When a strong field is applied, the individual nuclear spins align themselves either with or against the field. Alignment with the external field is a lower energy level than alignment against the field (B). The difference between energy levels is generally small compared to the thermal energy available at room temperature. Hence both levels are occupied, with only a very slightly larger population of the lower energy level. Diagram B shows 6 nuclear spins, 4 aligned with the field, 2 against the field. This is somewhat misleading. In an actual sample of about one million ^{1}H nuclei at room temperature, for example, the lower energy level population would outnumber the higher energy population by only 2 nuclei.

higher energy state
lower energy state
no magnetic field (A)
strong magnetic field (B)
strong magnetic field with rf energy added (C)

FIGURE 4.1: Orientation of nuclear magnets.

When a radiofrequency (rf) of the correct energy is absorbed by one of the lower energy state nuclei, it can change its spin orientation to that opposed to the applied field (C). This is known as nuclear magnetic resonance (NMR). In order for the resonance phenomenon to continue, the population excess of the lower energy state must be restored. This happens by a process known as spin lattice relaxation. This process occurs very slowly in solids and viscous liquids. Consequently such materials are usually examined when dissolved in an appropriate liquid solvent. An NMR spectrometer, diagramed in figure 4.3, determines the energy needed to excite the nuclei to this higher level.

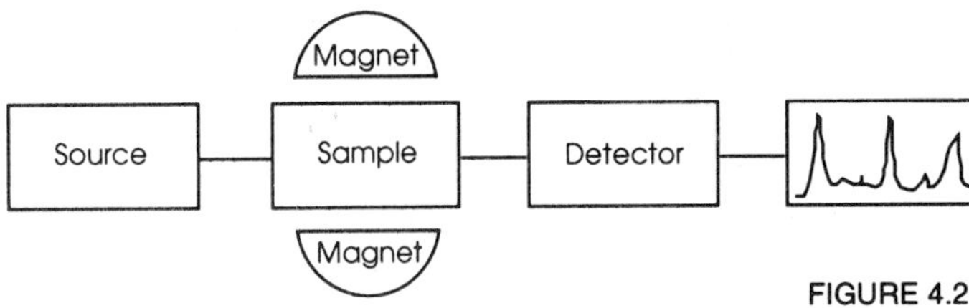

FIGURE 4.2: Schematic of an NMR spectrometer.

The sample is dissolved in a non-interfering solvent. For modern NMR spectroscopy this typically means using a deuterated solvent such as deuterated chloroform ($CDCl_3$). Deuterated solvents are those which have had their protons removed and replaced with deuterium. In practice, it is not possible to completely remove all the protons. Those molecules of solvent that have not had the protons removed will thus give rise to a signal corresponding to the proton. In deuterated chloroform, traces of $CHCl_3$ would give rise to a signal at 7.26 ppm in the proton NMR spectrum. Thus a signal at 7.26 ppm would appear in the spectrum of any compound for which deuterated chloroform had been used

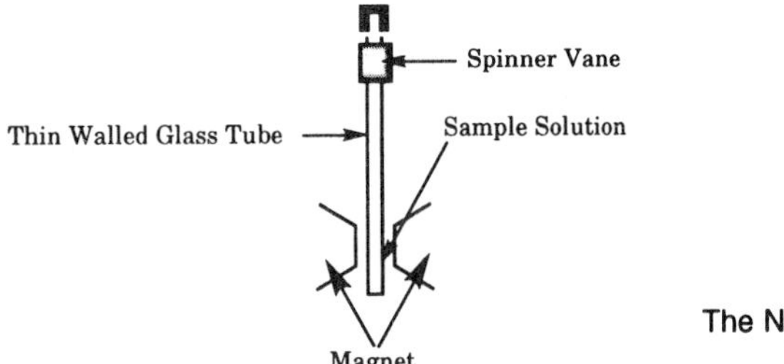

FIGURE 4.3: The NMR sample tube.

as the solvent. For carbon NMR, the solvents will typically have carbon atoms. For these solvents, isotopic replacement of the carbon atoms is not practical. Instead, we simply note that a signal attributable to the solvent may appear in the carbon NMR spectrum. For example, in many of the carbon NMR spectra in this chapter, we see several tightly clustered absorbance signals right at 77 ppm. This corresponds to the signal from the carbon atom in deuterated chloroform.

Once the sample is dissolved in an appropriate solvent, the solution is placed in a thin glass tube (Figure 4.3). This sample tube is placed between the poles of a powerful magnet. The tube is rapidly spun by air pressure directed against the spinner vane. The sample is spun in an effort to make the magnetic field it experiences homogeneous. The strength of the magnetic field at which the sample absorbs a fixed frequency radio wave is recorded. The signal may be processed in various ways, but in general the NMR spectrum is a plot of relative absorbance versus frequency of resonance.

The exact frequency of resonance is not a very useful unit, since it varies depending upon the strength of the individual magnetic field. Instead, relative frequency results are reported in units of δ (or parts per million, ppm), by which the frequency of the absorbance is shifted from that of an internal standard (usually tetramethyl silane, $(CH_3)_4Si$).

$$\delta = \frac{\text{observed shift from TMS (in Hz)}}{\text{spectrometer frequency (in MHz)}} \qquad \text{EQ. 4.1}$$

Much of the recent advance in NMR technique is the result of very high magnetic field strength made possible by superconducting magnets. The strength of the absorbance by NMR active nuclei is approximately proportional to the square of the applied magnetic field strength. Also, as the strength of the applied magnetic field increases, the spacing between energy levels for the alternate alignments of the nuclear spin increases, as does the difference in population between those levels. Thus resonance occurs at higher radiofrequencies. Since individual nuclei in the same chemical environment have constant δ values, equation 4.1 shows that increasing the spectrometer frequency increases the absolute value of the shift measured in Hz. This all means that better magnets have brought about stronger and more easily resolved signals from the nuclei. More advanced means of exciting the nuclei, and recording and processing the absorbance information have brought about Fourier Transform Nuclear Magnetic Resonance (FT-NMR). This technique allows multiple analyses to be run in a fraction of the time previous continuous wave analysis methods used. By rapidly accumulating and averaging numerous signals, FT-NMR provides much superior spectra for analysis.

4.2 MODES OF ^{13}C NMR ANALYSIS

There are two ways that ^{13}C NMR analyses are commonly conducted: normal mode (also known as spin uncoupled mode) and **spin coupled mode**. Both of these modes provide valuable clues about the structure of the carbon containing compound. In the normal mode of operation, a ^{13}C NMR spectrum provides a separate signal for each chemically unique carbon in the molecule. For example, the figure below is a normal mode FT-NMR spectrum of an alcohol.

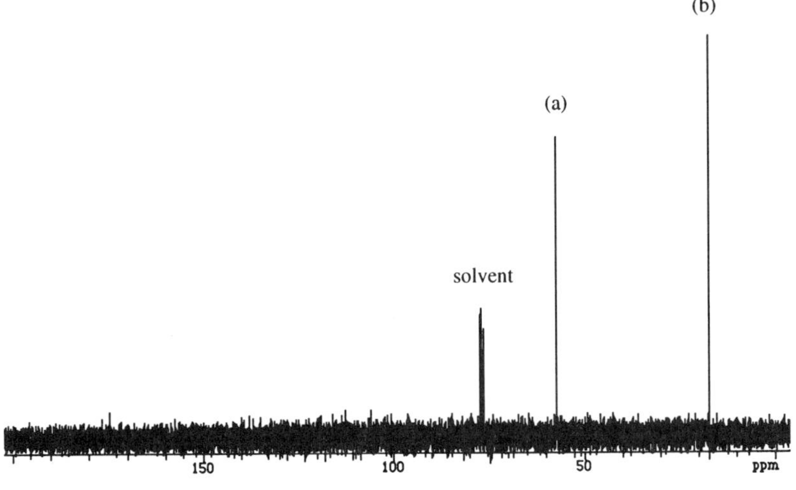

FIGURE 4.4(I): Normal mode ^{13}C FTNMR spectum of a simple alcohol.

Normal Mode

There are two separate signals in this spectrum, indicating two distinct carbon environments in the compound. One absorbance signal (listed a) has a δ value of about 58. The carbon nucleus giving rise to this signal is said to be "deshielded" or "shifted downfield" relative to the standard value of **TMS**, which is used to define 0 on the δ scale. The other carbon signal (listed b) occurs at about δ = 18. The carbon responsible for this signal is deshielded relative to carbon in TMS, but not nearly as deshielded as the carbon giving rise to signal (a). The relative amount of deshielding observed in a ^{13}C NMR spectrum may frequently be rationalized in terms of the electron environment around the carbon in question. The electrons have spin, and as they move around the nucleus, they set up a tiny magnetic field (H_{local}) in opposition to the applied field (H_o). The effective magnetic field is thus smaller than the applied field, according to the equation below:

$$H_{effective} = H_o - H_{local}$$ EQ. 4.2

When a carbon atom is bonded to an element of lower electronegativity than itself (for example, silicon in TMS), then the carbon keeps nearly full control of its electrons, and is maximally shielded from the applied magnetic field. When carbon is bonded to an atom that is more electronegative than itself, it loses control of some of its electrons. That carbon is **deshielded**, and will resonate at a higher δ value. The following table shows some common ^{13}C chemical shifts for different carbon environments.

TABLE 4.2: SOME COMMON CHEMICAL SHIFTS FOR ^{13}C NMR (TABLE 4.2)

TYPE OF CARBON	DELTA	STRUCTURE
RCH_3	10-30	R—C(H)(H)—H
RCH_2R	15-50	R—C(H)(H)—R
R_3CH	20-55	R—C(R)(R)—H
CH_3COOR	~20	H_3C—C(=O)—O—R

TYPE OF CARBON	DELTA	STRUCTURE
CH_3COR	~30	
RCH_2Br	25-50	
RCH_2NH_2	35-45	
RCH_2Cl	35-60	
RCH_2OH	50-65	
ArH	110-160	
RC≡CH	75-85	
RCH═CHR	110-140	
RC≡CH	65-70	
HC≡N	115-125	
$RCH{=}CH_2$	115-140	
RCOOR'	165-175	
$RCONR_2$	165-175	
RCOOH	175-185	
RCHO	195-210	
RCOR'	200-220	

TABLE 4.2

SPIN COUPLED MODE

In the spin coupled mode, the single line peaks we observed in the normal mode may split into more complicated patterns. Figure 4.4(I) is a spin uncoupled spectrum. The (a) signal in Figure 4.4(I) is split into 3 peaks (called a triplet), while the (b) signal is split into 4 peaks (called a quartet), as seen in Figure 4.4(II).

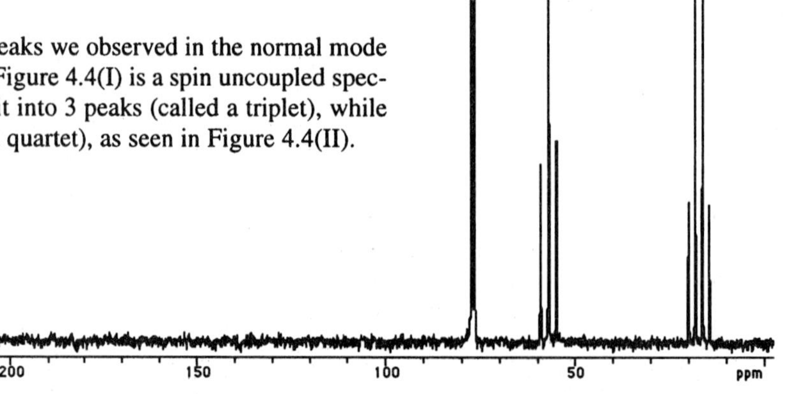

FIGURE 4.4(II): Spin coupled ^{13}C NMR spectrum of a simple alcohol.

This splitting is due to the local magnetic environment around the ^{13}C nucleus caused by any ^{1}H nuclei bonded to it. Remember that the ^{1}H nuclei are also magnetically active. The applied magnetic field strength, Ho, is altered at the ^{13}C site by bonded protons. Unlike the case of electron shielding, the local magnetic environment at the carbon atom can either be increased or decreased by the ^{1}H magnet, depending upon whether the ^{1}H is aligned with or against the applied field.

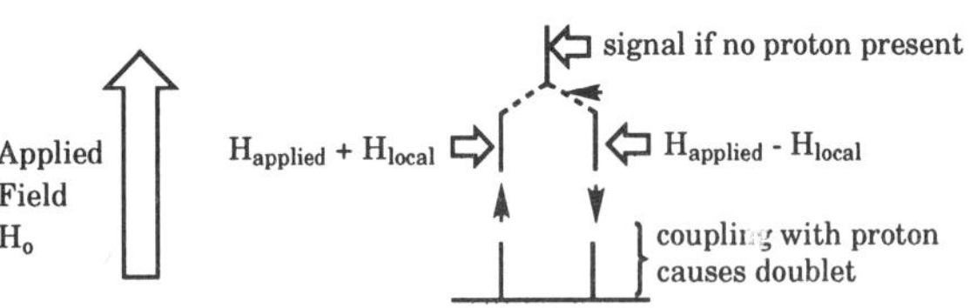

FIGURE 4.5:
^{13}C splitting due to two alternate ^{1}H alignments.

Since, as we saw earlier, these alternate alignments of a proton are essentially equally likely, the presence of a single ^{1}H bonded to a single ^{13}C will result in either increasing or decreasing the applied field H_0 a fixed amount. Of course, for individual atoms either one or the other occurs. But a single sample has numerous such molecules, and two equally intense absorbance bands are seen in the doublet. Splitting of a single ^{13}C peak into a doublet is evidence that the carbon has exactly one hydrogen bonded to it. When two protons are bonded to a carbon, there is one way they can both be aligned against the field, one way they can both be aligned with the field, and there are two possible ways the two can align in mutual opposition.

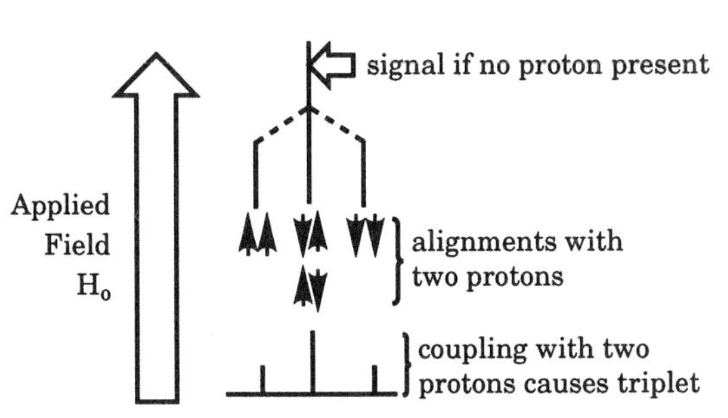

FIGURE 4.6(a):
^{13}C splitting due to three alternate alignments of two protons.

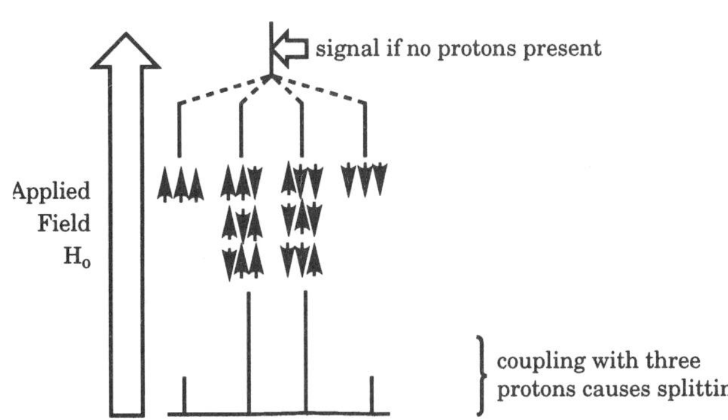

FIGURE 4.6(b):
^{13}C splitting due to four alternate alignments of three protons.

A carbon bonded to two protons produces a triplet in the spin coupled ^{13}C NMR spectrum, with an intensity ratio of 1:2:1. A carbon bonded to three protons produces a quartet, with an intensity ratio of 1:3:3:1. The number of peaks into which the ^{13}C signal is split depends upon the number of protons bonded to the carbon atom. The **"n + 1 rule"** is a simple way of remembering the splitting pattern: when "n" is the number of hydrogen atoms bonded to a carbon atom, the NMR signal resulting from that ^{13}C will be split into n + 1 peaks. This is shown in the next figure.

	n	peak shape			n	peak shape	
-C-	0	ǀ	singlet (s)	H-C-H	2	ılı	triplet (t)
-C-H	1	ǁ	doublet (d)	H-C-H (H)	3	ıllı	quartet (q)

FIGURE 4.7:
^{13}C NMR signal splitting due to n hydrogens on the carbon.

The intensity ratios of the components into which the signal is split are predicted by Pascal's Triangle, a triangular arrangement of digits in which each number is the sum of the two immediately adjacent digits in the previous line.

```
          1
        1   1
      1   2   1
    1   3   3   1
  1   4   6   4   1
1   5  10  10   5   1
```

FIGURE 4.8:
Pascal's Triangle.

The fact that in the alcohol ^{13}C NMR spectrum we observed a triplet and a quartet in spin coupled mode indicated a methyl group (-CH_3, with n = 3, thus a quartet) and a methylene group (-CH_2, with n = 2, thus a triplet). The methylene group was shifted farther downfield, as would be expected if it were attached to the -OH group of the alcohol. All the NMR spectral evidence is consistent with an interpretation of the alcohol as being ethanol, CH_3CH_2OH. The normal mode showed exactly two different types of carbon environments. The spin coupling mode showed a methyl and methylene carbon environment, consistent with the deshielding observed for the two signals in the normal mode, assuming that the methylene group was attached to the hydroxyl (OH group). Finally, note that there is only one structure of ethanol. Although we represent ethanol in various diagrams shown below, they each represent the same molecule. Remember that such two dimensional diagrams of three dimensional molecules must not be taken too literally:

```
                                        H
     H H           H  H                H:O:              H  OH
                   |  |                                  |  |
H:O:C:C:H       HO—C—C—H             H:C:C:H          H—C—C—H
                   |  |                                  |  |
     H H           H  H                H H               H  H


   H H             H  H                H H               H  H
                   |  |                                  |  |
 H:C:C:O:H       H—C—C—OH            H:C:C:H          H—C—C—H
                   |  |                                  |  |
   H H             H  H                H:O:              H  OH
                                        H
```

FIGURE 4.9:
Various representations of ethanol.

If the molecules represented in the figure above do not seem equal, it would be best to construct three-dimensional models to demonstrate their equivalence. The apparently different forms can be made by simply rotating the molecule. These are called rotational (or conformational) **isomers**. The word "isomer" remember, means "equal parts". Rotational (conformational) isomers are in fact the same molecule, either viewed from a different perspective, or generated by rotation around single bonds.

4.3 ^{13}C NMR AND STRUCTURAL ISOMERS

COMPOUNDS WITH THE FORMULA C_3H_8O

Next, let us consider the compound C_3H_8O. From the zero index of hydrogen deficiency value, it is clear that this compound has only single bonds and no rings. The oxygen can thus be part of either an ether or an alcohol. (Functional groups including C=O are not possible due to the molecular formula.) Infrared analysis indicates that the compound in question is an alcohol, as identified by the characteristic O-H stretch.

NMR analysis of an unknown alcohol with the formula C_3H_8O can produce either of the following NMR spectra:

COMPOUND I

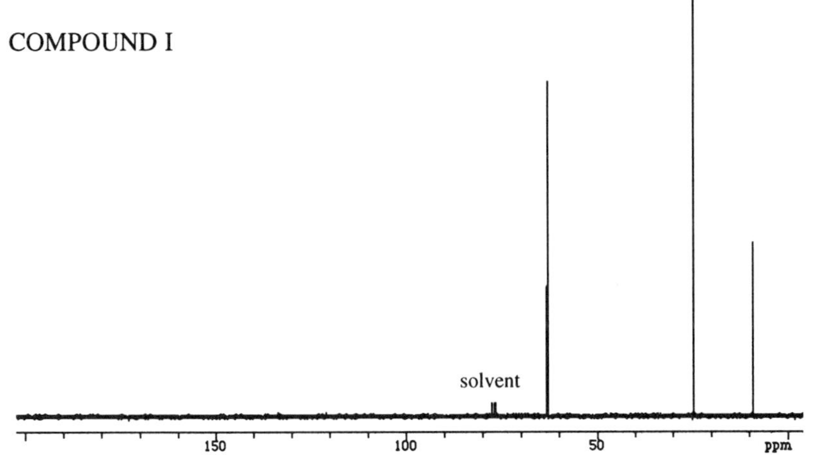

FIGURE 4.10(I): ^{13}C spectra of two alcohols, both with the formula C_3H_8O.

COMPOUND II

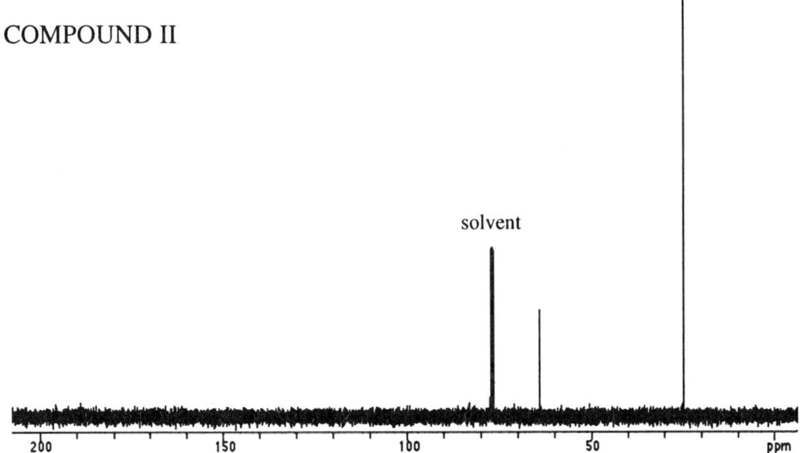

Compound I shows three signals, but compound II shows two signals. Both compound I and II have three carbons. The spectrum shows that Compound I has three distinct carbon environments, or one for each carbon. The spectrum for Compound II, however, has only two distinct carbon types.

In Chapter 3, we noted that propanol could be written in the form $HOCH_2CH_2CH_3$, or 1-propanol. In 1-propanol, carbon #1 is attached to the oxygen. Carbon #2 is one carbon-carbon single bond removed from the oxygen, and carbon # 3 is two carbon-carbon single bonds removed from the oxygen. All three carbons in 1-propanol have a unique environment, and thus 1-propanol should generate three unique ^{13}C NMR signals. 1-propanol has the C-NMR spectrum of Compound I. Again, note that placing the hydroxyl group "at the other end" of the carbon chain $CH_3CH_2CH_2OH$ produces the same three carbon environments; this is a rotational isomer of 1-propanol (i.e., it is the same molecule.) An alternate structure of propanol is shown in figure 4.11.

This alcohol is named 2-propanol, because the carbon to which the OH is attached is second from the end of the chain. The carbon in the CH_3 (methyl) group to the left of the middle carbon could be called carbon #1, and considered the start of the chain. The carbon in the methyl group to the right of the middle carbon, however, could equally well be considered the start of the chain. The two methyl groups are both connected to the same atom, and thus are termed equivalent. 2-propanol generates a ^{13}C NMR spectrum with only two signals: one signal for the central carbon attached to the OH group, and another signal for the equivalent methyl group carbons. 2-propanol has the NMR spectrum of Compound II.

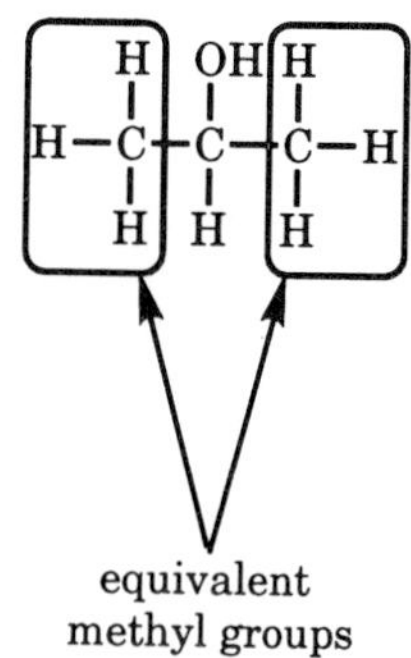

FIGURE 4.11:
Structure of 2-propanol.

Concept Check: Shielding and Chemical Shift

Figure 4.10 shows that the carbon NMR spectrum of 2-propanol has only two signals: one at 24 ppm, and another at 63 ppm. One signal is associated with the central carbon atom, the other signal corresponds to the two CH_3 carbons which are equivalent. Which carbon(s) are associated with which peaks?

Solution: In 2-propanol, the methyl carbons can be numbered #1 and #3, while the central carbon is #2. Both carbon environments are deshielded by the presence of oxygen, an electronegative element that removes electron density from nearby carbons. Since carbon #2 is closest to the oxygen atom, it is most deshielded. Thus carbon #2 is associated with the resonance at 63 ppm. The electron environment of the methyl carbons are less influenced by the oxygen, since they are farther removed. Consequently, their resonance is not shifted as much, and the signal associated with carbons #1 and #3 occurs at 24 ppm.

Concept Check: Spin-Spin Coupling in Carbon NMR

The ^{13}C NMR spectrum of 2-propanol in Figure 4.10 was run in the decoupled mode. What would be the multiplicity of each of the two signals if the spectrum had been run in the coupled mode?

Solution: Carbon #2 (directly bonded to oxygen)has just one proton bonded to it. Thus the signal at 63 ppm, corresponding to carbon #2, would be split into a doublet, according to the n + 1 rule. The signal at 24 ppm, corresponding to the equivalent carbon environments for carbons #1 and #3, is the resonance of carbon attached to three protons. Thus the n + 1 rule predicts that the signal at 24 ppm would be split into a quartet.

Compounds I and II are in fact different molecules, despite both being alcohols with the formula C_3H_8O. This is clearly demonstrated by the two different ^{13}C NMR spectra. These two compounds, 1-propanol and 2-propanol, are called **structural isomers**. They are isomers because they are built out of the same numbers of the same types of atoms. The isomers are designated structural because there is a structural difference between the two. Unlike **rotational iso-**

mers, which can be interconverted by rotating around bonds, structural isomers can be interconverted only by breaking and remaking bonds.

Compounds With the Formula C_4H_8O

Next, we consider several compounds with the formula C_4H_8O. Infrared spectra indicate a carbonyl group present in each of these compounds, which accounts for the index of hydrogen deficiency value of 1. In other words, the carbons must be singly bonded to each other, and must not form any rings. When ^{13}C NMR spectra of compounds that match these conditions are obtained, we find the three alternatives given in the following figure:

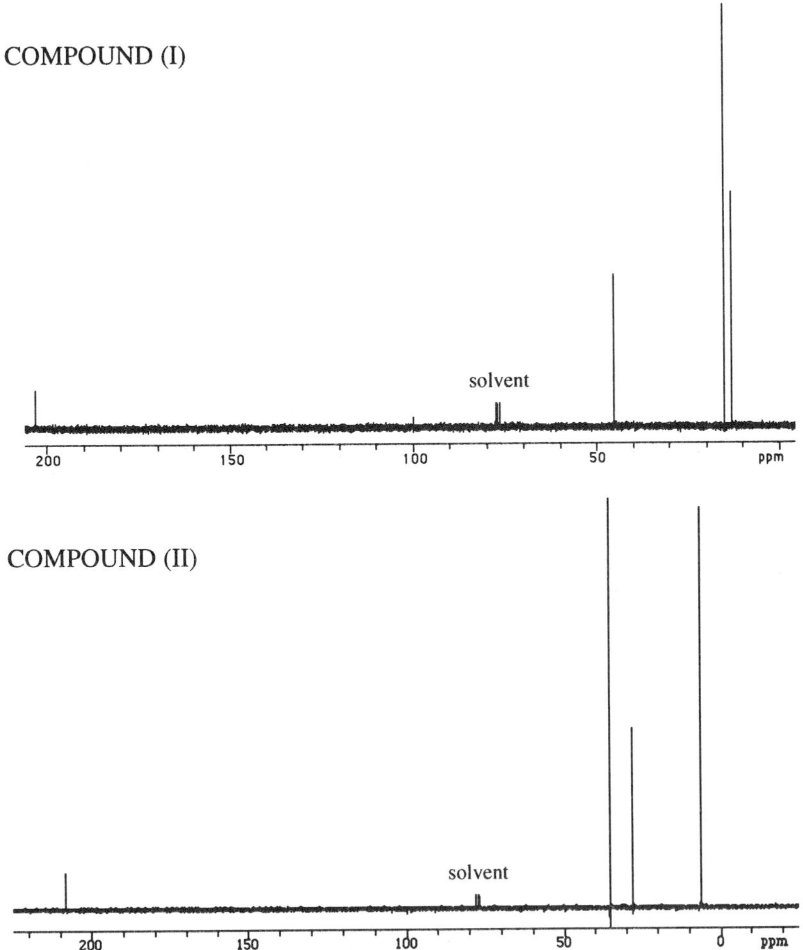

FIGURE 4.12: ^{13}C NMR spectra of three compounds, each C_4H_8O (cont'd on next page).

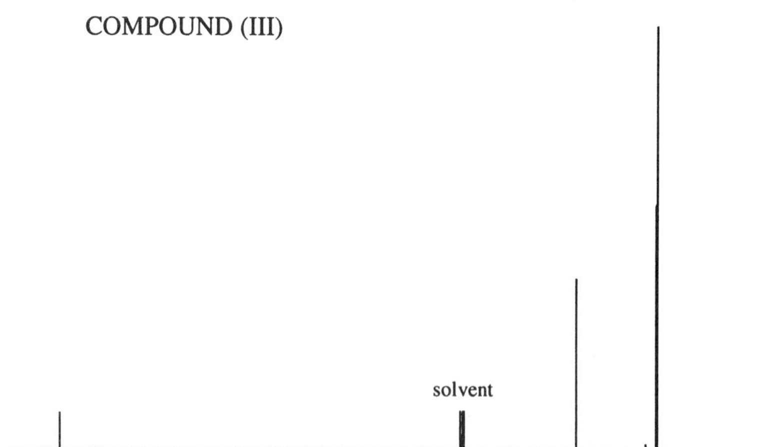

FIGURE 4.12(cont'd)

```
C—C—C—C
a  b  c  d
```

FIGURE 4.13: Positions on a four carbon straight chain.

Compounds I and II both show four carbon environments, one for each of the four carbons. Compound III, however, has just three environments; two of the carbons in compound III must be equivalent.

If we attempt to draw isomers for C_4H_8O, we can start by putting four carbons in a straight chain. Note that in the Figure 4.13, position (a) is equivalent to position (d), and position (b) is equivalent to position (c).

```
    H  H  H  O                  H  H  O  H
    |  |  |  ||                 |  |  ||  |
H—C—C—C—C—H          H—C—C—C—C—H
    |  |  |                     |  |      |
    H  H  H                     H  H      H
```

FIGURE 4.14: Structures of butanal and 2-butanone.

Thus, when the four carbons are in a straight chain, there are only two distinct positions for the oxygen to attach: on terminal carbon (a) or (d), forming an aldehyde, or on an internal carbon (b) or (c), forming a ketone. The two possible structures are shown in Figure 4.14:

The structures shown in the figure each have four distinct carbon environments. They must be associated with Compounds I and II, whose spectra are displayed in Figure 4.12. Which structure is related to which spectrum remains a question (until after the spin coupling spectra have been examined). A third structure must exist; one that has only 3 distinct carbon environments. Since we have exhausted the possibilities with the four carbons in a straight chain, and since rings are precluded based on IR and molecular formula, the backbone structure must be branched. There is only one possible way that four carbons can be branched, and that is shown in Figure 4.15.

```
      a
      C
      |
   C—C—C
   a  b  a
```

FIGURE 4.15: Branching of a 4 carbon chain.

Note that with only the carbons present, there are two carbon environments: the terminal carbons (a) and the central carbon (b). Doubly bonding the oxygen to this backbone is limited to the terminal carbons, because the central carbon already has three single bonds. Since carbon obeys the octet rule, it can form a maximum of four bonds. The central carbon in Figure 4.15 can not doubly bond to oxygen. Doubly bonding the oxygen to any of the three terminal carbons results in the same compound, shown in Figure 4.16.

```
    H  H  O
    |  |  ||
H—C—C—C—H
    |  |
    H  C
      /|\
     H H H
```

FIGURE 4.16: 2-Methylpropanal (isobutyraldehyde).

Note that this structure now has three distinct carbon environments, as shown in the Figure 4.17.

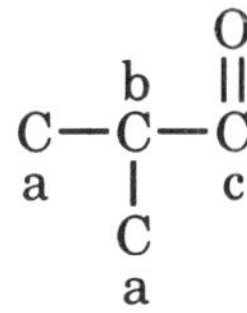

FIGURE 4.17: Carbon environments for isobutyraldehyde.

The two methyl groups (a) are equivalent. The carbons in these groups give rise to only one signal. Their assignment, along with carbons (b) and (c) are shown in the spectrum of Compound III.

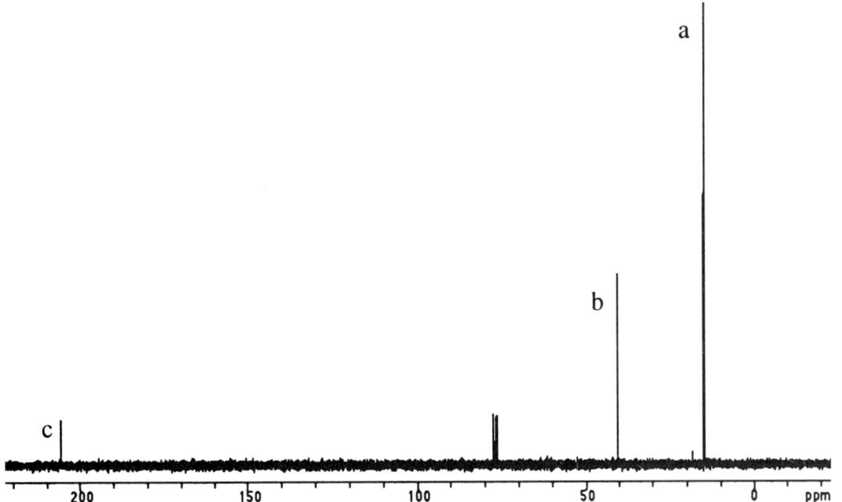

FIGURE 4.18: ^{13}C NMR spectrum of isobutyraldehyde, with peak assignments.

Note that the aldehyde carbon (c) is shifted farthest downfield. This carbon is significantly deshielded due to the electron withdrawing effect of the carbon oxygen double bond.

Spectra I and II in Figure 4.12 belong to either butanal or 2-butanone, but to determine which is which, the spin coupled ^{13}C NMR spectra are obtained.

FIGURE 4.19: ^{13}C NMR spin coupled spectra, compounds I and II (on next page).

COMPOUND (I)

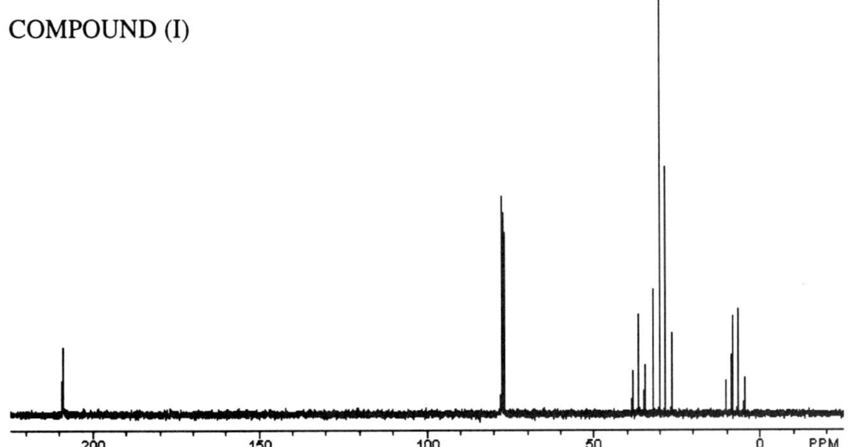

FIGURE 4.19(cont'd)

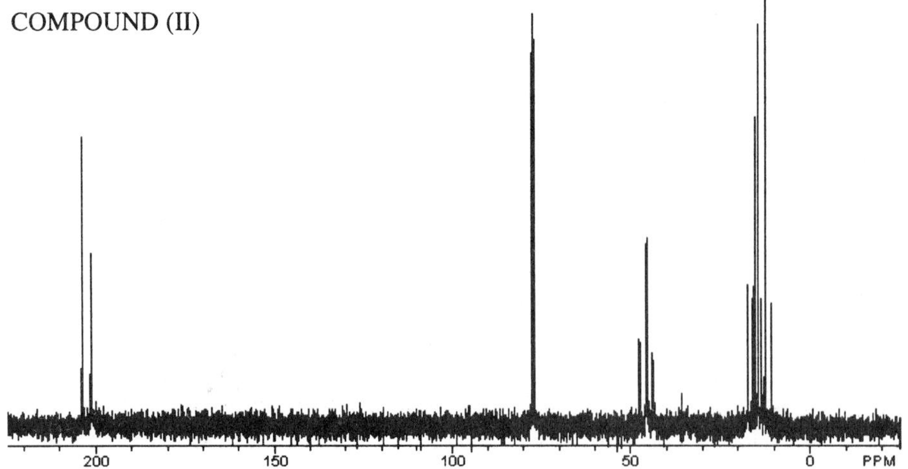

The four signals in the spectrum for Compound I form one doublet, two triplets, and a quartet. From the n + 1 rule, this indicates one carbon with one hydrogen attached, two carbons with two hydrogens attached, and one carbon with three hydrogens attached. Comparing the structures in Figure 4.14, we see butanal has a carbon that is bonded to just one hydrogen, but 2-butanone has no such structure. NMR evidence supports assigned butanal as Compound I.

The four signals from Compound II form one singlet, a triplet, and two quartets. Using the n + 1 rule, this implies one carbon unattached to any hydrogens, one carbon attached to two hydrogens, and two carbons attached to three hydrogens. Comparing the structures in Figure 4.14, we see 2-butanone has a carbon that is not bonded to any hydrogen. This gives rise to the singlet in Figure 4.19 (II). All the spectral evidence is consistent with assigning Compound II as 2-butanone.

A full explanation of the names of the three compounds, butanal, 2-butanone, and 2-methylpropanal, will be presented later. Note, however, that the "al" suffix indicates that Compounds I and III are aldehydes, while the "one" suffix indicates that Compound II is a ketone.

Compounds With the Formula C_5H_{12}

The twelve hydrogens in a five-carbon compound represent the maximum possible, when calculated as 2n + 2 (with n being the number of carbon atoms.) Thus we are dealing with a saturated pentane structure: all single bonds and no rings. Experimentation shows that there are three distinct compounds, all with this same formula. Let us try to predict ^{13}C NMR spectra based upon our ability to construct alternate structures on paper. The first approach might be to put all five carbons in a straight chain. This is known as n-pentane. The "n" prefix indicates a normal, or straight chain, structure. The "-ane" suffix indicates that it is an alkane. "Pent" is from the Greek, meaning five (e.g., a five-sided figure is called a pentagon). Thus the systematic stem for five carbons is "pent". The structure of n-pentane is shown at right.

```
  H  H  H  H  H
  |  |  |  |  |
H-C--C--C--C--C-H
  |  |  |  |  |
  H  H  H  H  H
```

FIGURE 4.20: Structure of n-pentane.

This structure predicts three distinct carbon environments for ^{13}C NMR: the end methyl carbons, the next to the end methylene carbons, and the central carbon. There are no other pentane structures with all five carbons in a row. We must look to branching the chain to provide the remaining two alternatives.

```
       H
     H | H
   H  \C/  H  H
   |   |   |  |
H—C——C——C——C—H
   |   |   |  |
   H   H   H  H
```

FIGURE 4.21(a): Structure of 2-methylbutane.

As we saw earlier when discussing Figure 4.13, a straight chain of four carbon atoms has just two positions: on the end carbon, or on the second carbon from the end. Placing the fifth carbon on an end carbon simply regenerates n-pentane, five carbons in a straight chain. The only alternative with four carbons in a row is to attach the fifth carbon to an interior carbon. This produces 2-methylbutane, shown at right.

Butane is the systematic name for a straight chain four carbon alkane. The next to the end carbon is numbered "2" (not "3", since that is a higher, but equivalent position number). Since the methyl group is attached to the #2 carbon on a four-carbon chain, the molecule is called 2-methylbutane. This is the only unique structure for C_5H_{12} with four carbons in a row. The ^{13}C NMR spectrum should show four signals, since the two carbons labeled (a) in the figure at right are equivalent.

```
      a
      C
      |
C—C—C—C
a  b  c  d
```

FIGURE 4.21(b): Carbon environments for 2-methylbutane.

The next alternative is to reduce the backbone chainlength to three carbons, and to attach the other two carbons to this chain. As we have seen earlier, with three carbons in a row there are only two positions: terminal carbon or center carbon. Placing one or both carbons on a terminal carbon brings us back to the four-carbon or five-carbon long chains previously examined. The only remaining alternative is to place both carbons onto the middle carbon in the three-carbon chain. This produces the third (and last) structural isomer of C_5H_{12} shown below. Again, it is named based upon the longest carbon carbon chain. Thus the compound is a derivative of propane, with the methyl groups attached to the middle carbon.

```
    a
    C
    |b
C—C—C
a   |   a
    C
    a
```

FIGURE 4.22: Structure of 2,2-dimethylpropane (neopentane).

Note that there are only two carbon environments. The four methyl carbons are equivalent, and produce just one ^{13}C NMR signal. The central carbon produces a second signal.

TABLE 4.3: SUMMARY OF C_5H_{12} STRUCTURAL INFORMATION

TABLE 4.3

NAME	STRUCTURE	# ^{13}C NMR SIGNALS
n-pentane	$CH_3CH_2CH_2CH_2CH_3$	3
2-methylbutane	CH_3 | $CH_3CHCH_2CH_3$	4
2,2-dimethylpropane	CH_3 | CH_3CCH_3 | CH_3	2

Thus the three structural isomers of pentane can be readily distinguished by the number of ^{13}C NMR signals associated with each compound. The naming of these and related compounds will be examined in greater detail later this chapter.

4.4 INTRODUCTION TO ^{1}H NMR SPECTRA

The nuclear magnetic resonance spectra we have seen so far have all been ^{13}C spectra, either in normal mode or spin coupled mode. In this section we examine the magnetic resonance associated with protons. The phenomenon is essentially the same as that addressed earlier. The proton spin is either aligned with or against the external magnetic field, radiofrequency is used to cause the spin to flip, and the frequency at which this jump takes place depends on the chemical environment of the proton. Unlike ^{13}C NMR, where absorbance could range from 0 to 200 δ, most proton NMR absorbances occur between 0 and 10 δ, as shown in the table below.

TABLE 4.4

TABLE 4.4: SELECTED H-NMR δ VALUES

TYPE OF HYDROGEN	DELTA	STRUCTURE
RCH_3	0.7-1.0	R—C(H)(H)—H
RCH_2R	1.3-1.5	R—C(H)(H)—R
R_3CH	1.5-2.0	R—C(R)(R)—H
$R_2C{=}CR'CH_3$	1.6-2.0	$R_2C{=}C(R)CH_3$
$RCOCH_3$	2.0-2.5	R—C(=O)—CH_3
$ROCH_3$	3.2-4.0	R—O—CH_3
$ArCH_3$	2.0-3.0	Ar—CH_3
$RNHCH_3$	2.0-3.0	R—N(H)—CH_3
RC≡CH	2.5-3.0	R—C≡CH
RCH_2X (Cl,Br,I)	3.5-4.5	RCH_2X
$RCOOCH_3$	3.5-4.0	R—C(=O)—O—CH_3
$RCH{=}CH_2$	4.5-5.0	$R(H)C{=}CH_2$
$RCH{=}CR_2$	5.2-5.8	$R(H)C{=}CR_2$
RNH_2	1-3	RNH_2
$ArNH_2$	3-5	$ArNH_2$
ROH	3-6	RO H
ArOH	4-8	ArO H
RCONHR	5-9	R—C(=O)—N(H)—R
ArH	6-8	Ar H
RCHO	9-10	R—C(=O)H
RCOOH	10-13	R—C(=O)OH

The absorbance due to protons is much stronger than that observed for ^{13}C, and consequently weaker magnetic fields can be used. NMR instruments capable of measuring ^{13}C spectra can measure ^{1}H spectra, but the reverse is not necessarily true.

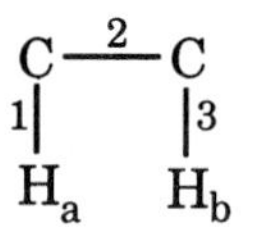

C —2— C
1| |3
H_a H_b

FIGURE 4.23: Spin-spin coupling for proton NMR.

Proton NMR spectra and ^{13}C NMR spectra also differ in the way that the signals can split. In the case of ^{13}C, the splitting was attributed to the number of protons directly bonded to the carbon in question. The n + 1 rule meant that if a carbon is bonded to no hydrogens, the signal is a singlet; bonded to 1 hydrogen atom, the signal becomes a doublet; bonded to two hydrogen atoms, the signal becomes a triplet. In the case of proton NMR signal splitting, the spin coupling phenomenon is somewhat different. The signal due to magnetic resonance of a proton is not influenced by the carbon to which it is attached. In proton NMR, the signal from one hydrogen is split by the adjacent hydrogens. The effect is over three bonds, as illustrated at right in Figure 4.23.

Here the signal from proton H_a is split into two peaks by H_b and the signal from proton H_b is split into two peaks by H_a. The n+1 rule used for hydrogen-coupled ^{13}C NMR applies here also. But remember, in proton NMR, n represents the number of equivalent hydrogens on an adjacent carbon.

HNMR for Ethanol

We began our discussion of ^{13}C NMR with ethanol. Let us continue our discussion of proton NMR with the same example. CH_3CH_2OH has hydrogen in three distinct environments: the methyl hydrogens (CH_3), the methylene hydrogens (CH_2), and the hydroxyl hydrogen. These give rise to three groups of signals, seen in the figure below.

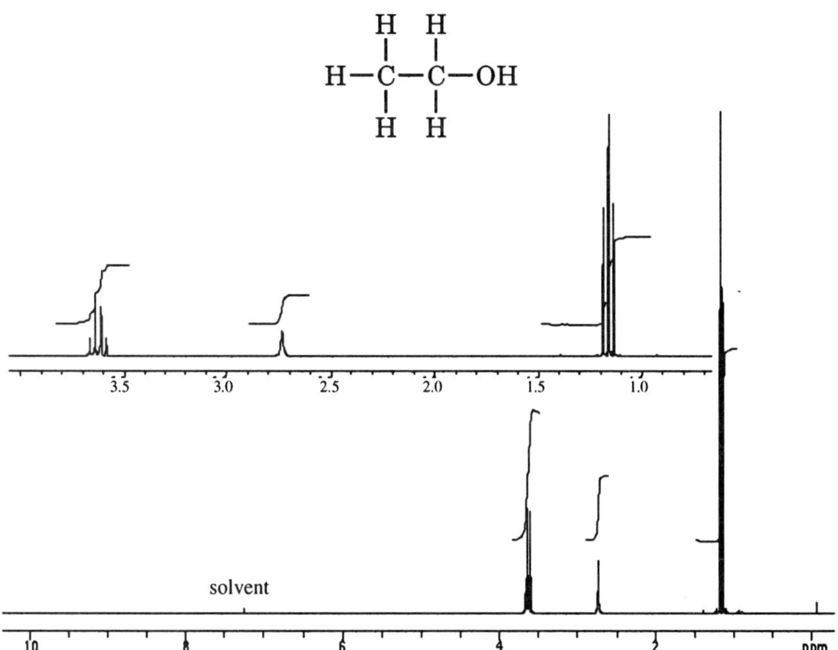

FIGURE 4.24: The structure and ^{1}H NMR spectrum of ethanol.

The signals in proton NMR can be integrated, which yields the area under the absorbance bands. This area is proportional to the number of equivalent hydrogen atoms giving rise to the signal. The area ratios for the three absorbance bands in Figure 4.24 are 2:1:3, for the quartet (q) at $\delta = 3.7$, the singlet (s) at $\delta = 2.7$, and the triplet (t) at $\delta = 1.2$ respectively.

The proton NMR spectrum for ethanol is explained as follows. The most shielded absorbance (lowest δ value) is for the CH_3- (methyl) protons. These hydrogens are furthest removed from the electron withdrawing power of the oxygen atom. The methyl hydrogen appears as a triplet because it is adjacent to a carbon with two protons attached. For proton NMR, the n + 1 rule uses n as the number of protons on the adjoining carbon atom. The integrated areas indicate three hydrogens were responsible for this absorbance. The -CH_2-(methylene) hydrogens are more deshielded due to the proximity of the oxygen atom, and they appear as a quartet at $\delta = 3.7$. These two protons appear as a quartet because they have an adjacent carbon atom with three hydrogens attached. Again, the n + 1 rule predicts a quartet for the hydrogen bonded to a carbon that has three protons three bonds away. The integrated areas indicate two hydrogens were responsible for this absorbance. Finally the singlet at $\delta = 2.7$ is due to the hydrogen on the oxygen. For reasons that will be addressed later, this hydrogen does not cause any splitting itself for adjacent protons, nor is its signal split. For our purposes, if one of the carbons in Figure 4.23 was replaced with an oxygen atom, no splitting (for either Ha or Hb) would occur. The integrated areas indicate one hydrogen was responsible for this absorbance.

When nuclei interact to mutually split one another, as the CH_3 protons interacted with the CH_2 protons in ethanol, we say spin-spin coupling occurs. The strength of the coupling is symbolized by J, and is measured in Hertz. The value of the **coupling constant J** is determined by the spacing (in Hz) between signals in a multiplet. In Figure 4.24, the spacing between the signals in the quartet corresponding to the CH_2 protons is equal to the spacing bewteen the signals in the triplet corresponding to the CH_3 protons. That these two multiplets have the same spacing (same coupling constant J) indicates that the protons responsible for those signals are in fact coupled and thus are adjacent. Also note that when very high resolution spectrometers are used, details of structure emerge in the spectrum that are beyond the scope of this course.

HNMR FOR 2-PROPANOL

The figure below shows the proton NMR spectrum for isopropanol.

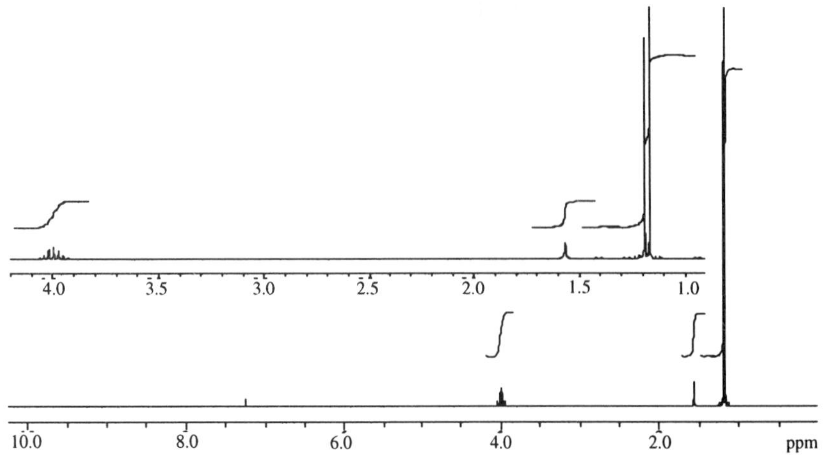

FIGURE 4.25: HNMR spectrum of 2-propanol CH_3-CH(OH)-CH_3.

Since the two methyl groups are equivalent, the six hydrogens on these two carbons are all equivalent, and they generate one signal. The signal for the methyl protons is a doublet (d) centered at δ = 1.2. Again, the relatively low δ value indicates these protons are relatively far removed from the deshielding effect of a bond with oxygen. This signal is a doublet because both methyl carbons are attached to only one carbon: the central carbon. The central carbon, in turn, is bonded to just one proton. Since the methyl protons are three bonds removed from the lone proton on the adjacent carbon, their peak is split according to the n + 1 rule into a doublet. The singlet at δ= 1.6 is due to the proton on oxygen. As mentioned earlier, hydrogens bonded through an oxygen are neither split, nor cause splitting in adjacent hydrogen signals. The septet at δ = 4.0 is, according to the n + 1 rule, split by six neighboring equivalent hydrogens. This is certainly the hydrogen on the central carbon. This signal is most deshielded, since the carbon to which the hydrogen is bonded is directly bonded to the oxygen atom. This proton has six adjacent protons from the two equivalent CH_3 groups bonded to the central carbon. Integration shows that the areas under the peaks at δ = 4.0, 2.0 and 1.2 are 1:1:6 respectively, in accordance with our assignment of protons responsible for each signal.

HNMR for 2-Butanone

The spectrum below is the proton NMR spectrum of 2-butanone, also known as methyl ethyl ketone.

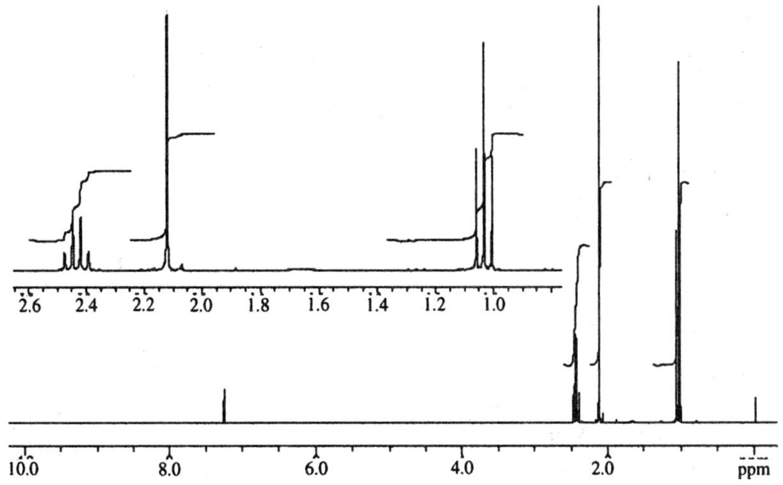

FIGURE 4.26: HNMR spectum of 2-butanone CH_3-CH_2-C(=O) CH_3.

The three proton environments are reflected in three signals. The singlet is due to the methyl protons immediately adjacent to the carbonyl group. The other methyl group (the #4 carbon, with oxygen on the #2 carbon) is split into a triplet by the two protons on the methylene (CH_2) group. The methylene protons, in turn, are split into a quartet by the adjacent methyl group's three equivalent carbons. Upon integration, the singlet, triplet, and quartet are shown to have areas in the ratio of 3:3:2, in accordance with their peak assignment.

Complex Splitting

In longer molecular chains, one may observe a more complicated form of splitting of HNMR signals. In the HNMR spectrum of 1-butanal, whose ^{13}C NMR spectrum was examined earlier, we see a less distinct splitting pattern.

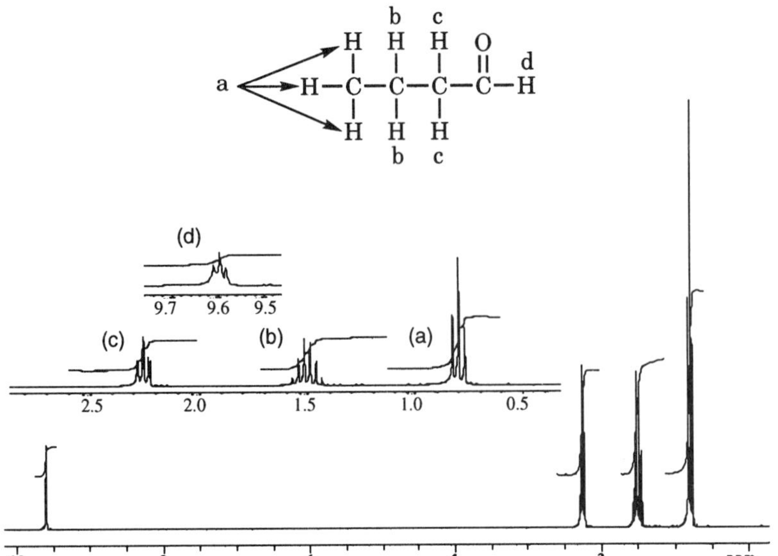

FIGURE 4.27: Structure and HNMR spectrum of butanal.

The terminal methyl group protons, labeled (a) in the structure in the figure above, are clearly split into a triplet by the two protons in the adjoining methylene group. The aldehyde proton (d) is split into a triplet by the adjoining methylene group at position (c). But the splitting of the signals from protons at positions (b) and (c) are more complex. The (b) position protons are split both by the three (a) methyl protons and the two (c) methylene protons. These five adjacent protons are not equivalent. The way the (a) protons and the (c) protons couple to the (b) protons are approximately equal in this case, so an (n + 1 = 6) sextet results. In other cases, the coupling from non-equivalent protons will not be equal, and no clear multiplet signal results. Such is the case for the (c) protons, which are unequally coupled to the one (d) aldehyde proton and the two (b) methylene protons. These three adjacent protons are not equivalent, so a standard n+1 quartet is not observed. Close examination of the multiplet structure for absorbance by the (c) protons shows that what may first appear to be a triplet is actually a triplet of doublets. The coupling of the (c) protons with the adjacent (b) protons is relatively strong, and distinct splitting of the (c) signal into triplets results. That is, the J value (the separation between the triplet peaks) is significant. These same (c) protons are also weakly coupled to the single (d) proton. This coupling is not as strong. A smaller J value results, which splits each of the three signals in the triplet into closely spaced doublets. Later, in Chapter 8, we will resume our discussion of coupling constants and the information they provide about the structure of the molecule.

4.5 BACKBONE STRUCTURES, ISOMERS, AND NOMENCLATURE

In the previous sections of this chapter, we have learned how molecular structure (especially backbone structure) is often determined by NMR. Our study of NMR spectra also introduced us to the concept of structural isomers, and some of the problems associated with naming those compounds. In this section we systematically examine hydrocarbon backbone structure. We will again address the issue of structural isomers, and outline a systematic approach to naming backbone components.

We begin with identifying the structures and names of the first ten straight chain alkanes in the following table.

TABLE 4.5

TABLE 4.5: C_1 THROUGH C_{10} STRAIGHT CHAIN ALKANES

CH_4	methane	CH_4	CH_4
C_2H_6	ethane	CH_3-CH_3	CH_3CH_3
C_3H_8	propane	CH_3-CH_2-CH_3	$CH_3CH_2CH_3$
n-C_4H_{10}	butane	CH_3-CH_2-CH_2-CH_3	$CH_3(CH_2)_2CH_3$
n-C_5H_{12}	pentane	CH_3-CH_2-CH_2-CH_2-CH_3	$CH_3(CH_2)_3CH_3$
n-C_6H_{14}	hexane	CH_3-CH_2-CH_2-CH_2-CH_2-CH_3	$CH_3(CH_2)_4CH_3$
n-C_7H_{16}	heptane	CH_3-CH_2-CH_2-CH_2-CH_2-CH_2-CH_3	$CH_3(CH_2)_5CH_3$
n-C_8H_{18}	octane	CH_3-CH_2-CH_2-CH_2-CH_2-CH_2-CH_2-CH_3	$CH_3(CH_2)_6CH_3$
n-C_9H_{20}	nonane	CH_3-CH_2-CH_2-CH_2-CH_2-CH_2-CH_2-CH_2-CH_3	$CH_3(CH_2)_7CH_3$
n-$C_{10}H_{22}$	decane	CH_3-CH_2-CH_2-CH_2-CH_2-CH_2-CH_2-CH_2-CH_2-CH_3	$CH_3(CH_2)_8CH_3$

Such straight chain alkanes are designated with the prefix "normal", or simply the symbol "n" prior to their name. Again, the alkane family connection is shown in the suffix "-ane". Alkanes with C5 through C10 (and higher) have a stem based on the Greek for the number of carbons. Do not erroneously substitute the Latin for six and seven (sex- and sept-) for the Greek (hex- and hept-).

The table above shows the structure of alkanes in several forms. Throughout this text we will show molecular structure, particularly hydrocarbon structure, in a variety of forms. For example, the structure of butane,C_4H_{10}, can be represented in the forms seen below.

```
  H H H H
H-C-C-C-C-H   CH3CH2CH2CH3     CH3(CH2)2CH3
  H H H H
    (I)            (II)            (III)          (IV)
```

FIGURE 4.28: Alternate structural representations of *n*-butane.

Form (I) is a representation of structure derived from a Lewis dot diagram. It provides the clearest picture of bonding between atoms. The condensed form (II) and the even more condensed form (III) convey the same structure, but do not directly show all the bonding. These forms do have the advantage of being easier to write (especially when working with a word processor). Form (IV), the line angle form, is the most succinct. Carbon atoms are symbolized by the ends of the lines and by the intersection of lines. No hydrogen atoms are explicitly shown. All these forms will be used throughout the text.

As we saw earlier in this chapter, a four-carbon alkane is the simplest that can have a branched (not straight chain) backbone. These are represented in the figure below.

CH_3-CH_2-CH_2-CH_3 or

CH_3-CH(CH_3)-CH_3 or

FIGURE 4.29: Structures of n-butane and isobutane.

Again, in the common name, the straight chain is indicated by the prefix "n ". When the branch occurs at the second carbon (next to the end) the prefix "iso" is used in the common name. The systematic (IUPAC) name is based upon the longest continuous carbon to carbon chain. For isobutane, the longest chain is three carbons long, so it is named as a derivative of propane. The methyl group substituent is placed on the second carbon, hence the systematic name of 2-methylpropane.

When C_5H_{12} is examined, the additional carbon expands the number of possible isomers to three, pictured below.

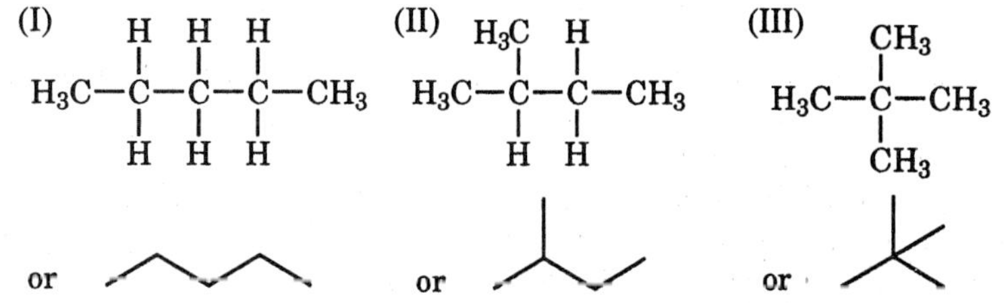

(I) H_3C—CH_2—CH_2—CH_2—CH_3 or

(II) H_3C—C(CH_3)H—CH_2—CH_3 or

(III) H_3C—C(CH_3)$_2$—CH_3 or

FIGURE 4.30: Structural isomers of C_5H_{12}.

Structure I is a five-carbon straight chain alkane, and thus is n-pentane. Structure II has the common name of isopentane, since the branch occurs next to the end. The systematic name of structure II is based upon the longest chain of carbons (four). Hence it is named as a methyl derivative of butane, or 2-methylbutane. Structure III, commonly called neopentane, has a systematic name based upon the three-carbon long chain: 2,2-dimethylpropane. The repetition of the numerical prefixes indicates both groups added to the chain attach to the second carbon. Prefixes such as "di" are used to indicate the number of identical groups added to the backbone. The table below shows the appropriate numerical modifiers.

TABLE 4.6

TABLE 4.6: SOME NUMERICAL PREFIXES

# OF IDENTICAL SUBSTITUENTS	PREFIX
2	di-
3	tri-
4	tetra-
5	penta-
6	hexa-

For example, the compound shown in the figure 4.31 is called 2,2,4-trimethylpentane. It is one of eighteen possible structures with the formula of C_8H_{18}.

```
    H3C  H  CH3
     |   |   |
H3C—C—C—C—CH3        or
     |   |   |
    H3C  H   H
```

FIGURE 4.31: Structure of 2,2,4-trimethylpentane.

The suffix "tri" indicates three methyl groups are attached to the backbone. That the backbone is five carbons long is reflected in naming it as a derivative of pentane. The "2,2,4" indicates that two of the methyl groups are attached to the second carbon, and one to the fourth carbon.

As the number of carbons in the alkane increases, the number of possible structural isomers increases dramatically. This is shown in the following table.

TABLE 4.7

TABLE 4.7: POSSIBLE STRUCTURAL ISOMERS OF ALKANES

FORMULA	NUMBER OF ISOMERS
C_4H_{10}	2
C_5H_{12}	3
C_6H_{14}	5
C_8H_{18}	18
$C_{10}H_{22}$	75
$C_{12}H_{26}$	355
$C_{14}H_{30}$	1,858
$C_{20}H_{42}$	366,319
$C_{30}H_{62}$	4,111,846,763
$C_{40}H_{82}$	62,491,178,805,831

In the face of such astronomical numbers of different molecules, it should be obvious that a systematic approach to nomenclature is required. The next section provides this systematic approach, repeating some of what we have already seen, and examining some new problems.

4.6 NOMENCLATURE RULES FOR ALKANES

In general, the naming of organic molecules follows a very predictable pattern. The name is composed of three parts: prefix, stem, and suffix. The stem name is related to the backbone hydrocarbon from which the molecule is formally derived. The suffix indicates the chemical family to which the backbone belongs. (Since we are primarily describing alkanes in this section, the suffix will be "-ane".) The prefix describes the identity, number and position of each substituent added to the backbone. Below is a list of rules to properly combine prefix, stem, and suffix to correctly name a wide variety of hydrocarbons.

RULE 1. Find the longest continuous chain of carbon atoms and use the name of this alkane as the backbone from which the alkane is derived.

Do not be misled by the way the structure is represented on the page. Consider the two hexanes in the following figure.

Although we might be tempted to name structure (a) as a propane derivative, and structure (b) as a butane derivative, redrawing the structures emphasizes that both are in fact derivatives of pentane.

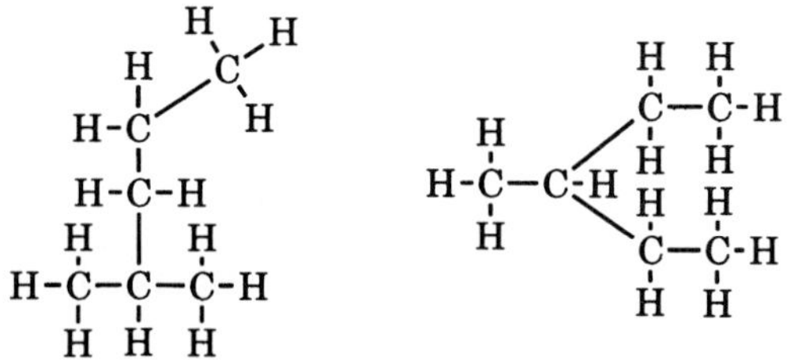

The key is to consider the various carbon-carbon chain options, and to find the longest one. Consider one more example, the naming of the branched alkane in figure 4.34.

While one might see this as a five-carbon chain (pentane derivative) or possibly a six-carbon chain (hexane derivative), the longest continuous chain has seven carbons (a heptane derivative.) The seven carbons of the heptane chain are boldfaced in figure 4.35.

(a)

(b)

FIGURE 4.32

Rule 2. Find and name all the groups that are branches off the longest continuous chain.

For alkanes, these substituents are commonly methyl, ethyl, propyl, and isopropyl groups, which were introduced back in Chapter 3. Longer hydrocarbon fragments attached to the backbone are named as described earlier: use the stem from the name for the alkane, drop the suffix "-ane" and add "-yl".

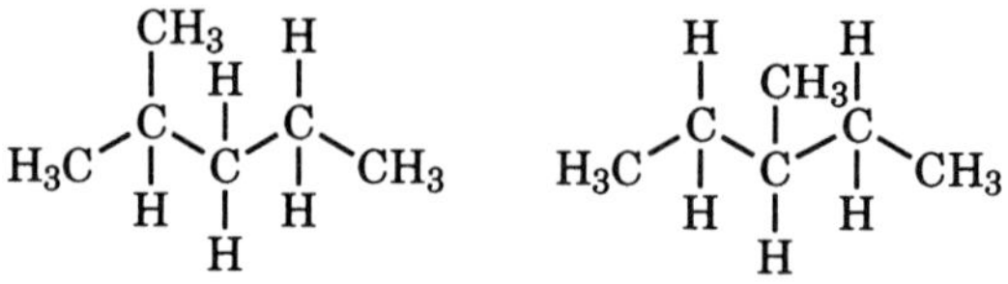

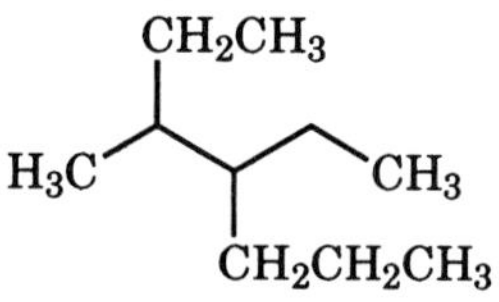

(a)

(b)

FIGURE 4.33: Redrawn structures from Figure 4.32.

The compound in Figure 4.35 is named 4-ethyl-3-methylheptane. The two-carbon ethyl group is associated with carbon number 4, and the one carbon methyl group is associated with carbon number 3. The order in which we name these substituents and the way in which the carbon numbers were determined are described in subsequent rules.

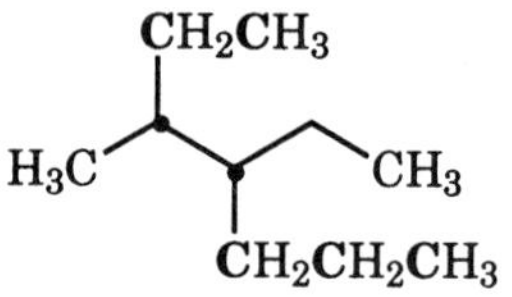

FIGURE 4.34: One possible structure for $C_{10}H_{22}$.

Rule 3. Assign each group a position on the longest continuous chain; number the chain so as to give the groups the lowest numbers.

An application of both rules 2 and 3 is shown below.

FIGURE 4.35: One possible structure for $C_{10}H_{22}$.

(I)	(II)
CH_3-CH_2-CH-CH_2-CH_3 (with CH_3 on the second carbon)	CH_3-CH_2-CH-CH_2-CH_3 (with CH_3 on the third carbon)
numbering right to left	numbering to left
5 4 3 2 1	5 4 3 2 1
Name: 4-methylpentane	3-methylpentane
numbering left to right	numbering to right
1 2 3 4 5	1 2 3 4 5
Name: 2-methylpentane	3-methylpentane

FIGURE 4.36: Carbon numbering.

For Compound (I), the choice is between 4-methylpentane and 2-methylpentane. The latter, representing the lower numerical value, is the correct name. In some instances, such as Compound (II), either numbering approach yields the correct name.

For multiple substituents, the position of each is indicated with a number. An example is shown in the figure at right.

CH_2CH_3

H_3C CH_3

CH_3

1 2 3 4 5 6 7

7 6 5 4 3 2 1

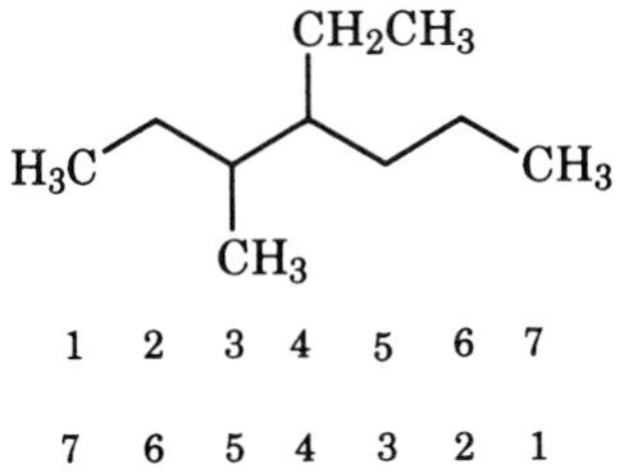

FIGURE 4.37: Numbering of the heptane chain for 4-ethyl-3-methylheptane.

The ethyl group is on the middle carbon in either system, and thus will be signified as "4-ethyl" in either approach. The right to left numbering puts the methyl group at carbon number 5, while left to right numbering places the methyl at the third carbon. The lower numbered alternative is used.

RULE 4. If there are several groups added to the backbone that are the same, use the prefixes di-, tri-, tetra- and numbers to indicate their positions.

Various isomers of octane are important in gasoline. The structure of one of those important octane molecules was shown earlier, in Figure 4.31. This compound has the systematic name of 2,2,4-trimethylpentane. The "tri" in the prefix indicates three equivalent substituents. Note that the structure is not named 2,4,4-trimethylpentane. Although that structure is equivalent, the numbering is always chosen to minimize the value of the numbers used, in accordance with rule 3.

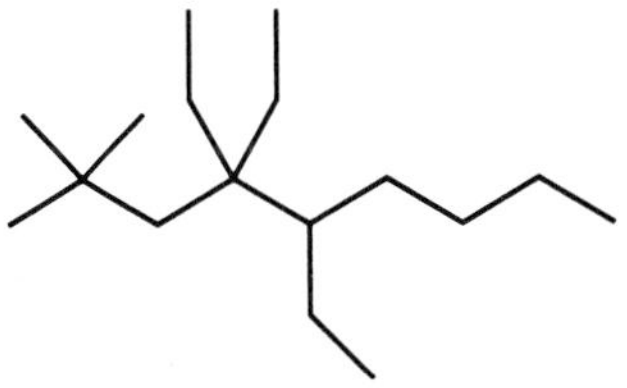

When multiple substituents are attached to the backbone, the number and position of each substituent must be indicated in the prefix. The following branched alkane has nine carbons in its longest chain. Thus it is named as a nonane derivative.

FIGURE 4.38: Structure of 4,4,5-triethyl-2,2-dimethylnonane.

RULE 5. List substituents in alphabetical order (ignoring prefixes di-, tri- tetra-, and italicized prefixes of n-, sec-, and tert-.) By convention, "iso" is considered part of the word it modifies. For example, isopropyl is alphabetized as a word beginning with the letter I.

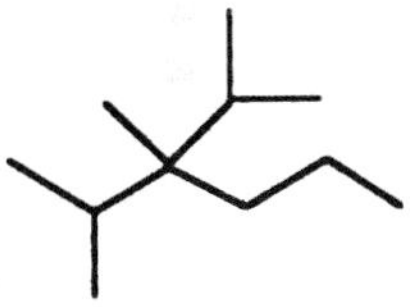

For example, in naming the structures in Figures 4.37 and 4.38, the ethyl substituents are named before the methyl substituents. Note that in the example 4,4,5-triethyl-2,2-dimethylnonane, ethyl precedes methyl, even though "tri" alphabetically follows "di".

FIGURE 4.39: Structure of 3-isopropyl-2,3-dimethylhexane.

The compound at right is named with the isopropyl substituent appearing in the name prior to the methyl substituents.

Some common terms, like isopropyl, find their way into systematic names. An alternate name for this compound would call the isopropyl substituent "1-methylethyl". Thus this compound in Fig. 4.39 could also be named 3-(1-methylethyl)-2,3-dimethylhexane.

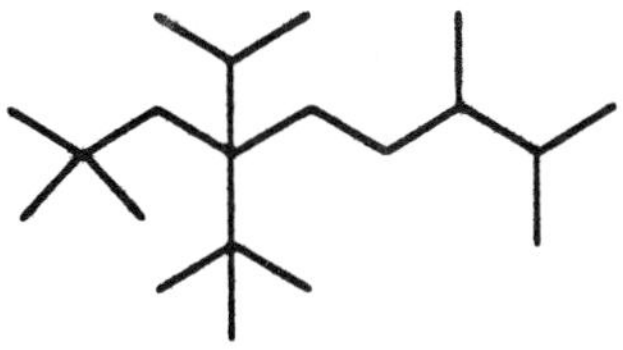

FIGURE 4.40: Structure of 4-tert-butyl-4-isopropyl-2,2,7,8-tetramethylnonane.

Another example is given in Figure 4.40. Note that the alphabetical ordering puts t-butyl first (as a B-word), puts isopropyl second (as an I-word), and puts tetramethyl last (as an M-word.)

We will find this rule for listing substituents alphabetically holds for non-hydrocarbon substituents as well. Consider the compound at right in Figure 4.41.

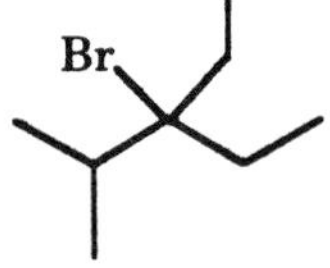

Br

FIGURE 4.41: Structure of 3-bromo-3-ethyl-2-methylpentane.

Note the naming of the bromo, ethyl, and methyl substituents in alphabetical order.

Concept Check: Hydrocarbon Structure, Nomenclature, Spectra

Table 4.7 shows that hexane, C_6H_{14}, can exist as five structural isomers. Draw and name each of these five isomers. How many unique carbon environments does each isomer have?

Solution:

(a) $CH_3CH_2CH_2CH_2CH_2CH_3$; n-hexane; 3 carbon environments; #1 and #6 identical; #2 and #5 identical; #3 and #4 identical

(b) $CH_3CH_2CH_2CH(CH_3)_2$; 2-methylpentane; 5 carbon environments; the two methyl carbons attached to carbon #2 are identitical.

(c) $CH_3CH_2CH(CH_3)CH_2CH_3$; 3-methylpentane; 4 carbon environments; #1 and #5 are identical; #2 and #4 are identical.

(d) $(CH_3)_2CHCH(CH_3)_2$; 2,3-dimethylbutane; 2 carbon environments; the four CH_3 carbons are identical; the two middle carbons are identical.

(e) $CH_3CH_2C(CH_3)_3$; 2,2-dimethylbutane; 4 carbon environments; the three CH_3 carbons attached to carbon #2 are identical.

4.7 NAMING HYDROCARBONS WITH MULTIPLE BONDS OR RINGS

Alkenes and Alkynes

Rule 1 states that the longest carbon chain is identified as the backbone, which provides the stem name. The one common exception to this approach may occur when the carbon backbone itself has multiple bonds; that is, for alkenes and alkynes. As with the alkanes, the longest continuous chain is chosen, provided it contains the double or triple bond. The suffix -ene for alkene or -yne for alkyne is added to the stem. The position of the multiple bond is designated by the lowest number possible. For example, consider the names of the following alkenes and alkynes at right.

Note that in the third example, the compound is named as a derivative of a five-carbon long chain. This is done to include the double bond, even though a longer, six-carbon chain is present. Alkenes and alkynes, like many small molecules, are often called by common (non-systematic) names.

```
   H  H  H  H  H
   |  |  |  |  |
H-C--C==C--C--C-H
   |        |  |
   H        H  H

   1  2  3  4  5
```
2-pentene (not 3-pentene)

```
   H  H           H  H  H
   |  |           |  |  |
H-C--C--C≡C--C--C--C-H
   |  |           |  |  |
   H  H           H  H  H

   1  2  3  4  5  6  7
```
3-heptyne (not 4-heptyne)

```
H          CH2CH2CH3
  \       /
   C == C
  /       \
H          CH2CH3
```
2-ethyl-1-pentene

FIGURE 4.42: Structure and names of some alkenes and alkynes.

Cycloalkanes

Alkanes are characterized as hydrocarbons with all carbons singly bonded. Although all alkanes are considered saturated, not all alkanes have a zero index of hydrogen deficiency. When alkanes wrap around to form cycles (rings), the number of hydrogen atoms is reduced from the maximum predicted of 2n+2 (where n is the number of carbon atoms.) **Cycloalkanes** are another class of alkanes wherein all carbons are singly bonded, but there is a hydrogen deficiency caused by one or more rings. The names and structures of several simple cycloalkanes are shown in the table below.

TABLE 4.8

TABLE 4.8: STRUCTURES AND NAMES OF SOME CYCLOALKANES

STRUCTURE	FORMULA	NAME
△	C_3H_6	Cyclopropane
□	C_4H_8	Cyclobutane
⬠	C_5H_{10}	Cyclopentane
⬡	C_6H_{12}	Cyclohexane

There are several ways of showing the formulas for cycloalkanes. Again, there is a tradeoff between ease of writing and degree of bonding information explicitly shown. The figure below shows several ways in which the formula of cyclohexane can be represented.

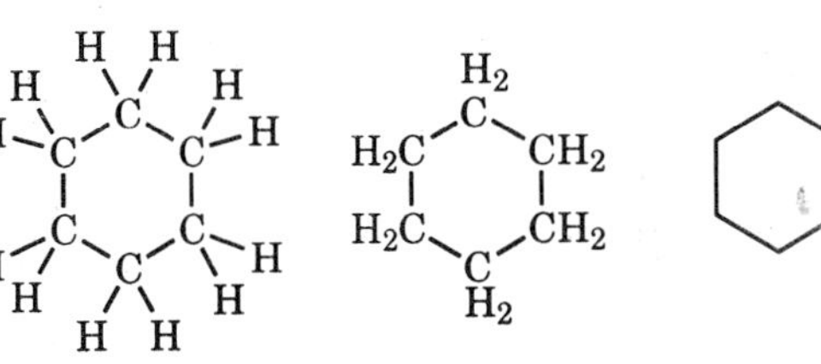

FIGURE 4.43: Structural formulas of cyclohexane (C_6H_{12}).

Positions of substituents on the ring are assigned numerically. The numbering occurs either clockwise or counterclockwise around the ring, whichever results in a lower number designation. Several examples are shown below, using isomers of dimethylcyclohexane.

Double bonds and rings prevent the free rotation around single bonds seen for all previous molecules. Later in this course we will see that this circumstance increases the number (and even the type) of isomers possible.

CONCEPT CHECK: RING STRUCTURES AND CARBON NMR SPECTRA

The structure of 1,1-dimethylcyclohexane is shown in Figure 4.44. How many distinct carbon environments should be observed in the uncoupled ^{13}C NMR spectrum of that compound?

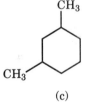

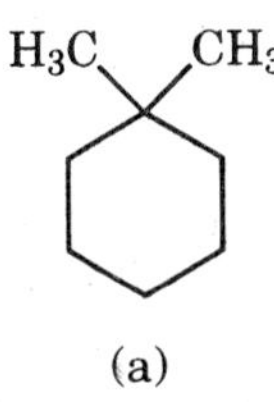

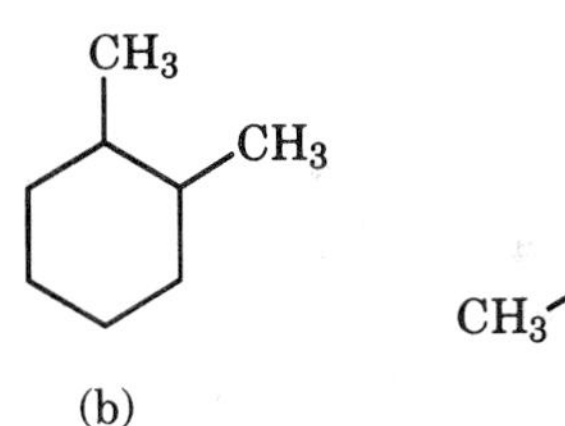

FIGURE 4.44: a) 1,1-dimethylcyclohexane, b) 1,2-dimethylcyclohexane, c) 1,3-dimethylcyclohexane.

SOLUTION: The two methyl groups are both attached to carbon #1 in the six-member ring, so these methyl carbons are equivalent. Numbering around the ring in a clockwise fashion yields positions called 2, 3, 4, 5, and 6 respectively. Thus carbons 2 and 6 are immediately adjacent to carbon #1, and carbons 2 and 6 are equivalent. Carbons 3 and 5 are both two bonds removed from carbon 1, and 3 and 5 are equivalent. Carbon 4, across the ring from carbon 1, is a unique carbon environment. Thus the carbon NMR of 1,1-dimethylcyclohexane should show a total of five signals: one signal for the two methyl carbons, one signal for carbon 1, one signal for carbons 2 and 6, one signal for carbons 3 and 5, and one signal for carbon number 4.

4.8 NAMING COMPOUNDS WITH OTHER FUNCTIONAL GROUPS

ALCOHOLS

As we have seen earlier, the suffix "-ol" is added to the stem to indicate the alcohol family. The stem is determined via the longest carbon chain. The position of the OH group on the backbone is indicated numerically, following the same rules listed earlier. When multiple OH substituents are attached, the -ol suffix is preceded by di, tri, tetra, etc. The table below shows the structures and names of some alcohols.

TABLE 4.9: STRUCTURES AND NAMES OF SOME ALCOHOLS

TABLE 4.9

COMMON NAME	SYSTEMATIC NAME	STRUCTURE
ethyl alcohol	ethanol	CH_3CH_2OH
isobutyl alcohol	2-methyl-1-propanol	$H_3C{-}C(H)(CH_3){-}CH_2{-}OH$
tert-butyl alcohol	2-methyl-2-propanol	$H_3C{-}C(CH_3)(OH){-}CH_3$
glycerol	1,2,3-propanetriol	$HO{-}CH_2{-}CH(OH){-}CH_2{-}OH$

AMINES

The suffix -amine is added to the stem named from the backbone structure. The position of the amine substituent is indicated according to the number of the carbon in the backbone to which it is attached. Common names for amine compounds often have an -ine suffix. When a compound is named for another functional group in addition to the amine, the NH_2 group is indicated as "amino" in the prefix to the stem. Substituted amines are indicated with "N-", as introduced earlier in Chapter 3.

HALIDES

Organic halides are simply named by specifying the number and position of the halogen atoms attached to carbons in the backbone chain.

TABLE 4.10: STRUCTURES AND NAMES OF SOME AMINES

TABLE 4.10

SYSTEMATIC NAME	STRUCTURE
ethanamine	$CH_3CH_2NH_2$
1,1-dimethylethanamine	$\begin{array}{c} CH_3 \\ \vert \\ H_3C-C-NH_2 \\ \vert \\ CH_3 \end{array}$
1,6-hexanediamine	$\begin{array}{c} \;\;\;\;\;\;H\;\;\;H\;\;\;H\;\;\;H\;\;\;H\;\;\;H \\ H_2N-C-C-C-C-C-C-NH_2 \\ \;\;\;\;\;\;H\;\;\;H\;\;\;H\;\;\;H\;\;\;H\;\;\;H \end{array}$
cyclohexanamine	NH_2
piperidine	$\begin{array}{c} H \\ \vert \\ N \end{array}$
2-aminoethanol	$\begin{array}{c} \;\;\;\;\;\;H\;\;\;\;H \\ H_2N-C-C-OH \\ \;\;\;\;\;\;H\;\;\;\;H \end{array}$
N-ethyl-N-methyl-3-pentanamine	$\begin{array}{c} H \\ \vert \\ H_3CH_2C-C-CH_2CH_3 \\ \vert \\ N \\ H_3CH_2C \;\;\; CH_3 \end{array}$

TABLE 4.11: STRUCTURES AND NAMES OF SOME ORGANIC HALIDES

TABLE 4.11

SYSTEMATIC NAME	STRUCTURE
1,1,1-tribromoethane	$\begin{array}{c} Br \\ \vert \\ Br-C-CH_3 \\ \vert \\ Br \end{array}$
2,3-dichlorobutane	$\begin{array}{c} \;\;\;\;\;\;Cl\;\;\;\;Cl \\ CH_3-C-C-CH_3 \\ \;\;\;\;\;\;H\;\;\;\;H \end{array}$
2,2-dibromo-1-chloro-3-methylpentane	$\begin{array}{c} \;\;\;\;H\;\;\;Br\;\;\;CH_3\;\;\;H \\ Cl-C-C-C-C-CH_3 \\ \;\;\;\;H\;\;\;Br\;\;\;H\;\;\;\;H \end{array}$

Concept Check: Organic Halide Nomenclature and NMR Spectra

A compound has the formula C_4H_9Br, and is known to be either 1-bromobutane or 2-bromobutane. The carbon NMR shows four carbon environments. Draw the structures of the two compounds. Can one specify the compound with this information?

Solution: 1-bromobutane ($CH_2BrCH_2CH_2CH_3$) has 4 unique carbon environments. 2-bromobutane ($CH_3CHBrCH_2CH_3$) also has 4 unique carbon environments. The information given concerning the uncoupled carbon NMR spectrum will not permit conclusive identification.

A proton NMR spectrum of the compound is given below. Can one specify the compound with this information?

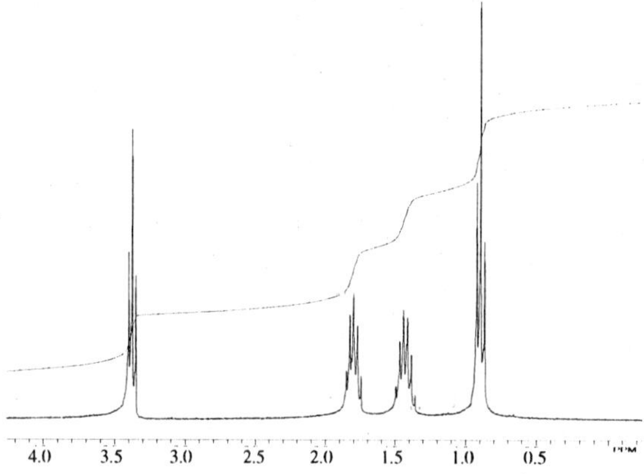

Solution: Integration of the spectrum shows that the triplet at 3.5 ppm corresponds to 2 protons; the pentet at 1.8 ppm corresponds to 2 protons; the sextet at 1.5 ppm corresponds to 2 protons; and the triplet at 1.0 ppm corresponds to 3 protons. The 2:2:2:3 ratio is sufficient to identify the compound as 1-bromobutane, since carbons 1, 2, and 3 all have two protons, and carbon 4 has 3 protons. Compare this to 2-bromobutane, whose carbons (from 1 to 4) would have protons in a 3:1:2:3 ratio.

Further evidence in support of the identification is also present in the spectrum. For 1-bromobutane, the protons on carbon 1 are most deshielded. This corresponds to the triplet at 3.5 ppm. This signal is a triplet since it is split by the two protons on carbon 2. The pentet at 1.8 ppm corresponds to the protons on carbon 2, which are the next most deshielded. The pentet results from nearly equal splitting by the two protons on carbon 1 and the two protons on carbon 3. The sextet at 1.5 ppm correspond to the protons on carbon 3. The sextet results from nearly equal splitting by the two protons on carbon 2 and the three protons on carbon 4. Finally, the most shielded signal comes from the protons farthest from the bromine, the protons on carbon 4. This is the triplet at 1.0 ppm. The triplet results from splitting by the two protons on carbon 3.

ALDEHYDES AND KETONES

The suffix -al signifies an aldehyde family, while the -one suffix indicates ketones. The naming of the stem, the location and the number of substituents follow the previous rules. As always, more than one functional group may be present.

TABLE 4.12: STRUCTURES AND NAMES OF ALDEHYDES AND KETONES

TABLE 4.12

COMMON NAME	SYSTEMATIC NAME	STRUCTURE
valeraldehyde	pentanal	$CH_3-CH_2-CH_2-CH_2-C(=O)-H$
acetone	2-propanone	$CH_3-C(=O)-CH_3$
acetylacetone	2,4-pentanedione	$CH_3-C(=O)-CH_2-C(=O)-CH_3$
benzaldehyde	benzaldehyde	$C_6H_5-C(=O)-H$
acrolein	2-propenal	$CH_2=CH-C(=O)-H$

CARBOXYLIC ACIDS AND ESTERS

The -oic acid suffix is used for organic acids. As explained in Chapter 3, the -ate ending of a complex anion may also be the characteristic ending of the name of an ester.

AMIDES AND NITRILES

As described earlier in Chapter 3, these families are designated with an "amide" or "nitrile" ending. The rules for naming the stem, and for designating the position of the functional group are the same. The N nomenclature developed under amines is also used for amides. If the nitrile group is named as an addition to another chemical family, the term "cyano" is used.

Compounds with more than one functional group were named in several of the previous tables. One functional group determines the family name, and any other functional groups are named as substituents. Which functional group determines the family name, and which are considered substituents is a matter of relative priority. The highest priority functional group determines the family name. The table below summarizes the priority of some functional groups, along with some example names.

TABLE 4.13: STRUCTURES AND NAMES OF SOME ACIDS AND ESTERS TABLE 4.13

COMMON NAME	SYSTEMATIC NAME	STRUCTURE
	3-methylpentanoic acid	$CH_3-CH_2-CH(CH_3)-CH_2-C(=O)-OH$
oxalic acid	ethanedioic acid	$HO-C(=O)-C(=O)-OH$
adipic acid	1,6-hexanedioic acid	$HO-C(=O)-(CH_2)_4-C(=O)-OH$
acrylic acid	propenoic acid	$CH_2=CH-C(=O)-OH$
lactic acid	2-hydroxypropanoic acid	$CH_2-CH(OH)-C(=O)-OH$
alanine	2-aminopropanoic acid	$CH_3-CH(NH_2)-C(=O)-OH$
benzoic acid	benzoic acid	$C_6H_5-C(=O)-OH$
ortho-phthalic acid	1,2-benzene dicarboxylic acid	$C_6H_4(COOH)_2$ (1,2-)
isopropyl butyrate	1-methylethyl butanoate	$H_3C-CH_2-CH_2-C(=O)-O-CH(CH_3)-CH_3$
ethyl benzoate	ethyl benzoate	$C_6H_5-C(=O)-O-CH_2-CH_3$

TABLE 4.14: STRUCTURES AND NAMES OF SOME AMIDES AND NITRILES TABLE 4.14

COMMON NAME	SYSTEMATIC NAME	STRUCTURE
formamide	methanamide	$HC(=O)-NH_2$
DMF	N,N-dimethylmethanamide	$H-C(=O)-N(CH_3)-CH_3$
benzamide	benzamide	$C_6H_5-C(=O)-NH_2$
acrylonitrile	2-propenenitrile	$CH_2=CH-C\equiv N$
	4-cyanobutanoic acid	$HOC(=O)-CH_2-CH_2-CH_2-C\equiv N$

TABLE 4.15: NOMENCLATURE PRIORITY OF FUNCTIONAL GROUPS

	STRUCTURE	FAMILY SUFFIX	SUBSTITUENT PREFIX	EXAMPLE NAME
Highest Priority				
COOH	—C(=O)—OH	-oic acid		
COOC	—C(=O)—O—	-ate	alkoxycarbonyl	2-methoxycarbonylethanoic acid
CONH	—C(=O)—N(H)—	-amide	amido	2-amidoethanoic acid
CN	—C≡N	nitrile	cyano-	4-cyano-1-butanoic acid
CHO	—C(=O)—H	-al	oxo-	3-oxo-1-propanoic acid
C—C(=O)—C		-one	oxo-	2-oxo-1-propanoic acid
COH	—C—OH	-ol	hydroxy-	2-hydroxy-1-propanoic acid
NH_2	—NH_2	amine	amino-	2-amino-1-propanoic acid
COC	—C—O—C—	ether	alkoxy-	2-ethoxy-1-ethanamine
C=C	C=C	-ene	-ene-	2-propenenitrile
C≡C	—C≡C—	-yne	-yne-	3-propynenitrile
others	methyl-, phenyl-, chloro-, nitro-, etc. named only as prefixes			
Lowest Priority				

Branched hydrocarbon: a carbon-hydrogen compound in which the carbon chain contains one or more carbon atoms that are bonded to more than two other carbons.

Coupling constant: symbolized by J, with units of Hertz; the strength of the interaction between nuclei, measured by the separation of peaks within a multiplet.

Cycloalkanes: saturated (singly bonded) hydrocarbons that have one or more ring structures.

Deuterated solvents: solvents typically used to dissolve samples for NMR analysis; these compounds have had the protons replaced by deuterium. Example: deuterated chloroform is $CDCl_3$.

Delta (δ): the frequency shift (in parts per million, ppm) of an NMR signal compared to the standard signal from TMS (at $\delta = 0$.)

Deshielding: the partial removal of electron density around a nucleus by a neighboring electronegative element. When the shielding effect of the electron density around the nucleus is reduced, the nucleus resonates at a higher delta value.

Integration mode: one mode of operating an NMR spectrometer that results in a measurement of the proportional number of nuclei giving rise to each signal.

Isomers: species with the same general formula, but different properties.

n + 1 rule: n + 1 is the multiplicity of the spin coupled signal. For ^{13}C NMR, n is the number of protons directly bonded to the carbon giving rise to the signal. For HNMR, n is the number of equivalent protons on an adjacent carbon atom (3 bonds removed from the proton giving rise to the signal.)

Nuclear magnetic resonance: an absorption of energy by a nucleus that realigns its direction of spin in a magnetic field.

Rotational isomers: the same molecule when viewed from different perspectives.

Spin coupled mode: one mode of operating an NMR spectrometer that results in splitting the regular signals according to the n + 1 rule.

Structural isomers: isomers with different molecular structure (i.e., different bonds).

TMS: tetramethylsilane, $(CH_3)_4Si$, the standard reference for NMR spectroscopy.

References for Nomenclature Questions

In this chapter we have established the basic rules of organic nomenclature. Compounds that can not be named by these rules will not occur frequently in this text. The reader interested in a more complete set of nomenclature rules is referred to the following references:

1. IUPAC's Nomenclature of Organic Chemistry, Pergamon Press, New York.

2. CRC Handbook of Chemistry and Physics, CRC Press, Boca Raton.

HOMEWORK Problems

MOLECULAR STRUCTURE, NOMENCLATURE, AND NMR SPECTRA

1. Write line-angle structures for:

$$\begin{array}{ccccc} & & CH_3 & & \\ & & | & & \\ H_3C & - & C & - & CH_3 \\ & & | & & \\ & & H & & \end{array} \qquad \begin{array}{ccccccccc} & & Cl & & H & & CH_3 & & \\ & & | & & | & & | & & \\ H_3C & - & C & - & C & - & C & - & C_2H_5 \\ & & | & & | & & | & & \\ & & H & & H & & CH_3 & & \end{array}$$

2. Why can't the following formulas be representative of real compounds?

$$\begin{array}{ccccc} & & H & & CH_3 \\ & & | & & | \\ H_3C & - & C & - & CH_2CH_3 \\ & & | & & \\ & & H & & \end{array} \qquad \begin{array}{ccccccc} & & H & & CH_3 & & \\ & & | & & | & & \\ H_4C & - & C & - & C & - & CH_3 \\ & & | & & | & & \\ & & H & & H & & \end{array}$$

3. Write structural formulas corresponding to the following names.
   (a) 1-butanol
   (b) 2-methyl-3-ethyl-2-pentanol
   (c) 2-chloro-3-methylhexanal

4. How many unique carbon environments will be represented in a CNMR spectrum of each of the compounds in problem #3?

5. Write structural formulas corresponding to the following names.
   (a) 5-bromo-3-ethylpentanal
   (b) 3-isopropyl-2-heptanone
   (c) 4-ethylnonane

6. How many unique carbon environments will be represented in a CNMR spectrum of each of the compounds in problem #5?

7. Write structural formulas corresponding to the following names.
   (a) N-isobutylpentylamine
   (b) 4-amino-3-isopropyldecane
   (c) 2-*sec*-butyl-3-methylheptanoic acid

8. How many unique carbon environments will be represented in a CNMR spectrum of each of the compounds in problem #7?

9. Write structural formulas corresponding to the following names.
   (a) dichloroacetic acid
   (b) isobutyl 3-methylpentanoate
   (c) phenyl acetate

10. How many unique carbon environments will be represented in a CNMR spectrum of each of the compounds in problem #9?

11. Write structural formulas corresponding to the following names.
   (a) pentamide
   (b) N-methylacetamide
   (c) N-methyl, N-ethylpropylamine

12. How many unique carbon environments will be represented in a CNMR spectrum of each of the compounds in problem #11?

13. Write structural formulas corresponding to the following names.
   (a) tertiarybutylamine
   (b) 3-pentanone
   (c) pentanal

14. How many unique proton environments will be represented in a HNMR spectrum of each of the compounds in problem #13?

15. Write structural formulas corresponding to the following names.
   (a) 2-propenenitrile
   (b) 3-oxo-1-propanoic acid
   (c) cyclohexanol

16. How many unique proton environments will be represented in a HNMR spectrum of each of the compounds in problem #15?

17. Write structural formulas corresponding to the following names.
   (a) N-methylpropanamide
   (b) 1,4-butanedioic acid
   (c) isopropylethanoate

18. How many unique proton environments will be represented in a HNMR spectrum of each of the compounds in problem #17?

19. Write structural formulas corresponding to the following names.
   (a) 2-aminopropanoic acid
   (b) cyclohexanone
   (c) 3-ethyl-2-pentene

20. How many unique proton environments will be represented in a HNMR spectrum of each of the compounds in problem #19?

21. Predict and draw the CNMR and the HNMR spectra of each of the four bromobutane isomers.

22. Name the following compounds.

a.
```
        H  CH3  H
        |   |   |
  H3C—C—C——C—CH2OH
        |   |   |
        H   H   H
```

b.
```
        CH3 H  O
         |  |  ||
  H3C—C——C—C—H
         |  |
         H  H
```

c.

```
      CH3 H   O  H  H
       |  |   ||  |  |
H3C—C——C—C—C—C—CH3
       |  |      |  |
       H  H      H  H
```

d.

```
      H  H
      |  |
H3C—C—C—NH2
      |  |
      H  H
```

e.

```
      H  H  H  O
      |  |  |  ||
H3C—C—C—C—C—OH
      |  |  |
      H  H  H
```

f.

```
      H  H  H  O
      |  |  |  ||
H3C—C—C—C—C—O—CH2CH3
      |  |  |
      H  H  H
```

g.

```
      H  H  H  O
      |  |  |  ||
H3C—C—C—C—C—NH2
      |  |  |
      H  H  H
```

h.

```
      H  CH3  H
      |  |    |
H3C—C—C——C—CH2OH
      |  |    |
      H  H    CH2CH3
```

i.

```
      H  Cl  O
      |  |   ||
H3C—C—C—C—H
      |  |
      H  H
```

j.

```
      H  H       O    H
      |  |       ||   |
H3C—C—C—C——C—CH3
      |  |            |
      H  CH2CH3  H
```

k.

```
      CH3 H  H  CH3
       |   |  |  |
H3C—C——C—N—C—CH3
       |   |      |
       H   H      H
```

l.

```
      H  Br   H  O
      |  |    |  ||
H3C—C—C——C—C—OH
      |  |    |
      H  CH3  H
```

m.

```
      H  H    H  O        H
      |  |    |  ||       |
H3C—C—C——C—C—O—C—Cl
      |  |    |           |
      H  CH3  H           CH3
```

n.

```
      H  Br   H  O
      |  |    |  ||
H3C—C—C——C—C—N—CH2CH
      |  |    |     |
      H  CH3  H     H
```

23. Predict the CNMR and HNMR spectra of

```
      Cl  H
      |   |
H3C—C—C—O—CH3
      |   |
      H   H
```

24. Suggest a plausible structure that would correspond to each of the following spectra

(a) $C_{10}H_{14}$

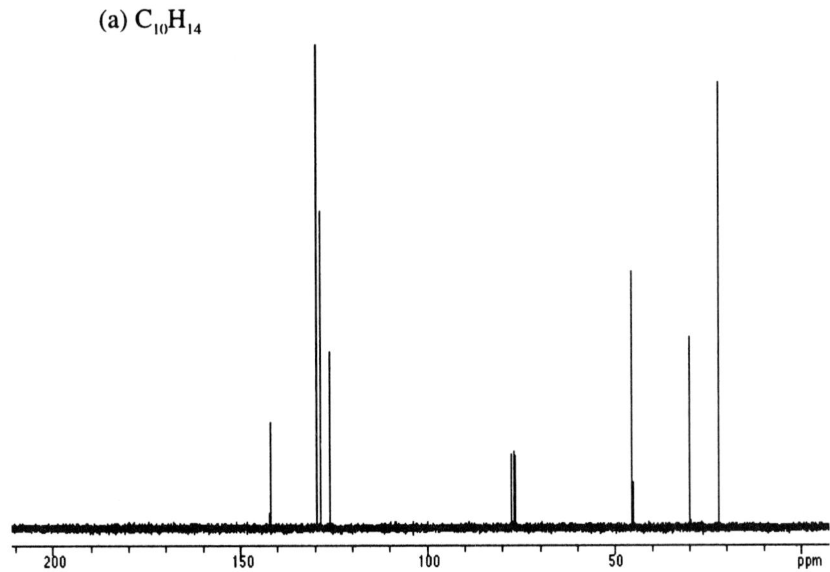

(b) $C_8H_8O_3$

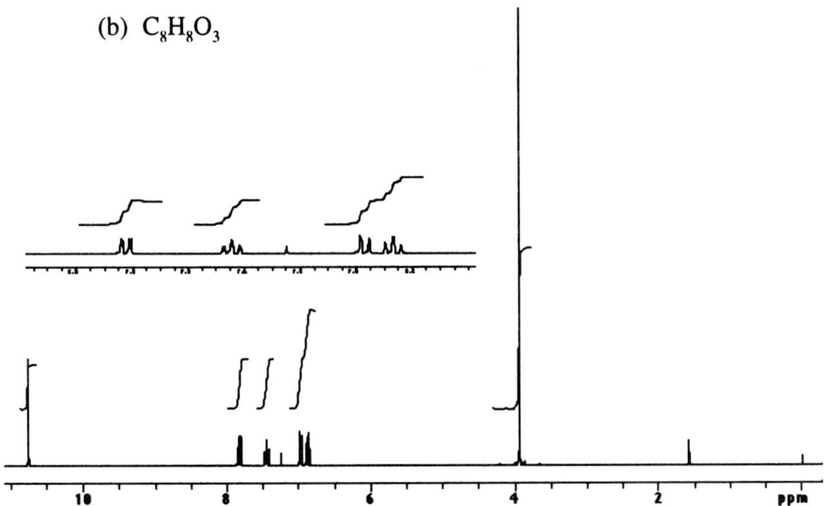

(c) $C_7H_{14}O$

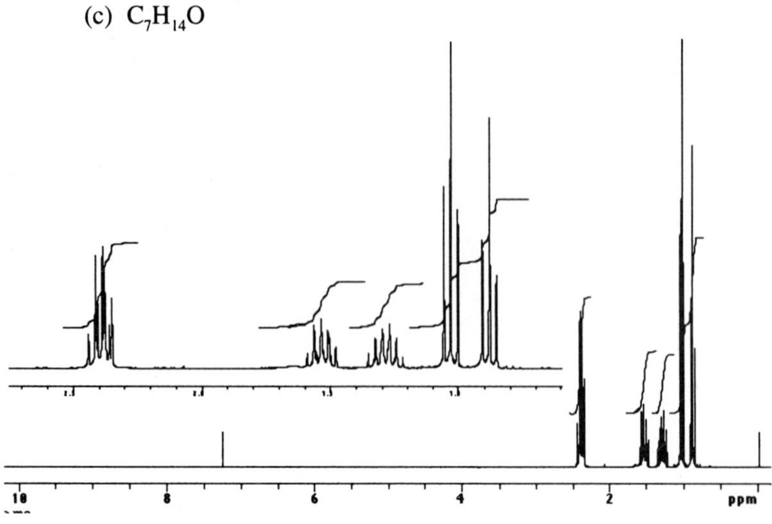

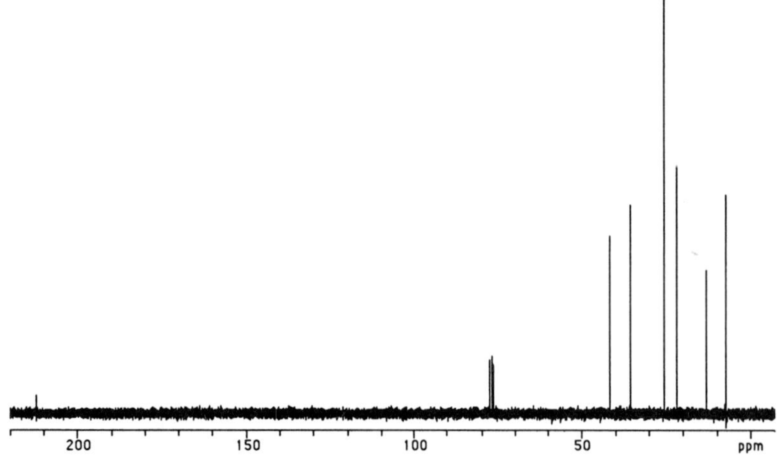

CHAPTER FOUR

OVERVIEW
Problems

25. A compound with the formula $C_5H_{10}O$ has a strong IR absorbance around 1700 cm^{-1}. The CNMR spectrum of this compound shows a signal with a chemical shift of about 200 ppm.

    (a) What is the IHD of this compound?

    (b) What functional group(s) is (are) possible with the indicated IHD and the IR and NMR information?

26. The compound in problem #25 shows just three unique carbon environments. Propose a structure for the compound, and name it.

27. A compound with the formula C_2H_5NO has a wide absorbance centered at 3500 cm^{-1}, and a strong absorbance around 1680 cm^{-1}.

(a) What is the IHD of this compound?

(b) What functional group(s) is (are) possible with the indicated IHD and the IR and NMR information?

28. The compound in problem #27 has a proton NMR that shows two signals. A broad peak around 6 ppm, corresponding to 2 protons, and a sharp singlet at 2.1 ppm, corresponding to 3 protons. The CNMR shows two signals: one at 173 ppm, and one at 22 ppm. Propose a structure for the compound and name it.

29. A compound with the formula C_3H_8O has a broad IR absorbance centered around 3350 cm^{-1}. The CNMR spectrum shows just two signals, one at 64 ppm, and the other at 25 ppm. The proton NMR shows three unique environments: a septuplet (corresponding to 1 proton) at 4 ppm, a singlet (corresponding to 1 proton) at 2.1 ppm, and a doublet (corresponding to 6 protons) at 1.2 ppm.

(a) Describe the significance of the IR absorbance

(b) What is the significance of two (not 3) peaks occuring in the CNMR spectrum?

(c) Draw the structure of the molecule

(d) Indicate which proton(s) is (are) responsible for each peak in the HNMR spectrum.

30. A compound with the formula $C_5H_{10}O_2$ has no absorbance above 3000 cm^{-1}, but a stong absorbance at 1720 cm^{-1}. The CNMR spectrum shows just 4 carbon environments, one of which is at 170 ppm. The HNMR spectrum shows three unique environments: a septuplet (corresponding to one proton) at 5.0 ppm, a singlet (corresponding to 3 protons) at 2.0 ppm, and a doublet (corresponding to 6 protons) at 1.3 ppm.

(a) Describe the significance of the IR spectrum

(b) What is the significance of four (not 5) peaks in the CNMR spectrum?

(c) Draw the structure of the molecule

(d) Indicate which proton(s) is (are) responsible for each peak in the HNMR spectrum.

CHEMICAL Reactions (I)

Some Reaction Types and the Mole Concept

CHEMICAL REACTIONS (I)

CHAPTER FIVE

Some Reaction Types and the Mole Concept

Since the time of alchemists, people have been fascinated by the transformation of one substance into another. The European alchemists were interested in studying methods for converting common metals into precious metals such as gold. The quest of Chinese alchemists involved the search for "elixirs" that would provide immortality. Chemists of today use their knowledge of chemical reactivity to prepare new drugs, new structural materials, and materials with novel properties. This chapter begins your introduction to chemical reactions and their applications.

It is convenient to classify chemical reactions into one of a small number of categories. The categories we choose are: (1) **combination reactions**; (2) **decomposition reactions**; (3) **metathesis**, or double replacement reactions; and (4) **redox**, or oxidation–reduction reactions. Learning this classification scheme will help you to predict the products of many inorganic and organic reactions. It is possible that a given reaction may fit into more than one of the categories. This should not, in practice, cause you trouble when predicting the products of those reactions. In this chapter, we will learn how to predict the products of the first three types of reactions, how to write balanced reaction equations, and how to quantify that information in a process known as stoichiometry. Finally, we will take a quick look at how one may adjust reaction conditions to favor the formation of a desired product.

Note that chemists usually represent reactions by showing reactants to the left of the arrow, products to the right, and any specific conditions (e.g., temperature, presence of a **catalyst**) over the reaction arrow. This approach is utilized when the balanced reaction is emphasized. Occassionally, chemists elect to show the synthesis of a compound by indicating the major reactant on the left, and the target molecule on the right; other reactants or products are listed over the reaction arrow. To the beginning chemist, this may cause some confusion. In this text, we will employ the following convention: if the species "X" appears over the reaction arrow, it may be assumed to be a catalyst.

$$C_2H_4 + H_2 \xrightarrow{Pt} C_2H_6$$

If "+ X" appears over the reaction arrow, X is assumed to be a reactant.

$$C_2H_4 \xrightarrow{+H_2} C_2H_6$$

If "- X" appears over the arrow, X is assumed to be a product.

$$C_2H_6 \xrightarrow{-\ H_2} C_2H_4$$

Note that in most other texts, the "+" or "-" will often be omitted. With practice, you will be able to see readily whether those authors mean the species over the arrow to be a catalyst, a reactant, or a product.

In chapter 6, we will continue our study of reactivity by examining the fourth category, redox reactions. We will also continue our discussion of reaction stoichiometry.

5.2 COMBINATION REACTIONS

As we begin the study of types of chemical reactions, the authors wish to remind the students that key terms presented in chapters one and two (e.g., metals, non-metals, ionic compounds, covalent compounds, weak and strong acids, weak and strong bases, soluble and insoluble salts, acidic anhydrides, basic anhydrides, polar and non-polar compounds) will be crucial in understanding reactivity. The formula and names of inorganic compounds, as summarized in Appendix E, and the structure and names of organic compounds , as summarized in chapters three and four, will also aid our understanding of chemical reactions.

Combination reactions are those reactions that involve two or more elements and/or compounds and that result in the formation of a single product. More specifically two (or more) elements may react to form a single compound, an element and a compound may react to form a new compound, or two compounds may react to form a third compound.

TWO ELEMENTS COMBINE TO FORM A COMPOUND

When two elements react to form a compound there is always a driving force accompanying the reaction. There is a release of energy due to the products being more stable than the reactants. While the exact details are more complicated than we can go into here, this will usually be the case when there is an electronegativity difference between the two elements involved. Since almost any two elements have different electronegativities, almost all elements will react with each other under some conditions. The most common types of reactions are those involving metals reacting with non-metals, and two non-metals reacting with one another. In the latter case one of the elements commonly involved is oxygen.

Prediction of the stoichiometry of the products that will form from a combination reaction involving two elements can usually be made by knowing the charges of the simple monatomic ions that would be formed by the elements involved. For this purpose you may assume that the more electronegative element will be the anion in any proposed compound and the more electropositive (less electronegative) element will be the cation of any proposed compound. While this method is not fool-proof, it will work many times to predict a reasonable product that could be formed between the two elements. The maximum positive charge on a cation is the group number of the element. Some elements, particularly non-metallic elements, also form compounds in which it would appear that they are in a cationic state that is two or four units of charge less than the group number. Sometimes these lower valent compounds are the more stable ones. See Appendix E for additional information related to common oxidation states of elements in compounds.

SIDEBAR: BALANCING CHEMICAL REACTIONS BY INSPECTION

To balance a chemical reaction, first write down the correct chemical formulas for the reactants and the products on either side of the chemical equation arrow. For instance, if we consider a reaction between aluminum oxide (Al_2O_3) and hydrochloric acid (HCl) to form aluminum chloride ($AlCl_3$) and water (H_2O), the beginning of the process would be to write:

$$Al_2O_3 + HCl \longrightarrow AlCl_3 + H_2O$$

At this point we start the balancing process by looking for atoms, except for H and O, which appear on both sides of the equation in only one compound. If such atoms are present, the balancing process usually goes better if one of those is chosen and balanced first. Once one of those is chosen, it is best to continue the process until only H and O are left. In many cases, when it is time to balance them they will already be in balance. In this case Al and Cl fit our criterion. Let us try Al first by putting a 2 in front of the aluminum chloride to balance the aluminum atoms:

$$Al_2O_3 + HCl \longrightarrow 2\ AlCl_3 + H_2O$$

We can now balance the chlorine atoms with a 6 in front of the hydrochloric acid:

$$Al_2O_3 + 6\ HCl \longrightarrow 2\ AlCl_3 + H_2O$$

Once the aluminum and chlorine are in balance, we find it necessary only to place a 3 in front of the water to complete the balance:

$$Al_2O_3 + 6\ HCl \longrightarrow 2\ AlCl_3 + 3\ H_2O$$

At this point it is always good to check the process by determining the number of each type of atom on both the reactant and product side:

ATOM	LEFT SIDE	RIGHT SIDE
Al	2	2
O	3	3
H	6	6
Cl	6	6

There are times when it is easiest to consider a group of atoms together as if they were a single atom. To illustrate, consider the equation for the reaction between calcium chloride and potassium phosphate to form calcium phosphate and potassium chloride:

$$CaCl_2 + K_3PO_4 \longrightarrow Ca_3(PO_4)_2 + KCl$$

Notice that the phosphorous atom only appears in this equation in conjunction with four oxygen atoms. It is easiest, therefore, to consider the PO_4 group as if it were a single species and not as separate atoms. In this situation we shall balance the Ca atoms, then the PO_4 group, and finally the K and Cl atoms. When all done the final equation is:

$$3\ CaCl_2 + 2\ K_3PO_4 \longrightarrow Ca_3(PO_4)_2 + 6\ KCl$$

Consider the reaction between metallic sodium and gaseous chlorine. In this situation we are dealing with a reaction between a metal and a non-metal. Based on the electronegativity of the two elements involved, the metal is obviously going to be the cation; from the fact that sodium is a Group I metal we know that its charge will be +1. Likewise we can predict that the non-metal (a halogen in this case) is going to be the anion. Since chlorine is in Group VII, the charge on the chloride anion will be the group number - 8, or 7 - 8 = -1. The product compound must be neutral. Thus the product must have a 1:1 ratio of Na^+ and Cl^- in the formula. The reaction will then be:

$$2\ Na(s) + Cl_2(g) \longrightarrow 2\ NaCl(s)$$ EQ. 5.1

Next consider the reaction between lithium metal and nitrogen. Again we have a metal reacting with a non-metal. For the same reasons given in the previous example the cation will be lithium and will have a charge of +1. The anion will be formed by the nitrogen (Group V) and will have a charge of 5 - 8 = -3. The formula of the product is again predicted on the basis of charge balancing. The overall reaction is:

$$6\ Li(s) + N_2(g) \longrightarrow 2\ Li_3N(s)$$ EQ. 5.2

When the reactants are both non-metals the situation becomes slightly more complicated. While the reaction of a metal with a non-metal almost always results in a truly ionic compound, the reaction of two non-metals results in a covalent compound. Although it is unrealistic to think of the reaction as involving ions, it nevertheless will aid in determining the stoichiometry of the product formed if we consider the compound formed to be ionic. (Note: this is only to determine the stoichiometry — the compounds that are formed are really covalent).

Consider the reaction of carbon with oxygen. Both of the elements are non-metallic but carbon is the more electropositive of the two. We should therefore consider it to be the "cation" in the proposed compound. If carbon were to form a monatomic cation, the charge on the ion would be +4. Since the oxygen will be the "anion" in the proposed compound, its charge would be -2. The **chemical reaction** is:

$$C(s) + O_2(g) \longrightarrow CO_2(g)$$ EQ. 5.3

In this case, however, there is also the possibility of another compound forming. The above reaction requires a ratio of 1:1 for carbon atoms and oxygen molecules. If there is less oxygen than this, for example if the ratio is 2:1, another reaction takes place with a different product formed. Its reaction is:

$$2\ C(s) + O_2(g) \longrightarrow 2\ CO(g)$$ EQ. 5.4

We can rationalize the stoichiometry of this compound by thinking of the carbon atom as an ion with a charge of +2. This is not too unusual for non-metallic species; often two elements will form more than one compound with the stoichiometry indicating differences of 2 units in the "charge" of the "cation".

Next consider the reaction between elemental phosphorous and sulfur. Using the methods described above, the phosphorous "cation" would have a charge of +5, and the sulfur "anion" would have a charge of -2. This would lead to the reaction:

$$2\ P(s) + 5S(s) \longrightarrow P_2S_5(s)$$ EQ. 5.5

Again there are other possible compounds formed between phosphorous and sulfur, the most important of which is P_2S_3, where phosphorous has a +3 oxidation state. Note that phosphorus in this compound has an oxidation state of the group number –2.

CONCEPT CHECK: COMBINATION REACTIONS BETWEEN ELEMENTS

Predict the product formed by the reaction between barium and arsenic.

SOLUTION: Barium, a Group IIA alkaline earth metal, is expected to have a +2 oxidation state in any compound. Arsenic, a Group VA nonmetal, should have a –3 charge as a monatomic anion. The compound barium arsenide is a neutral species and thus has the formula Ba_3As_2.

A COMPOUND AND AN ELEMENT COMBINE TO FORM A NEW COMPOUND

Up to this point we have considered only elements reacting with other elements. It is also possible for elements to react with compounds. For example, consider the reaction between sulfur dioxide and oxygen. These two species combine to form sulfur trioxide. This reaction is of particular interest for those concerned with air pollution and acid rain.

$$2\ SO_2(g) + O_2(g) \longrightarrow 2\ SO_3(g)$$ EQ. 5.6

It is possible to rationalize the stoichiometry of this product by remembering that many non-metals can assume more that one "charge" or **oxidation** state. In this particular example, we can think of the oxygen atom as having a "charge" of -2. This means that sulfur would be in the +4 oxidation state in sulfur dioxide, and in the +6 oxidation state in sulfur trioxide.

SIDEBAR: MOLECULAR FORMULA OF THE NON-METALLIC ELEMENTS

Some elements, such as Group VIIIA, the noble (or inert) gases, are always monatomic. Other non-metals are typically diatomic: H_2, N_2, O_2, F_2, Cl_2, Br_2, and I_2. Other elements may have a variety of molecular forms. Sulfur, for instance, is often observed in rings with 6, 8, 10, or 12 sulfur atoms, or as chains of variable numbers of sulfur atoms. Thus in writing the reaction between sulfur and oxygen, we denote sulfur simply as "S", but oxygen as the common diatomic molecule "O_2".

Organometallic compounds often form as a result of a reaction between an organic compound and a **metal**. The most famous of these compounds are, perhaps, the Grignard reagents. The reaction between an alkyl halide, typically an alkyl bromide, and magnesium produces the Grignard reagent. A simple example of this reaction is that of ethyl bromide reacting with magnesium in an ether solution to form ethyl magnesium bromide.

$$CH_3CH_2Br + Mg^0 \longrightarrow CH_3CH_2MgBr$$ EQ. 5.7

The stoichiometry of the Grignard reagent can be rationalized by realizing that magnesium in compounds typically is considered to have a +2 charge. We know that the bromine anion has a -1 charge and, therefore, that the alkyl group (CH_3CH_2) may be thought of as adopting a -1 charge as well. In fact, it will be convenient to think of the carbon atom in the alkyl group bonded to magnesium as carrying a partial negative charge, while the rest of the Grignard reagent, MgBr, carries a partial positive charge. An alkyl group that carries a negative charge is especially reactive. That is why the reaction in Eq 5.7 must be run in dry ether (i.e., without a trace of water.) In the presence of water or any other compounds that can donate a hydrogen ion (such as an alcohol) the Grignard reagent will rapidly decompose. The negatively charged alkyl group will combine with any available protons to form a more stable species. For example, the Grignard reagent shown in Eq 5.7 rapidly reacts with water to form C_2H_6 and HOMgBr.

Reactions of an organic compound with a non-metallic element resulting in the formation of a new compound are quite common. A good model for this type of reaction is the addition of hydrogen to an alkene, a process known as hydrogenation. You might notice that this process requires a catalyst to get the reaction to proceed in a reasonable amount of time. The nature of this catalyst will be discussed in more detail later. In the following example, hydrogen adds to ethene to produce ethane.

$$H_2C{=}CH_2 + H_2 \xrightarrow{\text{catalyst}} H_3C{-}CH_3$$ EQ. 5.8

Hydrogenation is of particular importance in the food industry. Many vegetable oils contain multiple alkene bonds and are, therefore, referred to as polyunsaturated. The liquid oils can be converted to semi-solid vegetable oil based spreads by the process of hydrogenation, which we examine in Chapter 6.

As you might expect, hydrogen can also add to alkynes. Consider the addition of hydrogen to 1-butyne. One of the products that you might expect to isolate from the reaction is 1-butene, but you might also expect to observe significant amounts of butane as a product of this reaction. The composition and relative amounts of products formed are a function of the type of catalyst used.

$$H_3CH_2C{-}C{\equiv}CH + H_2 \xrightarrow{\text{catalyst}} (H_3CH_2C)(H)C{=}C(H)(H) \dashrightarrow \text{Alkane}$$ EQ. 5.9

Another example of a reaction between an alkene and a diatomic molecule is the addition of a halogen (in this case Cl_2, Br_2, and I_2) to an alkene. Consider the reaction between 2-butene and bromine. Again, the diatomic molecule adds across the double bond. The product, in this case, is 2,3-dibromobutane.

$$CH_3CH{=}CHCH_3 + Br_2 \longrightarrow CH_3{-}CH(Br){-}CH(Br){-}CH_3$$ EQ. 5.10

Similar reactions can occur between alkynes and halogens. If only one molecule of the halogen adds, the expected product would, of course, be the di-halo compound. If another molecule of halogen were to add to this compound then the tetra-halo compound would result. Consider the addition of iodine to 2-butyne. Addition of one molecule of iodine results in the formation of 2,3-diiodo-2-butene.

$$H_3C{-}C{\equiv}C{-}CH_3 + I_2 \longrightarrow H_3C{-}C(I){=}C(I){-}CH_3$$ EQ. 5.11

This compound is still capable of further reaction with a halogen and/or hydrogen. If, for example, this compound then reacted with chlorine, the final reaction product would be 2,3-dichloro-2,3-diiodobutane.

Two Compounds Combine to Form a New Compound

Two inorganic compounds rarely combine unless one is an acid and the other is a base. For the purposes of this discussion the most common acids are the oxides of non-metals and the most common bases are the oxides of metals. When these react with one another they form simple ionic compounds, many of which are quite common. Consider the reaction of magnesium oxide (a base) and sulfur trioxide (an acid). The chemical reaction between a basic anhydride and an acidic anhydride forms a salt, in this case magnesium sulfate:

$$MgO(s) + SO_3(g) \longrightarrow MgSO_4(s)$$ EQ. 5.12

Alkenes are also capable of undergoing addition reactions with unsymmetrical binary compounds such as hydrogen bromide or water (it may be helpful for you to think of water as a proton and the hydroxide ion when thinking about these reactions). In both cases, one portion of the binary compound becomes attached to one of the alkene carbons while the other portion becomes attached to the other alkene carbon. Consider the addition of hydrogen bromide and of water to cyclohexene.

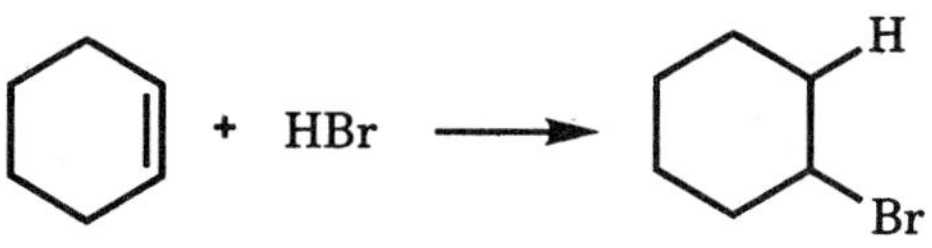

$$\text{cyclohexene} + HBr \longrightarrow \text{cyclohexane with H and Br}$$ EQ. 5.13

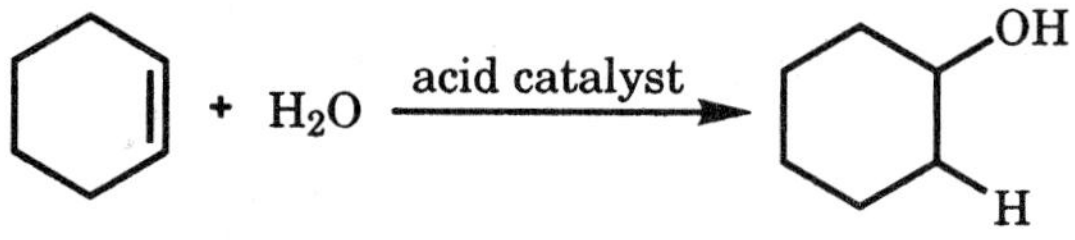

$$\text{cyclohexene} + H_2O \xrightarrow{\text{acid catalyst}} \text{cyclohexane with OH and H}$$ EQ. 5.14

The product of the first reaction is bromocyclohexane while the product of the second reaction is cyclohexanol.

You may have noticed that the first example involved the addition of a binary compound to a symmetrical alkene. What could happen if the binary compound were to add to an unsymmetrical alkene? In principle either part of the binary compound could add to either one of the alkene carbons. It should be possible, therefore, for two products to form in such a reaction. Consider the addition of hydrogen bromide to 1-methylcyclohexene. You might predict the two possible products to be 1-bromo-1-methylcyclohexane and 1-bromo-2-methylcyclohexane. In practice, however, very little, if any, of the 1-bromo-2-methylcyclohexane is formed.

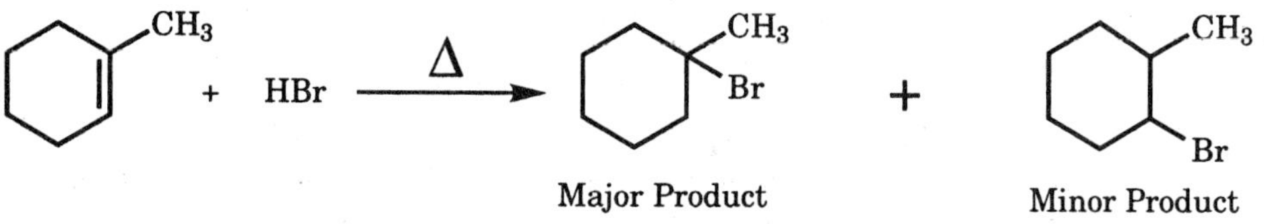

$$\text{1-methylcyclohexene } (CH_3) + HBr \xrightarrow{\Delta} \text{cyclohexane with } CH_3, Br + \text{cyclohexane with } CH_3 \text{ and } Br$$ EQ. 5.15

Major Product — Minor Product

These results are empirical evidence for **Markovnikov's Rule** that states that, when considering the addition of binary compounds that contain hydrogen to unsymmetrical alkenes, the hydrogen atom will add to the less substituted carbon. Another way of thinking about this is to remember the phrase "them that has gets". In this case the carbon that contains more hydrogen gets even more.

5.3 DECOMPOSITION REACTIONS

PYROLYSIS AND ELECTROLYSIS OF INORGANIC COMPOUNDS

Decomposition reactions are typically considered the reverse of combination reactions. Although a few decomposition reactions (those involving compounds that are inherently unstable) may proceed spontaneously, most decomposition reactions require some type of energy input to make the reaction go. All compounds will decompose into their elements if enough energy is supplied in the form of heat. For a thermal decomposition reaction (known as a pyrolysis) to occur under reasonable conditions, it is necessary that one of the products of the reaction be a small, stable, covalent molecule. The most common of these molecules are non-metal oxides formed when metal oxo-salts are heated. Examples of such reactions include:

$$CaSO_4(s) \xrightarrow{\Delta} CaO(s) + SO_3(g)$$ EQ. 5.16

and

$$MgCO_3(s) \xrightarrow{\Delta} MgO(s) + CO_2(g)$$ EQ. 5.17

A second class of decomposition reactions includes those which are promoted by the passage of an electrical current through a solution of the compound in water or through the molten compound itself. Products formed by the electrolysis of solutions can be difficult to predict and will not be covered here. Predicting the products of the electrolysis of molten binary salts is, however, not difficult at all. Consider the electrolysis of molten sodium chloride:

$$2\ NaCl(l) \xrightarrow{\text{electricity}} 2\ Na(l) + Cl_2(g)$$ EQ. 5.18

Most ternary, or higher, salts decompose (along the lines of equations 16 and 17) prior to melting and will, therefore, not undergo an electrolytic reaction.

PYROLYSIS OF ORGANIC COMPOUNDS

You may have been wondering where all the different hydrocarbon compounds originate. It turns out that hydrocarbons (as well as many other organic compounds) are often derived from crude petroleum and/or coal. These two materials are formed from the decay of plant and animal life over long periods of time. They can then be pyrolyzed to provide many simple hydrocarbon molecules.

As we saw earlier in Chapter Two, petroleum can be separated into individual components by a process know as distillation. In practice, when distilling petroleum, one often collects fractions that contain a mixture of compounds that boil within a given range. Mixtures of compounds derived from petroleum that you are already familiar with include gasoline and kerosene.

Other fractions, particularly those containing longer chain molecules, are often subjected to a process known as cracking. This is simply a pyrolytic technique that results in the decomposition of long chain alkanes into shorter alkanes and alkenes. Once these new compounds are formed they are separated by another distillation.

Coal is a good source of a variety of aromatic compounds. Upon heating coal under anaerobic (oxygen depleted) conditions, several products can be isolated including a liquid known as coal tar. Distillation of coal tar allows one to isolate various aromatic compounds including benzene, toluene, and naphthalene.

UNSATURATED HYDROCARBONS FROM ALCOHOLS AND ALKYL HALIDES

We saw in a previous section that it is possible for water to add across a double bond to form an alcohol. As you may have predicted, this process is reversible. Water can be eliminated from an alcohol to form an alkene as well. This process is know as **dehydration**. In both cases it is, however, necessary to use an acid catalyst. When 3-pentanol is treated with an acid catalyst and heat, water can be eliminated and 2-pentene can be isolated.

OH

$$\xrightarrow[\text{heat}]{H^+\text{ (cat.)}} \quad + H_2O$$ EQ. 5.19

This example was a simple case where the starting material was a symmetrical alcohol. What happens if we begin with an unsymmetrical alcohol? Consider the acid catalyzed dehydration of 2-butanol.

OH $\xrightarrow[\text{heat}]{H^+ \text{(cat.)}}$ Minor + Major EQ. 5.20

In this case it is possible to remove a proton from two different sites. The proton could be removed from the terminal carbon resulting in the formation of 1-butene, or the proton could be removed from carbon-3 resulting in the formation of 2-butene. Experimental results show that while both reactions occur, the major product of the reaction is 2-butene. This means that proton removal takes place preferentially from carbon-3. Alexander Saytzeff, a Russian chemist, first discovered this phenomenon. He found that in elimination reactions of this type, the major product will have the greatest number of alkyl groups on the alkene carbons. In the example above, 2-butene contains 2-alkyl groups on the alkene carbons whereas 1-butene contains only one.

Similar reactions are observed with alkyl halides. The one major difference between the two reactions is that alkyl halides require stoichiometric amounts of base to promote reaction. Consider for example, the reaction of potassium hydroxide and bromocyclohexane.

Br $\xrightarrow{KOH}$ + H_2O + KBr EQ. 5.21

The product of the reaction is cyclohexene and results from the loss of hydrogen bromide. This reaction is referred to as a **dehydrohalogenation** and is formally the reverse of the addition reaction of hydrogen bromide to an alkene. **Saytzeff's rule** also applies to reactions of this type. Consider the reaction between 1-bromo-2-methylcyclooctane and sodium hydroxide.

CH_3, Br + NaOH → CH_3 major + CH_3 minor + $NaBr + H_2O$ EQ. 5.22

Once again the major product is the product that contains the more substituted double bond, while the minor product contains the less substituted double bond.

Concept Check: Saytzeff's Rule

When 3-bromo-2-methylpentane is treated with KOH, two different alkenes are produced. Show the structure and name of both, and indicate which is expected as the major product.

Solution: The starting material, $(CH_3)_2CHBrCH_2CH_3$, undergoes a dehydrohalogenation reaction, eliminating HBr. With bromine on the #3 carbon, the hydrogen has to be removed from either the #2 or the #4 carbon, forming either:

(a) $(CH_3)_2C{=}CHCH_2CH_3$, 2-methyl-2-pentene, or....

(b) $(CH_3)_2CHCH{=}CHCH_3$, 2-methyl-3-pentene.

The major product is (a), 2-methyl-2-pentene, since the alkene carbon atoms are bonded to 3 other carbon atoms. This is a more substituted double bond than that present in 2-methyl-3-pentene, where the alkene carbons are connected to only 2 other carbon atoms.

Compounds containing two halogen atoms are sometimes capable of undergoing two dehydrohalogenation reactions to provide alkynes. Consider the following two reactions.

Br Br + KOH ⟶ Br + KBr + H_2O EQ. 5.23

Br + $NaNH_2$ ⟶ $CH_3CH_2\text{-}C{\equiv}C\text{-}CH_3$ + NaBr + NH_3

Br Br + KOH ⟶ Br + KBr + H_2O EQ. 5.24

Br + $NaNH_2$ ⟶ $CH_3CH_2\text{-}C{\equiv}C\text{-}CH_2CH_3$ + NaBr + NH_3

The first reaction involves two consecutive dehydrohalogenation reactions beginning with a **vicinal** dihalide (a compound that contains two halogen atoms on adjacent carbon atoms). The second reaction begins with a **geminal** dihalide (a compound that contains two halogen atoms on the same carbon atom). Both reactions proceed through vinylic halides (compounds that contain a halogen atom bonded to an alkene carbon). It is important to notice that a prerequisite for these reactions is the presence of sufficient hydrogen atoms on adjacent carbons. Potassium hydroxide, a strong base, is required to remove hydrogen bromide in the first half of both reactions. Sodium amide ($NaNH_2$) an even stronger base, is required to remove the second molecule of HBr to form the alkyne.

5.4 TWO COMPOUNDS REACT TO FORM TWO NEW COMPOUNDS

INORGANIC METATHESIS REACTIONS

Double replacement reactions always involve reactions between two materials that can be conveniently thought of as ionic. Unless a reactant is a weak acid, a weak base, or is insoluble, it should be written as totally ionized when in an aqueous solution. When predicting products for these reactions one must look for a driving force, i.e., some reason for the reaction to proceed to the right. There are three such driving forces that should be considered: (1) the formation of an insoluble substance; (2) the formation of a gas; and (3) the formation of a weakly ionized substance (typically a weak acid, weak base, or water).

To be successful you must know the list of strong acids and bases (a very short list) and the solubility rules presented earlier in Chapter Two. Once the reactants have been properly written, tentatively write a set of products in which the cations and anions have been switched. Next ask yourself if there is anything about the products that would give rise to a driving force for the reaction. That is, is there an insoluble substance among the products but not among the reactants? Has a gas been formed? Has a weakly ionized substance been formed? If the answer to any of these questions is 'yes,' you must only complete the equation by removing any species that appear on both sides of the reaction and you will have successfully predicted the products of the reaction.

Let us consider some examples:

Consider the reaction between aqueous solutions of sodium chloride and silver nitrate.

The solubility rules tell us that both sodium chloride and silver nitrate are soluble. The first step is to write out all of these species as completely ionized.

$$Na^+ + Cl^- + Ag^+ + NO_3^- \longrightarrow$$ EQ. 5.25

The next step is to consider whether the possible products, sodium nitrate and silver chloride, fit into any of the three categories mentioned above. The solubility rules tell us that silver chloride is insoluble! We must then finish the equation:

$$Na^+ + Cl^- + Ag^+ + NO_3^- \longrightarrow AgCl(s) + Na^+ + NO_3^-$$ EQ. 5.26

At this point we see that the sodium and nitrate ions appear on both sides of the equation (ions appearing in this fashion are termed "**spectator ions**") and can be removed from the equation if the question has requested the net ionic equation for the reaction. The final equation is:

$$Cl^-(aq) + Ag^+(aq) \longrightarrow AgCl(s)$$ EQ. 5.27

Next consider the reaction between calcium carbonate and hydrochloric acid.

The solubility rules tell us that calcium carbonate is insoluble and our knowledge of the list of strong acids tells us that hydrochloric acid is a strong acid. The reactants are written:

$$CaCO_3(s) + H^+ + Cl^- \longrightarrow$$ EQ. 5.28

If we now switch cations and anions the products would be calcium chloride ($CaCl_2$) and carbonic acid (H_2CO_3). Our knowledge of the solubility rules tells us that calcium chloride is soluble while our knowledge of the list of strong acids tells us that carbonic acid is not on the list and must, therefore, be a weak acid. We have found a driving force for this reaction! We write the reaction as:

$$CaCO_3(s) + H^+ + Cl^- \longrightarrow Ca^{2+} + Cl^- + H_2CO_3(aq)$$ EQ. 5.29

Notice that the chloride ion is a spectator ion. Knowing that carbonic acid is really a solution of carbon dioxide in water allows us to write the final, net ionic, equation:

$$CaCO_3(s) + 2\,H^+ \longrightarrow Ca^{2+} + H_2O(l) + CO_2(g)$$ EQ. 5.30

Finally consider the reaction between sodium hydroxide and acetic acid.

The list of strong acids does not contain acetic acid; therefore it must be a weak acid. The list of strong bases does include all soluble metal hydroxides; sodium hydroxide is soluble, and therefore is a strong base. The reactants are written:

$$CH_3COOH(aq) + Na^+ + OH^- \longrightarrow$$ EQ. 5.31

Again if we switch cation and anion to form tentative products we find those products to be sodium acetate and hydrogen hydroxide, better known as water. We have found our driving force! The final, net ionic, equation is:

$$CH_3COOH(aq) + OH^- \longrightarrow CH_3COO^- + H_2O$$ EQ. 5.32

Note that the reaction of aqueous solutions of acids and bases always form water and a salt as the products.

ORGANIC SUBSTITUTION REACTIONS

Many different types of organic reactions can be classified as **substitution reactions**. Some examples will be presented here and many of these will be studied in more detail in the future. Organic substitution reactions, like inorganic metathesis reactions, involve reactions between two different compounds and result in the formation of two new compounds. In many cases one of the reactants and one of the products may exist in ionic form. The driving force for these reactions is, however, not always as clear cut as those for the inorganic reactions.

Again, let us look at some examples.

Consider the reaction between ethanol and hydrogen bromide. You know that ethanol is certainly not ionic; but hydrogen bromide, a strong acid, forms ions in aqueous solution. The reaction, therefore, would be written in the following manner.

$$CH_3CH_2\text{–}OH + H^+ + Br^- \longrightarrow$$ EQ. 5.33

If we now think about the reactants switching partners, we may envision the hydroxyl group combining with the proton to form water and the bromide ion combining with the ethyl group to form ethyl bromide. The entire reaction can be written as follows.

$$CH_3CH_2\text{–}OH + H^+ + Br^- \longrightarrow CH_3CH_2\text{–}Br + H_2O$$ EQ. 5.34

Reactions that are the reverse of the above process are also quite common. The reaction between and alkyl halide and water to form an alcohol and the corresponding hydrogen halide may be encountered. For example, consider the reaction between t-butyl bromide and water.

$$CH_3\text{–}\underset{\underset{CH_3}{|}}{\overset{\overset{CH_3}{|}}{C}}\text{–}Br + H_2O \longrightarrow CH_3\text{–}\underset{\underset{CH_3}{|}}{\overset{\overset{CH_3}{|}}{C}}\text{–}OH + H^+ + Br^-$$ EQ. 5.35

In this case it is again convenient to think of water as a proton and a hydroxide ion. The combination of the hydroxide ion with the t-butyl group yields an alcohol.

The reaction might also be performed in an aqueous solution of potassium hydroxide. Consider the reaction between octyl bromide and potassium hydroxide. In this case the potassium hydroxide is clearly ionized. The reaction is written as follows.

$$CH_3(CH_2)_7Br + K^+ + OH^- \longrightarrow CH_3(CH_2)_7OH + K^+ + Br^-$$ EQ. 5.36

In this case, the hydroxide ion combines with the octyl group and the bromide ion combines with the potassium. The products of the reaction are, then, 1-octanol and potassium bromide.

In the preceding reactions, we have seen that substitution reactions can convert an alkyl bromide into an alcohol, or an alcohol into an alkyl bromide. Which reaction occurs (i.e., whether we expect alkyl halides to become alcohols, or alcohols to become alkyl halides) depends upon the type of reactants used, and this will be studied later in this course.

The formation and degradation of two other classes of organic compounds can also be considered substitution reactions. These compounds are the esters and the amides. Esters, which are composed of a carboxylic acid portion and an alcohol portion, are important compounds both commercially and biologically. Many simple esters are used as flavoring or fragrance agents. Polyesters, such as dacron, are of tremendous importance in the textile industry. In biochemical systems the ester functional group is present in many of the lipid molecules that make up lipid bilayer membranes. The amide, which is composed of a carboxylic acid portion and an amine portion, is the functional group that is commonly known as the peptide bond, which forms proteins. We will first look at the degradation of these compounds and then at their synthesis to begin to understand these reactions as substitution reactions.

Both amides and esters can be broken down into two components by a process known as **hydrolysis**, that is cleavage with water. Let us look at some examples. Consider the reaction of methyl salicylate (which, incidentally is oil of wintergreen) with water.

$$\text{HO-C}_6\text{H}_4\text{-C(=O)OCH}_3 + H_2O \longrightarrow \text{HO-C}_6\text{H}_4\text{-C(=O)OH} + CH_3OH$$ EQ. 5.37

As you can see, the products of the reaction are salicylic acid and methanol. It is often convenient to represent the pathway as shown below for the hydrolysis of methyl acetate.

$$H_3C\text{—}\overset{O}{\overset{\|}{C}}\text{—}O\text{—}CH_3 \quad HO\text{—}H \quad \rightarrow \quad CH_3COOH \quad + \quad CH_3OH$$ EQ. 5.38

This type of drawing is often referred to as the "lasso" representation of the mechanism. It is an attempt to show, in more detail, what is going on at the molecular level. In the case shown above, it is clear that one of the hydrogen atoms from water combines with the "CH_3O" group from methyl acetate to form methanol and that the "hydroxide portion" the water molecule combines with the acyl portion of the methyl acetate to form acetic acid.

Concept Check: Ester Hydrolysis and Ester Structure

An unknown ester has the following HNMR spectrum: a septet (corresponding to 1 proton) at 5.0 ppm; a singlet (corresponding to 3 protons) at 2.0 ppm; and a doublet (corresponding to 6 protons) at 1.2 ppm. When the unknown ester is hydrolyzed, the products of the reaction are acetic acid and isopropyl alcohol (2-propanol). Identify the unknown ester, and explain the HNMR spectrum by indicating which protons in the structure of the ester gave rise to which peaks.

Solution: Since the hydrolysis reaction yields CH_3COOH and $CH_3CH(OH)CH_3$, the ester would be formed by reacting acetic acid and isopropyl alcohol. This would yield isopropyl acetate, $CH_3COOCH(CH_3)_2$. The methyl (CH_3-) protons in the CH_3COO^- fragment would give rise to the singlet at 2.0 ppm. This signal is not split, since the adjacent carbon has no hydrogens bonded to it. The single proton in the isopropyl group (**CH**$(CH_3)_2$) is split by 6 equivalent protons on adjoining carbons. This gives rise to the septet at 5.0 ppm, with an integrated area corresponding to one proton. Finally, the 6 protons (two methyl groups) in the isopropyl fragment (CH(**CH$_3$**)$_2$) give rise to the doublet at 1.2 ppm. This is a doublet because these protons are are on carbons bonded to a carbon that has just one hydrogen atom.

Amides are hydrolyzed in the same way. For example, the following dipeptide is hydrolyzed into the two amino acids that make up the dipeptide.

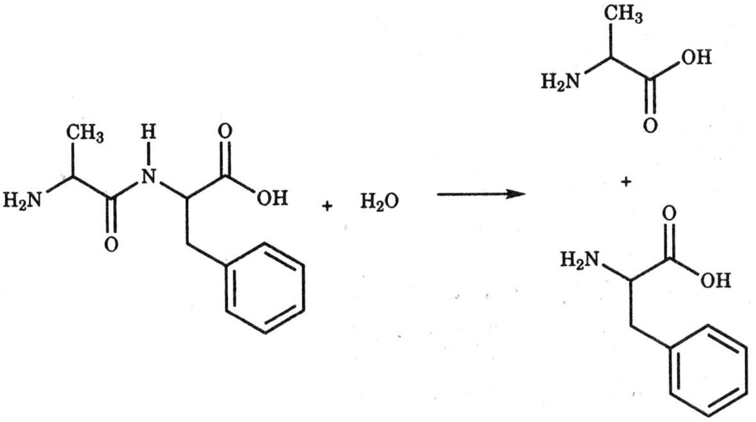

EQ. 5.39

In practice, the hydrolysis of both esters and amides are usually done under either acidic or basic conditions. At this point, however, it is enough for you to recognize the hydrolysis products of these reactions without worrying too much about reaction conditions.

As mentioned above, esters are simply the product of the reaction between a carboxylic acid and an alcohol. The following are a few examples of esterification reactions.

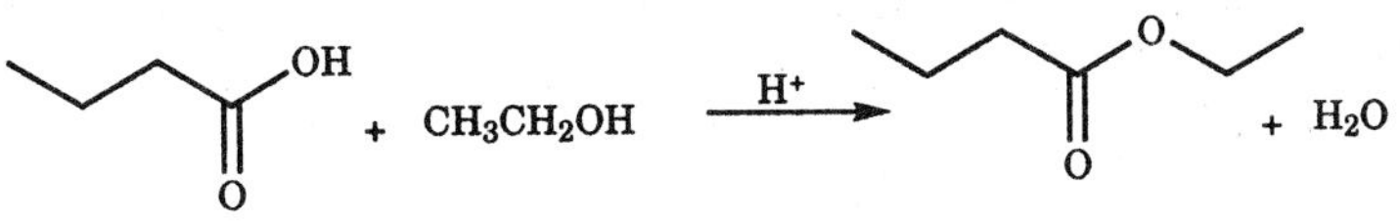

EQ. 5.40

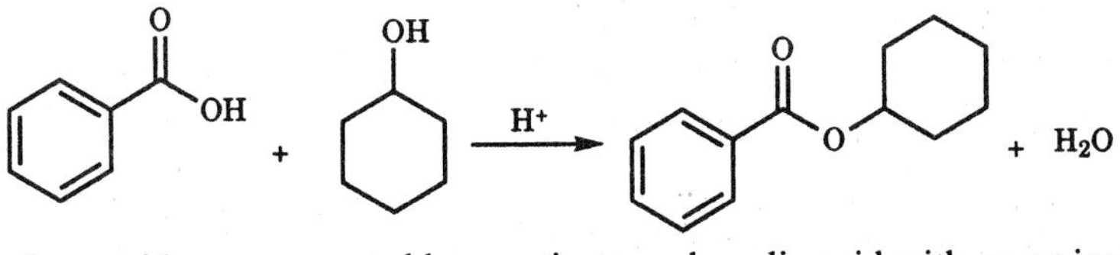

EQ. 5.41

Likewise, amides are prepared by reacting a carboxylic acid with an amine. The reaction is typically conducted in the presence of some "activating agent". For now it is sufficient that you are able to predict the products of these types of reactions. The following are representative of typical amide forming reactions.

$$CH_3CH_2\text{-}\overset{O}{\overset{\|}{C}}\text{—}OH + NH_3 \xrightarrow[\text{agent}]{\text{activating}} CH_3CH_2\text{-}\overset{O}{\overset{\|}{C}}\text{—}NH_2 + H_2O$$

EQ. 5.42

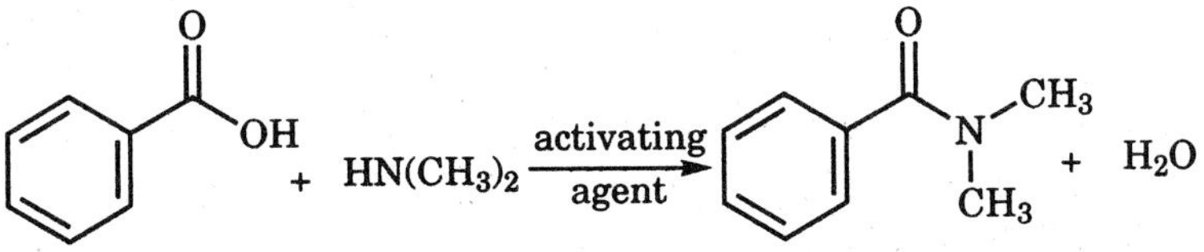

EQ. 5.43

Symmetrical ethers are prepared industrially by a similar dehydration reaction.

$$2\ CH_3CH_2OH \xrightarrow{H_2SO_4} CH_3CH_2\text{—}O\text{—}CH_2CH_3 + H_2O$$ EQ. 5.44

Sulfuric acid is used in the reaction since it facilitates the removal of water. Unsymmetrical ethers are prepared in another type of substitution reaction, combining an alkyl halide with a metal alkoxide.

$$CH_3I + CH_3CH_2ONa \longrightarrow CH_3OCH_2CH_3 + NaI$$ EQ. 5.45

Alkyl halides are also used to form nitriles, via reaction with an inorganic cyanide salt.

$$CH_3CH_2CH_2Br + NaCN \longrightarrow CH_3CH_2CH_2C{\equiv}N + NaBr$$ EQ. 5.46

Note that this reaction has special synthetic implications because the length of the carbon chain has been increased by one carbon atom.

Organic Addition Reactions

One final class of organic reactions that we wish to place under the general heading of two compounds reacting to form two new compounds is that of addition reactions to carbonyl compounds. While many different species react in this fashion, we will limit our discussion, at this point, to the Grignard reaction. You should recall that Grignard reagents are formed when an alkyl halide in an ether solution is allowed to react with elemental magnesium. These reagents show a special reactivity with carbonyl containing compounds. The carbonyl functional group has a polar double bond between carbon and oxygen, with oxygen being partially negative and carbon being partially positive due to oxygen's greater electronegativity. The Grignard reagent, which as we noted earlier has a partially negative alkyl portion and a partially positive MgBr portion, adds across the carbonyl bond in a specific fashion. The partially negative alkyl group from the Grignard reagent becomes attached to the partially positive carbonyl carbon. Conversely, the partially positive MgBr portion attaches to the partially negative carbonyl oxygen atom. This converts the carbonyl compound into an alkoxide salt of an alcohol. Consider, for example, the reaction between propanal and methyl magnesium bromide.

$$CH_3CH_2\text{—}\overset{\overset{\displaystyle O}{\|}}{C}\text{—}H + CH_3MgBr \longrightarrow CH_3CH_2\text{—}\underset{\underset{\displaystyle CH_3}{|}}{\overset{\overset{\displaystyle O^- MgBr^+}{|}}{C}}\text{—}H$$ EQ. 5.47

Protonation of this species results in formation of the alcohol and a magnesium salt. Notice that the reaction forms both organic and inorganic products. The net result, in terms of modifying the carbonyl containing compound, is the addition of an R group and an H group across the carbon-oxygen double bond. The general reaction of a carbonyl group with a Grignard reagent is an important reaction, and one that we will often utilize in the future.

$$CH_3CH_2\text{-}\underset{\underset{\displaystyle CH_3}{|}}{\overset{\overset{\displaystyle O^- MgBr^+}{|}}{C}}\text{-}H + NH_4Cl \longrightarrow CH_3CH_2\text{-}\underset{\underset{\displaystyle CH_3}{|}}{\overset{\overset{\displaystyle OH}{|}}{C}}\text{-}H + MgBrCl + NH_3$$ EQ. 5.48

Here is another example, wherein an aromatic group is added to the carbonyl carbon and an alcohol is formed.

$$\text{2-hexanone} + C_6H_5MgBr \longrightarrow \xrightarrow{+H^+} \text{2-phenyl-2-hexanol (OH)} + MgBr^+$$ EQ. 5.49

Again, an alkoxide anion (RO^-) is formed in the process leading to the alcohol. We note in passing that alkoxide ions can also be formed from the alcohol by reaction with a strong base like NaH, sodium hydride:

$$ROH + NaH \longrightarrow RO^- Na^+ + H_2$$ EQ. 5.50

APPLICATIONS OF CHEMICAL REACTIONS

5.5 CHEMICAL REACTIONS AS TOOLS FOR ANALYSIS

The study of chemical reactivity emerged from the shadow of alchemy and into the dawn of a new age in the late eighteenth century. Scientists began weighing the amounts of reactants used and products formed during reactions of the types described in this chapter. Their work led to the discovery of the quantitative nature of chemical reactions, and made possible the chemical analysis of compounds. Perhaps the greatest insights gained by use of these new tools were several empirical laws that are fundamental for our current understanding of the behavior of matter.

Among the persons active during this time were two Frenchmen, Antoine Lavoisier (1743–1794) and Joseph Louis Proust (1754–1826), and an Englishman by the name of John Dalton (1766–1844). Antoine Lavoisier was employed by Ferme Generale, a private firm that collected certain taxes for the government. This gave him access to the most accurate balances of that era. That equipment enabled him to pursue his scientific studies. Lavoisier conclusively demonstrated the **Law of Conservation of Mass** that states that matter is neither created nor destroyed during an ordinary chemical reaction. In one of his experiments he heated a mixture of liquid mercury and oxygen gas in an enclosed container and showed that the weight of the mercuric oxide formed was the same as the weight of the mercury and oxygen consumed during the reaction. Unfortunately for Lavoisier, his connection with the Ferme Generale caused him to be attacked by a number of radicals at the time of the Reign of Terror following the French Revolution. As a direct consequence he was executed by guillotine on May 8, 1794.

In 1799, shortly after the death of Lavoisier, Joseph Proust used decomposition reactions to break compounds into their constituent elements. By determining the mass of each element in comparison to the total mass of the compound, Proust determined the percent composition of many compounds. His careful measurements showed that nearly all compounds, regardless of their source, always contain the same percentage composition of elements. For example, the compound calcium carbonate, whether obtained from limestone, from seashells, or synthesized in the laboratory, always contains 40% calcium, 12% carbon, and 48% oxygen by weight. Proust therefore formulated his Law of Constant Composition (also called the **Law of Definite Proportions**) which states that compounds contain elements only in certain fixed proportions by weight regardless of the preparation method.

This work was rapidly followed by that of John Dalton. In 1803, Dalton reported his studies on several combination reactions. These studies showed that when two elements formed more than one compound (e.g., hydrogen and oxygen forming two different compounds known as water and hydrogen peroxide) they always do so in such a way that, for a fixed weight of one element in both compounds, the weights of the other element in the two compounds are in a small, whole number ratio. For example, 100.0 g of oxygen can react with 6.3 g of hydrogen to form hydrogen peroxide; the same 100.0 g of oxygen can react with 12.6 g of hydrogen to form water. For the fixed mass of 100.0 g of oxygen, the two possible masses of hydrogen (6.3 g and 12.6 g) are in a 1 to 2 ratio. This was formally stated as the **Law of Multiple Proportions**.

Continued work on the combining ratios of elements led Dalton to propose a series of atomic weights and finally to put forth an explanation of all the previous laws by proposing an atomic theory that states that elements can be subdivided repeatedly until there is a particle that cannot be further subdivided. This he termed the "atom", a term used by the Greek philosopher Democritus, as noted earlier in Section 1.2. For these insights, Dalton is often termed the "Father of Modern Chemistry."

Dalton's atomic theory had three major postulates:

1. All chemical elements are composed ot very small, indivisible, indestructible particles called atoms. These atoms are neither created nor destroyed during chemical reactions, merely rearranged.

2. All atoms of the same element are identical in mass and all other properties. Atoms of different elements differ from those of all other elements.

3. Chemical compounds are formed when atoms of different elements combine with one another in small, whole–number ratios.

While, as we will see, modern atomic theory differs from that put forth by Dalton in important ways, this insight of Dalton paved the way for approximately one century of work during which increasingly sophisticated measurements failed to show problems with the theory. It has only been since the discovery of radioactivity and subatomic particles by nuclear chemists and physicists that Dalton's first and second postulates have had to be revised to account for subatomic particles and isotopes of the same element with differing masses.

5.6 ATOMIC MASS AND THE MOLE CONCEPT

One of the important postulates of Dalton's atomic theory was that all atoms of a given element are identical in all properties, including weight. He was able, by additional further studies, to assign relative weights to each of the elements. For instance, it is possible to show that, in round numbers, 12 grams of carbon can combine with either 16 grams or 32 grams of oxygen to form two distinctly different compounds (an example of the Law of Multiple Proportions.) Dalton's reasoning for this was that these two compounds each contained one atom of carbon in the molecule with the first containing one atom of oxygen and the second containing two atoms of oxygen. With this information and much additional experimentation, Dalton was able to devise a table of relative atomic weights. The information presented above is powerful evidence that, whatever the actual weight of an oxygen atom, it must be 16/12 that of a carbon atom. Dalton was able to determine that the lightest atom was hydrogen and assigned 1 as its atomic weight. All other atoms were assigned weights relative to the atomic weight of the hydrogen atom.

Because, as we shall see, Dalton's theory was incorrect when it said that all atoms of the same element had the same weight, current atomic weights are the average weight of an atom in a naturally occurring collection of atoms of the element. By the currently accepted International Union of Pure and Applied Chemistry (IUPAC) convention, the current atomic weight scale was derived by assigning a value of exactly 12 to the ^{12}C isotope and measuring all other atomic weights relative to the weight of that isotope. The units of such a scale are called atomic mass units, or amu. The periodic table shows us that the atomic weight for carbon is listed as 12.011, that of hydrogen is 1.0079, that for fluorine as 18.9984, etc. These correspond to the mass, in amu, of the average atom of those elements.

While we can think of the atomic weight of an element as listed on the periodic table as the mass of one atom of the element, it can also be thought of as the weight, in grams, of a certain number of atoms of the element. This number, known as Avogadro's Number, after the Italian physicist and chemist Amadeo Avogadro (1776–1856), has a currently accepted value of 6.02205×10^{23}. Chemists use this number in much the same way that bakers use the dozen. That is, quantities of **reactants** and **products** in chemical reactions are measured in units of this number that is defined as the number of particles in the **mole** (abbreviated as mol). Since 1.0000 mol of carbon atoms weighs 12.011 grams and occupies a volume of slightly less than 6 cm^3, it must be obvious that a single carbon atom is extremely small — so small that it has not yet been possible to "see" one using even the most powerful microscopes yet designed. Of course, weighing a single atom of carbon (or any other element) is equally impossible using even the most sophisticated balances now available.

The fact that it is not truly possible to count the number of atoms in a mole of carbon atoms does not make that number less useful. It simply means that we must find other ways of measuring a mole of material other than by counting individual particles. Consider, for example, a situation in which we wish to determine the number of marbles in a paper bag filled with marbles all of which are the same size. We could do that if we knew the weight of a single marble and the total weight of the marbles in the bag. For instance, if a single marble weighs 1.35 grams and the total weight of marbles in the bag is 882.9 grams, then there must be:

$$\left(\frac{882.9 \text{ grams marbles}}{\text{bag}}\right)\left(\frac{1 \text{ marble}}{1.35 \text{ grams}}\right) = 654 \frac{\text{marbles}}{\text{bag}}$$ EQ. 5.51

Extending our analogy even further, if a certain number of marbles were defined as 1 elom of marbles, even if we did not know how many marbles were in an elom, we could still determine the number of eloms of marbles in the bag provided that we knew the weight of an elom of marbles. Using the present example, let us suppose that the weight of one elom of marbles was 22.95 grams. Therefore our bag must contain:

$$\left(\frac{882.9 \text{ grams marbles}}{\text{bag}}\right)\left(\frac{1 \text{ elom}}{22.95 \text{ grams}}\right) = 38.47 \frac{\text{eloms}}{\text{bag}}$$ EQ. 5.52

If we now picked up another bag of marbles of a different size and knew that an elom of these marbles weighed 26.69 grams, we could quickly determine that a weight of 1314.09 grams of marbles in this bag contained:

$$\left(\frac{1314.09 \text{ grams marbles}}{\text{bag}}\right)\left(\frac{1 \text{ elom}}{26.69 \text{ grams}}\right) = 49.24 \frac{\text{eloms}}{\text{bag}}$$ EQ. 5.53

Likewise it is possible to determine the number of moles of material provided that the total weight and the weight of one mole of the material are known. Let us look at several examples:

Example 1: A sample of pure iron weighs 23.45 grams. How many moles of iron does it contain?

$$\left(23.45 \text{ g Fe}\right)\left(\frac{1 \text{ mole}}{55.847 \text{ g Fe}}\right) = 0.4199 \text{ mole Fe}$$ EQ. 5.54

Example 2: How many atoms of silicon are contained in a sample of silicon weighing 4.57 g?

$$\left(4.57 \text{ g Si}\right)\left(\frac{1 \text{ mole}}{28.086 \text{ g Si}}\right)\left(\frac{6.022 \times 10^{23} \text{ atoms}}{1 \text{ mole}}\right) = 9.80 \times 10^{22} \text{ atoms Si}$$ EQ. 5.55

Example 3: What weight of palladium must be taken to obtain 0.156 moles of palladium?

$$\left(0.156 \text{ mole Pd}\right)\left(\frac{106.4 \text{ g Pd}}{1 \text{ mole Pd}}\right) = 16.6 \text{ g Pd}$$ EQ. 5.56

Since the concept of the mole as a unit of matter is so important in chemistry, it is extremely important that conversions between grams of material and moles of material become second nature and be easily accomplished.

5.7 CHEMICAL FORMULAS AND PERCENTAGE COMPOSITION

Using Dalton's atomic theory we understand that the formulas of chemical compounds indicate the atomic composition of that compound. For instance, the formula for carbon monoxide, CO, indicates that each molecule of carbon monoxide consists of one carbon atom bonded to one oxygen atom. In a similar fashion one molecule of carbon dioxide, CO_2, consists of one carbon atom bonded to two oxygen atoms. For ionic compounds, such as NaCl and $MgCl_2$, there are extended lattices, not discrete molecules. Formulas like NaCl and $MgCl_2$ convey the relative proportion of each ion in the lattice. While we can talk about individual molecules of carbon dioxide by the formula CO_2, we can not talk about an individual molecule of $MgCl_2$. Instead we refer to a **formula unit** of magnesium chloride, composed of one magnesium cation and two chloride anions. The symbolic representation of a **chemical formula** provides information about the number of each type of atom in a molecule, or the number of each type of ion in a formula unit. For salts and for simple molecules, this may be the only real function of the formula.

As we have seen in Chapter Three and Chapter Four, the formulas for more complicated molecules can be written in various ways. Some formulas convey more information about the structure of the molecule than others. For instance the formula for the molecule 2,4,6-trinitrotoluene, better known simply by its initials, TNT, could be written several different ways. The simplest of these formulas, written as $C_7H_5N_3O_6$, tells only the identity and number of atoms in the molecule. A more useful formula for some purposes is one that begins to indicate the manner in which the atoms are bonded together: $CH_3C_6H_2(NO_2)_3$. A yet higher degree of structural information can be provided by a diagrammatic representation of the formula, as seen at right in Figure 5.1:

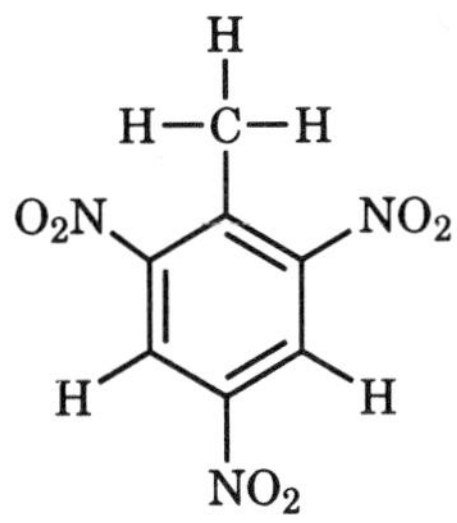

FIGURE 5.1: Structure of TNT.

If we know the chemical formula for a compound, then we can easily determine the mass corresponding to a mole of that compound, and the percentage by mass of each of its elemental constituents. For instance, 1.00 mole of the molecule TNT must contain 7.00 moles of C atoms, 5.00 moles of H atoms, 3.00 moles of N atoms, and 6.00 moles of O atoms. The total weight of 1.00 mole of TNT must then be:

$$\left(7.00 \text{ mole C}\right)\left(\frac{12.011 \text{ g C}}{1 \text{ mole C}}\right) = 84.\bar{0}77 \text{ g C} \quad \text{EQ. 5.57a}$$

$$\left(5.00 \text{ mole H}\right)\left(\frac{1.0079 \text{ g H}}{1 \text{ mole H}}\right) = 5.0\bar{3}95 \text{ g H} \quad \text{EQ. 5.57b}$$

$$\left(3.00 \text{ mole N}\right)\left(\frac{14.0067 \text{ g N}}{1 \text{ mole N}}\right) = 42.\bar{0}201 \text{ g N} \quad \text{EQ. 5.57c}$$

$$\left(6.00 \text{ mole O}\right)\left(\frac{15.9994 \text{ g O}}{1 \text{ mole O}}\right) = 95.\bar{9}96 \text{ g O} \quad \text{EQ. 5.57d}$$

$$84.\bar{0}77\text{g} + 5.0\bar{3}95\text{g} + 42.\bar{0}201\text{g} + 95.\bar{9}96\text{g} = 227.\bar{1}326 \text{ g} \quad \text{EQ. 5.57e}$$

The molecular weight of TNT would be 227.1 g/mol.

Since one mole of TNT weighs 227.1 g and contains 84.1 g of C, 5.04 g of H,

42.0 g of N, and 96.0 g of O, it is a simple matter to determine the mass percentage of each of these elements in TNT.

$$\left(\frac{84.\bar{0}7}{227.\bar{1}3}\right)(100) = 37.0\ \%\ \mathrm{C} \quad \text{EQ. 5.58a}$$

$$\left(\frac{5.0\bar{3}95}{227.\bar{1}3}\right)(100) = 2.22\ \%\ \mathrm{H} \quad \text{EQ. 5.58b}$$

$$\left(\frac{42.\bar{0}201}{227.\bar{1}3}\right)(100) = 18.5\ \%\ \mathrm{N} \quad \text{EQ. 5.58c}$$

$$\left(\frac{95.\bar{9}96}{227.\bar{1}3}\right)(100) = 42.3\ \%\ \mathrm{O} \quad \text{EQ. 5.58d}$$

Similar calculations will provide the percentage composition of ionic substances whose chemical formulas are known.

5.8 DETERMINING CHEMICAL FORMULAS

The percentage composition of a material can be empirically determined, that is, by experimental means. From the percentage by mass of each constituent element, a formula for the compound can be determined. Formulas for compounds determined from percentage composition are termed **empirical formulas**. Remember that the percentage composition of a material provides information about the relative weights of elements that make up the compound, but the chemical formula provides information about the relative number of atoms comprising the compound. To obtain an empirical formula from mass percentage, the mass of each element must be converted into moles. For example, consider a compound that is known to contain 40.0% C, 6.71% H, and 53.3% O. Since these are mass percentages, this means that, if we arbitrarily decomposed 100 grams of this compound into its constituent elements, we would obtain 40.0 g of carbon, 6.71 g of hydrogen, and 53.3 g of oxygen. If we are to obtain the chemical formula for this compound, we must convert these weight ratios into mole ratios. Hence, in 100 grams of this compound there are:

$$(40.0\ \mathrm{g\ C})\left(\frac{1\ \mathrm{mol}}{12.011\ \mathrm{g\ C}}\right) = 3.33\ \mathrm{mol\ C} \quad \text{EQ. 5.59a}$$

$$(6.71\ \mathrm{g\ H})\left(\frac{1\ \mathrm{mol}}{1.0079\ \mathrm{g\ H}}\right) = 6.66\ \mathrm{mol\ H} \quad \text{EQ. 5.59b}$$

$$(53.3\ \mathrm{g\ O})\left(\frac{1\ \mathrm{mol}}{15.9994\ \mathrm{g\ O}}\right) = 3.33\ \mathrm{mol\ O} \quad \text{EQ. 5.59c}$$

which tells us that the ratio of atoms in the compound is $C_{3.33}H_{6.66}O_{3.33}$. To convert this to a proper chemical formula that is a ratio of small whole numbers, it is necessary to divide each of the coefficients by the smallest:

$$C_{\frac{3.33}{3.33}}H_{\frac{6.66}{3.33}}O_{\frac{3.33}{3.33}} \text{ or } CH_2O \quad \text{EQ. 5.60}$$

If, after division by the smallest coefficient the result is not a series of whole numbers, it then becomes necessary to adjust coefficients by multiplying each by some number that will make them all whole. Consider a molecule that contains 83.2% C and 16.8% H.

$$\left(83.2\text{ g C}\right)\left(\frac{1\text{ mol C}}{12.011\text{ g C}}\right) = 6.93\text{ mol C}$$ EQ. 5.61a

$$\left(16.8\text{ g H}\right)\left(\frac{1\text{ mol H}}{1.0079\text{ g H}}\right) = 16.7\text{ mol H}$$ EQ. 5.61b

This would lead to a formula: $C_{\frac{6.93}{6.93}}H_{\frac{16.7}{6.93}} = C_1H_{2.41}$. The object now is to determine what value we must use to multiply both coefficients to have them become integers (within round–off error). A bit of serendipity or some trial and error may be required to do this. In the present situation, note that the .4 fraction on the hydrogen coefficient is 2/5. Hence we might first want to try the value 5. This will give us a formula $C_5H_{12.05}$. That is close enough to C_5H_{12} to stop our search. Sometimes it is difficult to know whether something is close enough to an integer to consider the process finished. The only true way to solve this problem is to have an estimate of the uncertainty in the percentage compositions. In many situations such problems can be solved by an independent determination of other properties of the compound, such as its molecular weight.

CONCEPT CHECK: DETERMINING EMPIRICAL FORMULAS

In his study of the effect of electric fields on living cells, Dr. Barnett Rosenberg and his coworkers noted decreased growth and cell division in a bacteria culture passed between normally inert platinum electrodes. The effect was traced to the formation of a compound, cisplatin, in the medium. Cisplatin has since proven to be very effective in combating certain types of cancer. Analysis of cisplatin shows that it is 23.63% chlorine, 2.02% hydrogen, 9.33% nitrogen, and 65.02% platinum by mass. What is the empirical formula of cisplatin?

SOLUTION: It is simplest to assume 100.0 grams of compound, then to relate the % by mass to actual weight in grams. The mass of each element in grams in then converted to the number of moles of that element in 100.0 g of compound.

$$23.63\text{ g Cl}\left(\frac{1\text{ mole Cl}}{35.453\text{ g}}\right) = 0.6665\text{ mole Cl}$$

$$2.02\text{ g H}\left(\frac{1\text{ mole H}}{1.0079\text{ g}}\right) = 2.00\text{ mole H}$$

$$9.33\text{ g N}\left(\frac{1\text{ mole N}}{14.007\text{ g}}\right) = 0.666\text{ mole N}$$

$$65.02\text{ g Pt}\left(\frac{1\text{ mole Pt}}{195.08\text{ g}}\right) = 0.3333\text{ mole Pt}$$

A relative ratio of moles is obtained by dividing the moles of each element by the term with the smallest number of moles

$$\frac{0.6665\text{ mole Cl}}{0.3333\text{ moles}} = 2.000\text{ mole Cl}$$

$$\frac{2.00\text{ mole H}}{0.3333} = 6.00\text{ mole H}$$

$$\frac{0.666 \text{ mole N}}{0.3333} = 2.00 \text{ mole N}$$

$$\frac{0.3333 \text{ mole Pt}}{0.3333} = 1.000 \text{ mole Pt}$$

The empirical formula of cisplatin is: $PtN_2H_6Cl_2$.

As indicated earlier, formulas determined from percentage composition are termed empirical formulas. The actual **molecular formula** may equal the empirical formula, or it may equal an integral multiple of the empirical formula. To see that this is true, note that the real compounds whose molecular formulas are C_2H_4, C_3H_6, C_4H_8, etc., all have percentage compositions of 85.6% C and 14.4% H. This would lead to an empirical formula CH_2. To be certain that the formula derived from percentage composition data is a molecular formula, it is necessary to have information about the molecular weight of the compound. Consider a compound containing 26.7% C, 2.24% H, and 71.1% O whose molecular weight is known to be approximately 90 g/mol.

$$(26.7 \text{ g C})\left(\frac{1 \text{ mol C}}{12.011 \text{ g C}}\right) = 2.22 \text{ mol C}$$ EQ. 5.62a

$$(2.24 \text{ g H})\left(\frac{1 \text{ mol H}}{1.0079 \text{ g H}}\right) = 2.22 \text{ mol H}$$ EQ. 5.62b

$$(71.1 \text{ g O})\left(\frac{1 \text{ mol}}{15.9994 \text{ g O}}\right) = 4.44 \text{ mol O}$$ EQ. 5.62c

This leads to an empirical formula CHO_2. If we compute the "molecular weight" of a compound whose formula is CHO_2, we find that it is 45 g/mol — exactly half of the 90 g/mol stated in the problem. Therefore the molecular formula of this compound must be $C_2H_2O_4$.

CONCEPT CHECK: MOLECULAR FORMULA TO MOLECULAR STRUCTURE

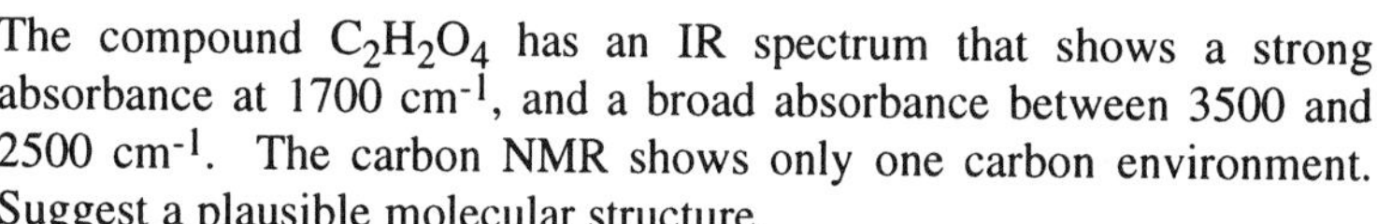

The compound $C_2H_2O_4$ has an IR spectrum that shows a strong absorbance at 1700 cm^{-1}, and a broad absorbance between 3500 and 2500 cm^{-1}. The carbon NMR shows only one carbon environment. Suggest a plausible molecular structure.

SOLUTION: The IR spectrum strongly suggests the presence of a caboxylic acid functional group, since both a carbonyl stretch and an OH stretch typical of acids are present. Since the two carbons are equivalent, one suspects a symmetric molecule of the form HOOC-COOH. (This compound is a dicarboxylic acid called oxalic acid)

5.9 CONTROLLING CHEMICAL REACTIONS: A PREVIEW OF EQUILIBRIUM

Throughout this chapter we have seen many chemical reactions. Some of these reactions were designed to synthesize a desired product from readily available reactants. Those are processes we wish to maximize. Chemistry is concerned not only with identifying reactions that can occur, but in controlling the extent of reactions.

Methods of controlling chemical reactions vary almost as widely as the reactions themselves. But again, a few generalizations are helpful. Every chemical reaction goes at least part way, and only part way. There are no 0% or 100% complete reactions. Instead, we end up with some mixture of reactants and products that are dynamically balanced. **Equilibrium** is reached when the rate at which reactants are forming products matches the rate at which the products are reforming reactants. Consider the generalized reaction:

$$A + B \rightleftharpoons C + D$$ EQ. 5.63

Here, the reaction has two arrows, indicating that both the forward reaction:

$$A + B \longrightarrow C + D$$ EQ. 5.63a

and the reverse reaction:

$$C + D \longrightarrow A + B$$ EQ. 5.63b

are occuring.

If C or D in Eq. 5.63 is a desired product, we wish to maximize the forward reaction, Eq. 5.63 (a), and minimize the reverse reaction, Eq. 5.63 (b). This can be done in several ways. The concentration of A and/or B can be increased, which increases the rate of Eq. 5.63 (a), and results in a higher concentration of C and D present at equilibrium. Another option is to remove C or D as it is formed. If Eq. 5.63 (a) continues while Eq. 5.63 (b) does not, reactants will be converted to products very efficiently.

Consider the aqueous phase reaction seen earlier as Eq. 5.25:

$$Na^+ + Cl^- + Ag^+ + NO_3^- \longrightarrow$$

According to the solubility rules, we predicted that sodium nitrate stays in solution, but that silver chloride would precipitate. By falling out of solution, silver chloride is removing itself from the reaction mixture. This removal of product is what drives the reaction forward. Similarly, we can increase the yield of solid silver chloride formed by increasing the concentration of sodium chloride or silver nitrate used.

When solid or liquid reactant(s) form a gaseous product which is not contained in a closed vessel, the gas escapes. Again, the removal of one product can drive a reaction towards completion. The conversion of calcium sulfate into calcium oxide seen earlier as Eq. 5.16 is an example of this:

$$CaSO_4(s) \longrightarrow CaO(s) + SO_3(g)$$

As sulfur trioxide escapes, the forward reaction continues, but the reverse reaction is greatly diminished due to the small amount of SO_3 present near the calcium oxide.

The reaction in Eq. 5.16 is also heated to promote the formation of products. If heat were to be considered part of this equation, we would show it as a reactant:

$$CaSO_{4(s)} + \text{heat} \longrightarrow CaO(s) + SO_{3(g)}$$

Running the reaction at an elevated temperature is roughly equivalent to increasing the concentration of a reactant. This increases both the rate at which product is formed, and the amount of product formed.

Heat can also be a product of a reaction, as is the case when hydrogen adds to ethene to form ethane as we saw in Eq. 5.8.

$$H_2C{=}CH_2 + H_2 \xrightarrow{\text{catalyst}} H_3C{-}CH_3 + \text{heat}$$

In this case, removing a product (heat) will ultimately result in a greater yield of the product ethane. Unfortunately, cooling a reaction makes it occur more slowly. Thus, while cooling a reaction that liberates heat increases the amount of product formed, it also greatly increases the time required to obtain the product. That is a problem. The reaction indicated in Eq. 5.8 is done using a catalyst, a compound that increases the rate of the reaction, to compensate for this.

To summarize, the equilibrium position of a reaction can be shifted either towards products or towards reactants. Adding reactants shifts the equilibrium towards more products. Removing a product shifts the reaction towards more products. Similarly, adding products, or removing reactants will shift the equilibrium towards the reactant side. Temperature can also be used as a controlling feature, since heat is nearly always a "reactant" or a "product". These rules were neatly summarized in 1884 by Henry-Louis Le Chatelier (1850 - 1936). **Le Chatelier's principle** states that an equilibrium responds to an applied stress in a way that minimizes that stress. Adding extra reactants to a system (a stress) shifts the reaction to make more products (using up reactants and thus minimizing the stress.) Removing a product (a stress) shifts the reaction to make more products (acting to minimize the stress.) Throughout this course we will examine issues related to chemical equilibrium, and Le Chatelier's principle will be our most important guide.

Catalyst: a substance that increases the rate of a chemical reaction, but is neither consumed nor produced during the course of the reaction.

Chemical formula: a means of indicating the ratios of the numbers of atoms in a molecule or formula unit of a compound.

Chemical reaction: a transformation of one or more species into different species

Combination reactions: those chemical reactions in which there is only one product

Decomposition reactions: those reactions in which one reactant produces more than one product

Dehydration: a reaction in which a water molecule is removed from a compound

Dehydrohalogenation: a reaction in which a hydrogen halide molecule is removed from a compound

Empirical formula: the simplest ratio of whole numbers of atoms within a compound.

Equilibrium: a dynamic balance between the forward reaction that makes products and the reverse reaction that reforms reactants

Formula unit: the simplest ratio of whole numbers of ions present in an ionic compound.

Formula weight: the sum of the weights (in amu) of all atoms appearing in one formula unit of an ionic compound. This is numerically equal to the grams of compound present per mole of formula units.

Geminal: describes substituents bonded to the same atom

Hydrogenation: a reaction in which molecular hydrogen is added to an unsaturated compound.

Hydrolysis: reactions in which water is used to react with another compound thereby breaking the bonds in the water molecule.

Law of Conservation of Mass: matter is neither created nor destroyed in an ordinary chemical reaction.

Law of Constant Composition: compounds contain elements only in certain definite proportions by weight.

Law of Multiple Proportions: the ratios of the weights of an element combining with a given weight of another element to form two different compounds are always small, whole numbers.

Le Chatelier's Principle: a system at equilibrium responds to a stress in a fashion that acts to counteract the stress

Markovnikov's rule: a rule that recognizes that when hydrogen containing binary compounds add to unsymmetrical alkenes, the hydrogen atom adds to the less substituted carbon

Metathesis reactions: those reactions in which two ionic materials react to form two new materials

Mole: the chemist's basic unit of amount — one mole of anything contains the same number of particles as one mole of anything else.

Molecular formula: see chemical formula

Molecular weight: the sum of the weights (in amu) of all atoms appearing in one molecule. This is numerically equal to the grams of compound present per mole of molecules of compound.

Product: one of the ending materials in a chemical reaction.

Pyrolysis: a decomposition reaction caused by heat, typically done in the absence of oxygen.

Reactant: one of the starting materials in a chemical reaction.

Redox reaction: a contraction of Reduction–Oxidation reaction: a reaction in which oxidation and reduction occur

Saytzeff's rule: a rule that recognized that elimination reactions proceed to form, as the major product, the alkene that contains the greatest number of alkyl groups

Spectator ion: an ion in solution that is chemically unchanged during a reaction; such ions are cancelled in determining the net ionic equation

Stoichiometry: the relative amounts (in moles, molecules or formula units) of reactants and products in a chemical reaction.

Substitution reactions: those reactions in which one group is substituted for another in an organic compound

Vicinal: describes substituents bonded to adjacent atoms

HOMEWORK Problems

PREDICTING THE PRODUCTS OF REACTIONS

1. Write chemical equations for the combination reactions that occur between the following reactants

(a) $Li(s) + Br_2(l) \longrightarrow$

(b) 2-butene + $H_2 \xrightarrow{\text{catalyst}}$

(c) $MgO(s) + CO_2(g) \longrightarrow$

(d) cyclohexene + $Br_2 \longrightarrow$

(e) 2-hexene + $H_2O \longrightarrow$

(f) 1-methylcyclopentene + HBr $\longrightarrow$

2. Write chemical equations for each of the following decomposition reactions:

(a) $BaCO_3(s) \xrightarrow{\Delta}$

(b) 2-bromocyclohexanone + KOH $\longrightarrow$

(c) $Ba(OH)_2(s) \xrightarrow{\Delta}$

(d) 2-methyl-2-butanol $\xrightarrow[\Delta]{H+}$

(e) KCl(l) $\xrightarrow{\text{electricity}}$

3. Write chemical equations to indicate the products of the following double replacement reactions (write net ionic equations where appropriate):

(a) $BaCl_2(aq) + K_3PO_4(aq) \longrightarrow$

(b) $NaOH(aq) + CH_3CH_2COOH \longrightarrow$

(c) Ethanol + HCl $\longrightarrow$

(d) 2-bromo-2-methylpentane + $H_2O \longrightarrow$

(e) Ethyl acetate + $H_2O \longrightarrow$

(f) $Ba(OH)_2(s) + H_2SO_4(aq) \longrightarrow$

(g) Acetic Acid + $CH_3OH \xrightarrow{H+}$

(h) $FePO_4(s) + HCl(aq) \longrightarrow$

(i) Benzoic Acid + $NH_3 \xrightarrow[\text{agent}]{\text{activating}} \xrightarrow{H+}$

(j) Hexanal + $CH_3MgBr \longrightarrow$

(k) Cyclohexanone + C_6H_5MgBr (phenylmagnesium bromide) $\rightarrow$ $\xrightarrow{H^+}$

4. Predict the products of the following by writing net ionic equations. If no reaction occurs, write N.R. Unless otherwise indicated, all reactions occur in aqueous solution with soluble reactants.

(a) $CuSO_4(aq) + H_2S(aq) \longrightarrow$

(b) $SrCO_3(s) \xrightarrow{\Delta}$

(c) $P_4(s) + O_2(g) \longrightarrow$

(d) $BaCO_3(s) + HCl(aq) \longrightarrow$

(e) $Na(s) + I_2(s) \longrightarrow$
(f) $HNO_3(aq) + Ca(C_2H_3O_2)_2(aq) \longrightarrow$
(g) $Cr(s) + Cl_2(g) \longrightarrow$
(h) $Pb(NO_3)_2(aq) + KBr(aq) \longrightarrow$
(i) $AgI(s) + HNO_3(aq) \longrightarrow$
(j) $CaO(s) + SO_3(g) \longrightarrow$

(k) [structure: benzene ring bearing OH and C(=O)OCH_2CH_3] $\xrightarrow{H_2O}$

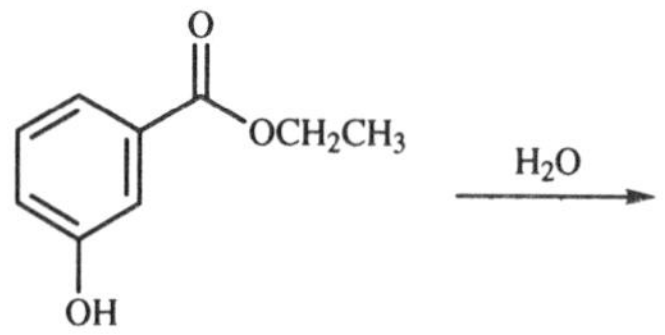

(l) [structure: HO–CH_2CH_2–CH(COOH)–NH–C(=O)–CH(NH_2)–CH_2–OH] $\xrightarrow{H_2O}$

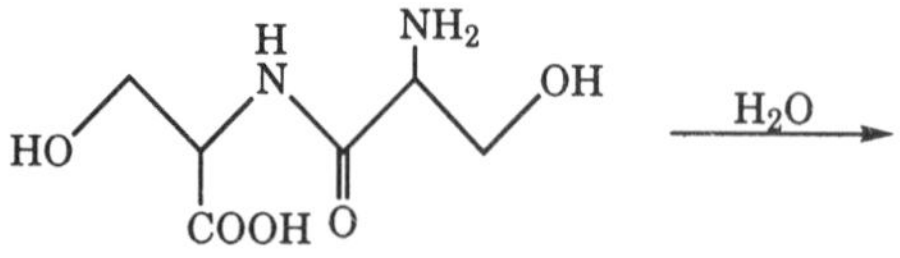

Mass, Moles, and % Composition

5. Calculate the number of moles of each substance contained in:

(a) 28.95 g of sodium metal
(b) 6.547 mg of uranium
(c) 12.89 kg of iron
(d) 5.87 g of nickel

6. How many atoms of each material are present in question 5?

7. Calculate the formula weight (g/mol) of:

(a) uranium hexafluoride (UF_6)
(b) sucrose ($C_{12}H_{22}O_{11}$)
(c) potassium ferricyanide [$K_3Fe(CN)_6$]
(d) acetone (CH_3COCH_3)

8. Calculate the percentage composition of each of the compounds in problem 7.

9. A hydrocarbon was found to contain 85.6% C and 14.4% H.

(a) Determine its empirical formula.
(b) Subsequent determination showed the molecular weight of the com pound to be approximately 57 g/mol. What is the molecular formula of the compound?

10. A compound, known to contain only C, H, and Cl was found to contain 37.23 %C, 7.81%H, and 54.94 % Cl. Determine its empirical formula.

11. A compound, known to contain no elements except carbon, hydrogen, and, perhaps oxygen, was found to contain 68.63% C and 8.64% H. Determine the empirical formula of the compound.

LeChatelier's Principle

12. Specify whether the equilibrium reaction indicated below is shifted towards reactants, towards products, or is not shifted in response to each of the following stresses:

$$\text{heat} + MnO_2(s) + 2\,Cl^- + 4\,H^+ \rightleftharpoons Mn^{2+} + Cl_2(g) + 2\,H_2O(l)$$

(a) adding hydrochloric acid
(b) using an open vessel, allowing the chlorine gas to escape
(c) adding $Mn(NO_3)_2$
(d) adding nitric acid
(e) adding $AgNO_3$
(f) increasing the temperature of the reaction

13. Specify whether the equilibrium reaction indicated below is shifted towards reactants, towards products, or is not shifted in response to each of the following stresses:

$$(CH_3)_3COH + HCl_{(aq)} \rightleftharpoons (CH_3)_3CHCl + H_2O_{(l)} + \text{heat}$$

(a) increasing the concentration of hydrochloric acid used
(b) separating and removing the water as it is formed
(c) increasing the temperature of the reaction
(d) adding $AgNO_3$

CHAPTER FIVE

OVERVIEW Problems

14. Carefully controlled fractional distillation of a mixture of alkanes yields three compounds with the same formula of C_5H_{12}, boiling at 9.5, 27.8 and 36.1 degrees C. Draw three different structures that have the formula C_5H_{12}.

15. How many unique environments for carbon atoms are there in each structure shown in your answer to #14? (i.e., how many signals does one expect in the uncoupled CNMR spectrum of each compound?)

16. When 1-butene (C_4H_8) reacts with HBr, the major product isolated (formula C_4H_9Br) has a HNMR spectrum that shows 4 unique proton environments. When integrated, these signals correspond to 1, 2, 3 and 3 protons respectively. Draw the structure of the major product, and show the 4 unique proton environments by labeling each type of proton in your structure.

17. When 1-butene (C_4H_8) reacts with HBr, the minor product isolated (formula C_4H_9Br) has a HNMR spectrum that shows 4 unique proton environments. When integrated, these signals correspond to 2, 2, 2 and 3 protons respectively. Draw the structure of the minor product, and show the 4 unique proton environments by labeling each type of proton in your structure.

18. When 2-methyl-1-propene reacts with water under certain conditions, the major product is $C_4H_{10}O$. The IR of this compound has a broad IR absorbance centered around 3350 cm^{-1}.

(a) What is the IHD of the product compound?
(b) What functional group is indicated by the IR spectrum?

19. The product of the reaction described in problem #18 has just two NMR signals: a singlet (corresponding to 1 proton) at 1.7 ppm, and a singlet (corresponding to 9 protons) at 1.3 ppm. Draw the structure of this product, and show the 2 unique proton environments by labeling each type of proton in your structure.

20. When bromocyclohexane is reacted with KOH, the organic product is isolated and examined by IR spectroscopy. The reactant (bromocyclohexane) had no IR absorbance above 3000 cm^{-1}, but the product (formula C_6H_{10}) shows a sharp peak at 3030 cm^{-1}.

(a) What is the IHD of the product C_6H_{10}?
(b) The sharp peak at 3030 cm^{-1} is attributable to which functional group?
(c) Draw the structure of the organic product of this reaction.

21. Based on your structure for the organic product formed by reacting bromocyclohexane with KOH, how many unique carbon environments are present in the product? (i.e., how many peaks are expected in the uncoupled CNMR spectrum?)

CHEMICAL Reactions (II)

Stoichiometry and Redox Reactions

CHAPTER SIX

Chemical Reactions (II)

Stoichiometry and Redox Reactions

In chapter five, we began our study of chemical reactivity by examining combination, decomposition, and metathesis reactions. We also saw how these reactions helped to establish the fundamentals of stoichiometry and ultimately led to Dalton's atomic theory.

In chapter six, we expand our examination of chemical stoichiometry to include reacting ratios for reactants and products. This leads to the concept of limiting reagents and to theoretical and actual yields. Next, we return to our study of chemical reactivity by examining the fourth class of chemical reactions: oxidation-reduction (redox) reactions. Since these reactions often are not readily balanced by simple inspection, we also learn a special procedure to follow in order to balance redox reactions. Then we conclude this chapter with several important applications of oxidation-reduction reactions.

6.2 CHEMICAL REACTION STOICHIOMETRY

In our study of stoichiometry last chapter, we saw how chemical reactions, especially combination and decomposition reactions, facilitated our determination of the composition of compounds. Chemical reactions also show relationships between compounds, either as two reactants, two products, or one reactant and one product. Consider the metathesis reaction we saw last chapter:

$$3\,CaCl_2 + 2\,K_3PO_4 \longrightarrow Ca_3(PO_4)_2 + 6\,KCl$$ EQ. 6.1

This chemical equation describes the reaction between calcium chloride and potassium phosphate to form calcium phosphate and potassium chloride. Specifically this equation tells us that three formula units (not three molecules, since these are ionic compounds) of calcium chloride react with two formula units of potassium phosphate to form one formula unit of calcium phosphate and six formula units of potassium chloride. Alternatively the equation also tells us that three moles of calcium chloride reacts with two moles of potassium phosphate to form one mole of calcium phosphate and six moles of potassium chloride. From this basic understanding, it is relatively easy to determine the answers to many questions involving the quantitative relationships among reactants and products.

Let us use this equation to answer the question: "How much calcium phosphate can be prepared from 5.00 grams of calcium chloride reacting with an excess of potassium phosphate?" To answer this question, we must know the **formula weights** of the materials and the balanced equation describing the reaction.

$$\left(5.00\text{g } CaCl_2\right)\left(\frac{1\text{ mol } CaCl_2}{110.99\text{ g } CaCl_2}\right)\left(\frac{1\text{ mol } Ca_3(PO_4)_2}{3\text{ mol } CaCl_2}\right)\left(\frac{310.18\text{ g } Ca_3(PO_4)_2}{1\text{ mol } Ca_3(PO_4)_2}\right)$$ EQ. 6.2

$$= 4.66\text{ g } Ca_3(PO_4)_2$$

The reacting ratio, $\left(\frac{1\text{ mol } Ca_3(PO_4)_2}{3\text{ mol } CaCl_2}\right)$, in the calculation is obtained by an inspection of the balanced chemical equation.

This process is easy to adapt to multi–step reactions. For example, consider the problem: "How much X can be made starting with Y g of Z if all other materials are present in excess?" An example of this type of problem is:

The total commercial synthesis of sulfuric acid (H_2SO_4) from sulfur (S_8) occurs by a series of reactions outlined below. Assuming complete conversion of all sulfur to sulfuric acid, what weight of sulfur must be mined per day to supply a plant with an ability to produce 50.0 metric tons (1 metric ton = 1000 kg) of sulfuric acid per day?

$$S_8 + O_2 \longrightarrow 8\,SO_2$$ EQ. 6.3a

$$2SO_2 + O_2 \longrightarrow 2\,SO_3$$ EQ. 6.3b

$$2\,SO_3 + H_2O \longrightarrow H_2S_2O_7$$ EQ. 6.3c

$$H_2S_2O_7 + H_2O \longrightarrow 2H_2SO_4$$ EQ. 6.3d

Note that this is, essentially, the same problem as the previous one. The only difference is the use of a series of reacting ratios, one for each chemical equation in the synthetic pathway. Also notice that the entire calculation can be set up at once simply by stringing together the stoichiometric factors obtained from each of the chemical equations in turn.

$$\left(\frac{50.0 \text{ tons } H_2SO_4}{\text{day}}\right)\left(\frac{1000 \text{ kg}}{\text{ton}}\right)\left(\frac{1000 \text{ g}}{\text{kg}}\right)\left(\frac{1 \text{ mol } H_2SO_4}{98.08 \text{ g } H_2SO_4}\right)\left(\frac{1 \text{ mol } H_2S_2O_7}{2 \text{ mol } H_2SO_4}\right)$$
$$\left(\frac{2 \text{ mol } SO_3}{\text{mol } H_2S_2O_7}\right)\left(\frac{2 \text{ mol } SO_2}{2 \text{ mol } SO_3}\right)\left(\frac{1 \text{ mol } S_8}{8 \text{ mol } SO_2}\right)\left(\frac{256.512 \text{ g } S_8}{\text{mol } S_8}\right)\left(\frac{1 \text{ kg}}{1000 \text{ g}}\right)\left(\frac{1 \text{ ton}}{1000 \text{ kg}}\right)$$
$$= \frac{16.3 \text{ metric tons } S_8}{\text{day}}$$ EQ. 6.4

In each of the previous examples the statement has been made "if all other materials are present in excess." For the first result, calculated in Eq 6.2, this means that there were more than 2/3 times as many moles of potassium phosphate present as there were moles of calcium chloride. In other words, the amount of product was limited by the amount of calcium chloride present at the start of the reaction. In such a situation calcium chloride is termed the limiting reagent and such was stated at the outset of the problem. What does one do if no such statement is made?

The situation is similar to asking how many automobiles can be made if one starts with 44 tires, 10 bodies, 12 steering wheels, etc. If we wished we could write an assembly "equation" that stated:

$$4 \text{ tires} + 1 \text{ body} + 1 \text{ steering wheel} \longrightarrow 1 \text{ automobile}$$ EQ. 6.5

There are several ways of attacking this type of problem. One could, for example, just pick one of the "reactants", and calculate the equivalent amount of another reactant. For example,

$$(44 \text{ tires})\left(\frac{1 \text{ body}}{4 \text{ tires}}\right) = 11 \text{ bodies}$$

The "reacting ratio" of 1 body to 4 tires is obtained from the balanced assembly equation. There are sufficient tires to react with 11 bodies, but that there are only 10 bodies present. This indicates that, when comparing just tires and bodies, tires are in excess, and bodies are limiting. We then switch our attention to automobile bodies, and their relationship to the other reactant, steering wheels:

$$(10 \text{ bodies})\left(\frac{1 \text{ steering wheel}}{1 \text{ body}}\right) = 10 \text{ steering wheels}$$

Again, the reacting ratio of 1 body to 1 steering wheel is given by the assembly equation. Since 10 bodies require 10 steering wheels, but 12 steering wheels are present, we say that steering wheels are in excess, and bodies are limiting. Thus, the final answer would be 10 automobiles since there are only 10 bodies present, and more than enough tires and steering wheels to make 10 automobiles. To solve this kind of problem, one must not only prove how much product can be made from a reactant, one must also prove that the reactant chosen is in fact the limiting reactant. One can readily recognize a limiting reagent problem by noting if information is provided about the amount present of more than one reactant.

The process outlined above is one way to solve limiting reagent problems. The authors recommend another approach to this kind of problem: the limiting reagent table method. The table method has several advantages. First, it organizes what can otherwise be a "hit or miss" approach to solving the problem. Second, the tabulated values allow one to answer directly any of a variety of questions, including how much of each reactant remains at the end of the reaction. Finally, the general approach of the table method will be used later in more complicated equilibrium problems; mastering the table now will have a long term benefit as well.

Consider the following limiting reagent problem: How many grams of calcium phosphate may be made from the reaction of 5.00 g of $CaCl_2$ and 5.00 g of K_3PO_4? Whatever method of solving limiting reagents is used, the first step is to write the balanced equation for the reaction:

$$3\,CaCl_2 + 2\,K_3PO_4 \longrightarrow Ca_3(PO_4)_2 + 6\,KCl$$ EQ 6.1

As was demonstrated earlier with car parts, one could attack the problem by guessing which of the reactants was limiting, then determining what equivalent amount of the other would react. This time, let us use the table method.

Across the top of the table we create a column for each reactant and each product shown in the equation.

TABLE 6.1: LIMITING REAGENT TABLE FOR EQ. 6.1 TABLE 6.1

	$3CaCl_2$	$2K_3PO_4$	$Ca_3(PO_4)_2$	$6KCl$
moles present	$\frac{5.00}{110.99} = 0.045\bar{0}5$	$\frac{5.00}{212.27} = 0.023\bar{5}5$	---	---
moles reacted	$3x$	$2x$	---	---
moles formed	---	---	x	$6x$
after reaction	$0.045\bar{0}5 - 3x$	$0.023\bar{5}5 - 2x$	x	$6x$
values of x	$0.045\bar{0}5 - 3x = 0$ $x = 0.015\bar{0}2$	$0.023\bar{5}5 - 2x = 0$ $x = 0.011\bar{7}8$ *		
moles left	0.00971	0	$0.011\bar{7}8$	$0.070\bar{6}6$

* Limiting Reagent

In the second row we place the number of moles of each of the substances given in the problem:

$$\text{moles } CaCl_2 = (5.00\text{ g } CaCl_2)\left(\frac{1\text{ mol } CaCl_2}{110.99\text{ g } CaCl_2}\right) = 0.045\bar{0}5$$ EQ 6.6

$$\text{moles } K_3PO_4 = (5.00\text{ g } K_3PO_4)\left(\frac{1\text{ mol } K_3PO_4}{212.27\text{ g } K_3PO_4}\right) = 0.023\bar{5}5$$ EQ 6.7

The third row contains the number of moles of each that react. Since this number is unknown we assign it a value of x times the coefficient for that substance in the balanced equation. The fourth row contains the number of moles of each of the products formed in terms of the same x value. In the fifth row we place the expression for the number of moles of material after reaction. This will be equal to the quantity present at the beginning minus what reacts plus what forms in the reaction. The next row contains possible values of x. This is computed by realizing that, if a reactant is the limiting reagent, none of it will remain after a complete reaction. Therefore the expression in the fifth row is set equal to zero and the equation solved for x. The limiting reagent is the one whose x value is the smallest and has been indicated by the asterisk in the table above. The values in the last line are computed by using the value of x determined for the limiting reagent in the expression in the fifth row. The last line in the table will be the actual number of moles of each material present after reaction. The answer to the question is determined from the value of moles left for calcium phosphate:

$$\left(0.011\bar{7}8 \text{ mol } Ca_3(PO_4)_2\right)\left(\frac{310.18 \text{ g } Ca_3(PO_4)_2}{1 \text{ mol } Ca_3(PO_4)_2}\right) = 3.65 \text{ g} \qquad \text{EQ. 6.8}$$

CONCEPT CHECK: LIMITING REAGENTS

Elemental phosphorus is produced on a large scale by heating calcium phosphate (a common mineral) with sand and coke. The net balanced reaction is shown below. Calculate the moles of all species present assuming a complete reaction when 50.0 g of $Ca_3(PO_4)_2$, 25.0 g of SiO_2, and 10.0 g of C are reacted.

$$2Ca_3(PO_4)_2 + 6\,SiO_2 + 10\,C \longrightarrow P_4 + 6\,CaSiO_3 + 10\,CO$$

SOLUTION: First one determines the moles of each reactant, and enters those values into the limiting reagent table:

$$\left(50.0 \text{ g } Ca_3(PO_4)_2\right)\left(\frac{1 \text{ mol}}{310.20 \text{ g}}\right) = 0.16\bar{1}18 \text{ mol}$$

$$\left(25.0 \text{ g } SiO_2\right)\left(\frac{1 \text{ mol}}{60.085 \text{ g}}\right) = 0.41\bar{6}07 \text{ mol}$$

$$\left(10.0 \text{ g C}\right)\left(\frac{1 \text{ mol}}{12.011 \text{ g}}\right) = 0.83\bar{2}57 \text{ mol}$$

	$2Ca_3(PO_4)_2$ +	6 SiO_2 +	10 C	P_4 +	6 $CaSiO_3$ +	10 CO
moles						
initial	0.1612	0.4161	0.8326	0	0	0
react	2 x	6 x	10 x	-	-	-
form				x	6 x	10 x
left?	0.1612-2x = 0	0.4161-6x = 0	0.8326-10x = 0	-	-	-
x = ?	0.0806	0.06935*	0.08326			
left	0.1612 - 2(.06935) = 0.0225	0.4161 - 6(.06935) = 0	0.8326 - 10(.06935) = 0.1391	0.06935 0.06935	6(.06935) 0.4161	10(.06935) 0.6935

* Indicates limiting reagent (SiO_2)

The moles left are: calcium phosphate, 0.022 moles; silicon oxide, 0; carbon, 0.139 moles; phosphorus, 0.0694 moles; calcium silicate, 0.416 moles; and carbon monoxide, 0.694 moles.

6.3 PERCENT YIELD

The above calculations allow the computation of the amount of material that should be obtained when a chemical reaction occurs. For many reactions, especially those involving simple ionic rearrangements, these calculations are extremely good predictors of the quantity of products formed. Other reactions, and in particular more complicated organic reactions, can give a variety of products because of a variety of side reactions that are possible. For these reactions we are able to calculate only a **theoretical yield** of product using the methods above. The **actual yield** from a reaction is something that must be measured. The ratio of the two, expressed as a percentage, is termed the **percentage yield**.

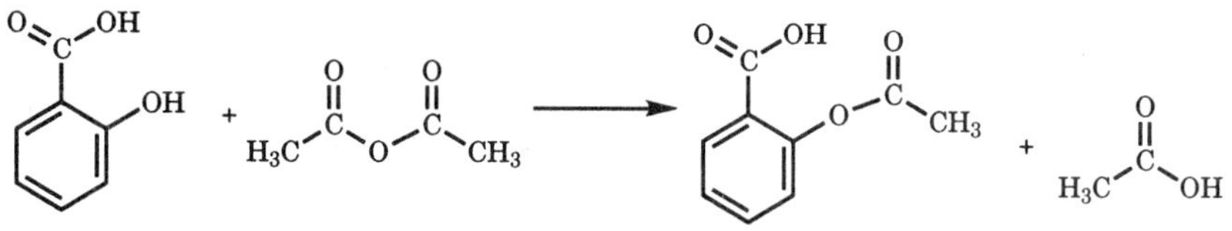

SA AS

EQ. 6.9

For instance, consider the synthesis of aspirin (AS) from salicylic acid (SA) and acetic anhydride. This reaction is often performed in beginning chemistry laboratory courses. A student isolated 1.43 g of aspirin after starting with 1.50 g of salicylic acid and excess acetic anhydride. Let us calculate the theoretical and **percent yields** for this reaction.

Theoretical Yield =

$$\left(1.50\text{ g SA}\right)\left(\frac{1\text{ mol SA}}{138.12\text{ g SA}}\right)\left(\frac{1\text{ mol AS}}{1\text{ mol SA}}\right)\left(\frac{180.17\text{ g AS}}{1\text{ mol AS}}\right) = 1.96\text{ g AS}$$

EQ. 6.10

Percent Yield =

$$\left(\frac{1.43\text{ g actual}}{1.96\text{ g theoretical}}\right)\left(100\%\right) = 73.0\%$$

EQ. 6.11

SIDEBAR: TYPES OF STOICHIOMETRIC PROBLEMS

We have now seen three common types of numerical problems that involve chemical reactions. Let us review these three types:

The first type is a simple stoichiometric problem, wherein information is given about one species in a chemical reaction, and information is requested about another. Example: "How many grams of HBr can be synthesized from 20.0 g of H_2 and excess Br2?" The recommended approach is a simple conversion of units from grams of hydrogen to moles of hydrogen to moles of bromine to grams of bromine.

The second type is a limiting reagent problem, wherein information is given about more than one reactant, and information is requested about reactants and/or products. Example: "How many grams of HBr can be synthesized from 10.0 g of H_2 and 10.0 g of Br_2?" The recommended approach is to construct the limiting reagent table, which allows the determination of the amount of hydrogen, bromine, and hydrogen bromide present at the end of the reaction.

The third type is a % yield problem, which is a common addition to either of the first two types of problems. After being given information that allows the determination of the mass of product theoretically possible, information is given related to the actual yield. Example: "Calculations show that 10.13 g of HBr is the maximum amount of HBr that could be formed. The actual mass of product recovered is 9.54 g. What is the percent yield?"

6.4 OXIDATION REDUCTION REACTIONS

OXIDATION NUMBERS AND THEIR UTILITY

The last major category of types of chemical reactions we address in this text is redox reactions. The key to understanding a given redox reaction will be a knowing the "normal" oxidation states of elements in stable compounds, and then finding one or more elements in oxidation states that are either more positive or more negative than is usual for stable compounds. (See Appendix E for information regarding common oxidation states.) One then predicts that the element whose oxidation state is too positive will be reduced (i.e., its oxidation state will decrease) and the element whose oxidation state is too negative will be oxidized (i.e., its oxidation state will increase) in the reaction. A working definition of **oxidation** is, then, the loss of electrons, while **reduction** is defined as the gain of electrons. Oxidation and reduction must both occur in a reaction; one cannot occur without the other.

When part of a compound, the "normal" oxidation states for the representative metals are positive the Group number (i.e., for sodium, the normal oxidation state is +1; for magnesium it is +2, etc.). Toward the bottom of the periodic table, representative metals in Groups III-VI can take on an additional oxidation state differing from the group number by 2 units. For instance thallium has oxidation states of +1 and +3, lead has oxidation states of +2 and +4, and bismuth has oxidation states of +3 and +5. Any oxidation state other than a positive oxidation state for a metal in a compound is very uncommon and makes the metal a very strong **reducing agent**. Elemental metals (in the zero oxidation state) can often serve as reducing agents, since those metals may be easily oxidized. Common stable oxidation states for the transition metals are +2 and +3 with, in general, oxidation states above +3 being strong oxidizing states. Like the representative metals, transition metals do commonly show oxidation states up to, and including, the Group number (except Group VIIIB). Unlike the representative metals, these oxidation states do not need to vary by 2 units.

Non-metals commonly show negative oxidation numbers equal to the Group number minus eight. Thus oxygen, a group VI non-metal, shows a -2 oxidation state in nearly all of its compounds. As a very electronegative element, oxygen will almost always have a negative oxidation number. The exception to that is a compound of oxygen with fluorine, the only element more electronegative than oxygen. In OF_2, fluorine will be in a -1 oxidation state, and oxygen in a +2 oxidation state. The only other compounds where oxygen is not in a -2 oxidation state are the uncommon peroxides and superoxides. Sulfur, another group VI non-metal, also exhibits a common oxidation number of -2. When combined with a more electronegative element, especially oxygen, a non-metal like sulfur can exhibit oxidation states up to, and including, positive the Group number. These oxidation numbers usually vary by 2 units. Hence sulfur exhibits oxidation numbers of -2, 0, +2, +4, and +6. Chlorine, a group VII non-metal, has a common oxidation number of -1; it also can have higher oxidation states, varying in units of 2: +1, +3, +5, and +7 besides the oxidation number of 0 exhibited by all elements in their uncombined state. Non-metals are most "happy" in negative oxidation states and tend to be strong **oxidizing agents** in positive oxidation states. This is especially true of the halogens, less so of some of the other elements, especially sulfur and phosphorous.

CONCEPT CHECK: OXIDATION STATES FOR INORGANIC PEROXIDES AND SUPEROXIDES

(a) What is the oxidation state of oxygen in a peroxide such as Na_2O_2 or BaO_2?
(b) What is the oxidation state of oxygen in a superoxide such as CsO_2?

SOLUTION:
(a) Based on charge balance, it is apparent that ionic peroxides have the anion O_2^{2-}. E.g., two Na^+ + O_2^{2-} yields the neutral compound. Thus in peroxides, the individual oxygen atom has a formal oxidation state of –1. Since this is higher than the typical value of –2 for oxygen, peroxides are typically very good oxidizing agent.

(b) Based on charge balance, it is apparent that superoxides contain the O_2^- ion. E.g. Cs^+ + O_2^- yields the neutral compound. Since the combination of two oxygen atoms has a charge of –1, each oxygen atom is in a formal oxidation state of – 0.5. Superoxides are quite powerful oxidizing agents.

When called upon to predict the products of oxidation-reduction reactions, you should remember what oxidation state the elements are "happiest" in and write the products accordingly. Remember that elements in oxidation states that are too positive are good oxidizing agents and will be in less positive oxidation states as products and vice versa. To precisely predict the products of such reactions requires experience that you will gain as you continue studying chemistry.

Let us now consider a reaction of potassium permanganate ($KMnO_4$) with ferrous chloride ($FeCl_2$) that occurs in acid solution. Since both compounds are metal salts and soluble, they can be written as completely ionized:

$$K^+ + MnO_4^- + Fe^{2+} + Cl^- \longrightarrow$$ EQ. 6.12

For compounds, the sum of the oxidation numbers of the constituent elements must sum to zero. For ions, the sum of the oxidation numbers must equal the charge on the ion. Potassium in the ion K^+ is in the +1 oxidation state. The sum of the oxidation states for the manganese atom and the four oxygen atoms in permanganate must equal -1. Although there are exceptions (such as peroxides and superoxides) it is best to assume that each oxygen atom in a compound is in a -2 oxidation state. Thus the four oxygen atoms in MnO_4- contribute 4(-2) = - 8 to the overall -1 charge. The oxidation state of manganese in permanganate ion is thus +7. Thus we see two elements in the equation above that are candidates for being in "unhappy" oxidation states: Mn is in the +7 state and Fe in the +2 state. Positive 7 is a very high oxidation state for manganese and the permanganate ion should then be a very strong oxidizing agent. While +2 is a stable state for the iron atom, we remember that +3 is another common oxidation state for iron. The ferrous ion is, therefore, a weak reducing agent. Having decided this, we postulate that this reaction is a good candidate for an oxidation reduction reaction (we have previously ruled out a double displacement reaction because the potential products are both soluble and it does not look promising for either a combination or decomposition reaction). We have only to determine the "right" products.

It is not difficult to decide upon the right product for the iron — it is ferric ion (the +3 state). We are reasonably certain of that because it is another quite common species. We have seen the ferric ion often, and know that it is stable. Besides, iron is a transition metal and +2 and +3 states are quite common for transition metal ions.

The "right" oxidation state for the manganese atom is more difficult to predict. We know that it will be more negative than +7 but that leaves practically everything between +7 and 0 (in practice +2 because it is very rare that a solution reaction yields an element as a product). The correct answer is the +2 oxidation state — a fact that we would know only if we had: (1) looked it up, (2) seen the reaction before, (3) asked an upper-class chemistry major or the instructor (let's hope they know!), or, (4) made a lucky guess. The final net ionic reaction is:

$$MnO_4^- + 8\,H^+ + 5\,Fe^{2+} \longrightarrow Mn^{2+} + 5\,Fe^{3+} + 4H_2O$$ EQ. 6.13

OXIDATION/REDUCTION OF ORGANIC COMPOUNDS

When dealing with organic compounds it is often easier to look for oxidation or reduction by inspection than to determine the oxidation numbers of every atom in each organic compound. Specifically, one can simply look for a change in the number of oxygen or hydrogen atoms as an organic molecule is modified during a chemical reaction. Oxidation can be seen either as the gain of oxygen or the loss of hydrogen. Reduction can be seen either as the loss of oxygen or the gain of hydrogen. Let us consider a series of compounds containing three carbon atoms. If we draw the compounds in order from most reduced to most oxidized, the following series results.

$$CH_3CH_2CH_3 \quad CH_3CH{=}CH_2 \quad CH_3CH_2CH_2OH \quad CH_3CH_2{-}\overset{O}{\overset{\|}{C}}{-}H \quad CH_3CH_2{-}\overset{O}{\overset{\|}{C}}{-}OH$$

least oxidized ⟶ most oxidized

FIGURE 6.1: Carbon in increasingly oxidized states in related compounds.

As you can see by inspection, the above working definitions apply to this series of compounds. Of course the ultimate oxidation product of any organic species is carbon dioxide, which is formed by the combustion of organic compounds.

Next we will consider several oxidation/reduction reagents and their utility in the laboratory.

OXIDIZING AGENTS

One of the most common oxidizing agents is, of course, oxygen itself. You have all seen examples of oxygen at work as an oxidizing agent. The formation of rust on iron objects is an oxidation reduction reaction. Iron, in the presence of oxygen and water, forms iron(III) oxide as a hydrate. This material is commonly known as rust.

$$4\,Fe^{2+} + O_2(g) + 4\,H_2O(l) + 2x\,H_2O \longrightarrow 2Fe_2O_3 \cdot xH_2O(s) + 8\,H^+$$ EQ. 6.14

Oxygen can also act on organic compounds. Aldehydes are prone to easy air oxidation. It is not uncommon to find stock bottles of aldehydes contaminated with the corresponding carboxylic acid. For example, benzaldehyde is easily oxidized to benzoic acid in the presence of oxygen.

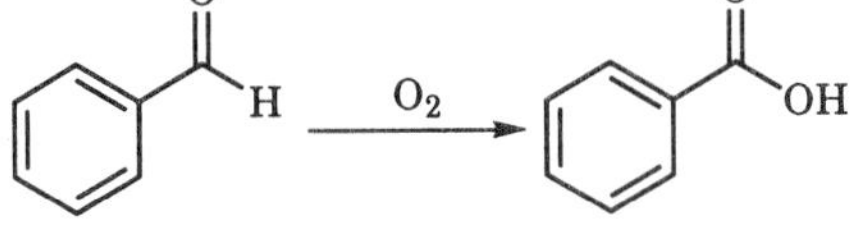

$$C_6H_5{-}CHO \xrightarrow{O_2} C_6H_5{-}COOH$$ EQ. 6.15

This process is easy to detect since benzaldehyde is a clear liquid with the odor of almond and benzoic acid is a white odorless solid.

Ozone, a close relative of oxygen, is an even more potent oxidizing agent. As you may know, ozone depletion is a serious environmental concern. Ozone by itself slowly decomposes to form oxygen. In the presence of nitric oxide, a common pollutant, ozone undergoes the rapid oxidation reduction reaction seen below.

$$O_3(g) + NO(g) \longrightarrow NO_2(g) + O_2(g)$$ EQ. 6.16

As a synthetic reagent, ozone can be used to convert alkenes into carbonyl compounds. A general reaction for this process, known as **ozonolysis,** is provided below.

$$\text{WXC=CYZ} \xrightarrow[\text{2. workup}]{\text{1. } O_3} \text{WXC=O} + \text{O=CYZ}$$ EQ. 6.17

The product composition is dependent upon both the groups on the double bond and the nature of the workup. Consider the reaction between ozone and 2-methyl-2-pentene.

$$(H_3C)_2C=CH(CH_2CH_3) \xrightarrow{\text{1. } O_3} \begin{cases} \xrightarrow{\text{2. oxidative workup}} (H_3C)_2C=O + O=C(OH)CH_2CH_3 \\ \xrightarrow{\text{2. reductive workup}} (H_3C)_2C=O + O=C(H)CH_2CH_3 \end{cases}$$ EQ. 6.18

This tri-substituted olefin provides access to different products depending upon the workup conditions employed. In the oxidative workup conditions, H_2O_2 and NaOH are used to provide further mildly oxidizing conditions. The products of this oxidative workup are a ketone and a carboxylic acid. On the other hand, reductive workup conditions include Zn or dimethyl sulfide to prevent further oxidation of the initial ozonolysis products. The results from a reductive workup are a ketone and an aldehyde.

Peroxycarboxylic acids are also potent oxidizing agents. They have the general formula

$$R-\overset{\overset{\displaystyle O}{\|}}{C}-O-O-H$$

and are usually prepared by treating a carboxylic acid with hydrogen peroxide.

$$R-\overset{\overset{\displaystyle O}{\|}}{C}-O-H + H_2O_2 \longrightarrow R-\overset{\overset{\displaystyle O}{\|}}{C}-O-O-H + H_2O$$ EQ. 6.19

These compounds are synthetically useful in two ways. First, they are used to convert alkenes into oxiranes (epoxides) in a process known as epoxidation. Consider the reaction of 3-chloroperoxybenzoic acid and 2-butene.

$$H(H_3C)C=C(H)CH_3 + 3\text{-}ClC_6H_4C(=O)OOH \longrightarrow H(H_3C)C\overset{O}{-}C(H)CH_3 + 3\text{-}ClC_6H_4C(=O)OH$$ EQ. 6.20

The products of the reaction are the corresponding oxirane and 3-chlorobenzoic acid. The peroxyacid supplied one of its oxygen atoms to the alkene.

The peracids are also useful for converting ketones into esters, a process known as the Baeyer-Villiger oxidation. The reaction has proven to be most useful for the conversion of cyclic ketones into lactones (cyclic esters). For example performic acid can convert cyclopentanone into the lactone of 5-hydroxypentanoic acid.

$$\text{cyclopentanone} + \text{H-}\overset{\text{O}}{\overset{\|}{\text{C}}}\text{-O-O-H} \longrightarrow \text{lactone} + \text{H-}\overset{\text{O}}{\overset{\|}{\text{C}}}\text{-O-H} \quad \text{EQ. 6.21}$$

So far we have encountered methods for oxidizing alkenes that involve cleavage of both bonds (ozonolysis) and cleavage of only one bond (epoxidation). Another reagent that is capable of oxidizing alkenes while cleaving only one bond is a combination of osmium tetroxide (OsO_4) and hydrogen peroxide. For example, the reaction of cyclohexene with this mixture results in the formation of 1,2-cyclohexane diol.

$$\text{cyclohexene} + H_2O_2 \xrightarrow{OsO_4} \text{cyclohexane-1,2-diol (OH, OH)} \quad \text{EQ. 6.22}$$

The same type of transformation can also be accomplished with $KMnO_4$. Consider the reaction between 3-hexene and the permanganate ion.

$$\text{3-hexene} + MnO_4^- \longrightarrow \text{3,4-hexanediol (OH, OH)} + MnO_{2(s)} \quad \text{EQ. 6.23}$$

Notice that in the reaction Mn starts out in the +7 oxidation state, but ends up in the +4 oxidation state. Manganese has, therefore, been reduced. Note that permanganate is reduced to MnO_2 under neutral or slightly basic conditions. Earlier we saw permangante formed Mn^{2+} under acidic conditions. Clearly the organic species has been oxidized. This reaction is also interesting to observe. As you may already know, $KMnO_4$ is a bright purple solution. The MnO_2, on the other hand, is a brownish solid that precipitates in the reaction vessel. The reaction, in effect, contains its own indicator. When the purple color has been replaced with a brown precipitate, the reaction is done.

Potassium permanganate is also commonly used for the quantitative analysis of a variety of inorganic ions that undergo redox reactions. An acidic solution containing arsenic, for example, can be conveniently titrated with potassium permanganate of known concentration to determine the amount of arsenic present. The reaction is shown below. During the titration the arsenic present in solution is oxidized from the +3 oxidation state to the +5 oxidation state.

EQ. 6.24

$$5\,H_3AsO_3 + 2\,MnO_4^- + 6\,H^+ \longrightarrow 5\,H_3AsO_4 + 2\,Mn^{2+} + 3\,H_2O$$

Another species that has great utility in both quantitative analysis and synthetic chemistry is chromium. Several reagents exist in which chromium has an oxidation number of +6. During the course of a redox reaction, the chromium in these reagents is reduced to chromium (III) while at the same time the reducing agent becomes oxidized. One of the most common uses of chromium (VI) in quantitative analysis is for the determination of iron (II). The redox reaction proceeds as follows.

$$6Fe^{2+} + Cr_2O_7^{2-} + 14H^+ \longrightarrow 6Fe^{3+} + 2\,Cr^{3+} + 7\,H_2O$$ EQ. 6.25

In this case the iron (II) was oxidized to iron (III), while chromium was reduced from a +6 to a +3 oxidation state. Balancing redox reactions like those above can be difficult. Later in this chapter, we will learn a protocol to follow that will make balancing these reactions fairly easy.

Synthetically, the chromium (VI) reagents are also used to oxidize alcohols to carbonyl compounds and/or aldehydes to carboxylic acids. Common oxidizing agents include the dichromate ion in aqueous acidic solution or complexes of chromium trioxide with pyridine in non-aqueous solutions. These two reagent combinations are complimentary. The first can be used for water soluble molecules that are not prone to reaction with aqueous acid. The second can be useful for those reactions that involve compounds that are not water soluble and/or would be susceptible to reaction with aqueous acid. Finally, the chromium trioxide pyridine complex is often a useful reagent because oxidation of primary alcohols, using this reagent, stops at the aldehyde oxidation state rather than progressing to the carboxylic acid as is common when using acidic dichromate solutions. Let us look at some examples of these types of reactions.

OH $\xrightarrow[H_2SO_4]{K_2Cr_2O_7}$ O, OH EQ. 6.26

OH $\xrightarrow[H_2SO_4]{K_2Cr_2O_7}$ O EQ. 6.27

OH $\xrightarrow{CrO_3/pyridine}$ O EQ. 6.28

OH $\xrightarrow{CrO_3/pyridine}$ O, H EQ. 6.29

Concept Check: Organic Oxidation-Reduction Reaction

What alcohol and what chromium (VI) species would one select to prepare 2-methylpropanal? Is the organic molecule being oxidized or reduced in this reaction?

Solution: To prepare 2-methylpropanal, $(CH_3)_2CHCHO$, one would start with 2-methylpropanol, $(CH_3)_2CHCH_2OH$. If acidic dichromate was used, the carboxylic acid would be formed. Instead, reaction with CrO_3 in pyridine is more likely to yield the desired aldehyde. The alcohol is said to be oxidized when converted to an aldehyde. (Organic oxidation is readily interpreted as either the gain of oxygen or, in this case, the loss of hydrogen from the molecule.)

Reducing Reagents

Our discussion of reducing reagents will be limited to those that are used most often for synthetic transformations. One of the most common reducing agents, and one that we have already looked at when we examined addition reactions, is hydrogen. The addition of hydrogen across a multiple bond is a reduction. The conversion of an alkene to an alkane or an alkyne to an alkene can, therefore, be classified as a reduction.

While the reduction of carbon-carbon multiple bonds is often important in synthetic organic chemistry, the reduction of carbon-oxygen double bonds is probably even more significant. Typically when we think about the reduction of compounds that contain carbonyl bonds we consider the reduction of carboxylic acids or their derivatives to alcohols (or amines in the case of amide reduction) and the reduction of aldehydes and ketones to alcohols. As we will see later, it is difficult to stop a reduction reaction at the aldehyde oxidation stage, but it can be done with the appropriate reagents under suitable conditions.

In organic synthesis, two most commonly encountered reducing reagents are lithium aluminum hydride ($LiAlH_4$) and sodium borohydride ($NaBH_4$). It is often convenient to think of these two reagents as sources of hydride (H^-). The subtle distinction between the two reagents is reactivity. $LiAlH_4$ is the more reactive of the two reducing agents and, as such, is capable of reducing all carbonyl containing compounds. Sodium borohydride, on the other hand, will reduce only aldehydes and ketones. Here again we begin to see the concept of selectivity and its importance. Some examples of typical reductions of this type are provided here.

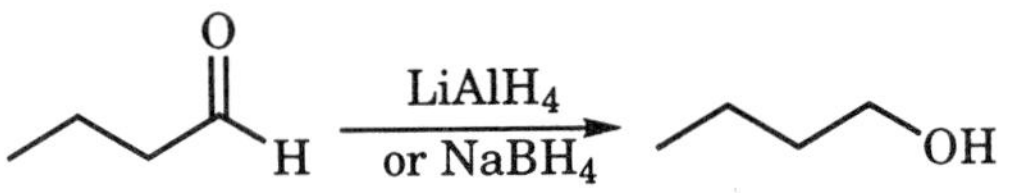

O, H —$LiAlH_4$ or $NaBH_4$→ OH EQ. 6.30

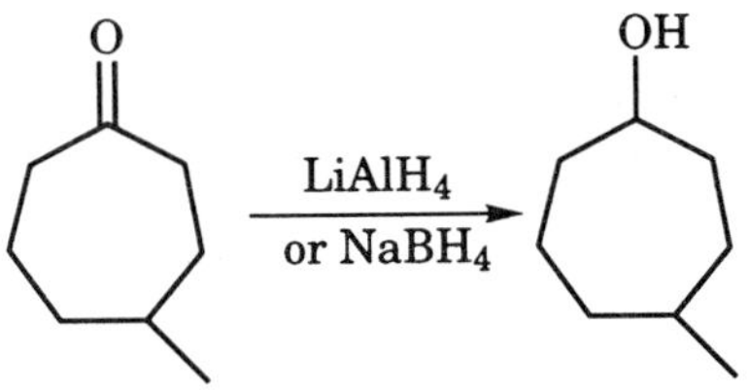

O —$LiAlH_4$ or $NaBH_4$→ OH EQ. 6.31

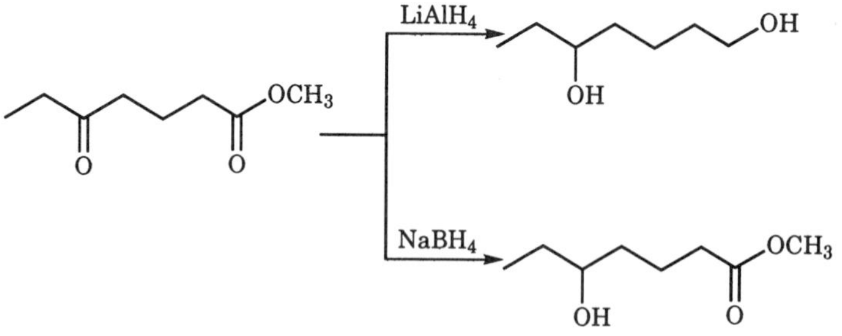

$$\text{methyl 4,7... keto ester} \xrightarrow{LiAlH_4} \text{diol (OH, OH)}$$ EQ. 6.32

$$\xrightarrow{NaBH_4} \text{hydroxy ester (OH, } OCH_3\text{)}$$ EQ. 6.33

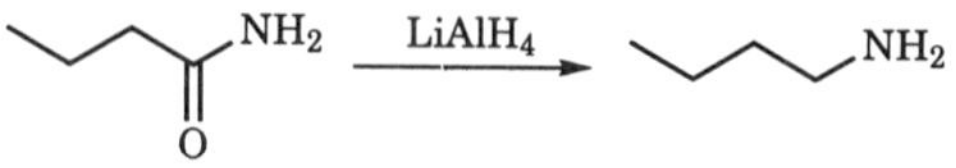

$$\text{amide } (NH_2) \xrightarrow{LiAlH_4} \text{amine } (NH_2)$$ EQ. 6.34

Lithium aluminum hydride is also used to reduce nitriles into primary amines.

$$CH_3C{\equiv}N \xrightarrow{LiAlH_4} CH_3CH_2NH_2$$ EQ. 6.35

Reducing agents such as $LiAlH_4$ and $NaBH_4$ are appropriate for small scale redox reactions in laboratory synthesis. But on an industrial scale, the cost of such reagents is prohibitive. Catalytic hydrogenation is often used as a more cost-effective means of reduction. Unsaturated vegetable oils (coconut, cottonseed, peanut, safflower, or soybean oils) have low melting points. Reducing the alkene groups present in such molecules increases their melting points, turning the oils into fats which are semi-solids (margarine, shortening.)

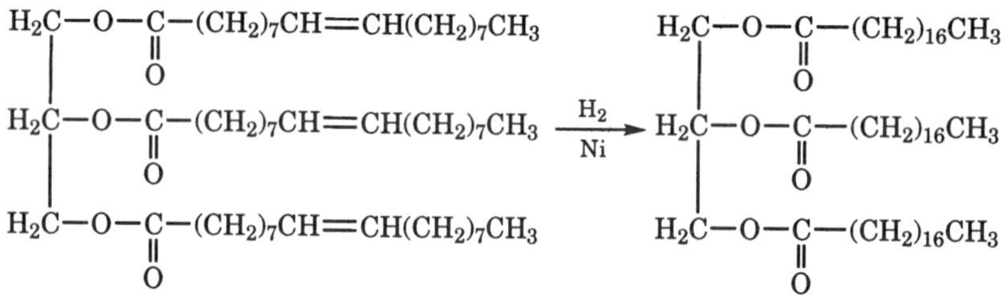

$$\begin{array}{l} H_2C{-}O{-}C({=}O){-}(CH_2)_7CH{=}CH(CH_2)_7CH_3 \\ H_2C{-}O{-}C({=}O){-}(CH_2)_7CH{=}CH(CH_2)_7CH_3 \\ H_2C{-}O{-}C({=}O){-}(CH_2)_7CH{=}CH(CH_2)_7CH_3 \end{array} \xrightarrow[Ni]{H_2} \begin{array}{l} H_2C{-}O{-}C({=}O){-}(CH_2)_{16}CH_3 \\ H_2C{-}O{-}C({=}O){-}(CH_2)_{16}CH_3 \\ H_2C{-}O{-}C({=}O){-}(CH_2)_{16}CH_3 \end{array}$$ EQ. 6.36

Triolein, an oil (liquid) — Tristearin, a fat (solid)

6.5 BALANCING OXIDATION-REDUCTION REACTIONS

Balancing oxidation-reduction reactions is even easier than balancing other types of reactions if a few simple rules are followed. These rules are easily laid out as a series of steps that can be followed to get a balanced reaction when all steps are followed to completion.

Let us take for our first example, the reaction between ferrous ion and dichromate ion in acidic solution. Earlier we have seen that the ferrous ion may be readily oxidized, and that the dichromate ion (with chromium in a +6 oxidation state) is a good oxidizing agent.

The first step is to determine the identity of the products.

In the first reaction the products are ferric ion (Fe(III)) and the chromium ion (Cr(III)).

The skeleton of the reaction should then be written.

$$\text{I.}\quad Cr_2O_7^{2-} + Fe^{2+} \longrightarrow Cr^{3+} + Fe^{3+}$$

The next step is to identify those elements in each reaction that are undergoing oxidation or reduction and to separate them into two half-equations. In this example, these elements are iron and chromium.

$$\text{Ia.}\quad Fe^{2+} \longrightarrow Fe^{3+}$$

$$\text{Ib.}\quad Cr_2O_7^{2-} \longrightarrow Cr^{3+}$$

The next step is to balance each half-reaction for the element undergoing a change in oxidation state. This requires that reaction Ib. be rewritten, balancing chromium:

$$\text{Ib.}\quad Cr_2O_7^{2-} \longrightarrow 2\ Cr^{3+}$$

The next step is to balance oxygen atoms by the addition of water. This requires that reaction Ib be rewritten:

$$\text{Ib.}\quad Cr_2O_7^{2-} \longrightarrow 2\ Cr^{3+} + 7\ H_2O$$

Next hydrogen atoms are balanced by the addition of hydrogen ions. This requires the rewriting equation Ib:

$$\text{Ib.}\quad Cr_2O_7^{2-} + 14\ H^+ \longrightarrow 2\ Cr^{3+} + 7\ H_2O$$

The next step is to balance the charge by the addition of electrons. This will require rewriting both half-equations:

$$\text{Ia.}\quad Fe^{2+} \longrightarrow Fe^{3+} + e^-$$

$$\text{Ib.}\quad Cr_2O_7^{2-} + 14\ H^+ + 6\ e^- \longrightarrow 2\ Cr^{3+} + 7\ H_2O$$

The next step is to add the two half-reactions together in such a way as to cancel the electrons. This will often mean that they will need to be multiplied by some factor before being added together. In our first example, the iron oxidation reaction had to be multiplied by 6 in order to produce the 6 electrons that are consumed in the dichromate reduction reaction. This process yields:

$$\text{I.}\quad 6\ Fe^{2+} + 14\ H^+ + Cr_2O_7^{2-} \longrightarrow 6\ Fe^{3+} + 7\ H_2O + 2\ Cr^{3+}$$

The next step is to cancel anything that appears on both sides of the equation. There is nothing (except the electrons) in reaction I that needs to be cancelled

To summarize the process so far: the skeleton reaction is broken into half reactions. Mass is balanced first, beginning with the element undergoing a change in oxidation state, then other elements, then oxygen is balanced (using water), and finally hydrogen is balanced using protons. Once mass is balanced, charge is balanced by adding electrons to one side of each reaction. The balanced half reactions are combined in a way that conserves the number of electrons.

At this point everything (atoms and charge) is in balance in our first example. This is the end of the balancing procedure if the reaction occurs in acid solution or if there are no hydrogen ions appearing in the equation for a reaction said to occur in basic solution.

Next, let us consider an example balancing a redox reaction that occurs in base: the reaction between the Cr(III) ion and hydrogen peroxide to yield the chromate ion (CrO_4^{2-}) and water. As before, we begin by splitting the skeleton reaction into half reactions:

IIa. $Cr^{3+} \longrightarrow CrO_4^{2-}$

IIb. $H_2O_2 \longrightarrow H_2O$

The chromium is balanced in IIa and the other reaction involves only oxygen and hydrogen, which are balanced in subsequent steps.

IIb. $H_2O_2 \longrightarrow 2\ H_2O$

Next, oxygen is balanced in both reactions by the addition of water.

IIa. $Cr^{3+} + 4\ H_2O \longrightarrow CrO_4^{2-}$

IIb. $H_2O_2 \longrightarrow 2\ H_2O$

The mass balancing is completed by adding protons to balance hydrogen.

IIa. $Cr^{3+} + 4\ H_2O \longrightarrow CrO_4^{2-} + 8\ H^+$

IIb. $H_2O_2 + 2\ H^+ \longrightarrow 2\ H_2O$

The charge is balanced in both half reactions by adding electrons:

IIa. $Cr^{3+} + 4\ H_2O \longrightarrow CrO_4^{2-} + 8\ H^+ + 3\ e^-$

IIb. $H_2O_2 + 2\ H^+ + 2\ e^- \longrightarrow 2\ H_2O$

Before combining the half reactions, the number of electrons must be conserved. Multiplying the chromium half reaction by 2 and the peroxide half reaction by 3 results in 6 electrons being consumed and produced.

IIa. $2\ Cr^{3+} + 8\ H_2O \longrightarrow 2\ CrO_4^{2-} + 16\ H^+ + 6\ e^-$

IIb. $3\ H_2O_2 + 6\ H^+ + 6\ e^- \longrightarrow 6\ H_2O$

The half reactions are now combined.

II. $2\ Cr^{3+} + 8\ H_2O + 6\ H^+ + 3\ H_2O_2 \longrightarrow 2\ CrO_4^{2-} + 16\ H^+ + 6\ H_2O$

Since protons and water appear on both sides, they are cancelled to yield a balanced simplified equation.

$$\text{II. } 2\,Cr^{3+} + 2\,H_2O + 3\,H_2O_2 \longrightarrow 2\,CrO_4^{2-} + 10\,H^+$$

Because reaction II was said to occur in basic solution and hydrogen ions do appear in it, we must go through one additional step to get rid of the hydrogen ions. This is done by adding exactly the same number of hydroxide ions to both sides of the equation as there are hydrogen ions in the equation. This means that we must add ten hydroxide ions in our example. This will give the following equation:

$$\text{II. } 2\,Cr^{3+} + 2\,H_2O + 3\,H_2O_2 + 10\,OH^- \longrightarrow 2\,CrO_4^{2-} + 10\,H^+ + 10\,OH^-$$

The hydrogen and hydroxide ions on the same side of the equation combine to form water that will then cancel some of the water on the opposite side of the equation. Once this is done the final equation is:

$$\text{II. } 2\,Cr^{3+} + 10\,OH^- + 3\,H_2O_2 \longrightarrow 2\,CrO_4^{2-} + 8\,H_2O$$

This completes the balancing process. It is definitely worthwhile to pause at the end of problems requiring the balancing of a redox reaction to verify that the number of atoms of each element and the total charge are the same on both the reactant and product side.

Also note that the skeleton reactions at the start and the final reactions at the end were written in net ionic form (omitting spectator ions). One could also be given a redox problem with the reactants and products written as compounds:

$$\text{I. } FeCl_2 + K_2Cr_2O_7 \longrightarrow FeCl_3 + CrCl_3$$

$$\text{II. } CrCl_3 + H_2O_2 \longrightarrow K_2CrO_4 + H_2O$$

It is far simpler to reduce this problem to a net ionic reaction before starting, balancing as before, then adding the counterions to show compounds reacting and forming:

$$\text{I. } 6\,FeCl_2 + 14\,HCl + K_2Cr_2O_7 \longrightarrow 6\,FeCl_3 + 7\,H_2O + 2\,CrCl_3 + 2\,KCl$$

$$\text{II. } 2\,CrCl_3 + 10\,KOH + 3\,H_2O_2 \longrightarrow 2\,K_2CrO_4 + 8\,H_2O + 6\,KCl$$

If one wished, however, one could follow the same balancing procedures starting with the compounds instead of the net ions; the result, more tediously obtained, is the same.

Occasionally, an element in a compound that does not ionize in solution enters a redox reaction. In this case, the species should be written as it actually exists, and the balancing procedure followed as described. Consider the unbalanced net ionic redox reaction in acid:

$$\text{III. } Cr_2O_7^{2-} + Hg \longrightarrow Cr^{3+} + Hg_2I_2$$

for which the skeleton half reactions are:

IIIa. $Cr_2O_7^{2-} \longrightarrow Cr^{3+}$

IIIb. $Hg \longrightarrow Hg_2I_2$

Half reaction IIIa. was balanced above to yield

IIIa. $Cr_2O_7^{2-} + 14H^+ + 6e^- \longrightarrow 2\,Cr^{3+} + 7\,H_2O$

The other half reaction, IIIb., has liquid mercury metal going to mercurous iodide. Our solubility rules tell us the mercurous iodide is insoluble in water, and thus exists as a solid in the net ionic reaction. This half reaction is balanced according to the rules. First the species undergoing redox (i.e., mercury) is balanced, with phases (other than ions in aqueous solution) indicated:

IIIb. $2\,Hg(l) \longrightarrow Hg_2I_2(s)$

Then species other than oxygen and hydrogen are balanced. In this case, the iodide on the product side is balanced by iodide on the reactant side:

IIIb. $2\,Hg(l) + 2\,I^- \longrightarrow Hg_2I_2(s)$

This half reaction is mass balanced. It is charge balanced by adding electrons:

IIIb. $2\,Hg(l) + 2\,I^- \longrightarrow Hg_2I_2(s) + 2e^-$

To conserve electrons, reaction IIIb. is multiplied by 3, and added to IIIa. to yield:

III. $6\,Hg(l) + 6\,I^- + Cr_2O_7^{2-} + 14\,H^+ \longrightarrow 3\,Hg_2I_2(s) + 2\,Cr^{3+} + 7\,H_2O$

This is now a balanced, net ionic redox reaction.

Until now, the redox reactions we have balanced have all been inorganic. Oxidation reduction reactions also play a very important part in organic reactions as well. Calculating the oxidation state of a carbon atom in an organic compound can be challenging, but the method of redox balancing we use does not require exact knowledge of the oxidation states involved, as long as one can recognize if oxidation or reduction has occurred. Remember that organic oxidation reduction reactions can often be recognized by a few tell-tale signs. For example, if the number of oxygen atoms in an organic molecule increases, or if the number of hydrogen atoms decreases, the species is being oxidized; if the number of oxygen atoms decreases, or if the number of hydrogen atoms increases, the species is being reduced. Recall that Figure 6.1 showed this trend: propane was oxidized first to propene, then to 1-propanol, then to propanal, then to propanoic acid.

Note that in organic redox reactions, the carbon backbone is generally not attacked. While it is possible to break the carbon backbone of an organic molecule in a redox reaction (e.g., burning it in an oxygen atmosphere) the organic redox reactions of interest will generally use reactants and conditions that modify functional groups, but do not attack the carbon backbone.

To conclude this section on balancing redox, let us do an example using the oxidation of cyclohexene to form 1,2-cyclohexanediol using permanganate ion. This is performed under basic conditions with the following unbalanced skeleton reaction:

$$\text{IV. } MnO_4^- + C_6H_{10} \longrightarrow MnO_2 + C_6H_{12}O_2$$

Note that adding two hydrogens (each with a +1 oxidation state) and two oxygens (each with a -2 oxidation state) means that the oxidation number of the carbons involved must increase to make the sum of oxidation numbers equal zero. Cyclohexene is being oxidized, and manganese is being reduced. We begin according to our rules by writing skeleton half-reactions:

$$\text{IVa. } MnO_4^- \longrightarrow MnO_2$$

$$\text{IVb. } C_6H_{10} \longrightarrow C_6H_{12}O_2$$

Next the species undergoing oxidation/reduction are balanced. These are manganese and carbon, and they are already in balance. Since only oxygen and hydrogen remain, we next balance oxygen by adding water:

$$\text{IVa. } MnO_4^- \longrightarrow MnO_2 + 2\,H_2O$$

$$\text{IVb. } C_6H_{10} + 2\,H_2O \longrightarrow C_6H_{12}O_2$$

To complete the mass balance, H^+ is added to balance hydrogen:

$$\text{IVa. } MnO_4^- + 4\,H^+ \longrightarrow MnO_2 + 2\,H_2O$$

$$\text{IVb. } C_6H_{10} + 2\,H_2O \longrightarrow C_6H_{12}O_2 + 2\,H^+$$

Then charges are balanced by adding electrons:

$$\text{IVa. } MnO_4^- + 4\,H^+ + 3\,e^- \longrightarrow MnO_2 + 2\,H_2O$$

$$\text{IVb. } C_6H_{10} + 2\,H_2O \longrightarrow C_6H_{12}O_2 + 2\,H^+ + 2\,e^-$$

Electrons are conserved, multiplying IVa. by 2, and multiplying IVb. by 3.

$$\text{IVa. } 2\,MnO_4^- + 8\,H^+ + 6\,e^- \longrightarrow 2\,MnO_2 + 4\,H_2O$$

$$\text{IVb. } 3\,C_6H_{10} + 6\,H_2O \longrightarrow 3\,C_6H_{12}O_2 + 6\,H^+ + 6\,e^-$$

Combining the balanced half reactions and canceling common terms yields:

$$\text{IV. } 2\,H_2O + 2\,H^+ + 2\,MnO_4^- + 3\,C_6H_{10} \longrightarrow 2\,MnO_2 + 3C_6H_{12}O_2$$

Since this reaction occurs in base, the last step is to neutralize the 2 H^+ by adding 2 OH^- to both sides of the reaction, and canceling common terms:

$$\text{IV. } 4\,H_2O + 2\,MnO_4^- + 3\,C_6H_{10} \longrightarrow 2\,MnO_2 + 3\,C_6H_{12}O_2 + 2\,OH^-$$

6.6 BIOCHEMICAL EXAMPLES OF REDOX REACTIONS

Many of the reactions that take place during metabolism are oxidation/reduction reactions. The process of electron transport also involves redox processes. It is, therefore, important to have a grasp of redox chemistry before trying to apply that understanding to biological systems. The previous sub-sections have served as a brief overview of several oxidation/reduction reactions. This section serves as your guide to how these phenomena apply to biological systems. Three compounds are chosen to illustrate biological redox processes because the method of electron flow through each compound is different. These compounds are nicotinamide dinucleotide (NAD+ along with NADH), the cytochromes, and flavin adenine dinucleotide (FAD along with $FADH_2$).

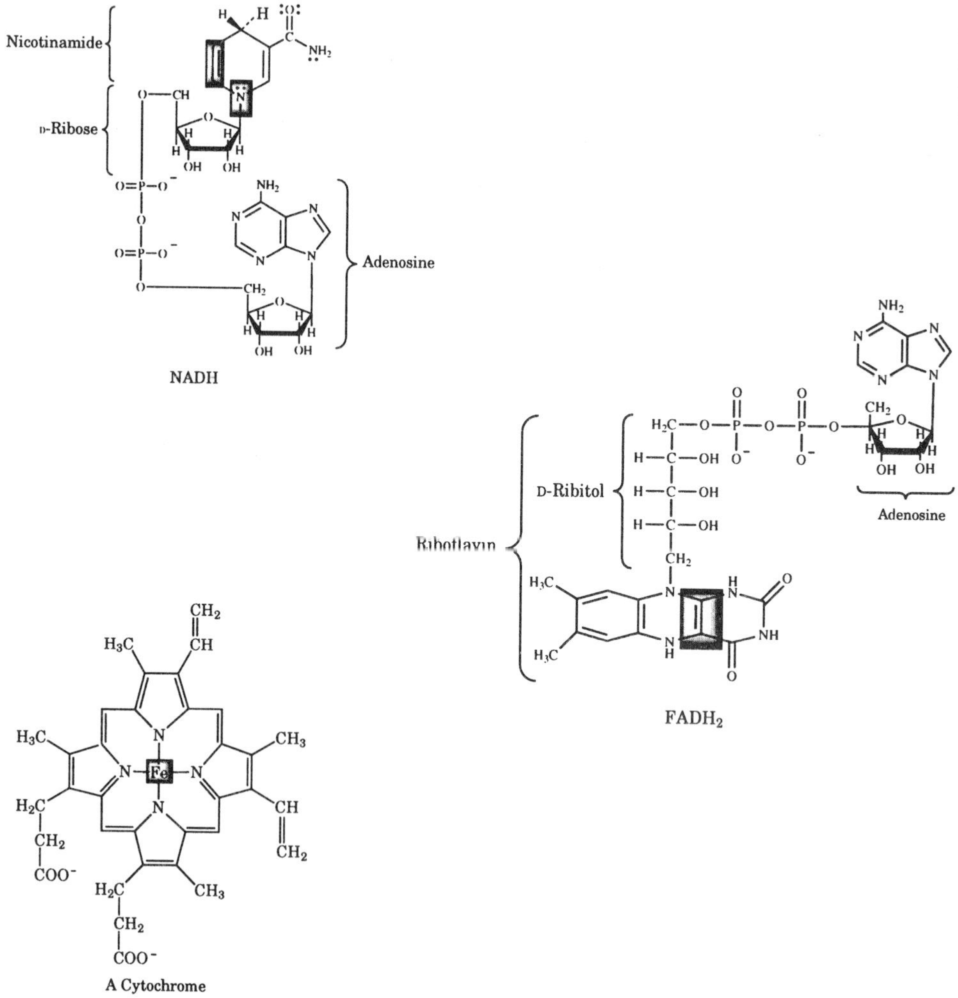

Nicotinamide

D-Ribose

Adenosine

NADH

D-Ribitol

Riboflavin

Adenosine

$FADH_2$

A Cytochrome

FIGURE 6.2: Drawings of NADH, $FADH_2$, and cytochromes, with emphasis on sites of electron transfer.

Electron transfer can be accomplished most simply by transferring free electrons from one species to another. This method of electron transfer is employed by the cytochromes during electron transport. As you may have noticed, each cytochrome contains an iron atom within the heme group. The iron atom, then, acts as the electron transfer agent by oscillating between the +2 and +3 oxidation state.

Another method of biochemical electron transfer is to have the transfer made from an electron donor to an electron acceptor as hydride ($H:^-$). The NAD^+/NADH pair serves to donate hydride (NADH) or to accept hydride (NAD^+) in one step and is, therefore, somewhat analogous to $LiAlH_4$ or $NaBH_4$. This electron transfer is shown schematically below.

NADH → NAD^+

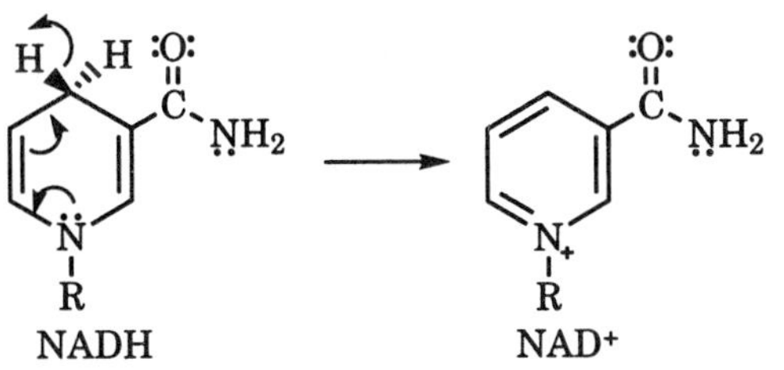

FIGURE 6.3: Nicotinamide ring with and without hydride.

A final form of biological electron transfer is that of transfer as a hydrogen atom. Since hydrogen atoms contain both 1 proton and 1 electron, the addition or removal of a hydrogen atom from a molecule also results in the addition or removal of 1 electron from that same molecule. This is the method of electron transfer that is employed by the FAD/$FADH_2$ redox pair. Again, this type of transport is shown schematically below.

$FADH_2$ ⇌ FADH + H• ⇌ FAD + H•

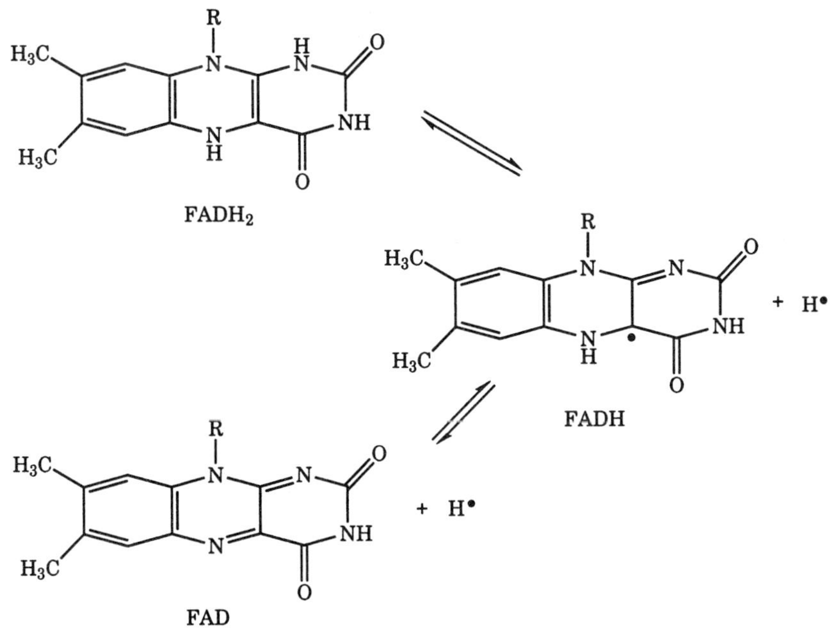

FIGURE 6.4: Electron transfers in FAD/$FADH_2$ pair.

6.7 ELECTROCHEMICAL CELLS

Redox reactions play an important role in the biochemistry of energy production, storage, and retrieval. In a similar fashion, the commercial storage and retrieval of electrical energy by chemical means is commonly done using redox reactions. This occurs in devices known as electrochemical cells, or batteries. In principle, any redox reaction can be carried out in an electrochemical cell in such a way that the energy from the reaction is captured and used for performing some useful work. As a specific example, let us look at a reaction that was one of the earliest studied of such reactions: the reaction between copper ions and zinc metal. This combination is known as a Daniell cell, and it was invented in 1836 by John Frederic Daniell (1790–1845). The overall reaction is:

$$Zn(s) + Cu^{2+} \longrightarrow Cu(s) + Zn^{2+}$$ EQ. 6.37

If this reaction is carried out in a beaker, for instance by placing some metallic zinc into a solution of copper sulfate, the electron exchange occurs directly between the zinc atoms and the copper ions. The energy released when this reaction occurs then appears in the solution as heat.

In the 1800s, a means of preventing this direct transfer of electrons was devised with the invention of the voltaic pile by Alessandro Volta (1745–1827). Further developments resulted in the appearance of what we now formally call the electrochemical cell and know informally by a number of different terms, the most common of which is battery. An electrochemical cell that uses equation 6.37 as the basis of its operation is illustrated in Figure 6.5. One of the beakers contains a solution of copper cations with virtually any anion as the counter ion. Copper sulfate is the most common material used. The other beaker contains a solution of zinc cations, usually provided as a solution of zinc sulfate. A strip of the metal, called an electrode, is then placed into each solution. The metal is often the same metal used for the cations in the solution. The strips of metal are connected by a wire to carry the electrical current between the electrodes. The circuit is completed by carefully placing a U–shaped tube containing a very concentrated solution of an electrolyte, for instance KCl, between the beakers. This tube is known as a salt bridge and serves the purpose of completing the electrical circuit by permitting the flow of electrical charge (as the ions move through the salt bridge between the beakers) but preventing the mixing of the copper and zinc solutions at least for the duration of the experiment.

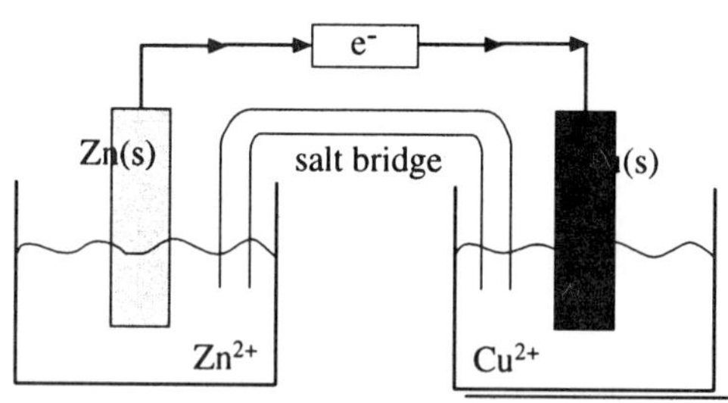

FIGURE 6.5: Simple copper zinc cell.

For the reaction to occur, the zinc atom must give up its electrons to the wire. After these electrons travel across the wire and do work (perhaps helping to turn an electrical motor), they arrive at the copper electrode where they are given up to a copper ion. The circuit is completed when two potassium ions migrate from the salt bridge into the copper solution and a zinc ion migrates into the salt bridge from the zinc solution, or two chloride ions migrate into the zinc solution from the salt bridge and a sulfate ion migrates into the salt bridge from the copper sulfate solution, or some other migration of ions that achieves the same balance of charge.

The voltage provided by such a cell is dependent upon the identity and concentration of the reactants in a manner that will be studied in detail in later chemistry courses. Our current purpose is to discuss some of the practical examples of electrochemical cells and their use in everyday life. Note that we began this section by saying that, "in principle it is possible ...". In practice it is not always easy to prevent the direct contact of the reactants and there are far fewer electrochemical cells in use than the number of possible different combinations of reducing and oxidizing agents. The table below lists a number of commercially available "dry" cells that can be used to provide electrical energy for small appliances such as flashlights, cameras, etc. In all of these cells, the indicated chemical reaction is one that produces energy when it proceeds.

TABLE 6.2: DRY CELLS COMMONLY AVAILABLE TABLE 6.2

COMMON NAME	CHEMICAL REACTION
Dry Cell	$2\ MnO_2(s) + 2\ NH_4Cl(s) + Zn(s) \longrightarrow Mn_2O_3(s) + H_2O(l) + Zn(NH_3)_2Cl_2(s)$
Alkaline Battery	$2\ MnO_2(s) + Zn(s) \longrightarrow Mn_2O_3(s) + ZnO(s)$
Mercury Battery*	$Zn(s) + HgO(s) \longrightarrow ZnO(s) + Hg(l)$
Silver Battery	$Cd(s) + Ag_2O(s) \longrightarrow CdO(s) + 2\ Ag(s)$
Zinc–Air Battery	$2\ Zn(s) + O_2(g) \longrightarrow 2\ ZnO(s)$

* Because of the harm to the environment caused by the disposal of mercury batteries, the production of these batteries has recently been banned in the U.S.

All the "batteries" listed in this table are designed for "throw–away" use because there are no practical means of regenerating the reactants. Cells in which the electrochemical reaction produces energy are known as "galvanic cells" or "voltaic cells" in honor of two scientists who contributed much to the understanding of electrical phenomena, Alessandro Volta and Luigi Galvani (1737–1798). Of course the reason these cells eventually have to be thrown away is that they "go dead" because they run out of reactants and the reaction must cease. Another type of galvanic cell that is designed to never run out of reactants is termed the fuel cell. In this type of cell reactants are continuously fed into the chemical reaction and products removed so that the cell continues to generate electricity. The best known of these fuel cells are the hydrogen–oxygen fuel cells used in the space program. They operate via a chemical reaction that indirectly combines hydrogen and oxygen gases at a low temperature to form water:

$$2\ H_2(g) + O_2(g) \longrightarrow 2\ H_2O(l)$$ EQ. 6.38

This cell uses very expensive catalysts to promote the reaction and, except for such uses as the space program where the expense can be justified for a low–weight, inexhaustible (as long as hydrogen and oxygen gases are available) source of electrical energy. Of even greater potential interest is the development of fuel cells that will use organic compounds such as hydrocarbons and/or alcohols as reducing agents and atmospheric oxygen as the oxidizing agent. Low temperature fuel cells are far more efficient in converting chemical energy into useful work than are the internal combustion engines used for powering automobiles. Fuel cells are also less polluting, since they do not have the by–products associated with high–temperature reactions in an internal combustion engine. Fuel cells may one day help to eliminate the problem of air pollution that now plagues many cities.

If energy is provided from an outside source, it is also possible to reverse the direction of the reaction in an electrochemical cell. Such a cell is termed an "electrolytic cell." Certain types of electrochemical cells have been designed that permit use in both directions — as a galvanic cell for the production of energy and as an electrolytic cell for the recharging of the cell. Technically the word "battery", or more completely "storage battery" should be used to refer to this type of cell only. I.e., the electrochemical cells in Table 6.2 should not be called batteries. Table 6.3 includes a number of storage batteries in common use.

TABLE 6.3: STORAGE BATTERY SYSTEMS

TABLE 6.3

NAME	CHEMICAL REACTION (discharge cycle — galvanic cell)
NiCad Battery	$Cd(s) + 2\,NiO(OH)(s) + H_2O(l) \longrightarrow Cd(OH)_2(s) + Ni(OH)_2(s)$
Lead–Acid Storage Battery	$Pb(s) + PbO_2(s) + 2\,H_2SO_4(aq) \longrightarrow 2\,PbSO_4(s) + 2\,H_2O(l)$

Electrolytic cells are also used in the electroplating industry where chromium, silver, and gold are often deposited on a backing of a less–expensive metal to produce the shine and/or corrosion–resistance of the more expensive metal. Electroplating is often carried out by making the electrode at which oxidation occurs the metal to be deposited and making the object to be covered the electrode at which reduction occurs. Both electrodes are then dipped into a solution containing the ions of the metal being plated. Performed in this manner there is no change in concentration of the reactive species as the plating proceeds because the ions are continually replaced. To illustrate, if we consider a silver–plating operation, the oxidation half–reaction would be:

$$Ag(s) \longrightarrow Ag^+ + e^-$$ EQ. 6.39

while the reduction half–reaction would be:

$$Ag^+ + e^- \longrightarrow Ag(s)$$ EQ. 6.40

Of course this leads to no net reaction but the complete process does transfer silver atoms from one electrode to the other which is what we wanted to accomplish in the electroplating operation. Figure 6.6 illustrates just such an electroplating operation.

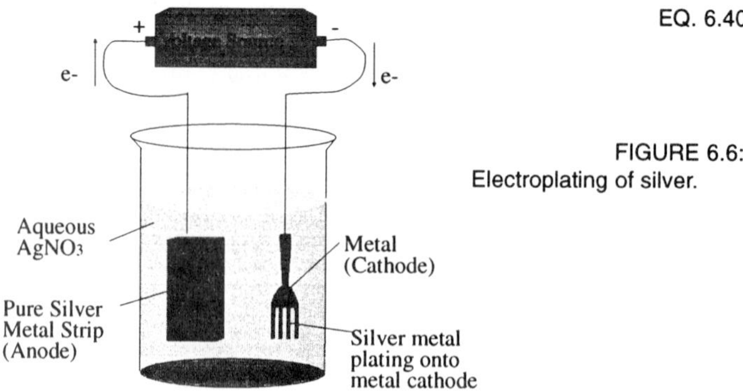

FIGURE 6.6: Electroplating of silver.

6.8 CORROSION

Corrosion occurs by an oxidation reaction in which a metal is oxidized and another material, usually oxygen gas from the atmosphere, is reduced. The most common type of corrosion is the rusting of iron that can be expressed as a chemical reaction as:

$$x\,\mathrm{Fe(s)} + y/2\,\mathrm{O_2(g)} \longrightarrow \mathrm{Fe}_x\mathrm{O}_y\mathrm{(s)}$$ EQ. 6.41

where the stoichiometric coefficients are left as variables because a variety of different products are obtained from this reaction — FeO, Fe_2O_3, Fe_3O_4, and mixtures are all possible. Besides being a cosmetic problem, the corrosion of iron creates a structural problem because the iron oxides (rust) formed do not adhere tightly to the metal, flake off the underlying metal, and weaken the article. Eventually, if the corrosion process is not stopped, the entire thickness of iron is attacked and a hole develops.

Corrosion is not a problem with some other metals, e.g., aluminum, because the oxide formed adheres tightly to the underlying metal and protects it from further attack by oxygen in the atmosphere. The chemical equation for the corrosion reaction of aluminum is shown in equation 6.42. Indeed, this reaction

$$\mathrm{Al(s)} + 3/2\,\mathrm{O_2(g)} \longrightarrow \mathrm{Al_2O_3(s)}$$ EQ. 6.42

is even more strongly favored than the oxidation of iron even though aluminum is normally considered to be a metal that does not corrode. It is interesting to take a sharp knife and scratch the dull surface of a natural aluminum object such as a storm door. Looking quickly you can see the bright shiny aluminum metal under the scratch where the overlay of aluminum oxide had been scraped away. This surface very quickly becomes dull because the oxygen in the atmosphere attacks the fresh surface very rapidly. Nevertheless aluminum objects never develop the structural problems than iron objects develop unless the aluminum is subject to some other type of chemical attack that dissolves the aluminum oxide film.

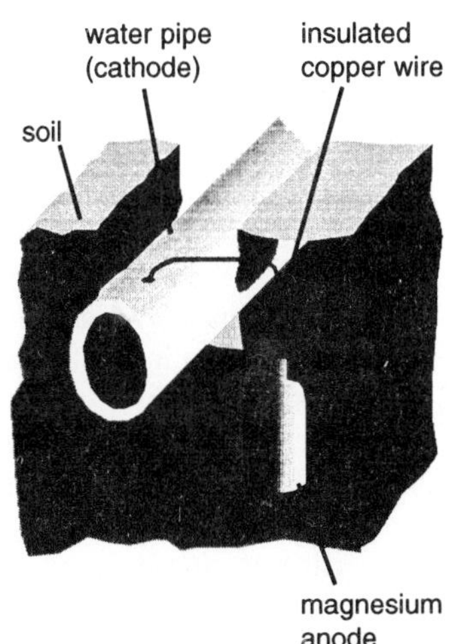

FIGURE 6.7: Sacrificial anode at work.

An interesting application related to electrochemical cells is used to protect normally corrodable objects that would be difficult or expensive to replace. In this application a relatively large amount of a more active (more easily oxidized) metal is placed in electrical contact with the metal being protected. This electrical contact between the two metals ensures that any electrons lost in the oxidation reaction will come from the more active metal. That large mass of metal is then slowly corroded away leaving the other metal unchanged. Water heaters and underground pipes are specific examples of metallic objects often protected by what is termed a sacrificial **anode**. The metal often used for the sacrificial anode is the very active alkaline earth metal, magnesium.

6.9 STOICHIOMETRY AND ELECTROLYSIS

In electrolytic cells, electrical energy is supplied to drive a reaction. Chemically, the reactions may appear to have electrons as a reactant species in the reduction half reaction, and as a product in the oxidation half reaction. Consider the electrolysis of molten sodium chloride, pictured in Figure 6.8:

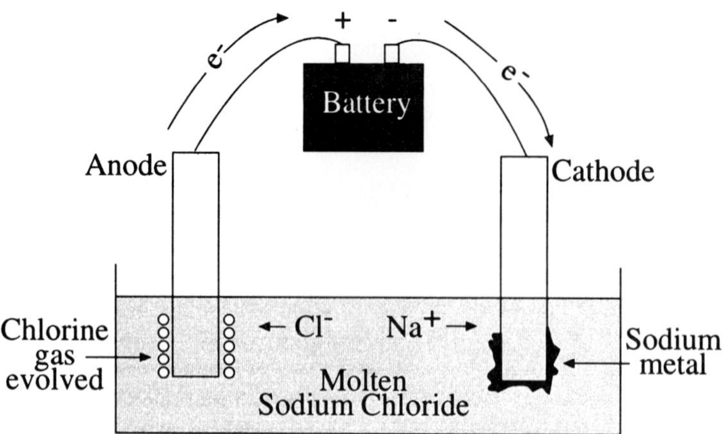

FIGURE 6.8:
Electrolysis of molten sodium.

The net reaction occuring in the cell is

$$NaCl(l) \longrightarrow Na(s) + \tfrac{1}{2}Cl_2(g)$$ EQ. 6.43

Which is the combination of the two half reactions:

$$Na^+ + e^- \longrightarrow Na(s)$$ EQ. 6.44

$$Cl^- \longrightarrow \tfrac{1}{2}Cl_2(g) + e^-$$ EQ. 6.45

While the electrons cancel in the net reaction, one electron is associated with the reduction of one sodium ion and the formation of one sodium atom. Similarly, the amount of chlorine produced is related to the number of electrons used in the electrolysis. While the amount of sodium or chlorine involved in the reaction can be determined by mass, the same is not true for electrons. The number of electrons involved in electrolysis reactions are determined in another fashion.

Each electron carries a specific amount of charge. One mole of electrons (6.022×10^{23} electrons) carries a combined charge of 96,485 C, where C (coulombs) is the unit of charge. Electric current is the flow of charge per unit time. Electric current is measured in amperes (A). An ampere is defined as a coulombs per second. Thus the value of the measured current multiplied by the time during which the current flows yields the total charge moved:

$$(\text{current})(\text{time}) = \text{charge}$$ EQ. 6.46

$$(\text{coulombs/sec})(\text{sec}) = \text{coulombs}$$ EQ. 6.47

Michael Faraday (1791-1867) first worked out the relationship between the amount of electric charge and the extent of chemical reaction. A mole of electrons is called a Faraday in his honor, and the constant used to convert between electrical charge and the number of electrons (96,485 C/mole of electrons) is known as Faraday's constant.

Faraday's results showed that the amount of material liberated at an electrode during an electrolysis procedure was proportional to the charge used. An apparatus used to demonstrate this result is given in the figure below. A battery delivers a current through three aqueous solutions: silver nitrate, copper (II) nitrate, and indium (III) nitrate. By measuring the current and the time the current flows, the total charge moving through each compartment is known. By weighing each cathode (the electrode at which the metal ion is reduced) before and after the current flows, Faraday was able to quantitatively establish the stoichiometry of the electrolysis reactions.

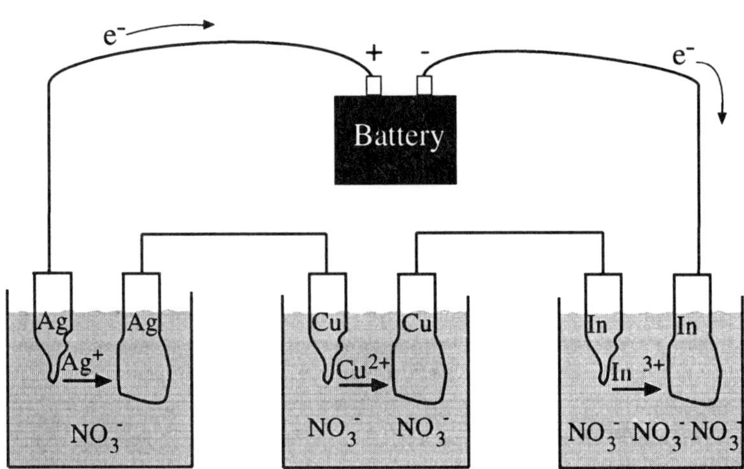

FIGURE 6.8: Electrodeposition of metals from salt solution.

For example, when a current of 1.000 ampere (A) flowed for 96,485 seconds, the amount of metal deposited at the cathode was about 107.9 g for silver, 31.77 g for copper, and 38.27 g for indium. Increasing the current, or increasing the time of electrolysis by a certain percentage would increase the mass of metal deposited at each cathode by that same percentage. When 1.000 ampere flows for 96,485 seconds, the total charge moved through each compartment is 96,485 coulombs, or one mole of electrons. The balanced half reactions are:

$$Ag^+ + e^- \rightarrow Ag(s)$$ EQ. 6.48

$$Cu^{2+} + 2\ e^- \rightarrow Cu(s)$$ EQ. 6.49

$$In^{3+} + 3\ e^- \rightarrow In(s)$$ EQ. 6.50

One can readily determine from the masses that 1.000 moles of silver, 0.5000 moles of copper, and 0.3333 moles of indium were deposited.

Concept Check: Stoichiometry of Electrolysis

In Figure 6.8, it was shown that a current of 1.000 amperes run for 96,485 seconds would deposit about 107.9 g of silver, 31.77 g of copper, or 38.27 g of indium. If a compartment containing aluminum metal and $Al^{3+}(aq)$ was added, how long would the current of 1.000 amperes have to run to deposit 14.50 g of aluminum (the mass of a typical 12 ounce soft drink can?)

Solution: The balanced half reaction is: $Al^{3+} + 3\ e^- \rightarrow Al(s)$. That is, 3 moles of electrons are required to produce 1 mole of aluminum. The mass of aluminum required can be converted to Faradays by:

$$\left(14.50\text{ g Al}\right)\left(\frac{1\text{ mol Al}}{26.982\text{ g Al}}\right)\left(\frac{3\text{ mol e}^-}{\text{mol Al}}\right) = 1.612\text{ Faradays}$$

The time required can be calculated by multiplying the moles of electrons required (Faradays) by Faraday's constant, and dividing by the current of 1 ampere:

$$\frac{\left(1.612\text{ F}\right)\left(96{,}485\ ^{C}\!/\!_{F}\right)}{1\ ^{C}\!/\!_{s}} = 1.555 \times 10^5\text{ seconds}$$

Note that it takes much more electricity to obtain a given weight of aluminum than any of the other metals examined in this example. Since aluminum is obtained from its ore by an electrolytic process, that process is rather expensive on a cost/gram basis. That is why recycling centers will pay for aluminum cans, but not other cans (mostly iron.) Iron can be produced in a more cost effective manner than aluminum.

Actual yield: the quantity of a product that was obtained when a reaction was run.

Anode: the electrode at which oxidation occurs

Battery: a storage device for electrochemical energy

Cathode: the electrode at which reduction occurs

Electrochemical cell: a combination of an anode, a cathode, and the material undergoing a redox reaction.

Electrolytic cell: an electrochemical cell that uses externally provided electrical energy to drive a chemical reaction

Faraday: a mole of electrons

Formula weight: the sum of the weights (in amu) of all atoms appearing in one formula unit of an ionic compound. This is numerically equal to the grams of compound present per mole of formula units.

Fuel cell: a galvanic cell in which the reactants (the fuel) are continuously supplied

Galvanic cell (also known as a voltaic cell)**:** an electrochemical cell that uses a chemical reaction to produce an electric current.

Limiting reagent: the reactant that is consumed while other reactants remain in excess.

Oxidation: the loss of electron(s) by a species in a chemical reaction

Oxidizing agent: a chemical species which causes another to be oxidized

Ozonolysis: an ozone induced cleavage of a carbon-carbon multiple bond

Percent yield: the ratio of the actual yield to the theoretical yield expressed as a percent.

Reducing agent: a chemical species that causes another to be reduced.

Reduction: chemically, the gain of electrons by an atom during a chemical reaction.

Sacrificial anode: a piece of readily oxidized metal electrically connected to other material that it protects from corrosion

Stoichiometry: the relative amounts (in moles, molecules or formula units) of reactants and products in a chemical reaction.

Theoretical yield: the quantity of a product that would be obtained if the reaction went to completion as written.

HOMEWORK Problems

OXIDATION-REDUCTION

1. Calculate the oxidation number of each element in the following:

   (a) $NaBrO_3$
   (b) Na_2O_2
   (c) SO_3^{2-}
   (d) $XeOF_4$
   (e) K_2CrO_4
   (f) KO_2
   (g) $(NH_4)_2S_2O_3$
   (h) I_3^-

2. Predict the products of the following reactions:

(a) H, O — $\xrightarrow[\text{time}]{O_3}$

(b) CH_2 — $\xrightarrow[\text{2. Oxidative Workup}]{\text{1. } O_3}$

(c) O — $\xrightarrow{LiAlH_4}$

(d) O, O, OCH_3 — $\xrightarrow{NaBH_4}$

(e) $\xrightarrow{KMnO_4}$

(f) O — $\xrightarrow{CH_3CO_3H}$

(g)

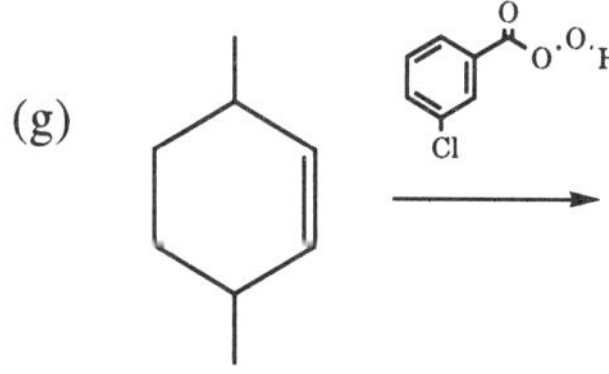

3. Explain, in your own words, how biochemical redox agents work.

4. When 0.4856 g of a compound, known to contain only C, H, and O, was burned in excess oxygen, 1.2122 g of CO_2 and 0.5953 g of H_2O were found. Determine the empirical formula of the compound.

5. Balance the following chemical equations (where appropriate write the equations in the form of net ionic reactions):

   (a) $Fe(s) + S_8(s) \longrightarrow Fe_2S_3(s)$

(b) $FeCO_3(s) + HCl(aq) \longrightarrow FeCl_2(aq) + H_2O(l) + CO_2(g)$
(c) $Ba(NO_3)_2(aq) + H_2SO_4(aq) \longrightarrow BaSO_4(s) + HNO_3(aq)$
(d) $C_8H_{18}(l) + O_2(g) \longrightarrow CO_2(g) + H_2O(l)$

6. Balance the following redox equations in the indicated medium (write net ionic equation where appropriate):

(a) $V^{2+} + H_2O_2(aq) \longrightarrow VO^{2+}$ (basic solution)
(b) $S_2O_8^{2-}(aq) + NH_3(aq) \longrightarrow NO_2(g) + SO_4^{2-}(aq)$ (acid solution)
(c) $SnCl_4^{2-}(aq) + HgCl_2(aq) \longrightarrow Hg_2Cl_2(s) + SnCl_6^{2-}(aq)$
(Note: $HgCl_2(aq)$ is unionized in solution) (HCl solution)
(d) $Cu(s) + HNO_3(aq) \longrightarrow Cu^{2+}(aq) + NO(g)$ (acid solution)

7. Balance the following redox equations:

(a) $CH_3CH_2CH_2OH + MnO_4^- \longrightarrow CH_3CH_2COOH + Mn^{2+}$ (acid solution)
(b) $CH_3CH(OH)CH_3 + MnO_4^- \longrightarrow CH_3COCH_3 + MnO_2$ (basic solu tion)
(c) acetaldehyde + $Cr_2O_7^{2-} \longrightarrow$ acetic acid + Cr^{3+} (acid solu tion)
(d) 2-butanone + $NaBH_4 \longrightarrow$ 2-butanol + H_3BO_3 (acid solu tion)

8. Balance the following redox reactions:

(a) cyclohexanone + $LiAlH_4 \longrightarrow$ cyclohexanol + $Al(OH)_4^-$ (base)
(b) 1-butanol + $CrO_3 \longrightarrow$ butanal + $Cr(OH)_3$ (acid)
(c) 2-methyl-2-butene + $MnO_4^- \longrightarrow$ acetaldehyde + acetone + Mn^{2+} (acid)

9. How many grams of carbon dioxide are produced by the complete combustion of 43.0 g of n–hexane?

10. What weight of sulfur dixode can be removed from an exhaust stream by 4.30 metric tons (1 metric ton = 1000 kg) of calcium oxide? How much calcium sufite would be produced?

11. What weight of the amino acid phenylalanine ($C_9H_{11}NO_2$) is produced by the complete hydrolysis of 50.0 g of Nutrasweet ($C_{14}H_{18}N_2O_5$)? (The reaction produces 1 mole of phenylalanine per mole of Nutrasweet hydrolyzed.)

12. Hydrogen fluoride (HF) is produced by the treatment of calcium fluoride (CaF_2) with hot, concentrated phosphoric acid (H_3PO_4). How much hydrogen fluoride can be produced by the reaction of 500.0 g of phosphoric acid with 300.0 g of calcium fluoride? How much calcium phosphate is produced? Which reactant is in excess and by how much? The (unbalanced) chemical reaction is

$$CaF_2 + H_3PO_4 \longrightarrow HF + Ca_3(PO_4)_2$$

13. Mesitylene (1,3,5-trimethylbenzene [C_9H_{12}]) can be made by the reaction of methyl bromide (CH_3Br) with benzene (C_6H_6) via the following, (unbalanced) reaction:

$$CH_3Br + C_6H_6 \longrightarrow C_9H_{12} + HBr$$

What weight of mesitylene can be made by the reaction of 80.0 g of methyl bromide with 80.0 g of benzene if the reaction goes exactly as the balanced reaction indicates?

14. When the reaction in problem 17 was actually run, a total of 25.85 g of mesitylene were recovered. What was the percent yield for the reaction?

15. The reaction of 85.0 g of sodium fluoride with 60.0 g of calcium nitrate produced 27.30 g of calcium fluoride. What was the percent yield for the reaction?

CHAPTER SIX

OVERVIEW Problems

16. When butanal reacts with $LiAlH_4$, an organic product is isolated. The IR spectrum of this product has a broad absorbance cetered aroun 3350 cm^{-1}. The strong absorbance at 1700 cm^{-1} that had been present in the spectrum of the reactant (butanal) is no longer present in the IR spectrum of the product.

(a) The IR absorbance at 1700 cm^{-1} in the spectrum of the reactant was due to which functional group?

(b) The IR absorbance centered around 3350 cm^{-1} in the spectrum of the product is due to which functional group?

(c) The butanal molecule was (oxidized/reduced)

(d) $LiAlH_4$ was the (oxidizing/reducing) agent.

17. The organic product isolated from the reaction between butanal and $LiAlH_4$ was examined by HNMR spectroscopy. The spectrum showed 5 unique proton environments. The integrated signal showed that these were due to 1, 2, 2, 2, and 3 protons respectively. Draw the structure of this product, and show the 4 unique proton environments by labeling each type of proton in your structure.

18. When 1-propanol reacts with MnO_4^-, an organic product is isolated that has a broad IR absorbance between 3500 and 2500 cm^{-1}, and a strong absorbance at 1720 cm^{-1}. The CNMR of the organic product has three signals, one of which occurs at 182 ppm. The HNMR spectrum shows 3 environments: a singlet (corresponding to 1 proton) at 11.2 ppm, a quartet (corresponding to 2 protons) at 2.4 ppm, and a triplet (corresponding to 3 protons) at 1.2 ppm.

(a) Compared to 1-propanol, the organic product had more oxygen atoms and fewer hydrogen atoms. Consequently in this reaction of 1-propanol, the #1 carbon (the carbon previously bonded to the hydroxyl group) is said to be (oxidized/reduced).

(b) The two absorbances noted in the IR spectrum of the product are consistent with what single functional group?
(c) The CNMR signal at 182 ppm is due to what type of carbon?
(d) Draw the structure of the organic product.
(e) Interpret the HNMR spectrum, indicating which protons shown in *(d)* corresponds to which signal .

19. When 2-butanol reacts with $Cr_2O_7^{2-}$, an organic product is isolated that has no IR absorbance above 3000 cm^{-1}, but does have a strong absorbance at 1720 cm^{-1}. The CNMR of the organic product has four signals, one of which occurs at 209 ppm. The HNMR spectrum shows 3 environments: a quartet (corresponding to 2 protons) at 2.4 ppm, a singlet (corresponding to 3 protons) at 2.1 ppm, and a triplet (corresponding to 3 protons) at 1.1 ppm.

(a) Compared to 2-butanol, the organic product had fewer hydrogen atoms. Consequently in this reaction of 2-butanol, the #2 carbon (the carbon previously bonded to the hydroxyl group) is said to be (oxidized/reduced).
(b) The absorbance noted in the IR spectrum of the product is consistent with what functional group?
(c) The CNMR signal at 209 ppm is due to what type of carbon?
(d) Draw the structure of the organic product.
(e) Interpret the HNMR spectrum, indicating which protons shown in *(d)* corresponds to which signal .

20. In a hydrolysis reaction, 50.0 g of ethyl acetate (MW = 88.10) react with excess water.

(a) Write the balanced reaction.
(b) What is the maximum mass of acetic acid produced?
(c) What is the maximum mass of ethanol produced?

21. When 10.0 g of 2-methylpropene react with excess HCl, an addition product is formed.

(a) Write the two possible products of this reaction.
(b) Indicate which of the two products is favored.
(c) Calculate the theoretical yield of the addition product (i.e., the maximum mass of the favored product that could be pro duced.)

22. In a dehydrohalogenation reaction, 10.000 g of bromocyclohexane (MW = 163.06) are reacted with 3.000 g of potassium hydroxide (FW = 56.10).

(a) Write the balanced reaction.
(b) Using a limiting reagent table, determine the maximum mass of cyclohexene produced.

23. Meta-chloroperbenzoic acid, a useful organic oxidizing agent, is synthesized by reacting meta-chlorobenzoic acid ($C_6H_4ClCOOH$, MW = 156.56) with hydrogen peroxide (H_2O_2, MW = 34.015). The synthesis procedure starts with 10.00 g of meta-chlorobenzoic acid.

(a) Write the balanced reaction.

(b)What mass of 30.0% (by mass) aqueous hydrogen peroxide solution must be added to the benzoic acid to assure a stoichiometric amount of each?

24. Elemental analysis shows that compound X is 77.75% carbon, 7.46% hydrogen, and 14.80% oxygen by mass. The molecular weight of the compound, determined by mass spectroscopy, is about 108 amu.

(a)What is the empirical formula of X?

(b)What is the molecular formula of X?

25. Compound X in the previous problem has a strong IR absorbance between 3500 and 3200 cm^{-1}, between 3150 and 3000 cm^{-1}, and between 3000 and 2800 cm^{-1}, but no absorbance between 2700 and 1500 cm^{-1}. The proton NMR of X shows a singlet at 7.4 ppm (corresponding to 5 protons), a singlet at 4.7 ppm (corresponding to 2 protons), and a singlet at 2.5 ppm (corresponding to 1 proton).

(a) Draw the structural formula of X.

(b) Explain the significance of the IR absorbance information.

(c) Label the protons in the structural formula of X in relation to the peaks in the proton NMR spectrum.

26. When compound X is treated with hydrogen peroxide, organic compound Y is formed. Compound Y is 79.22% carbon, 5.70% hydrogen, and 15.08% oxygen by mass. Mass spectroscopy indicates a molecular weight of about 106 amu.

(a) What is the empirical formula of Y?

(b) What is the molecular formula of Y?

27. Compound Y in the previous problem has no strong absorbance between 3500 and 3200 cm^{-1}, but it does absorb strongly at 1700 cm^{-1}. The proton NMR spectrum shows a singlet at 10.0 ppm (corresponding to one proton), and a complex splitting pattern between 7.4 and 7.9 ppm (corresponding to five protons).

(a) Draw the structural formula of Y.

(b) Explain the significance of the IR absorbance information.

(c) Label the protons in the structural formula of Y in relation to the peaks in the proton NMR spectrum.

(d) Balance the redox reaction: $X + H_2O_2 \longrightarrow Y + H_2O$

28. Compound V was analyzed to be 66.63% carbon, 11.18% hydrogen, and 22.19% oxygen by mass. The IHD of the molecular formula is 1.0. The IR spectrum revealed a strong absorbance at 1700 cm^{-1}.

(a) What is the empirical formula of V?

(b) What functional group(s) are likely in compound V?

29. The proton NMR spectrum of Compound V from above showed a quartet at 2.5 ppm (corresponding to 2 protons), a singlet at 2.2 ppm (corresponding to 3 protons) and a triplet at 1.1 ppm (corresponding to 3 protons).

(a) What is the structural formula of V?

(b) Label the protons in the structural formula of V in relation to the peaks in the NMR spectrum.

30. Compound V reacts with $NaBH_4$ to form organic compound W.

Compound W has a strong IR absorbance between 3600 and 3200 cm^{-1}, but no absorbance between 2800 and 1500 cm^{-1}. Compound W has a singlet at 4.0 ppm (corresponding to 1 proton), a sextet at 3.7 ppm (corresponding to 1 proton), a complex splitting pattern around 1.5 ppm (corresponding to 2 protons), a doublet at 1.2 ppm (corresponding to 3 protons), and a triplet at 0.9 ppm (corresponding to 3 protons).

(a)Draw the structural formula of W.

(b) Explain the significance of the IR absorbance information.

(c) Label the protons in the structural formula of W in relation to the peaks in the proton NMR spectrum.

(d) Balance the redox reaction: $V + NaBH_4 \longrightarrow W + H_3BO_3$

CHAPTER seven

MOLECULAR Structure

MOLECULAR Structure

We have a variety of ways of symbolizing molecules. We can use the word "acetone", we can give the formula C_3H_6O, or we give the formula as CH_3COCH_3 in an effort to emphasize what atoms are bonded together and in what order. But to really "see" the molecule, we build models or create images that reflect acetone's three-dimensional geometry.

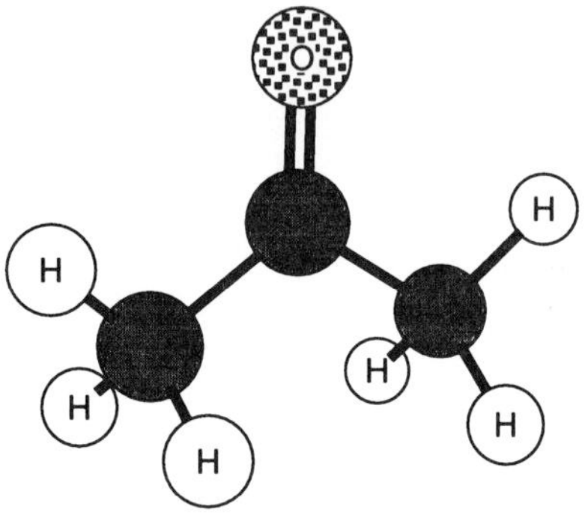

Molecular geometry, i.e., the position of individual atoms relative to one another, is of paramount importance in determining the properties of the molecule. Enzymes, for example, are large organic molecules that play critical roles in biological processes. Every enzyme has a particular shape and function. Even a minor alteration in the shape of the enzyme may render it non-functional and prove fatal to the organism. Another example is the design of drugs. It is becoming more common that such efforts are based at least in part upon how the structure of a molecule (yet to be synthesized) matches the geometric parameters thought to determine the drug's effectiveness. In this chapter we build a theory to predict molecular structure, that is, where atoms are positioned relative to one another. We will find that the keys to this determination are the electrons.

FIGURE 7.1: Structure of acetone.

7.2 VALENCE ELECTRONS: WHERE THE ACTION IS!

Earlier, we described the energy levels occupied by an atom's electrons using terms like $1s^2\ 2s^2\ 2p^6\ 3s^2\ 3p^6$ etc. We have also noted that these electrons occupy either the outermost shell (called the valence shell) or else are **core electrons**, which are closer to the nucleus and at lower energy than the valence electrons. The valence electrons, being outermost, are the electrons that are shared when atoms covalently bond to one another.

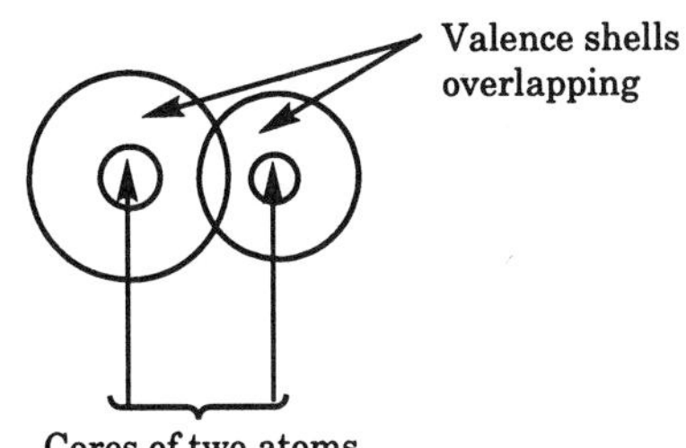

FIGURE 7.2: Orbital overlap.

Similarly, when two atoms interact to become ions, the electron is typically transferred from the valence shell of a metal to an available opening in the valence shell in a non-metal.

It is the electron distribution that dictates the physical and chemical properties of an atom or group of atoms. In particular, the structure of groups of atoms depends upon the positioning of valence electrons around the atoms.

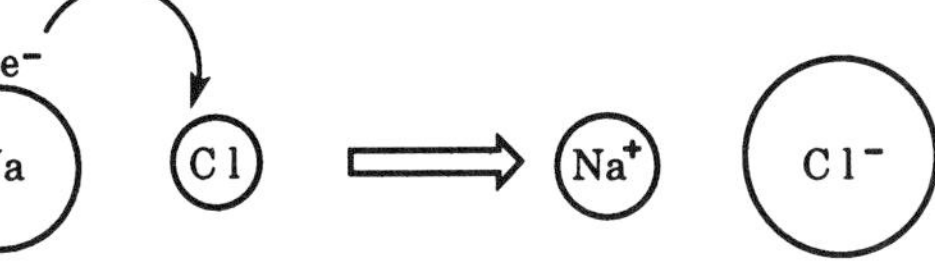

FIGURE 7.3: Formation of ionic NaCl.

7.3 TWO BY TWO

By drawing Lewis dot diagrams, we pictured the number of electrons around each atom in a group of atoms bonded together. Electrons around an individual atom may be pictured as belonging entirely to that atom (lone pairs) or pictured as being shared between two atoms (bonding pairs.)

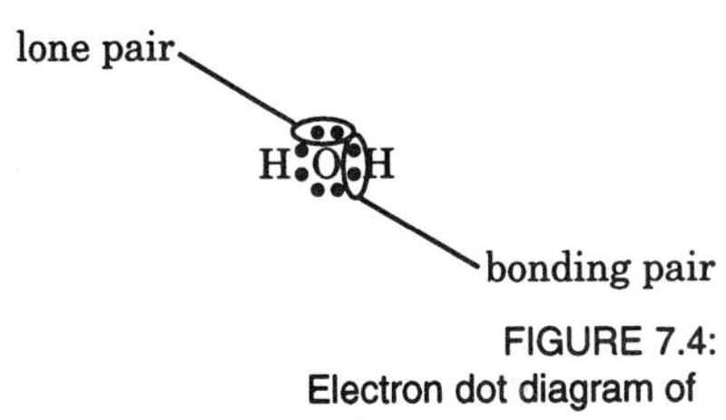

FIGURE 7.4: Electron dot diagram of water.

We typically refer to such electrons as pairs, although on occasion unpaired electrons can occur in Lewis dot diagrams. This happens when an odd number of electrons are available. Nitric oxide, NO, is an example of an odd electron species.

:N::O:

FIGURE 7.5: Electron dot diagram of NO.

The odd electron is shown as belonging to nitrogen. Since oxygen is more electronegative than nitrogen, oxygen will achieve its octet at the expense of nitrogen.

Electrons in dot diagrams are typically treated as pairs (with a single electron as a necessary exception.) When multiple bonding occurs, such as in the nitrogen molecule, we refer to three pairs of electrons, not a pair of electron triplets.

:N:::N:

FIGURE 7.6: Electron dot diagram of N_2.

Pairs predominate in Lewis dot diagrams for the same reason that pairs predominated in electron configurations. We write $1s^2\ 2s^2\ 2p^6$ (actually $2p_x^{\ 2}\ 2p_y^{\ 2}\ 2p_z^{\ 2}$) because no more than two electrons can occupy the same position in space. An electron can join another electron in the same position only if it has the opposite spin (e.g., positive instead of negative.) This is known as **Pauli's Exclusion Principle**. Once a pair of electrons occupies a position, it repels any additional electrons.

7.4 VALENCE SHELL ELECTRON PAIR REPULSION (VSEPR)

Lewis dot diagrams should not be interpreted too literally. As we previously noted with the specific example of water, a Lewis dot diagram does not indicate molecular geometry. While water can be represented by either Lewis dot diagram below in Figure 7.7, the experimentally observed angle formed by the hydrogen oxygen hydrogen is neither 180° nor 90°.

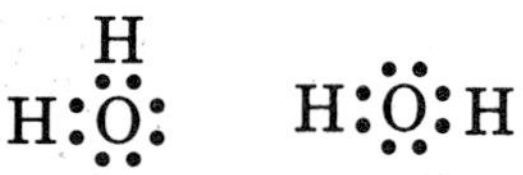

H
H:Ö: H:Ö:H

FIGURE 7.7: Electron dot diagram of water.

Dot diagrams are, after all, two-dimensional, while molecules exist in a three-dimensional universe. While triatomic molecules by definition will be planar, molecules with 4 or more atoms may not be planar.

Valence shell electron pair repulsion theory (VSEPR) is an attempt to predict the geometric arrangements of groups of atoms. It is based upon the repulsion that pairs of electrons around the same atom experience.

> **SIDEBAR: VIOLATION OF THE OCTET RULE— ELECTRON DEFICIENT SPECIES**
>
> As mentioned in Chapter One, the octet rule is sometimes violated. For example, species with an odd number of electrons must violate the rule. Other violations occur when a species is, for some reason, apparently content to have less than an octet of electrons. Electron deficient species often have a central atom containing a Group IIA or Group IIIA element, such as beryllium or boron. When we examine the structure of compounds like $BeCl_2$ and BCl_3, it is determined that the central atom has fewer than eight valence electrons around it. Application of the NAS approach to Lewis dot diagrams would indicate that the octet rule could be satisfied by forming one or more double bonds between chlorine and the central atom. Multiple bonds do not form between chlorine and either beryllium or boron because of a large electronegativity difference. Chlorine would be contributing 3 of the 4 electrons in a double bond between itself and either beryllium or boron. Since chlorine is much more electronegative than beryllium or boron, it does not share any extra electron density. We shall see additional justification for this violation of the octet rule when we study formal charge in Chapter 13.

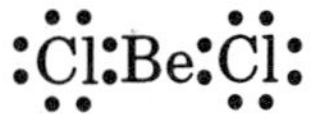

:C̈l:Be:C̈l:

FIGURE 7.8: Electron dot diagram of $BeCl_2$.

Consider the case of the gas phase molecule $BeCl_2$, whose dot diagram is shown in Figure 7.8.

The chlorine atoms are held to the beryllium atom by bonding pairs of electrons. The **electronic geometry** around beryllium (i.e., where in space the two pairs of electrons are located) determines where the chlorine atoms will be, and thus determines the molecular geometry. The repulsive forces between these two pairs of electrons around beryllium will act to maximize the separation. The farthest apart the electron pairs can locate themselves and still be in beryllium's valence shell is 180° apart, illustrated in Figure 7.9.

:Be:

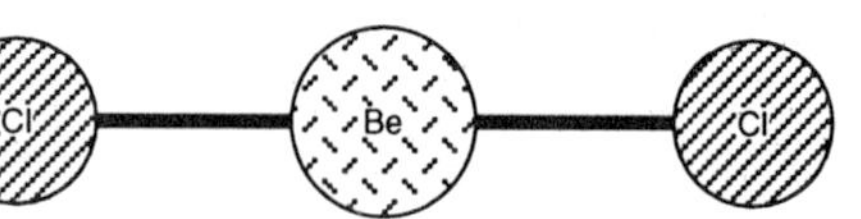

FIGURE 7.9: electronic and molecular geometry of $BeCl_2$

When the chlorine atoms are added to the picture, it is clear that the repulsion between the two pairs of electrons around beryllium results in a linear molecule of $BeCl_2$. Experimental evidence confirms that the $BeCl_2$ molecule is linear.

7.5 VSEPR AND 2 REGIONS OF ELECTRON DENSITY

H:C:::N:

FIGURE 7.10: Electron dot diagram of HCN.

$BeCl_2$ is determined to be linear based upon the following process: (1) drawing the Lewis dot diagram; (2) determining the number of regions of electron density around the central atom; and (3) maximizing the distance of separation of the valence electron pairs. The "regions of electron density" in step (2) means more than the number of electron pairs, although in the case of $BeCl_2$, those terms mean the same thing. Consider, however, the case of hydrogen cyanide, HCN. We begin with the Lewis dot diagram in Figure 7.10.

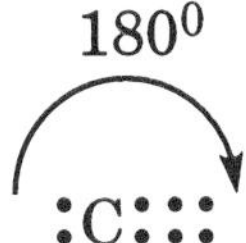

FIGURE 7.11: Electron geometry around C.

Note that the three bonding pairs of electrons between the carbon and nitrogen atoms do not really occupy the same region of space. That would violate the Pauli Exclusion Principle. They are drawn there by convention. (Remember the advice not to interpret Lewis dot diagrams too literally.) What is indicated in the Lewis dot diagram are two groupings of electrons around the carbon atom. These Lewis groupings are loosely termed "regions of electron density". This is a term strictly related to Lewis dot diagrams and VSEPR, and likewise should not be interpreted too literally. The two regions of electron density shown around the carbon atom in Fig 7.10 act as if they were two pairs of electrons. These groupings minimize their repulsion by maximizing their separation. As was the case for $BeCl_2$, the electronic geometry of these groupings around the carbon atom is described as linear:

H:C:::N:

FIGURE 7.12: electronic and molecular geometry of HCN.

Completing the picture by adding the peripheral hydrogen and nitrogen atoms to the bonding pairs of electrons yields a linear molecule as seen in figure 7.12.

Two regions of electron density are observed whenever a carbon atom forms a triple bond. Alkynes, given by the generic formula R—C≡C—R', have a Lewis dot diagram shown in Figure 7.13 .

:C:::C:

FIGURE 7.13: Electron dot diagram of an alkyne.

Both carbons engaged in the triple bond have two regions of electron density, and VSEPR predicts a linear arrangement. Acetylene, H—C≡C—H , is in fact a linear molecule. When R or R' represent one or more additional carbon atoms in the molecule, the molecule as a whole may not be linear. The **local molecular geometry** (i.e., of the 2 triply bonded carbon atoms and the X atoms immediately adjacent: X—C≡C—X will be linear.

7.6 VSEPR AND 3 REGIONS OF ELECTRON DENSITY

FIGURE 7.14: Electron dot diagram of B in BCl_3.

Three regions of electron density (or in a simple case, three electron pairs) around an atom minimize their repulsion and maximize the distance of separation by a triangular arrangement around the central atom. BCl_3 is an example of such a case. The boron atom is in a plane along with the three bonding pairs of electrons as shown in figure 7.14.

The electronic configuration around the boron atom is described as trigonal planar. Adding the chlorine atom to each bonding pair of atoms results in four coplanar atoms, with the three chlorine atoms corresponding to the points of an equilateral triangle circumscribing the boron atom. This molecular geometry is referred to as trigonal planar. Although figure 7.15(a) shows an equilateral triangle (i.e., with three equal sides,) note that the chlorine atoms are not actually bonded to each other. The lines in the first diagram are representative of a geometric relationship only. The actual bonding is shown in figure 7.15(b) and (c).

The Cl-B-Cl **bond angle** is measured to be exactly 120°, as predicted by VSEPR.

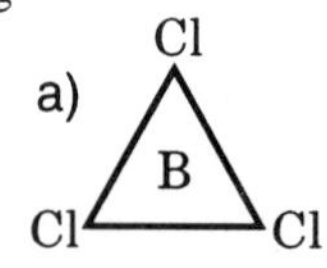

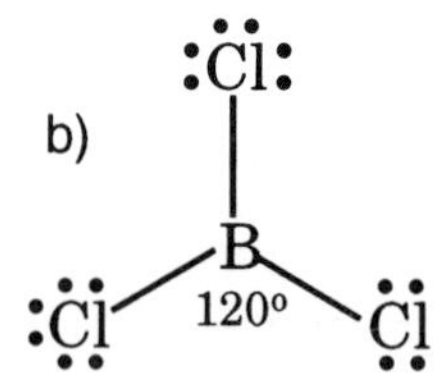

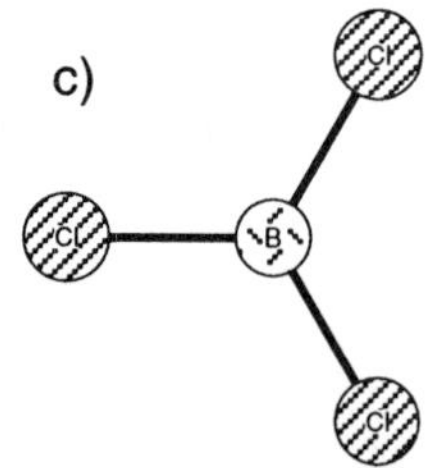

FIGURE 7.15: Molecular geometry of BCl_3.

The nitrate ion, NO_3^-, has three regions of electron density surrounding the central nitrogen atom, as shown in Figure 7.16 .

VSEPR predicts that the three regions of electron density will be situated at the corners of an equilateral triangle, and that the O-N-O bond angles will be exactly 120°. Again, experimental results confirm the prediction.

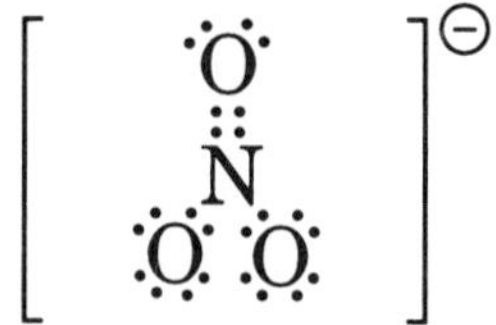

FIGURE 7.16: Electron dot diagram of NO_3^-.

A similar case would be formaldehyde (methanal), CH_2O. The Lewis dot diagram in Figure 7.18 indicates three regions of electron density around the carbon atom.

The electronic geometry around the central carbon atom would be trigonal planar, with the oxygen atom and the two hydrogen atoms representing the corners of a triangle. Since the three constituent atoms forming the imaginary triangle are not truly identical, the triangle is not equilateral, and the bond angles are only approximately 120°. (The H-C-H bond angle is about 118°, the H-C-O bond angle about 121°.)

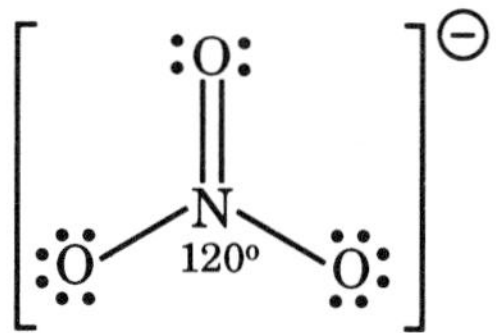

FIGURE 7.17: Molecular structure of nitrate ion.

As noted earlier, VSEPR can predict local molecular structure within larger molecules. The first example in this chapter was acetone, and the electronic geometry around its carbonyl carbon is trigonal planar, as indicated by the three regions of electron density around the central carbon in the Lewis dot diagram:

The trigonal planar electronic geometry around the central carbon atom indicates that the two terminal carbons and the oxygen atom will be situated at the corners of a triangle around the central carbon:

The local molecular geometry around the carbonyl carbon is also trigonal planar. The entire molecule can not, however, be described as planar. The terminal carbon atoms each have four regions of electron density, and as such the hydrogen atoms will not be co-planar. We develop this idea in the next section.

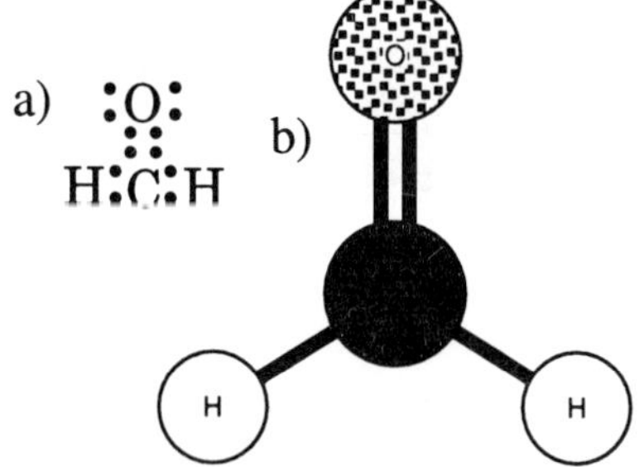

FIGURE 7.18: Formaldehyde as a) dot diagram, and b) ball and stick model.

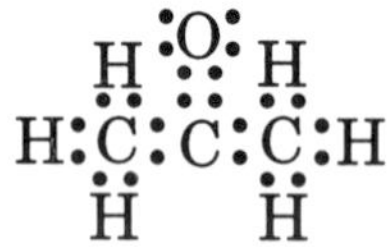

FIGURE 7.19: Electron dot diagram of acetone.

FIGURE 7.20: Local molecular geometry of CCOC.

7.7 VSEPR AND 4 REGIONS OF ELECTRON DENSITY

Methane, CH_4, has a simple Lewis dot diagram shown in Figure 7.21 The dot diagram by convention shows four pairs of electrons in a square around the central atom. This planar arrangement of the five atoms would predict a 90° H-C-H bond angle. VSEPR predicts that the repulsion between these four pairs of electrons will push them as far apart as possible from each other while keeping them within the valence shell of the carbon atom. While dot diagrams are restricted to 2 dimensions, electrons in the real world are not. Instead of a square planar electronic configuration, the electrons find a maximum separation in a non-planar geometry. These alternative locations of four electron pairs are shown in Figure 7.22 .

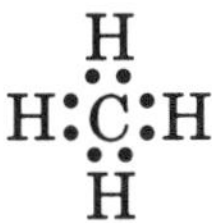

FIGURE 7.21: Electron dot diagram of methane.

A regular tetrahedron, pictured at right, is composed of four equilateral triangular faces. The four electron pairs attain maximum separation by occupying the vertices of the tetrahedron in Figure 7.23.

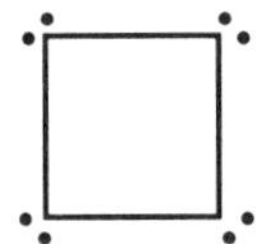

FIGURE 7.22: Square planar and tetrahedral electronic geometry.

The angle formed between any two corners of the tetrahedron and its center is about 109.5°. If carbon is pictured as being in the center of a tetrahedron and one of the four bonding pairs of electrons being on each corner, then the electronic geometry is described as **tetrahedral**. The repulsion between pairs of electrons at 109.5° is less than at 90°. Being farther apart represents more room and less repulsion. The tetrahedral orientation of electron pairs is preferred. The representation of molecular geometry is completed by adding the hydrogen atoms to the bonding electron pairs (Figure 7.24 .)

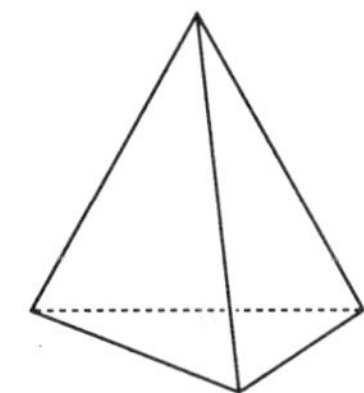

FIGURE 7.23: Solid tetrahedron.

Each of the carbons in alkanes form four single bonds. The requisite four bonding pairs of electrons exhibit tetrahedral electronic geometry. Thus the local molecular geometry around any carbon in an alkane is also tetrahedral. Consider the central carbon in n-pentane:

The third (or middle) carbon has four regions of electron density; two are shared pairs of electrons forming bonds with hydrogen atoms, and two are shared pairs of electrons forming bonds with adjacent carbon atoms. The local molecular geometry around the central carbon atom is tetrahedral, but since different groups are bonded, the exact internal angle characteristic of a regular tetrahedron will not be observed. Overall, however, the local molecular geometry will match the electronic geometry predicted by VSEPR whenever all electron pairs around the atom in question are bonded to another atom. The situation becomes more complicated when the regions of electron density include lone pairs in addition to bonding pairs of electrons. We shall examine this further in section 7.10.

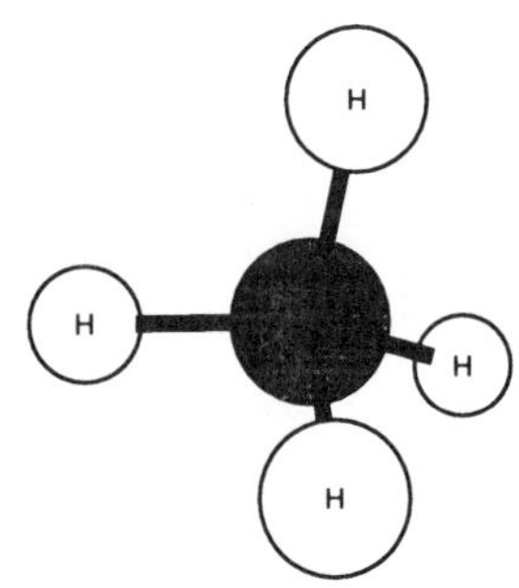

FIGURE 7.24: Molecular geometry of methane.

SIDEBAR: VIOLATION OF THE OCTET RULE— EXPANDED OCTETS

The octet rule will also fail for some compounds that have elements with more than 8 electrons in their valence shell. These will be elements from the third period or lower, and they will be the central atom in the compound. The NAS approach to constructing Lewis dot diagrams will not work for these compounds, and a different approach is required. Consider the compound AsF_5. The more metallic element is

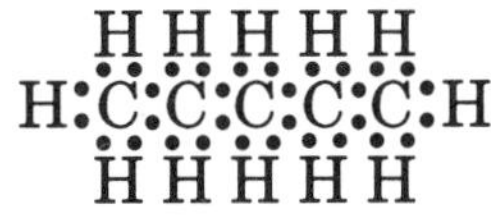

FIGURE 7.25: Lewis dot diagram of pentane.

assumed to be the central atom, with the other atoms arranges symmetrically around it. Arsenic is thus placed in the center and forms 5 bonds, one to each fluorine. Arsenic obviously will have more than an octet of electrons. The arsenic atom contributes 5 valence electrons, each fluorine contributes one unshared electron, so drawing a Lewis dot diagram is relatively straight forward.

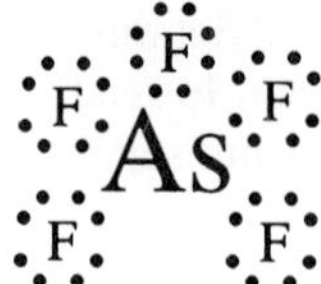

Now consider the compound XeO_2F_2. Again, the traditional approach of satisfying octets will not work. Instead, we begin by determining A, the number of available valence electrons. Xenon, the most metallic element (lowest electronegativity) is placed in the center, with two oxygen atoms and two fluorine atoms around it. Next each atom is connected to the central atom with a pair of electrons. At this stage, the dot diagram is

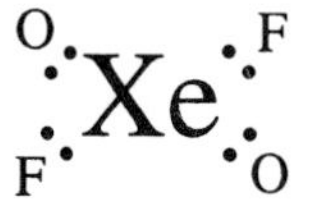

Next, the octet is completed for all peripheral atoms.

If any available electrons have not yet been included in the diagram, they are now added to the central atom as pairs of electrons. For our example, A = 34, and only 32 electrons have appeared in the diagram so far. Thus one additional pair of electrons is added to xenon.

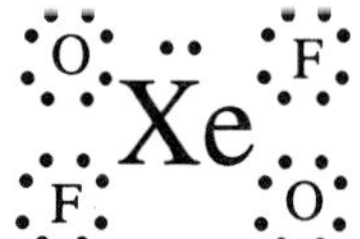

At this stage, the diagram is complete. We note that xenon has 5 pairs of electrons in its valence shell. In the following sections, we will learn the geometric consequences of expanded octets.

7.8 VSEPR AND 5 REGIONS OF ELECTRON DENSITY

The octet rule is based upon second period elements wanting a $2s^2$ $2p^6$ valence electron configuration, for a total of 8 electrons. When elements from the third or subsequent periods are considered, there are more than s and p electron orbitals. Beginning in the third period, d orbitals are accessible and more than 8 electrons can be accommodated. (Alternatively, one could say that as one goes from the second period to the third, fourth, fifth, etc., the size of the atom increases and more electrons can "fit" around the atom.) Consequently, species like PF_5 are observed, where the central phosphorous atom, a third period element, is bonded to five fluorine atoms. Since 5 bonds means 10 electrons around phosphorous, this is said to be an example of an **expanded octet**. In trying to arrange those 5 regions of electron density around the central atom, one discovers that there is no geometry that both maximizes distance between electron pairs and has all 5 sites in equivalent positions. Instead, one finds that the best compromise between those two demands is an arrangement called **trigonal bipyramidal**. This name conveys the structure of two triangle-based pyramids fused at the base.

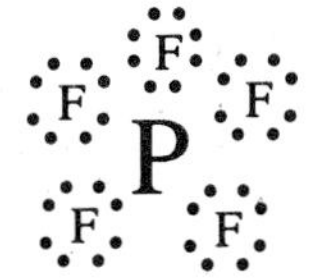

FIGURE 7.26: Lewis dot diagram of PF_5.

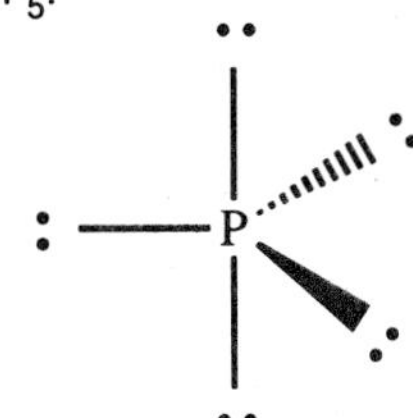

FIGURE 7.27: Trigonal bipyramidal electronic geometry.

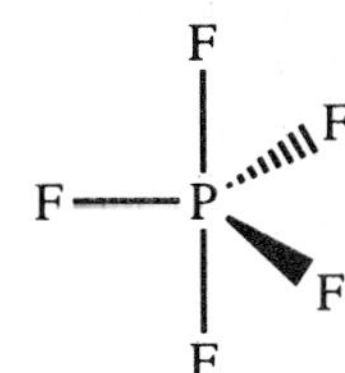

FIGURE 7.28: Molecular geometry of PF_5.

The five positions at the vertices of the trigonal bipyramid are not all equal, but are in two separate types. The three vertices pictured as lying along the "equator" of the structure are called **equatorial**, while the two vertices lying along the "pole" of the structure are called **axial**. This difference becomes important when not all 5 pairs of electrons are bonding pairs. If the central atom has one or more lone pairs around it in a trigonal bipyramidal electronic geometry, a variety of molecular geometries are possible.

7.9 VSEPR AND 6 REGIONS OF ELECTRON DENSITY

As we noted in the previous section, more than 8 valence electrons may be used around an element from the third or subsequent period. For example in the compound SF_6, sulfur, a third period element, is bonded to 6 fluorine atoms. This involves 6 pairs of electrons. In the case of 6 regions of electron density, there is a single structure that both maximizes the distance each electron pair is from each other and maintains all 6 sites as equivalent.

FIGURE 7.29: Lewis dot diagram of SF_6.

The geometric arrangement of 6 regions of electron density is referred to as octahedral because the geometry is that of the geometric figure with 8 triangular faces, or two square-based pyramids fused at the base. The electron pairs are maximally separated when placed on the 6 vertices of the octahedron. In the next section, we pursue the structure of molecules when all 6 sites are not bonding pairs of electrons.

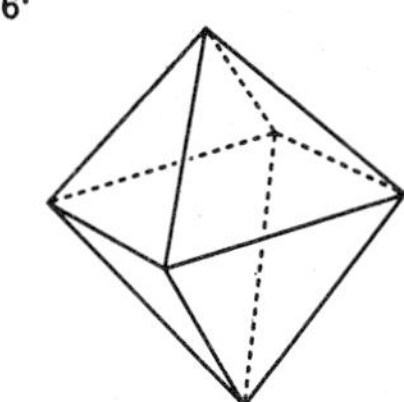

FIGURE 7.30: An octahedron.

FIGURE 7.31: Octahedral electronic geometry.

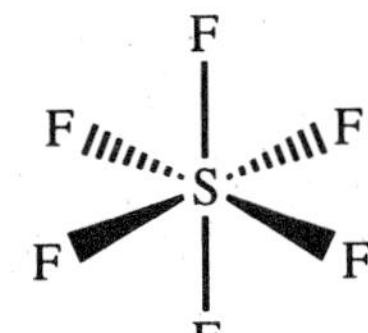

FIGURE 7.32: Molecular geometry of SF_6.

7.10 ELECTRONIC GEOMETRY VS. MOLECULAR GEOMETRY

Electronic geometry around the central atom of a molecule determines the molecular geometry, but the two are not necessarily equal. The regions of electron density around the central atom (A) may either be bonding pairs of electrons that attach other atoms (X) to A, or they may be lone pairs on A, symbolized by E. When there are one or more lone pairs of electrons around the central atom, the molecular geometry may not equal the electronic geometry.

Let us first consider the case of two regions of electron density, such as was the case around beryllium in $BeCl_2$. We have previously seen that both the electronic and molecular geometry are predicted to be linear. Now consider the nitrogen molecule, with the Lewis dot diagram shown in Figure 7.6. Either nitrogen atom has two regions of electron density, which are predicted to be situated 180° apart, or linear. Only one of these regions is involved in bonding; the other region is a lone pair. A diatomic molecule is, by definition, still linear, so there is no apparent difference between electronic and molecular geometry in the case of nitrogen molecule.

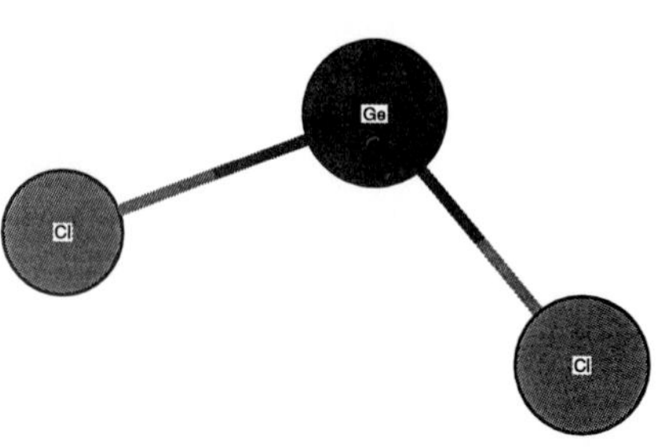

:Cl:Ge:Cl:

FIGURE 7.33:
Electron dot diagram and electronic geometry of $GeCl_2$.

Next consider three regions of electron density, such as was the case around boron in BCl_3. Both the electronic geometry around boron and the molecular geometry were previously shown to be trigonal planar, with a characteristic 120° angle. But consider the geometry of $GeCl_2$, shown in the figure at right:

Two pairs of electrons around germanium are bonding pairs, while one is a lone pair. This is an example of the general formula AX_2E. The electronic geometry is predicted by VSEPR to be trigonal planar. But the molecular geometry describes where the atoms of germanium and chlorine are. The chlorine atoms must be positioned along the direction of the bonding pairs of electrons, and thus the molecular geometry is described as "bent". There are only two choices for a triatomic molecule: linear or bent. In the case of linear electronic geometry, the molecular geometry will be linear. In the case of trigonal planar molecular geometry, three atoms will have a bent molecular geometry.

Methane, as we noted earlier, has tetrahedral electronic geometry and tetrahedral molecular geometry. This occurred because each pair of electrons around the carbon atom was associated with a hydrogen atom. The atoms are simply located "at the end" of each bonding pair of electrons, and the geometry is preserved. Here again, the presence of one or more lone pairs of electrons around an atom causes a difference between the electronic geometry and the local molecular geometry, since the former describes the location of the electrons, and the latter describes the location of the atoms.

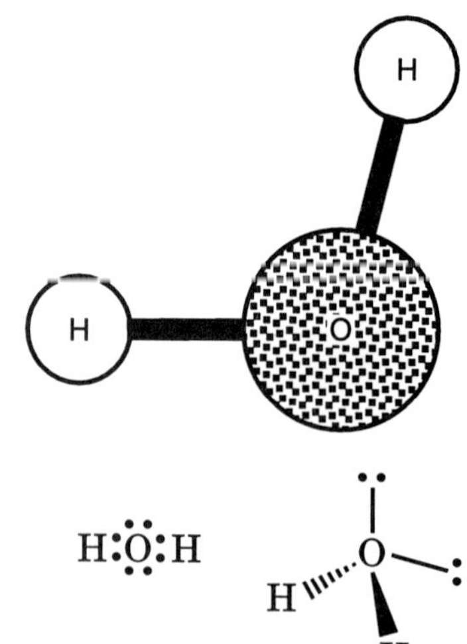

H:O:H

FIGURE 7.34:
Electron dot diagram and electronic geometry of water.

Consider water, which has two lone pairs among the four regions of electron density. The electronic geometry is clearly tetrahedral.

The molecular geometry of water describes the location of the three atoms in space. As noted earlier, this can only be linear or bent. Since the hydrogens have to be located at the end of pairs of electrons which are tetrahedrally positioned around the oxygen atom, it is impossible for water to be a linear molecule. Its molecular geometry is described as bent. Again, the tetrahedral electronic geometry (location of regions of electron density) determines, but is not necessarily identical to the molecular geometry (location of the atoms.) Additional examples are given in Table 7.1

TABLE 7.1

TABLE 7.1: RELATIONS BETWEEN ELECTRONIC AND MOLECULAR GEOMETRY

REGIONS OF e- DENSITY	BONDING REGIONS	LONE PAIRS	ELECTRONIC GEOMETRY	MOLECULAR GEOMETRY	GENERAL FORMULA	EXAMPLE
2	2	0	linear	linear	AX_2	$BeCl_2$
2	1	1	linear	linear	AXE	N_2
3	3	0	trig. planar	trig. planar	AX_3	BCl_3
3	2	1	trig. planar	bent	AX_2E	$GeCl_2$
4	4	0	tetrahedral	tetrahedral	AX_4	CH_4
4	3	1	tetrahedral	trig. pyramidal	AX_3E	NH_3
4	2	2	tetrahedral	bent	AX_2E_2	H_2O
4	1	3	tetrahedral	linear	AXE_3	HF
5	5	0	trig. bipyramidal	trig. bipyramidal	AX_5	PF_5
5	4	1	trig. bipyramidal	see-saw	AX_4E	SF_4
5	3	2	trig. bipyramidal	T-shaped	AX_3E_2	ClF_3
5	2	3	trig. bipyramidal	linear	AX_2E_3	XeF_2
6	6	0	octahedral	octahedral	AX_6	SF_6
6	5	1	octahedral	square pyramidal	AX_5E	BrF_5
6	4	2	octahedral	square planar	AX_4E_2	XeF_4

Next consider the case of AX_5, where A is the central atom around which 5 pairs of bonding electrons are arranged. We find trigonal pyramidal electronic geometry and trigonal bipyramidal molecular geometry. This was seen earlier for PF_5. When there are four bonding pairs of electrons and one unshared pair of electrons around the central atom, this is symbolized by AX_4E, where E represents an unshared pair of electrons. Since there are two distinct position in this geometry, there are two distinct molecular structures for AX_4E, pictured at right:

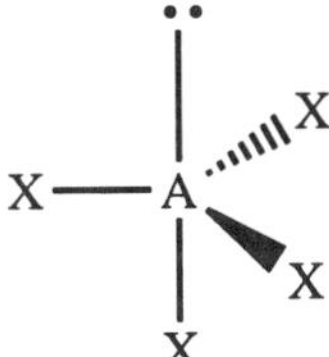

FIGURE 7.35: Molecular structure for axial placement of lone pair.

VSEPR assumes that lone pair electrons are more repulsive than shared (bonding) pairs of electrons, since only one nucleus attracts a lone pair, but two nuclei attract a bonding pair of electrons. We also note that there are two different angles between adjacent electron pairs in a trigonal bipyramidal structure: 90° and 120°. The interaction between electrons 90° apart is far more significant than the interaction at 120°. We can use these principles to decide which of the two possible structures for AX_4E is preferred. VSEPR predicts that the repulsion of the lone pair electrons is strongest, and thus it demands the most room. The structure in Fig 7.35 has the lone pair in an axial position, giving it 90° interactions with 3 bonding pairs of electrons. The structure in Fig 7.36 has the lone pair in an equatorial position, giving it 90° interactions with just 2 bonding pairs of electrons. Thus the equatorial placement of the lone pair shown in Fig 7.35 gives the preferred molecular geometry, referred to as a "teeter-totter" shape. SF_4 is an example of a molecule with this molecular geometry.

FIGURE 7.36: Molecular structure for equatorial placement of lone pair.

Concept Check: Predicting Molecular Structure from VSEPR

What geometry does one expect in a molecule AX_3E_2?

Solution: The question is asking where the one A and 3 X atoms are situated in space. We begin with the electronic geometry. With 3 bonding pairs and two lone pairs around A, we predict a trigonal bipyramidal electronic geometry. The question becomes where the lone pairs are situated in relation to the overall structure. Placing both lone pairs in axial positions would give three 90° interactions with the equatorial bonding pairs for each lone pair. Placing just one lone pair axial and one equatorial creates a 90° interaction between lone pairs, which is highly disfavored. If both lone pairs are placed in equatorial positions, the interaction between lone pairs is at a comfortable 120°, and there are only two 90° interactions for each lone pair. The predicted molecular geometry, described as "T-shaped" is shown in the Figure at right:

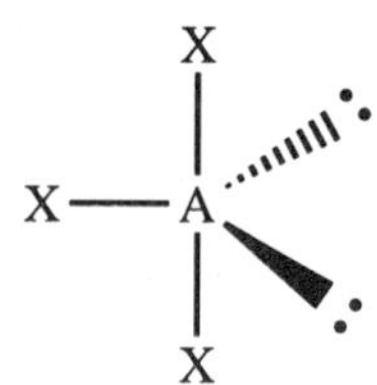

FIGURE 7.37: Molecular structure for AX_3E_2.

In the case of molecules with the formula AX_6, such as SF_6, both the electronic and molecular geometry were predicted to be octahedral. When an unshared pair of electron replaces a bonding pair, the molecule is represented by AX_5E. Unlike the case for trigonal biyramidal structure, all six sites in octahedral geometry are equivalent. (That is, if you try to draw a second structure, a simple rotation of the molecule in space reproduces the original structure.) The molecular geometry of AX_5E is described as square pyramidal, and is seen in the figure at right. The compound BrF_5 has this structure.

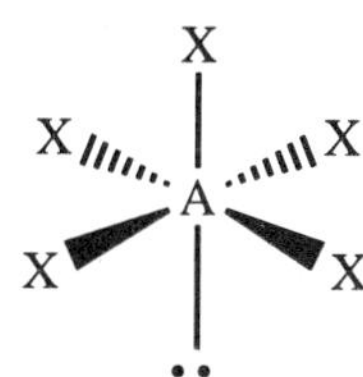

FIGURE 7.38: Square pyramidal molecular geometry.

For a molecule of the form AX_4E_2, the two lone pairs should be separated as much as possible. The choice between a 90° interaction between the lone pairs or a 180° interaction between the lone pairs is easily made, and results in the structure at right, described as square planar. Xenon tetrafluoride, XeF_4, is such a molecule.

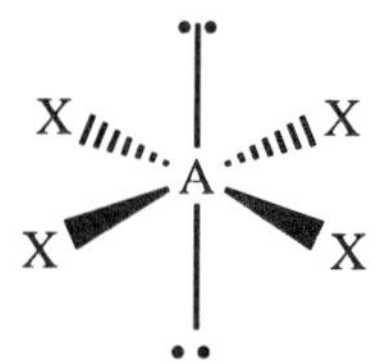

FIGURE 7.39: Square planar molecular geometry.

7.11 LONE PAIRS AND BOND ANGLES PREDICTED BY VSEPR

VSEPR theory predicts that electron pairs tend to maximize their distribution in space (spread themselves out.) This tendency is inhibited not only by the presence of other electron pairs, but by the presence of a positively charged nucleus. A positive nucleus exerts an attractive force on an electron pair, which acts to minimize the space it occupies. A **lone pair of electrons** is controlled by one nucleus. A shared pair of electrons finds itself between two atoms, and thus between two positive nuclei. A lone pair of electrons, under control of only one atom, occupies more space than a bonding pair of electrons, which is under the control of two atoms. The larger size of lone pair electrons results in a larger repulsive force than that for shared pairs of electrons. One can simply picture lone pairs as demanding (and getting) more room within the valence shell than bonding pairs.

It is expected that the repulsive forces are greatest between two lone pairs of electrons, and least between two bonding pairs of electrons. Repulsion between a lone pair (lp) and a bonding pair (bp) are intermediate between those extremes. This is expressed symbolically in the following inequality:

$$\text{lp-lp} > \text{lp-bp} > \text{bp-bp}$$

Consider the Lewis dot diagram of ammonia, Figure 7.40.

H
H:N:H

FIGURE 7.40: Electron dot diagram of NH_3.

There are clearly four regions of electron density around the central nitrogen atom. Three pairs of electrons are shared, and one pair of electrons is unshared (a lone pair.) VSEPR correctly predicts an electronic geometry that is roughly tetrahedral, as the four electron pairs repel each other and maximize their separation. The electronic geometry is roughly tetrahedral, and the molecular geometry (describing the positions of the four atoms) is described as trigonal pyramidal. The electronic geometry is not exactly tetrahedral because there are two different types of electron pairs: bonding pairs and lone pairs. Since the lone pair on nitrogen is larger, it demands more room than an equal division of space provides. As we observed for methane, an equal distribution of space resulted in the tetrahedral angle of 109.5° for the H-C-H bond. For ammonia, the extra repulsion of the lone pair reduces the H-N-H bond angle to 107.3°.

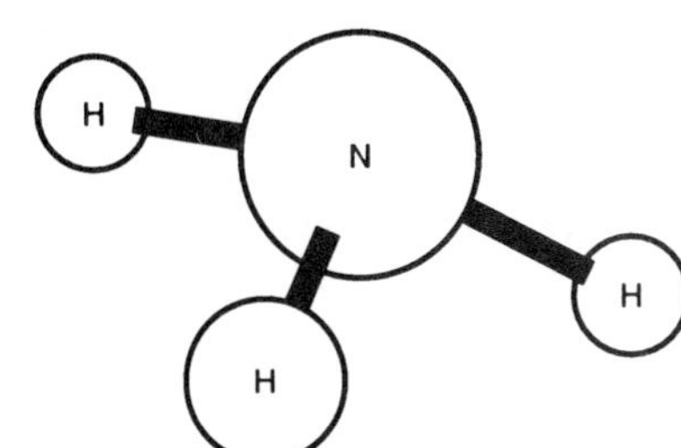

Earlier we showed the Lewis dot diagram of water to be H:Ö:H . The four regions of electron density around the oxygen lead us to predict roughly tetrahedral electronic geometry. The extra repulsion by the two lone pairs of electrons forces the two bonding pairs of electrons closer together than expected for a pure tetrahedral arrangement of electron pairs. As a result, the H-O-H bond angle is 104.5°. The various types of electron pair repulsion are shown in the Figure below:

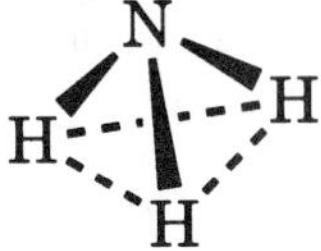

FIGURE 7.41: Molecular geometry of ammonia.

Note that the pure tetrahedral angle of 109.5° in methane was decreased to 107.3° in ammonia by the presence of one lone pair, and was further decreased to 104.5° in water by the presence of two lone pairs.

Lone Pair - Lone Pair Repulsions

Lone Pair - Bonded Pair Repulsions

Lone Pair - Bonded Pair Repulsions

O

H H

Bonded Pair - Bonded Pair Repulsions

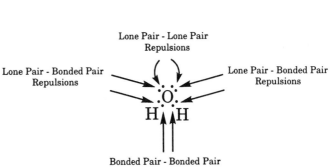

FIGURE 7.42: Electron pair repulsion in the water molecule.

7.12 POLAR BONDS AND POLARITY IN MOLECULES

Earlier in Chapter Two we noted the polarity of some solvents, as well as the importance of polarity in determining solubility. Discussion of polarity was delayed until now, when molecular geometry could be discussed.

A molecule is called polar if its measured value of **dipole moment** (usually in units of debyes) is non-zero. A dipole moment occurs due to a separation of charge in a molecule. The charge separation is a result of differences in electronegativity of atoms in the molecule. First consider two molecules: F_2 and HF. In diatomic fluorine, the molecule is, of course, linear. In Fig 1.8, we see that fluorine is the most electronegative element, meaning that fluorine atom bonded to another atom has a very strong pull on shared electrons. But in F_2, the pull in the opposite direction (due to the other fluorine atom) is just as strong. As a result there is equal sharing of the electron pair, and the center of electron density is exactly midway between the two atoms. The center of positive charge (due to the protons in the nuclei) is at the same location. Since there is no separation of charge, the molecule F_2 has a zero dipole moment, and is called nonpolar. The diatomic molecule HF is also linear. While the bond between hydrogen and fluorine is covalent, it differs from the bond in F_2. In F_2, the two atoms had a completely equal share of the electron density due to equal electronegativity. In the molecule HF, the electron pair between the atoms is shared, but not equally. Fluorine has a much greater share of the electron pair than does hydrogen due to fluorine's greater electronegativity. This is shown in the figure at right, where the arrow points in the direction of greater electron density. By convention, the other end of the arrow has a plus sign, indicating the direction of greater positive charge in the dipole.

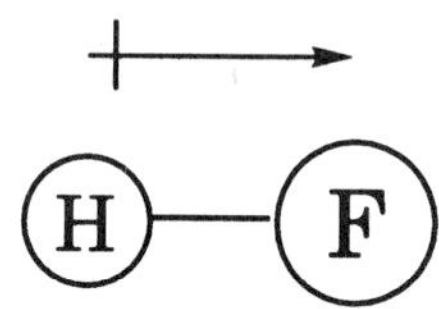

FIGURE 7.43: Dipole moment in HF.

As we saw earlier in Chapter One, electronegativity decreases going down a group. Thus we predict that for the binary compounds of hydrogen with the halogens, the dipole moment will decrease as the electronegativity decreases. This trend is observed in the following table.

TABLE 7.2: ELECTRONEGATIVITY DIFFERENCE AND DIPOLE MOMENT FOR HX

COMPOUND	Δ EN	μ (DEBYES)
HF	1.9	1.82
HCl	0.9	1.08
HBr	0.7	0.82
HI	0.4	0.44

TABLE 7.2

Next, consider the polarity of triatomic molecules. Carbon and oxygen differ in electronegativity, as do hydrogen and oxygen. Yet the molecule carbon dioxide is nonpolar while water is polar. This, also, is due to molecular geometry. The Lewis Dot representations of both molecules are shown at right.

O::C::O   H:O:H

FIGURE 7.44: Lewis dot diagrams of CO_2 and H_2O.

Carbon has two regions of electron density, and thus is predicted to have linear electronic and molecular geometry. While the bonds between carbon and oxygen are polar, they are equal and opposite. Electron density is drawn from carbon to oxygen, but in exactly opposite directions. The center of both positive and negative charge is the carbon atom. The molecule is nonpolar. This can be seen by imagining the vector addition of the dipole arrows along the bonds in carbon dioxide.

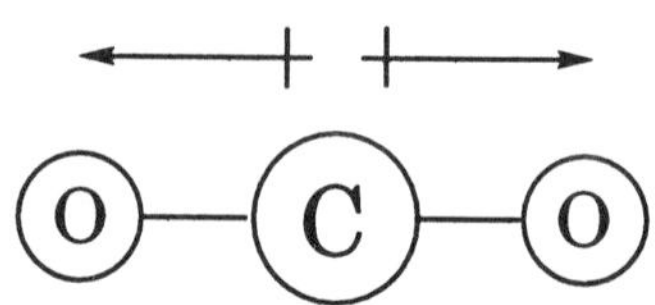

FIGURE 7.45: Dipole moment cancellation in CO_2.

As seen earlier, water has four regions of electron density, a tetrahedral electronic geometry and thus a bent (non-linear) molecular geometry. The vector sum of dipole moments along the two bonds in the molecule (adding arrows head to tail) results in a net dipole for the molecule of water.

The polarity of polyatomic molecules is determined in the same fashion. In Chapter Two we noted that carbon tetrachloride had a zero dipole moment, while dichloromethane had a dipole moment of 1.60 debye. Again, this is a consequence of the molecular geometry as well as the electronegativity differences between carbon and chlorine, or carbon and hydrogen. Carbon tetrachloride is a non-**polar molecule** despite the **polar covalent bonds** between carbon and chlorine because of the cancellation of charge separation upon vector addition.

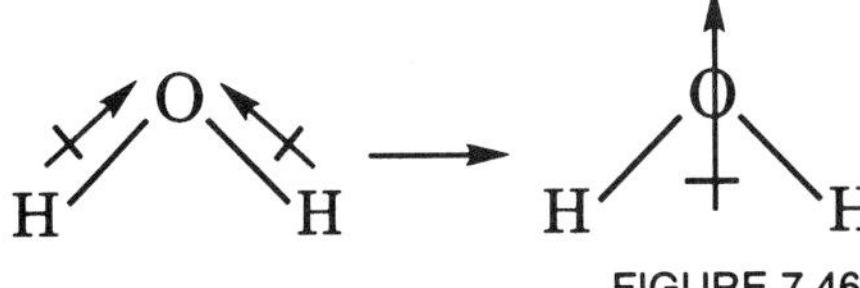

FIGURE 7.46: Dipole addition for H_2O.

A quick inspection of the Lewis dot diagram of dichloromethane in Figure 7.47 might tempt one to imagine that the molecule is nonpolar due to a similar cancellation. But viewing the molecules in their proper tetrahedral orientation (Figure 7.48) shows that, while the vectors cancel for carbon tetrachloride, there is no possible vector cancellation for dichloromethane. Thus dichloromethane is polar.

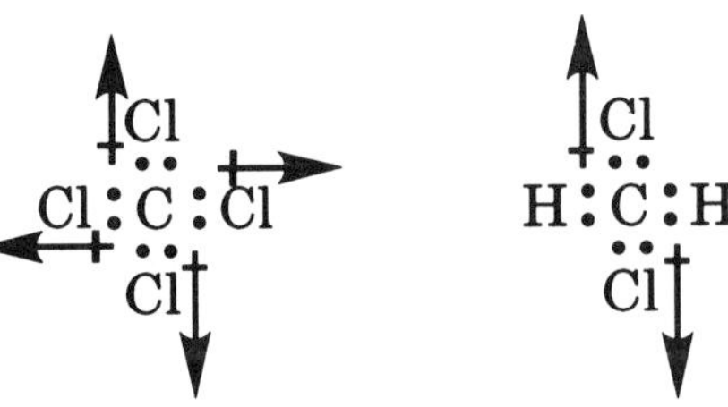

FIGURE 7.47: Lewis dot representation of dipole addition.

CONCEPT CHECK: ELECTRONIC GEOMETRY, MOLECULAR STRUCTURE, AND POLARITY

SCl_2 and CS_2 have similar formulas. One is polar, and the other is not. Explain.

SOLUTION: First, the Lewis dot diagrams of the two compounds are determined via NAS.

SCl_2: N = 3(8) = 24; A = 6 + 2(7) = 20; S = (24-20)/2 = 2

:Cl:S:Cl:

With 4 regions of electron density, the sulfur in SCl_2 has tetrahedral electronic geometry. Thus the molecular geometry of SCl_2 is bent, and the molecule is polar.

CS_2: N = 3(8) = 24; A = 4 + 2(6) = 16; S = (24-16)/2 = 4

:S::C::S:

With two regions of electron density, the carbon in CS_2 has a linear electronic geometry. Thus the molecular geometry of CS_2 is linear, and the molecule is non-polar.

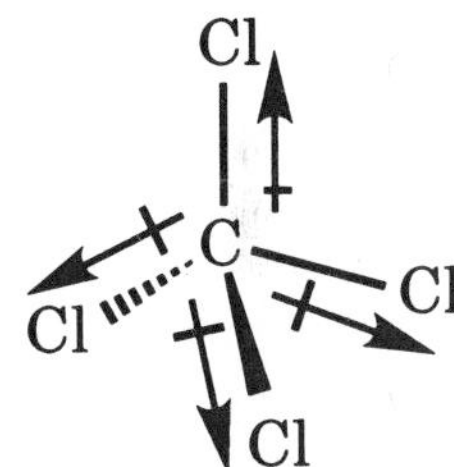

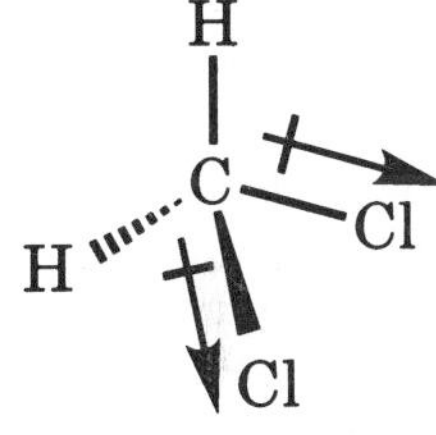

FIGURE 7.48: Three-dimensional representation of dipole addition.

7.13 SOME LIMITATIONS OF VSEPR

VSEPR theory has been successful in moving our thinking from flat Lewis dot diagrams to three dimensional molecular geometry. VSEPR works very well for molecules containing second period elements including carbon, nitrogen and oxygen. It correctly predicts whether certain molecules will be polar or nonpolar. As you continue to explore chemistry, you will find VSEPR to be a useful tool. But, like every theory, VSEPR has its limitations. One limitation depends upon the relative size of the atoms involved, and the range over which electronic repulsion are effective. For example, VSEPR correctly predicts the polarity and approximate geometry of water, H_2O; VSEPR even explains why the bond angle deviates from the predicted regular tetrahedral angle. But if we attempt to apply VSEPR to the closely related compound, H_2S, we find our predictions fail miserably. Instead of an angle near 109.5°, experiment yields an H-S-H bond angle much nearer to 90°. Part of the difficulty lies in the fact that sulfur is a much bigger atom than oxygen, and four electron pairs in its valence shell have more room (less repulsion) than four electron pairs in oxygen.

There are significant fundamental problems in basic **VSEPR theory**. One is related to how electrons get from electron orbitals with designated geometric orientations (*e.g.*, $2p_x$, $2p_y$, $2p_z$) to freely move around the circumference of the valence shell. Another problem deals with multiple bonding. Since the four electrons in a double bond, or the six electrons in a triple bond, are not in fact occupying the same region in space (due to Pauli's exclusion principle) then how can they be lumped together in a grouping called a region of electron density? And where exactly are the second and third electron pairs, if not in the same space as the first? Answers to these questions will come as we examine chemical bonding more fully in Chapter 11. Until then, we approach VSEPR as any other theory: useful in making certain correct predictions, but not necessarily the final word.

The final words of this chapter detail various means by which molecular geometry can be expressed.

7.14 REPRESENTATIONS OF 3-DIMENSIONAL GEOMETRY IN 2-DIMENSONS

We have now discovered how to predict both electronic and molecular geometry. For example we know that the geometry about carbon in a molecule of methane is tetrahedral. The model of methane shown below allows us to visualize this easily:

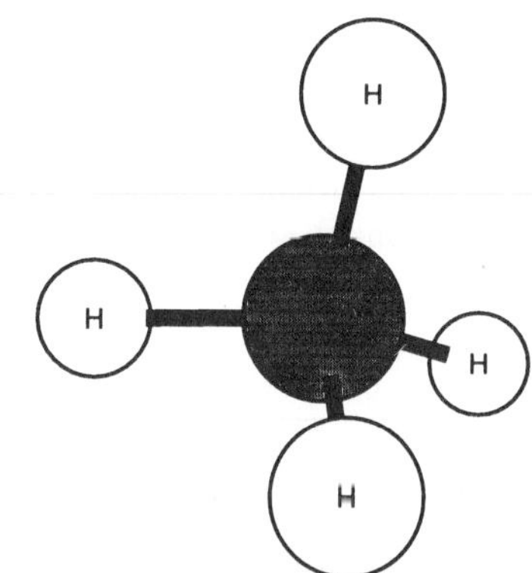

FIGURE 7.49: Ball and stick representation of methane.

The problem is, however, that most people would not want to draw this representation of methane all the time. The question is, then, how can we represent, on paper, the tetrahedral nature of the carbon atom in methane? The answer is quite simple. We need to add the feeling of perspective to the drawing. Perspective is added to chemical drawings by using wedges and dashes. Wedges are used to indicate to the reader that a particular atom or group of atoms is projecting out of the plane of the paper. Dashes are used to indicate that an atom or group of atoms is projecting back into the plane of the paper. Thus we can draw a methane group as follows: two of the hydrogen atoms lie in the plane of the paper, one projects in front of the plane of the paper, and one projects behind the plane of the paper. You may find it helpful to experiment with molecular models when studying these concepts.

The same type of drawing can be done for other compounds as well. Consider, for example, 3,3-dimethylpentane. By focusing on carbon-3 as the central atom, the following three-dimensional drawing can be produced. In the first example the ethyl-groups are placed in the plane of the paper. In the second, the

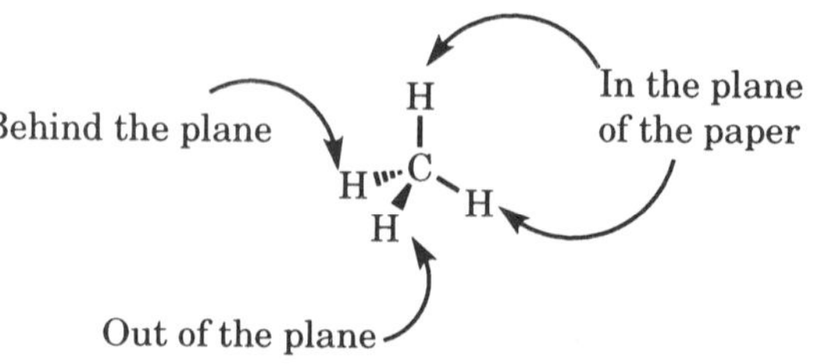

FIGURE 7.50: Dash and wedge representation of methane.

methyl-groups are placed in the plane of the paper. It is important to realize that both drawings are 3-dimensional drawings of the same molecule.

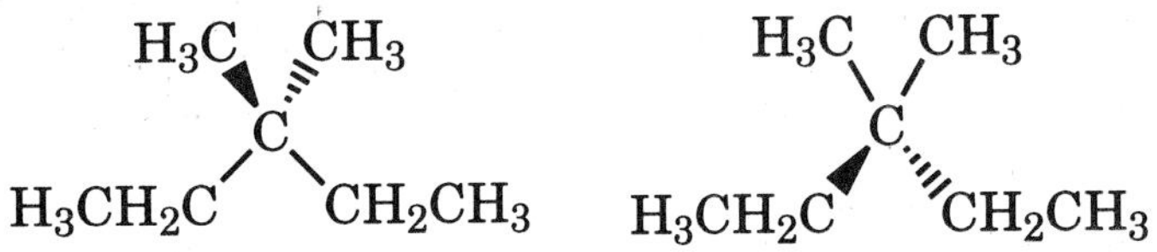

FIGURE 7.51: Two perspectives of 3,3-dimethylpentane.

We may also want to emphasize the tetrahedral nature of a particular carbon atom in an extended chain. The tetrahedral nature of carbon-4 of 4-bromoheptane is highlighted in the following drawing. In this case the molecule is drawn such that the bromine is projecting out of the plane of the paper and the hydrogen is projecting behind the plane of the paper.

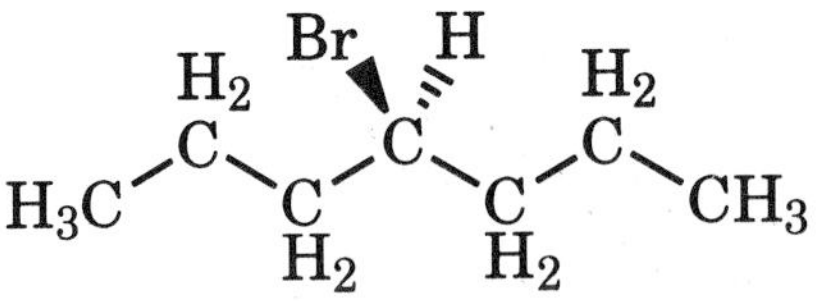

FIGURE 7.52: 4-bromoheptane.

Let us use *n*-butane to look at a variety of different representations of the same molecule. First we can look at the ball and stick model of this compound.

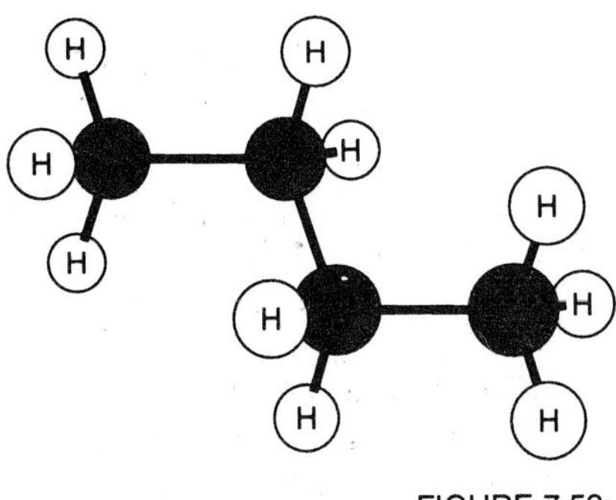

FIGURE 7.53: Butane.

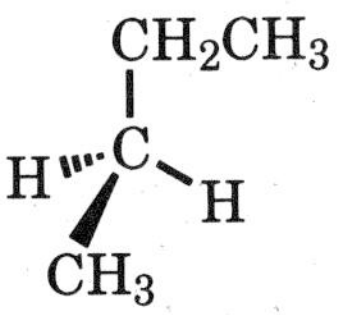

FIGURE 7.54: 3-D drawing of butane using dashes and wedges.

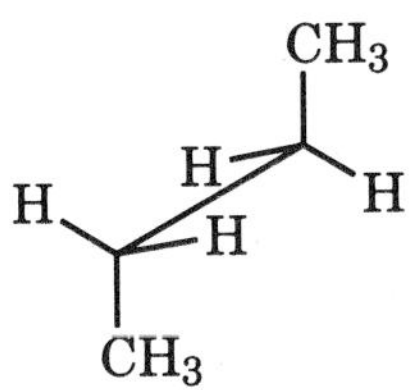

FIGURE 7.55: Sawhorse drawing of butane.

Again, it would be inconvenient for us to draw this each time we wanted to draw butane. We can, however, draw the 3-dimensional representation.

Sometimes it is also helpful to use the sawhorse representation of certain compounds. The sawhorse representation of n-butane is shown in Figure 7.55. In this case, perspective is given by the angles used in the drawing.

Try to imagine what you would see if you could actually hold a molecule of n-butane in your hand and look down the bond between the number two and number three carbons. You would see something shown by the following drawing, Figure 7.56.

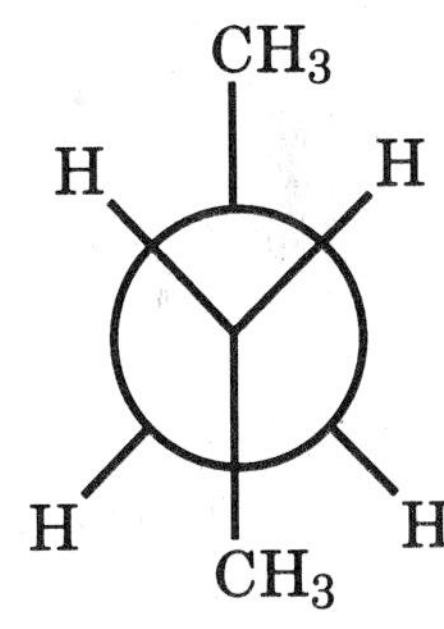

FIGURE 7.56: Newman projection of C_2-C_3 bond of butane.

This representation of n-butane is called a **Newman projection**. As you will see in subsequent chapters, both the **sawhorse diagram** and the Newman projection will be useful when trying to describe certain reaction pathways.

Now that you have an understanding of electonic and molecular geometry, we will begin to explore, in the next chapter, the consequence of geometry on the properties of certain types of compounds.

Concept Check: Sawhorse Diagrams and Newman Projections

Name the molecule pictured in the following sawhorse representation. Then convert the sawhorse representation into an equivalent Newman projection.

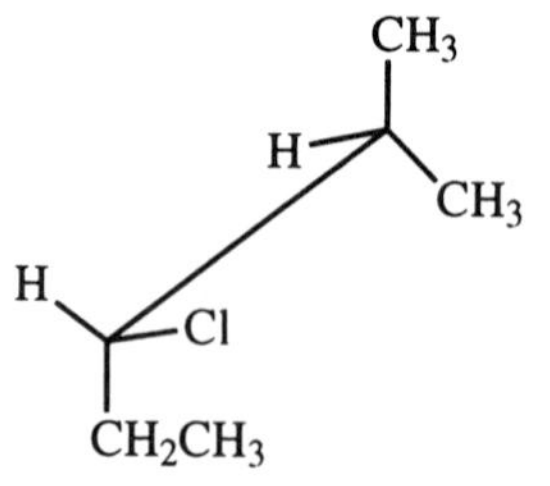

CH_3

H

CH_3

H

Cl

CH_2CH_3

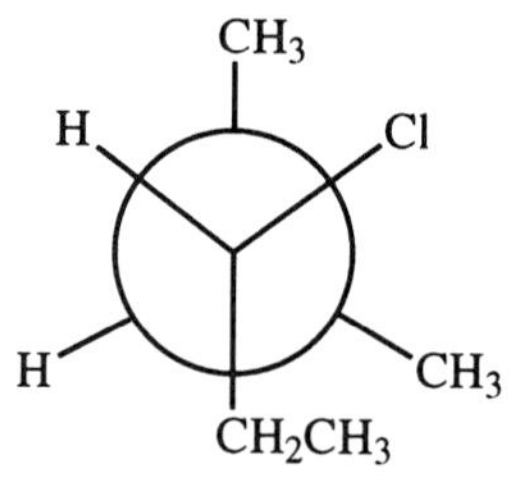

CH_3

H

Cl

3-chloro-2-methylpentane

H

CH_3

CH_2CH_3

Axial: in trigonal bipyramidal geometry, the positions above and below the triangular plane of the structure.

Bond angle: the difference in direction, measured in degrees, between two lines intersecting at an atom. These lines in turn connect that atom to two other atoms.

Bonding pair of electrons: two electrons that are shared between two atoms.

Core electrons: electrons in filled or closed shells, closer to the nucleus and at a lower energy than valence electrons.

Dipole moment: a measure of the separation of charge (polarity) in a molecule, measured in units of debyes.

Electron deficient: refers to an atom (in a compound) that has fewer than 8 electrons in its valence shell

Electronic geometry: a description of the spatial distribution of electron pairs around an atom.

Equatorial: in trigonal bipyramidal geometry, the positions corresponding to the three vertices coplanar with the central atom.

Expanded octet: refers to an atom (in a compound) that has more than 8 electrons in its valence shell.

Local molecular geometry: a description of the spatial distribution of those atoms immediately bonded to a specified atom in a molecule.

Lone pair of electrons: two electrons that are associated with only one atom.

Molecular geometry: a description of the spatial distribution of the atoms within a molecule.

Newman projection: a representation of a molecule in which the chemist is sighting down a carbon-carbon bond.

Octahedral: a geometry corresponding to the placement of the 6 vertices of an octahedron.

Pauli Exclusion Principle: a theory that posits that no two electrons in the same atom can occupy the same position unless they have opposite spins. Since only two spin states are possible, a maximum of two electrons can occupy a given region of space around an atom.

Polar covalent bond: an unequal sharing of a pair of electrons between two nonmetals of differing electronegativity.

Polar molecule: a molecule with a non-zero dipole moment. The centers of positive and negative charge in a polar molecule do not coincide.

Sawhorse diagram: a representation of a molecule that indicates perspective by the way it is drawn.

Tetrahedral: a description of the electronic (or molecular) geometry in which four regions of electron density (or four atoms) are situated around a central atom. The central atom is imagined at the center of a tetrahedron, and the four electron pairs (or atoms) at the corners.

Trigonal bipyramidal: a geometry that occurs when two traingle-based pyramids are fused at the base. The structure has 5 vertices.

Valence electrons: those outermost, highest energy electrons in an atom.

VSEPR Theory: Valence shell electron pair repulsion theory, which proposes that the shapes of molecules can be predicted based upon maximizing the separation of pairs of valence electrons around the atom.

HOMEWORK Problems

Lewis Dot Diagrams

1. Specify the values of N, A and S, then draw Lewis dot diagrams for each of the following:

   (a) NF_3
   (b) CN^-
   (c) PF_4^+
   (d) $AlCl_4^-$
   (e) $HgBr_2$

2. For each of the species in problem one, indicate the electronic geometry around the specified atom:

   (a) around N
   (b) around C
   (c) around P
   (d) around Al
   (e) around Hg

3. For each of the species in problem one, specify the overall geometry of the molecule or ion.

4. Specify the values of N, A and S, then draw Lewis dot diagrams for each of the following:

   (a) O_3
   (b) OF_2
   (c) SO_3^{2-}
   (d) H_3O^+
   (e) CdI_2

5. For each of the species in problem four, indicate the electronic geometry around the specified atom:

   (a) around middle oxygen
   (b) around O
   (c) around S
   (d) around O
   (e) around Cd

6. For each of the species in problem four, specify the overall geometry of the molecule or ion.

7. The species IF_4^+ has a central iodine atom that has 7 valence electrons. Each fluorine contributes one of its seven valence electrons to make a bond with iodine. But one of the 11 valence electrons around iodine is lost due to the positive charge. Draw the Lewis dot diagram of IF_4^+ and predict its electronic and molecular geometry.

8. The species IF_2^- has a central iodine atom that has 7 valence electrons. Each fluorine contributes one of its seven valence electrons to make a bond with iodine. In addition to those 9 valence electrons around iodine, there is an additional electron due to the negative charge. Draw the Lewis dot diagram of IF_2^- and predict its electronic and molecular geometry.

9. The species TeF_5^- has a central tellurium atom with 6 valence electrons. Each fluorine contributes one of its seven valence electrons to make a bond with tellurium. In addition to those 11 valence electrons around tellurium, there is an additional electron due to the negative charge. Draw the Lewis dot diagram of TeF_5^- and predict its electronic and molecular geometry.

10. The species ICl_4^- has a central iodine atom that has 7 valence electrons. Each chlorine contributes one of its seven valence electrons to make a bond with iodine. In addition to those 11 valence electrons around iodine, there is an additional electron due to the negative charge. Draw the Lewis dot diagram of ICl_4^- and predict its electronic and molecular geometry.

MOLECULAR GEOMETRY AND POLARITY

11. BCl_3 and PCl_3 have the same general type of formula, but different molecular geometries. Explain.

12. BeF_2 and SF_2 have the same general type formula, but different molecular geometries. Explain.

13. Which of the following has the largest O-N-O bond angle? Which has the smallest? Explain. (a) NO_2 (b) NO_2^+ (c) NO_2^-

14. Which of the following has the largest H-N-H bond angle? Which has the smallest? Explain. (a) NH_3 (b) NH_4^+ (c) NH_2^-.

15. BeF_2 and OF_2 have similar formulas. One is polar, the other is not. Explain.

16. BF_3 and AsF_3 have similar formulas. One is polar, the other is not. Explain.

17. Give the values of N, A and S, draw the Lewis dot structures, state the electronic geometry around the specified atom, state the molecular or ionic geometry, and show a three dimensional representation (using dashes and wedges when appropriate) for each of the following:

(a) Cl_2SO specified atom: the central sulfur atom
(b) PO_4^{3-} specified atom: the central phosphorus atom
(c) $CHCl_3$ specified atom: the central carbon atom
(d) PF_3 specified atom: the central phosphorus atom

18. Many polyatomic molecules have no single central atom. Nevertheless electronic geometry and local molecular geometry around a specified atom can be determined from Lewis dot diagrams and VSEPR. First, draw the complete Lewis dot diagram for each of the following. Then state the electronic and local molecular geometry around the specified atom.

(a) acetyl chloride, CH_3COCl
(specified atom: the carbon attached to oxygen and chlorine)

(b) 1,1,1-trichloroethane, CCl_3CH_3
(specified atom: carbon #1, attached to three chlorines)

(c) 2-butanone, $CH_3COC_2H_5$
(specified atom: carbon #2, attached to oxygen)

(d) 1-propyne, $H{-}C{\equiv}C{-}CH_3$
(specified atom: carbon #1, attached to one hydrogen atom)

19. Using dashes and wedges, draw as many unique representations, about the indicated atoms, as possible for the following compounds.

(a) C-3 of 3-bromopentane

(b) C-4 of 2-pentanone

20. Given the following sawhorse drawings, name each compound and draw the corresponding Newman projections.

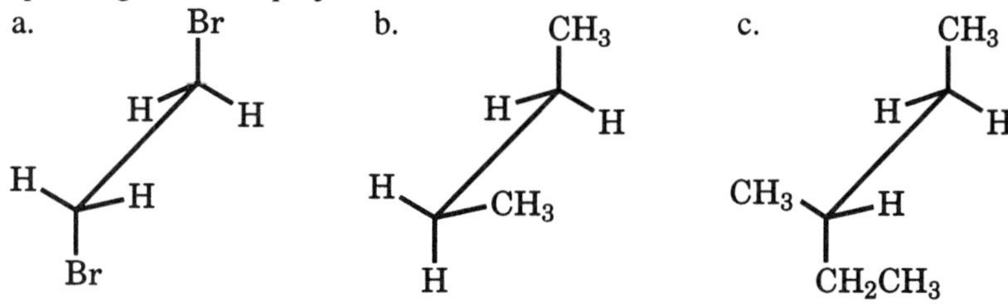

21. Given the following Newman projection drawings, name each compound and draw the corresponding sawhorse drawing.

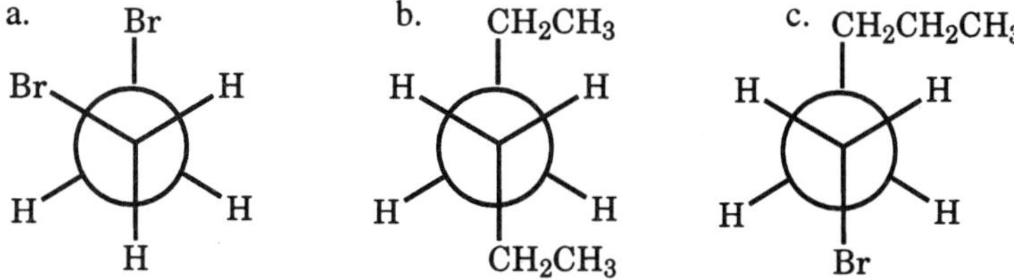

OVERVIEW Problems

22. Gaseous diatomic chlorine reacts in basic aqueous solution to form chloride anion and chlorate anion (ClO_3^-).
    a. Balance this reaction.
    b. Draw the Lewis dot diagram of the chlorate anion.
    c. From your Lewis dot diagram, what approximate O-Cl-O bond angles do you predict?

23. A reaction between iodine (I_2) and chlorine (Cl_2) in aqueous solution yields iodic acid, HIO_3, and HCl.
    a. Balance this reaction.
    b. Draw the Lewis dot diagram of iodic acid.
    c. From your Lewis dot diagram, what approximate O-I-O bond angles do you predict?

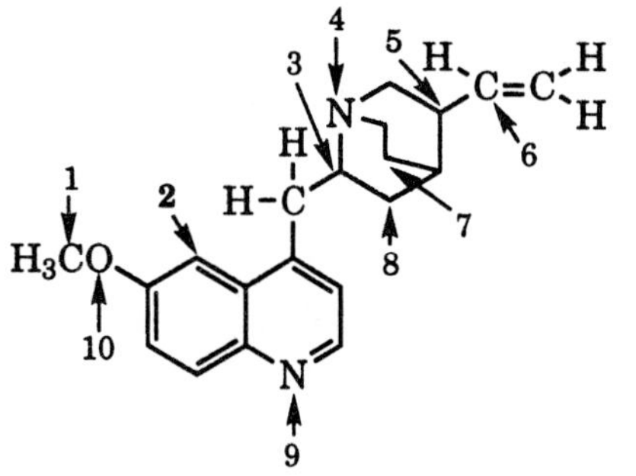

24. The figure above is that of quinine. Determine the molecular formula (i.e., determine the values for the subscripts w,x,y,z in $C_wH_xN_yO_z$.) Identify the electronic geometry about each of the ten atoms indicated in the figure of quinine shown above.

25. Ammonia and hypochlorite ion (OCl^-) react to yield chloride and hydrazine, N_2H_4.
    a. What is the oxidizing agent?
    b. What is the reducing agent?
    c. Write the balanced reaction that forms hydrazine.
    d. Draw the Lewis dot diagram of hydrazine.
    e. Is the bond between nitrogen atoms a single, double, or triple bond?
    f. What do you estimate the H-N-H bond angle to be?

26. Hydrazine, formed in the previous problem, is used as a rocket fuel. It reacts with oxygen to form nitrogen gas and water vapor. One mole of hydrazine reacting this way liberates 534.3 kJ of energy.
    a. Write the balanced combustion reaction of hydrazine.
    b. What mass of water is formed if 100.0 g of hydrazine react with excess oxygen?
    c. What amount of energy is released when 100.0 g of hydrazine is combusted?

27. Hydrazine, N_2H_4, reacts with nitrous acid, HNO_2, to form water and hydrogen azide, HN_3.

a. Write the balanced reaction that forms hydrogen azide.
b. What formal oxidation number would a nitrogen atom in HN_3 have?
c. Write an acceptable Lewis dot diagram of hydrogen azide.

28. The space shuttle has small rocket engines that react one mole of a hydrazine derivative, $(CH_3)_2NNH_2$ (dimethylhydrazine) with two moles of nitrogen tetroxide, N_2O_4. The products of the reaction are water, nitrogen gas, and carbon dioxide.

a. Is N_2O_4 the oxidizing agent or the reducing agent, and why?
b. What mass of N_2O_4 is needed to completely react with 100.0 kg of dimethylhydrazine? What mass of each product is formed?
c. Draw the Lewis dot diagram of N_2O_4.

29. Acrylonitrile has the formula C_3H_4N. Acrylonitrile has an alkene functional group and a nitrile functional group.

a. Draw a Lewis dot diagram for acrylonitrile.
b. What is the approximate C-C-C bond angle?
c. What is the approximate C-C-N bond angle?

30. Antimony trifluoride is prepared by reacting antimony (III) oxide with hydrofluoric acid, and removing the water that appears as a co-product.

a. Write the balanced reaction forming SbF_3 and water.
b. Write the Lewis dot diagram for SbF_3.
c. Would you predict SbF_3 to be very water soluble?
HINT: is it polar/non-polar?

31. Carbon disulfide, CS_2, is miscible with carbon tetrachloride, but less than 0.3 g of CS_2 will dissolve in 100 g of water at 20°C.

a. Draw the Lewis dot diagram for carbon disulfide.
b. Is CS_2 polar or non-polar?
c. How does one rationalize the solubility of CS_2 in water versus CCl_4?

32. Compound A has a molecular weight of about 168 and is composed of 14.31% carbon, 1.20% hydrogen, and 84.49% chlorine by mass. NMR spectra indicate that there is only one hydrogen environment and one carbon environment.

a. Name compound A.
b. Draw a sawhorse representation of compound A.
c. Draw a Newman projection of compound A.

33. Compound B is an isomer of compound A in the previous problem. That is, they have the same molecular weight, and the same elemental composition. NMR spectra indicate that while there is still only one hydrogen environment for compound B, there are two distinct carbon environments.

a. Name compound B.
b. Draw a sawhorse representation of compound B.
c. Draw a Newman projection of compound B.

STEREOCHEMISTRY
Molecules and Mirrors

STEREOCHEMISTRY—Molecules and Mirrors

As you examine the world around you, you will likely notice the existence of handedness. For example, you are either left-handed or right-handed, meaning that you prefer to do most tasks with your left or right hand, respectively. Likewise, as we will see in this chapter, molecules can also be right-handed or left-handed. Even more impressive, as we will discover later, is that biological macromolecules can distinguish between the right and left-hand forms of molecules so well that only one form of a molecule may be biologically active.

8.1 ISOMERISM

STRUCTURAL ISOMERS: A BRIEF REVIEW

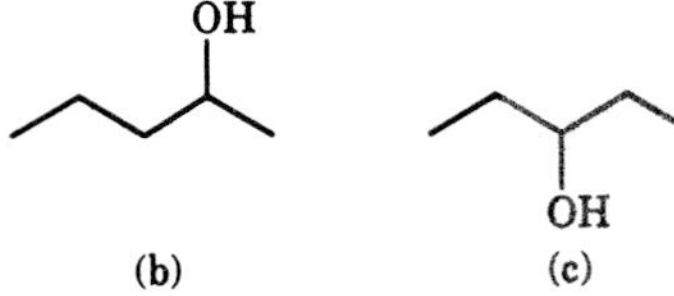

FIGURE 8.1:
a) 1-pentanol, b) 2-pentanol, and c) 3-pentanol.

As you have already learned, one type of isomerism involves structural isomers. Structural isomers (sometimes referred to as constitutional isomers) are those compounds that have the same molecular formula, but are put together differently. That is, they have different connectivity. For example 1-pentanol, 2-pentanol, and 3-pentanol all have the same molecular formula, $C_5H_{12}O$, but are connected differently.

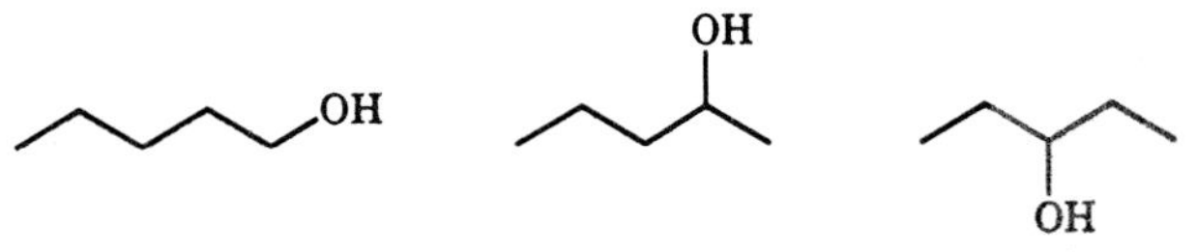

STEREOISOMERS

Stereoisomers, on the other hand, have the same atoms connected in the same order, but they differ in their orientation in space. Consider your hands. When you hold your right hand up to a mirror, you see your left hand as a reflection, and when you hold your left hand up to a mirror you see your right hand as a reflection.

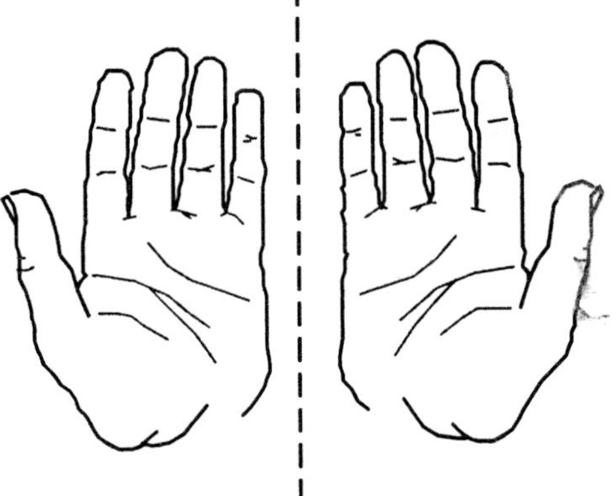

FIGURE 8.2:
Picture of a hand and the reflection.

If you were to try to superimpose the mirror image on the hand, you would find that it is not possible. In other words the right hand and its mirror image are non-superimposable, and in the jargon of the chemist are called chiral. **Enantiomers** are molecules that are non-superimposable on their mirror image. Not all stereoisomers exist as enantiomers. Those that are not related as enantiomers are called **diastereomers**.

8.2 GEOMETRIC ISOMERS: ONE TYPE OF DIASTEREOMER

Geometric isomerism can exist when rotation about a bond(s) is restricted in some way that leads to a defined arrangement of the groups attached to that bond(s). A common example of this is found when looking at the alkenes. Since, as you recall, the local molecular geometry around such bonds is planar, the groups on each end of the double bond are locked into place with respect to each other. In other words, rotation around single bonds is expected, but rotation around multiple bonds is prohibited. If we look at 2-butene, we can see that the methyl groups at each end of the molecule can be oriented either on the same side of the double bond, or on opposite sides of the double bond.

H H $C=C$ CH_3 CH_3 (a)

CH_3 H $C=C$ H CH_3 (b)

FIGURE 8.3:
a) *cis*- and b) *trans*-2-butene.

When the methyl groups are on the same side of the double bond, we refer to the molecule as *cis*-2-butene and when the methyl groups are on opposite sides of the double bond, we refer to the molecule as *trans*-2-butene. Designation of a structure as *cis* or *trans* works well if the groups on each end of the double bond are the same, or if each alkene carbon is bonded to one hydrogen atom and one R group.

As we mentioned in Chapter 4, coupling constants are useful for determining more than just the number of neighboring protons. In fact, they can also provide useful information about the location of these protons when rotation about carbon-carbon bonds is restricted. This is just the situation we encounter with geometric isomers about carbon-carbon double bonds. In this particular case the rotation is restricted because of the double bond, and protons on the adjacent carbon atoms are locked into place with respect to each other. In general, the value of the coupling constant for protons that are *trans* to each other is larger (12-18 Hz) that the coupling constant for protons that are *cis* to each other (6-12 Hz). Thus NMR spectroscopy is a good tool to determine whether a particular alkene has the *trans* configuration or the *cis* configuration.

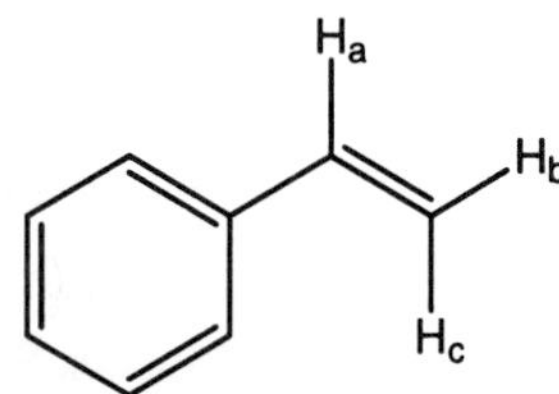

FIGURE 8.4: Structure of styrene.

In order to examine this phenomenon we will look at the proton spectrum of styrene. The structure of styrene is shown in Figure 8.4. As you can see, the proton NMR spectrum of styrene contains essentially four different types of protons with an integration of 5:1:1:1. Therefore, it is probable that all of the aromatic protons are responsible for the multiplet that appears from about 7-7.5 ppm, and that the other three peaks (appearing from 5 to 7 ppm) result from each of the three olefinic protons. This may seem odd since two of the olefinic protons are attached to the same carbon and we usually think of protons on the same carbon as being equivalent. These protons are, however, magnetically non-equivalent. This means that each of the protons has a slightly different magnetic environment as a result of the restricted rotation about the double bond. This magnetic non-equilvalence is a result of the different relationships (or interactions) each of these protons has with the groups attached to the other olefinic carbon. The coupling constant for protons attached to the same carbon is generally quite small (usually around 2 Hz). By examining the proton NMR spectrum of styrene we should be able to determine which proton give rise to each of the three signals.

If we look carefully at the signals for each of the three olefinic protons we note that each signal appears as a doublet of doublets. The two signals between 5 and 6 ppm each show one relatively large coupling and a very small coupling. Thus, it is quite likely that these peaks result from the two terminal alkene protons (H_b and H_c). Both of these protons are split by H_a and then again by each other. The coupling to each other should of course be equal, and in this case is around 1.5 Hz. We can also be fairly certain that the signal at around 5.25 ppm is the signal for H_b since it has the smaller of the two coupling constants (around 11 Hz). This would indicate the *cis* relationship between H_b and H_a. The trans relationship between H_a and H_c is shown by the larger coupling constant (around 17 Hz) in the signal at 5.8 Hz. Finally, the signal for H_a, at 6.75 ppm, shows couplings constant of both 11 Hz and 17 Hz since H_a is coupled to both H_b and H_c. Detailed splitting diagrams for these couplings are presented in Figures 8.6-8.8.

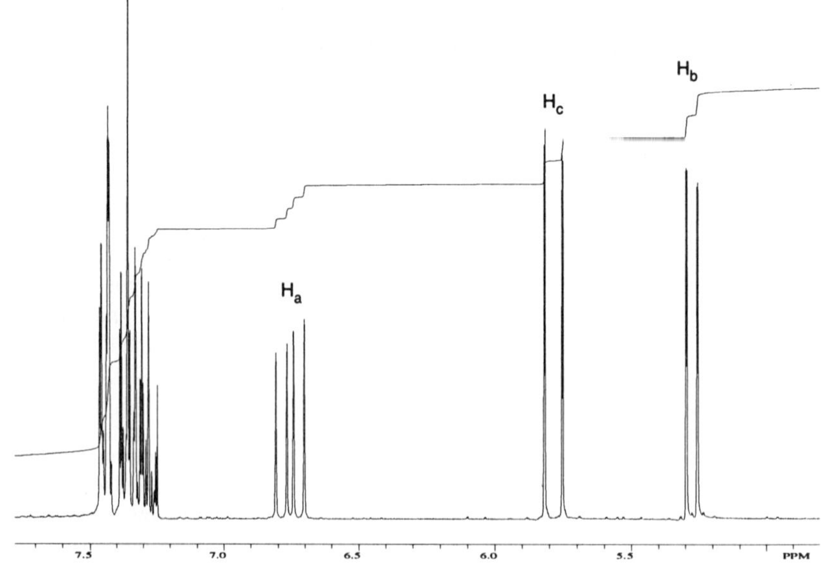

FIGURE 8.5: ¹H NMR spectrum of styrene.

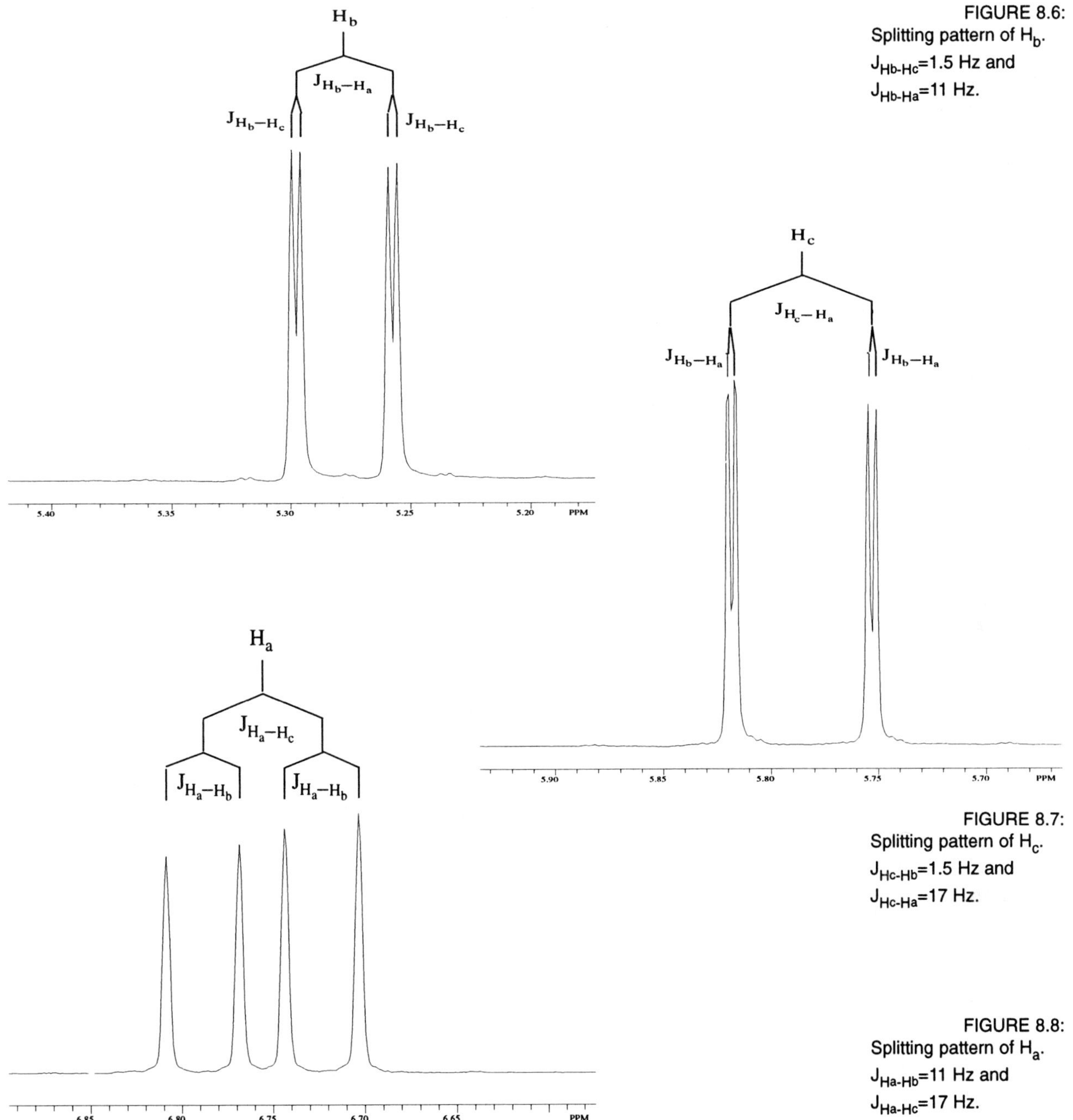

FIGURE 8.6: Splitting pattern of H_b. $J_{Hb\text{-}Hc}$=1.5 Hz and $J_{Hb\text{-}Ha}$=11 Hz.

FIGURE 8.7: Splitting pattern of H_c. $J_{Hc\text{-}Hb}$=1.5 Hz and $J_{Hc\text{-}Ha}$=17 Hz.

FIGURE 8.8: Splitting pattern of H_a. $J_{Ha\text{-}Hb}$=11 Hz and $J_{Ha\text{-}Hc}$=17 Hz.

Another nomenclature convention is applied for those cases when *cis* or *trans* designation is ambiguous. It is also correct to use this other system for simple alkenes that can be readily named using the *cis-trans* convention. Consequently we begin our explanation of this nomenclature system with the familiar 2-

butene example. We will then apply it to more complex systems. If, in the case of 2-butene, both methyl groups are on the same side of the double bond, as in *cis*-2-butene, we can name the molecule (*Z*)-2-butene (the *Z* is taken from the German word zusammen, meaning together). If, on the other hand, the methyl groups are on opposite sides of the double bond, as in *trans*-2-butene, we can name the molecule (*E*)-2-butene (the *E* is taken from the German word entgegen, meaning opposite).

Now let us look at a more complex example and try to determine a correct name for a compound using the *E*/*Z* nomenclature system. Consider the molecule in Figure 8.9.

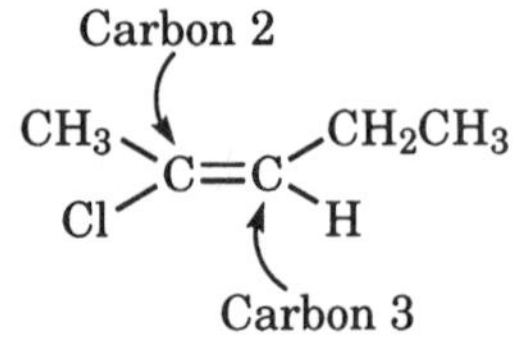

The longest carbon chain that contains the double bond consists of five carbon atoms; therefore we know this compound is a pentene derivative. We also know that the double bond begins at carbon-2, however mere inspection of this compound does not allow us to easily determine an *E*/*Z* stereochemical designation. We need to examine the molecule and apply the Cahn, Ingold, Prelog priority rules to name this compound correctly.

FIGURE 8.9: A more complex alkene.

> **RULE 1. Priority is determined for the atoms directly attached to the carbon atoms making up the double bond. The atom with the higher atomic number has higher priority.**

Application of Rule 1 to this molecule leads us to the following priority designations. Of the two atoms attached to carbon number two, the chlorine atom has a higher atomic number, and thus a higher priority, than the carbon in the methyl group. For carbon number 3, the two atoms attached are a hydrogen atom and a carbon atom. The carbon atom has a higher priority

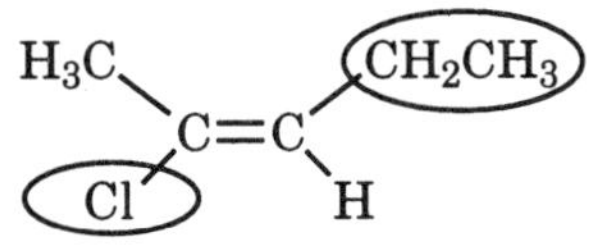

FIGURE 8.10: (*E*)-2-chloro-2-pentane with higher priority groups circled.

Since the higher priority groups on each of the alkene carbons are on the opposite sides of the double bond, we name this compound, (*E*)-2-chloro-2-pentene.

In the previous example, it was relatively easy to determine the priority of the groups on each of the alkene carbons. Now let us consider another, slightly more complex, example by trying to name the compound in Figure 8.10.

In this case, we begin by determining that the longest carbon chain that contains the double bond consists of seven carbon atoms, and that we are naming a heptene derivative. The priority of the groups attached to carbon-3 is readily determined from Rule 1. Carbon-4, however, is singly bonded to 2 carbon atoms, as methylenes. Thus Rule 1 does not allow assignment of priority. Then Rule 2 is invoked.

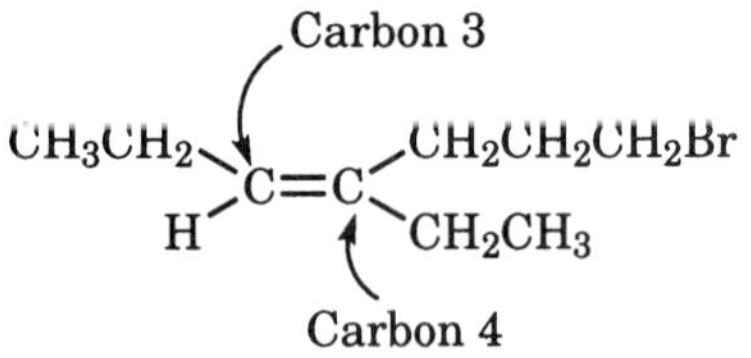

FIGURE 8.11: Another alkene to name.

> **Rule 2. If priority can not be established by the atomic number of the atoms directly bonded to the olefinic carbons, the chains are followed until a point of difference is found.**

In our example, the two carbons attached to carbon four have equal priority. Thus we must go one carbon further to determine priority. Here also the next atom in line is a carbon. But, carbon 6 is attached to a bromomethyl group, while the ethyl group ends with a hydrogen atom. Thus higher priority is assigned to the bromopropyl group.

In this case, the higher priority groups are on the same side of the double bond, and the correct name is (*Z*)-7-bromo-4-ethyl-3-heptene.

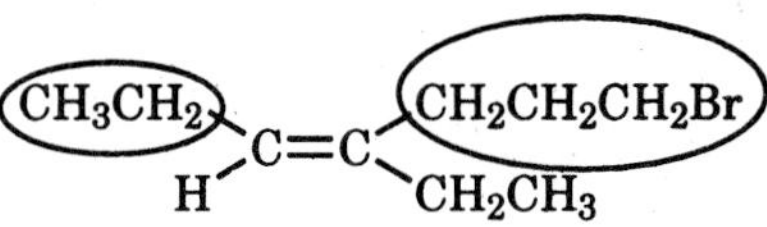

FIGURE 8.12: (*Z*)-7-bromo-4-ethyl-3-heptene with higher priority groups circled.

A final rule applies in cases of double and triple bonds.

> **Rule 3. In the case of double and triple bonds, priority is assigned based on doubling or tripling both atoms as illustrated at right in Figure 8.13.**

FIGURE 8.13: Priority assignments for double and triple bonds.

We also encounter cis/trans isomerism in certain inorganic compounds, such as the square planar complexes of platinum. Such planar complexes are common despite the fact that VSEPR theory predicts that such compounds would be tetrahedral (you will see the origin of this geometry at a later point). If we consider the compound diamminedichloroplatinum, we see that two diastereomers of this compound exist: the *cis*-isomer and the *trans*-isomer.

It is interesting to note that the *cis*-isomer has been identified as a potent chemotherapeutic agent, while the *trans*-isomer is inactive against cancer cells.

We also see *cis* and *trans* isomerism in octahedral complexes. When we extend VSEPR to 6 regions of electron density, we predict octahedral electronic geometry. An octahedron is simply two square-based pyramids stacked together at the bases. An example of an octahedral compound is sulfur hexafluoride.

If the central atom is bonded to two ligands of one type and four ligands of another type, cis/trans isomers are again possible. Consider the coordination compound formed by Co^{2+}, four ammonia molecules, and two chloride anions. In this case the chloride anions can be cis to one another, or they can be trans to one another.

FIGURE 8.14: a) *cis-* and b) *trans* platinum complexes.

Note that the structures (a) and (a') in Figure 8.16 are the same molecule, seen from different perspectives. Simlarly, (b) and (b') are identical.

FIGURE 8.15: Molecular structure of SF_6.

FIGURE 8.16: a) *cis-* and b) *trans*-isomers of a cobalt complex.

CONCEPT CHECK: NAMING GEOMETRIC ISOMERS

PROBLEM: Determine if the following compound is *E* or *Z*.

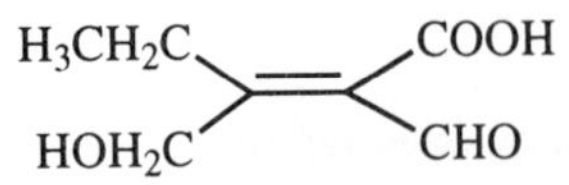

H_3CH_2C $COOH$

HOH_2C CHO

SOLUTION: One way to approach this problem is the divide the alkene right down the center with an imaginary plane as shown below.

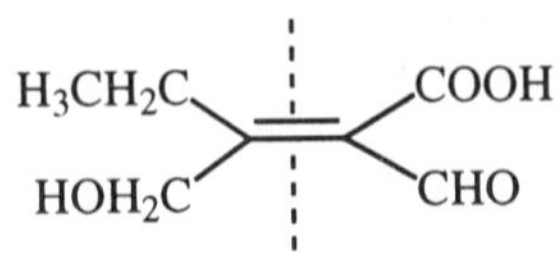

H_3CH_2C $COOH$

HOH_2C CHO

Now we need to examine the substituents on both of the alkene carbons. If we consider the alkene carbon on the left side we will be comparing the following two groups: $-CH_2CH_3$ and $-CH_2OH$. Comparing these two substituents according to the Cahn-Ingold-Prelog rules, we determine that the $-CH_2OH$ group has a higher priority than the $-CH_2CH_3$ group. If we compare the two substituents on the right side we find that the carboxylic acid (-COOH) has a higher priority that the aldehyde (-CHO). It might be helpful at this point to draw an ellipse around the higher priority group on each alkene carbon. This is shown below.

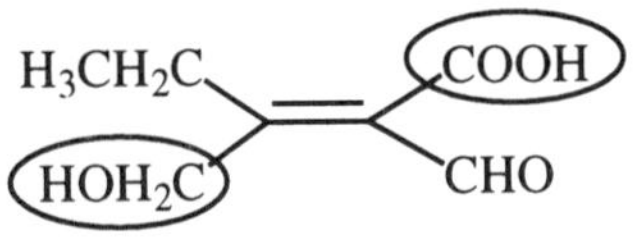

H_3CH_2C $COOH$

HOH_2C CHO

Examining the structure again, we note that the higher priority substituents are on opposite sides of the double bond. Thus, we can conclude that this alkene has the *E* configuration.

8.3 CHIRALITY AND ISOMERS

Chirality, as you saw earlier, has to do with whether an object is superimposable on its mirror image. We used a hand as our first example of a chiral object. As you well know, your right hand is not superimposable on your left hand and it is your right hand that you see when you hold your left hand up to a mirror. The fact that these images are non-superimposable is due to a lack of symmetry.

Molecules can also be chiral. A chiral molecule is not superimposable on its mirror image. Let us look at a few molecules and see if we can determine what structural characteristics are found in some chiral molecules. Let us consider some of the molecules we used as examples of structural isomers: 2-pentanol and 3-pentanol. First consider a molecule of 3-pentanol and the mirror image of this molecule.

H_3CH_2C / H / H_3CH_2C / OH | CH_2CH_3 / H / HO / CH_2CH_3

= C_2H_5

= OH

= H

FIGURE 8.17: A molecule of 3-pentanol and its mirror image.

As we can see, it is possible to superimpose the mirror image on the original molecule by simply rotating the mirror image. (Note: it is often helpful to build actual molecular models to determine if two molecules are superimposable.)

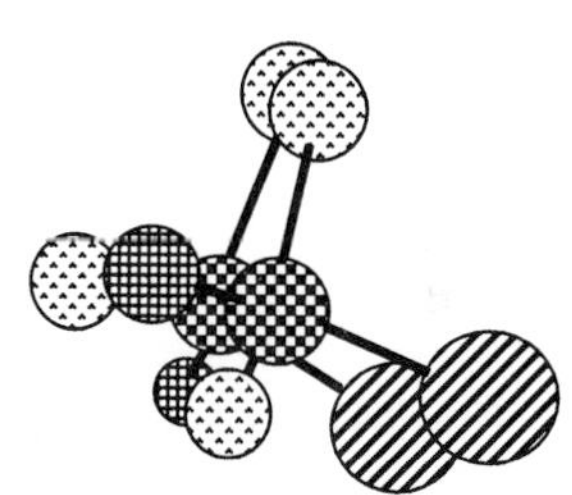

FIGURE 8.18: Superimposed images of 3-pentanol.

If we try the same exercise with a molecule of 2-pentanol, we find the mirror images are non-superimposable and the molecules are related as enantiomers.

There is something special about carbon-2 in this molecule. If you look at it carefully you will notice that this carbon is surrounded by four different substituents. A carbon that is surrounded by four different substituents is called a stereocenter or a chiral carbon atom. A characteristic of many chiral compounds is that they contain a chiral carbon atom.

$CH_2CH_2CH_3$ / H / H_3C / OH | $H_3CH_2CH_2C$ / H / HO / CH_3

= CH_3

= OH

= C_2H_5

= H

FIGURE 8.19: A molecule of 2-pentanol and its mirror image.

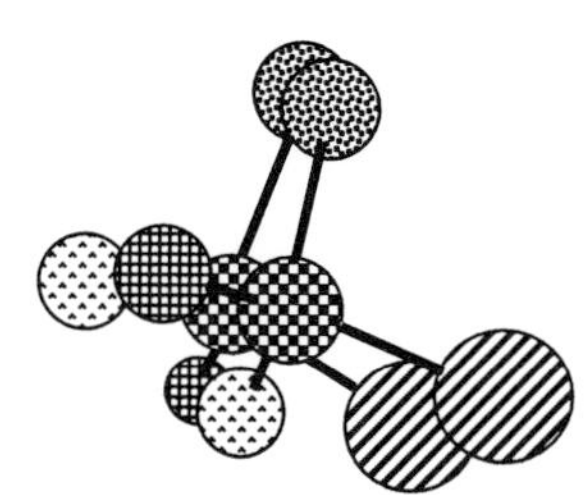

FIGURE 8.20: Non-superimposable images of the two enantiomers of 2-pentanol.

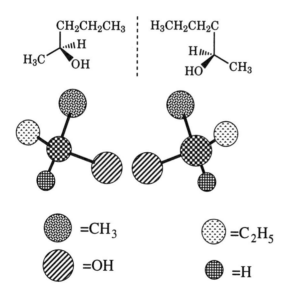

8.4 NOMENCLATURE AT THE STEREOCENTER: IS IT *R* OR IS IT *S*?

In order for a chemist to know what other chemists were talking about, a nomenclature system for chiral compounds was developed to designate the stereochemistry at chiral atoms. Let us begin by looking at the 2-pentanol example again. Shown at right are the two enantiomers of 2-pentanol. In order to name these compounds, including the sterochemical designation, a few rules need to be followed. These rules are similar to the rules we used when naming **geometric isomers**.

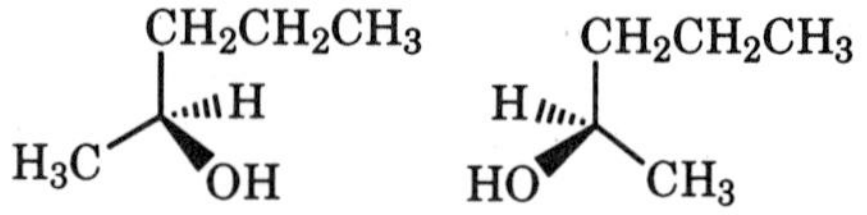

FIGURE 8.21: Enantiomeric 2-pentanol structures.

First, priority must be assigned to each group surrounding the **chiral center.** The basis for assigning priority is the same as that used when naming geometric isomers (See section 8.2). Thus in the case of 2-pentanol, the priority of the different groups would be as follows. The ranking is from high priority to low priority.

$$-OH > -CH_2CH_2CH_3 > -CH_3 > -H$$

After priority is assigned, the molecule is oriented in such a manner that the lowest priority substituent is toward the back, away from the observer. If we orient our 2-pentanol molecules in this manner we would see the following in Figure 8.22.

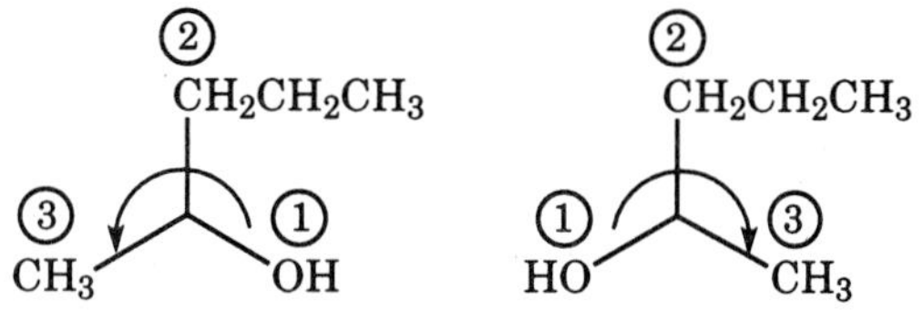

In the first case, the direction of priority, from highest priority (1) through medium priority (2) to lowest priority (3), is counter-clockwise and this stereoisomer is said to have the *S* configuration (after the Latin, *sinister*, meaning "left"). The other stereoisomer, in which the direction of priority from high to low is clockwise, is said to have the *R* configuration (after the Latin, *rectus*, meaning "straight or right"). Let us look at another example. Consider the ketone in Figure 8.23 and try to determine the correct stereochemical designation.

FIGURE 8.22: Priority assignments for the enantiomers of 2-pentanol.

In this case, the chlorine atom has the highest priority, the acetyl group has the second highest priority, the methyl group has the third highest priority, and of course the hydrogen atom has the lowest priority. If you were to orient the molecule so that the hydrogen atom pointed away from you, the direction in going from high to low priority would be in the counterclockwise direction. Thus, this is a drawing of (*S*)-3-chloro-2-butanone.

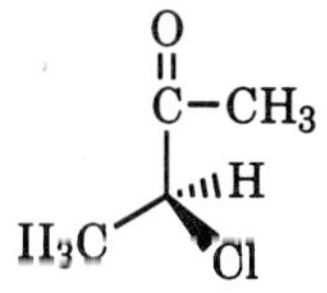

FIGURE 8.23: A chiral ketone.

CONCEPT CHECK: R OR S DESIGNATION AND CAHN INGOLD PRELOG RULE #3

PROBLEM: Name the following compound, including the stereochemical designation at the chiral carbon.

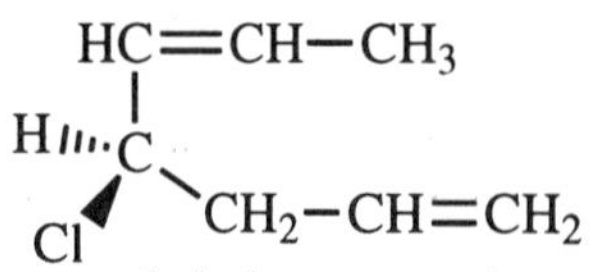

SOLUTION: The longest chain is seven carbons, and this contains two double bonds. The backbone is 1,5-heptadiene. Carbon number 4 is chiral, since 4 different groups are attached; H, Cl, $CH{=}CHCH_3$, and $CH_2CH{=}CH_2$. The procedure is to rotate the lowest priority group (H) away from the observer, and prioritize the other 3 groups. Chlorine, as the highest atomic weight atom attached to the chiral carbon, is easily assigned first priority. The choice between $CH{=}CHCH_3$ and $CH_2CH{=}CH_2$ for second priority is less obvious. Both are a three carbon chain. But the Cahn, Ingold, Prelog priority rule #3 indicates that the alkene carbons should be treated as if each had an additional carbon atom bonded to it. (These are shown as C*.)

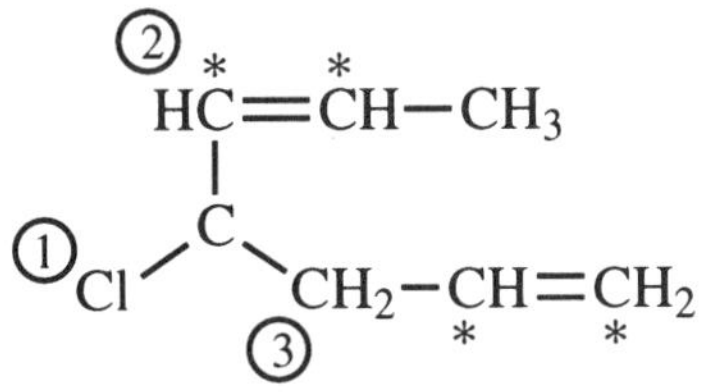

```
 (2)  *    *
     HC=CH—CH3
      |
      C
(1) Cl/  \CH2—CH=CH2
      (3)     *    *
```

The chain $C^*H{=}C^*HCH_3$ has a higher priority than $CH_2C^*H{=}C^*H_2$. Thus the compound is named (*R*)-4-chloro-1,5-diheptadiene.

8.5 OPTICAL ACTIVITY

You might be wondering at this point whether such subtle differences between enantiomers results in any difference in properties. As we see in Table 8.1, using 2-pentanol as our example again, almost every physical property is the same. The physical property that is different is the sign of the **optical rotation**; enantiomers have optical rotations that are equal in magnitude, but opposite in sign. Therefore we can use optical activity as a way of distinguishing between enantiomers. As we will see shortly, we can also use optical activity as a way of determining the enantiomeric purity of a given sample.

TABLE 8.1: SOME PROPERTIES OF THE ENANTIOMERS OF 2-PENTANOL

TABLE 8.1

	(*R*)-2-PENTANOL	(*S*)-2-PENTANOL
Boiling Point (°C)	119	119
Density at 20°C (g/ml)	0.8103	0.8103
Solubility in water at 20°C (g/100 ml H_2O)	16.6	16.6
Specific Optical Rotation Values $[\alpha]^{25}_{D}$	-13°	+13°

WHAT IS OPTICAL ACTIVITY?

Quite simply, optical activity is the ability of a chiral molecule to rotate plane polarized light. Light, as you know, consists of electromagnetic waves that oscillate in an infinite number of planes. If we were to look head on at the electrical portion of an approaching beam of light we would see an infinte number of orientations of electric oscillations, all perpendicular to the direction the light is travelling. This is illustrated diagramatically in Figure 8.24.

FIGURE 8.24: Electric field of normal light.

Plane polarized light is formed by passing this normal light through a polarizing filter. The polarizing filter removes all tne waves except one. If we were to view an approaching beam of plane polarized light we would see oscillation of the electric field in only one direction. This is illustrated diagramatically in Figure 8.25.

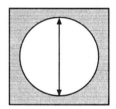

FIGURE 8.25: Electric field of plane polarized light.

Optical activity is measured using a polarimeter. The basic components of any polarimeter include a light source, a polarizing filter, a sample tube, and a detection device to measure the amount of rotation. When plane polarized light passes through a solution of a compound that is optically inactive, no rotation is observed. On the other hand, when plane polarized light passes through a solution of an optically active compound, the plane polarized light undergoes a rotation. This rotation is called the optical rotation which is designated by the Greek letter α and measured in degrees.

If the molecule rotates the light in a clockwise direction it is said to have a (+) sign of rotation that is also called **dextrorotatory** and if the molecule rotates the light in a counterclockwise direction it is said to have a (-) sign of rotation that is also called **levorotatory** (these terms are derived from the Latin; *dexter* = right-hand and *levo* = left-hand). It is important to note that there is no correlation between the direction that a given molecule rotates plane polarized light and the *R/S* designation of that particular molecule.

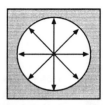

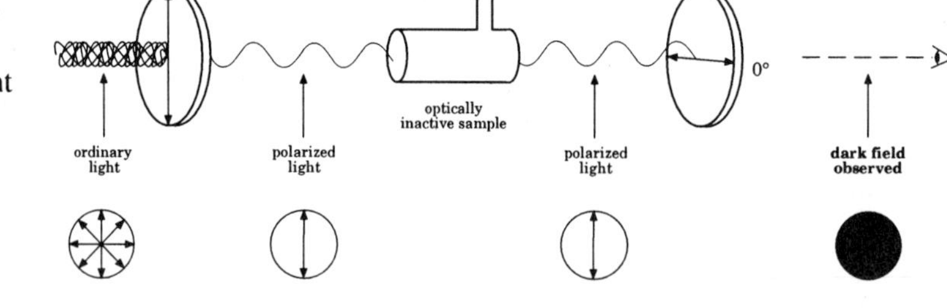

Specific Rotation

As you have probably guessed by now, the angle that the plane polarized light rotates when passing through a solution of a chiral material depends on how many of the chiral molecules interacted with the light. This quantity is in turn dependent upon the concentration of that species and the length of the tube in which the solution is contained. The rotation will also depend upon the wavelength of light that is passed through the sample and the temperature. Since chemists use optical rotation data when preparing known compounds, as a check on optical purity, it is necessary to standardize the way that rotation data are reported. Thus **specific rotation** has been defined as a basis for standardization. Specific rotation can be calculated according to the following formula.

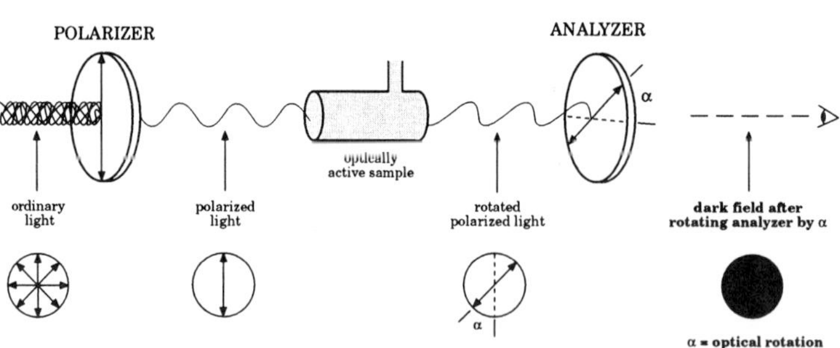

FIGURE 8.26: Rotation of light polarization vector.

$$[\alpha]^T = \frac{\alpha}{lc}$$ EQ. 8.1

where [α] = specific rotation
α= observed rotation
c= concentration in grams/milliliter (or density for neat liquids)
l= length of the polarimeter tube in decimeters
T= temperature at time of observation

Thus when specific rotations are reported all of this information must be included so that other chemists can duplicate the data if they prepare the same compound. Let us look at our familiar example, 2-pentanol, and see how the specific rotation for the two enantiomers is reported in Figure 8.27.

The specific rotations of these compounds were determined from neat (pure liquid) samples of each compound, but remember that in most cases the concentration and solvent used must also be specified after the numerical value.

$CH_2CH_2CH_3$, H, H_3C, OH	$CH_2CH_2CH_3$, H, HO, CH_3
(S)-2-pentanol	(R)-2-pentanol
$[\alpha]^{25}$= +13° [neat]	$[\alpha]^{25}$= -13° [neat]

FIGURE 8.27: Specific rotation values for neat samples for the enantiomers of 2-pentanol.

CONCEPT CHECK: DETERMINATION OF SPECIFIC ROTATION

PROBLEM: A 0.100 molar solution of codeine ($C_{18}H_{21}NO_3$, MW = 299.36) dissolved in ethanol is used to fill a cell with pathlength 20.0 cm. Analysis shows an observed rotation of - 8.13° at 15.0°C. Calculate the specific rotation of codeine under these circumstances.

SOLUTION: First, the concentration must be converted from molarity to grams of solute per ml of solution.

$$\left(\frac{0.100 \text{ mole codeine}}{1 \text{ Liter solution}}\right)\left(\frac{1 \text{ Li t}}{1000 \text{ ml}}\right)\left(\frac{299.36 \text{ g codeine}}{1 \text{ mole}}\right) = 0.02994 \text{ g}$$

The pathlength l, in decimeters, is:

$$l = (20.0 \text{ cm})\left(\frac{1 \text{ dm}}{10 \text{ cm}}\right) = 2.00 \text{ dm}$$

Substituting into equation 8.1 yields:

$$[\alpha]^{15} = \frac{\alpha}{l\,c} = \frac{-8.13°}{(2.00 \text{ dm})(0.02994 \text{ g/mL})} = -136°$$

OPTICAL PURITY AND ENANTIOMERIC EXCESS

Often we refer to a particular compound as being optically pure. Quite simply, this means that the sample in question contains only one of the stereoisomeric forms of the compound; that is, the compound is all of the *R* form or all of the S form. As you might expect, a sample of a compound that is optically pure will exhibit maximum rotation. If the sample is contaminated by any of the other enantiomeric form the rotation will be reduced, by rotation of the light in an equal, but opposite direction. Therefore, if we know the specific rotation of an optically pure sample of a compound, we can calculate the enantiomeric excess, that is how much one isomer is in excess over another isomer in a given mixture, in any other sample by determining the specific rotation of this new sample.

We know that the specific rotation of (*R*)-(-)-2-pentanol, at 25°C, is -13°. If a particular sample contained some of the *S*-isomer this rotation would be reduced and we could calculate how much excess of the *R* isomer is present in that sample according to the following equation.

$$\left(\frac{\text{measured specific rotation of mixture}}{\text{specific rotation for the pure enantiomer}}\right)(100) = \%\ \text{enantiomeric excess} \qquad \text{EQ. 8.2}$$

If we determined that the rotation of a mixture of the *R* and the *S* isomers rotated plane polarized light -3.25°, we could determine that the enantiomeric excess of the *R* isomer in that sample is 25%.

$$\left(\frac{-3.25^\circ}{-13^\circ}\right)(100) = 25\% \qquad \text{EQ. 8.3}$$

What do you suppose would happen to the specific rotation if we had an equal amount of both the *R* isomer and the *S* isomer in a given sample? You should predict that this sample would have zero rotation. For every rotation to the left from a molecule of (*R*)-2-pentanol, an equal but opposite rotation to the right from a molecule of (*S*)-2-pentanol would occur and, therefore, the net rotation would be zero. An equimolar mixture of the *R* and *S* enantiomers is called a **racemic mixture**.

8.6 MOLECULES THAT CONTAIN MORE THAN ONE STEREOCENTER: ANOTHER TYPE OF DIASTEREOMER

Let us now consider a molecule, 2,3-dibromopentane, that contains two stereocenters. Since the configuration at each stereocenter can be either *R* or *S*, we find that there are four possible stereoisomers of this compound; the (2*R*,3*R*) isomer, the (2*S*,3*S*) isomer, the (2*R*,3*S*) isomer, and the (2*S*,3*R*) isomer.

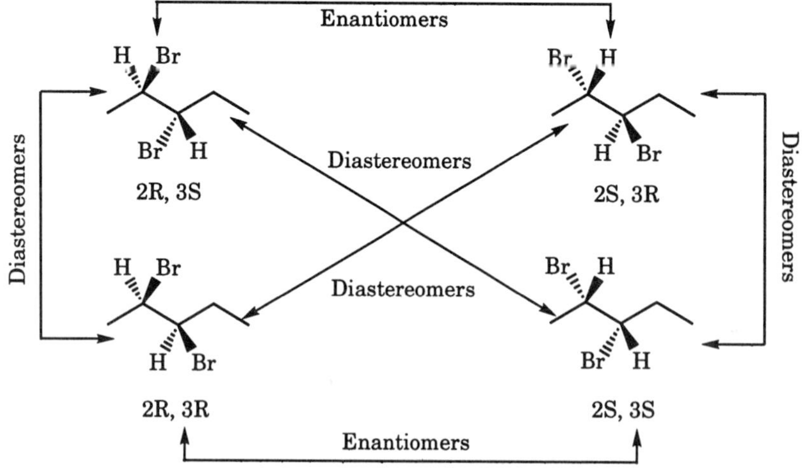

FIGURE 8.28: The four stereoisomers of 2,3-dibromopentane.

As you can see, 2,3-dibromopentane can exist as two pairs of enantiomers and four pairs of diastereomers. We now see two important principles. First, the maximum number of stereoisomers that are possible for a particular compound is equal to 2^n, where n is equal to the number of stereocenters present in that compound. Second, pairs of stereoisomers are enantiomers only if they have a different configuration at all the stereocenters in the molecule.

Diastereomers, unlike enantiomers, have different physical properties; they have different boiling points, melting points, and solubilities and therefore may be separated based upon these differences. You should also note that the specific rotations of diastereomers are not related in any particular manner. Both compounds may, for example, rotate plane polarized light to the right. This is, of course, in contrast to enantiomers that rotate plane polarized light in equal, but opposite directions. Thus you can not calculate the composition of a solution of diastereomers from optical rotation data, as was done for a mixture of enantiomers in equation 8.2.

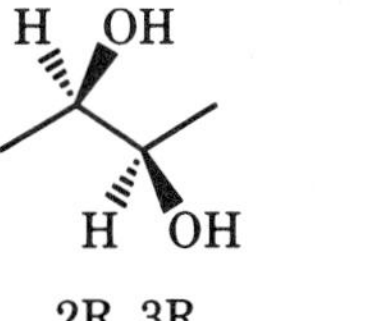

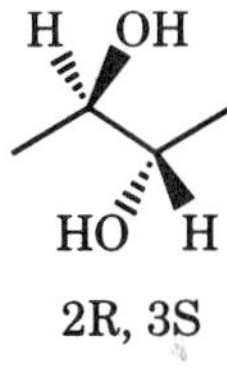

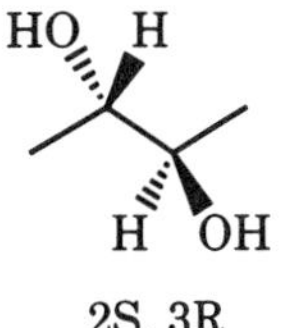

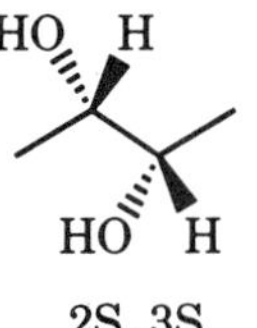

FIGURE 8.29:
2,3-butanediol isomers.

If we consider another compound, 2,3-butanediol, we discover that simply because a molecule contains two stereocenters does not mean that it will always have four stereoisomers. Each stereocenter could have either the *R* configuration or the *S* configuration.

In this case, the (2*R*,3*R*) isomer and the (2*S*,3*S*) isomer are a pair of enantiomers. However, the (2*R*,3*S*) isomer and the (2*S*,3*R*) isomer are not a pair of enantiomers. While they are mirror images, simply rotating one of the isomers turns it into the other isomer; in other words, they are the same molecule.

FIGURE 8.30:
Superimposition of meso-2,3-butanediol isomers.

This compound is called meso-2,3-butanediol. A **meso compound** is a compound that contains stereocenters, but is not chiral and, therefore, does not rotate plane polarized light. You should be sure not to confuse meso compounds with racemic mixtures. The former is achiral, while the latter is a 50:50 mixture of two chiral substances. It is usually quite simple to determine if a compound is *meso*; it has an internal plane of symmetry.

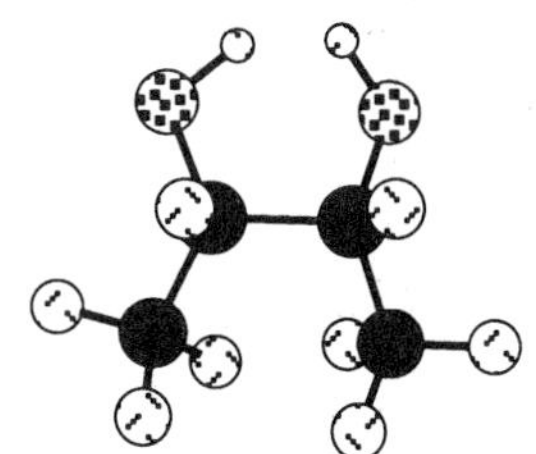

FIGURE 8.31:
Internal plane of symmetry in 2,3-butanediol isomers.

8.7 FISCHER PROJECTIONS

Fischer projections, named after the famous German carbohydrate chemist, Emil Fischer, are two dimensional representations of three dimensional molecules. It is often useful, and many times easier, to use these representations when drawing and looking at chiral molecules. Let us begin by looking at the Fischer projection of our simple example, 2-pentanol. Instead of looking at the molecule in the manner that we use to assign configuration, picture the molecule as though two of the substituents are pointing away from you, back into the plane of the paper, and two of the substituents are pointing toward you, out of the plane of the paper.

The Fischer projection is drawn as shown on the right. The chiral carbon atom is represented by crossing the two lines, the substituents on the vertical axis are extending behind the plane of the paper and the substituents on the horizontal axis are extending above the plane of the paper. You should note that the convention is to keep the longest carbon chain along the vertical axis.

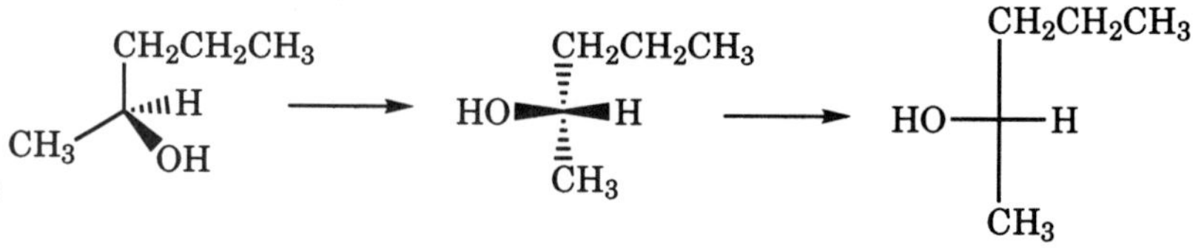

FIGURE 8.32: Constructing a Fischer projection of 2-pentanol.

It is also possible to assign configuration from the Fischer projection. You must, however, keep several "rules" in mind when doing so.

1. **A Fischer projection may be rotated 180° and the resulting Fischer projection is the same molecule.**

2. **Rotation of a Fischer projection by 90° produces the enantiomer of the original.**

3. **A Fischer projection lifted out of the plane of the paper and flipped over also produces the enantiomer.**

4. **Any three groups may be exchanged and the resulting structure will have the same configuration.**

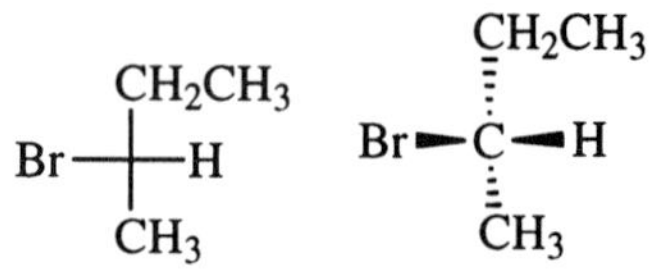

FIGURE 8.33: A Fischer projection of 2-bromobutane and its dash and wedge representation.

Consider the following compound in Figure 8.33 and let us try to assign the configuration about the stereocenter.

After the hydrogen atom is rotated directly behind the stereocenter, the remaining three groups are evaluated for priority.

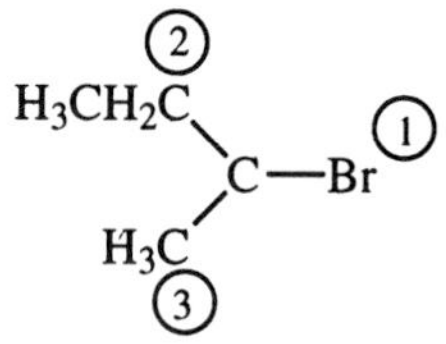

FIGURE 8.34: Assigning priorities in (*S*)-2-bromobutane.

Since the movement from priority one to two to three is in a counter-clockwise direction, the compound has the "*S*" configuration.

Rule One indicates that rotating the Fischer projection 180° does not change the stereochemistry. In the Figure 8.35, we redraw the compound in Figure 8.33 with the Fischer projection rotated 180°. We also show the dash and wedge representation.

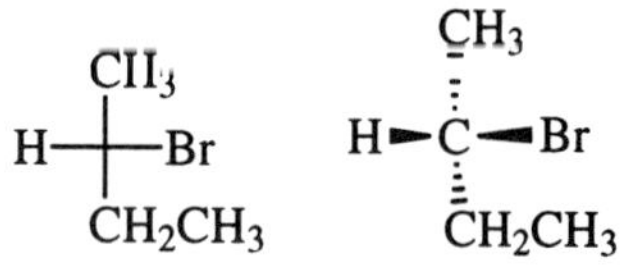

FIGURE 8.35: Fischer projection and dash and wedge representation.

In Figure 8.36, the hydrogen is again rotated behind the stereocenter, and relative priority is assigned. Again, the counter-clockwise progression of priority indicates an "S" configuration. Thus rotation by 180° did not change the configuration.

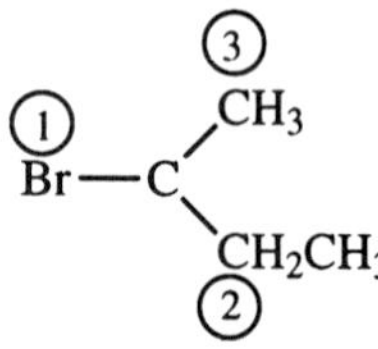

If the Fischer projection in Figure 8.33 is rotated 90°, the projection in Figure 8.37 results.

Rotating the hydrogen directly behind the stereocenter results in Figure 8.38.

FIGURE 8.36: Assigning priorities in (*S*)-2-bromobutane.

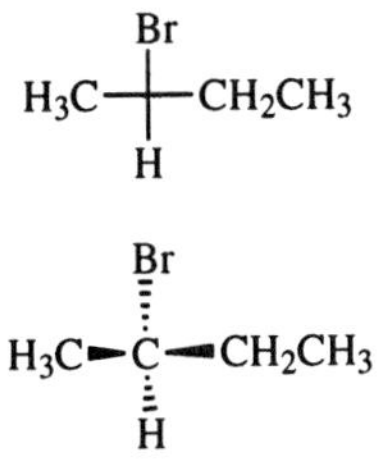

FIGURE 8.37: A Fischer projection and dash and wedge representation of 2-bromobutane.

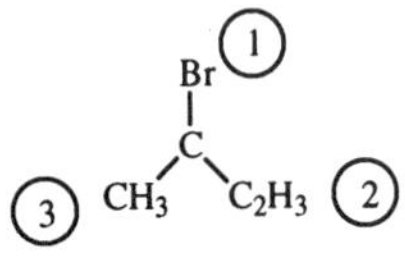

FIGURE 8.38: Assigning priorities in (*R*)-2-bromobutane.

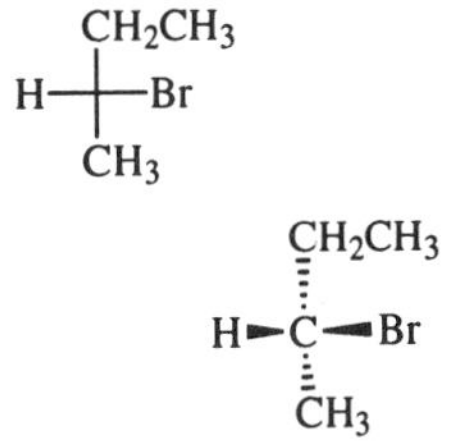

FIGURE 8.39: A Fischer projection and dash and wedge representation of 2-bromobutane.

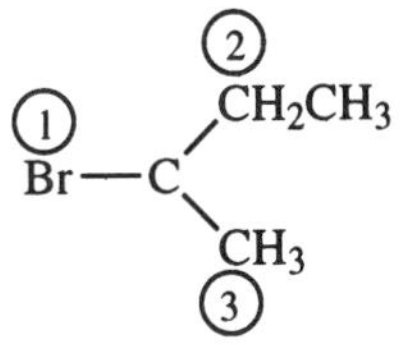

FIGURE 8.40: Assigning priorities in (*R*)-2-bromobutane.

Note that moving around the stereocenter in order of priority is a clockwise movement, and has changed the compound to the "*R*" configuration. Thus rotating the Fischer projecton by 90° produces the enantiomer of the original.

To examine Rule Three, the Fischer projection in Figure 8.33 is lifted out of the plane and flipped over, resulting in the projection and dash and wedge representation seen in Figure 8.39

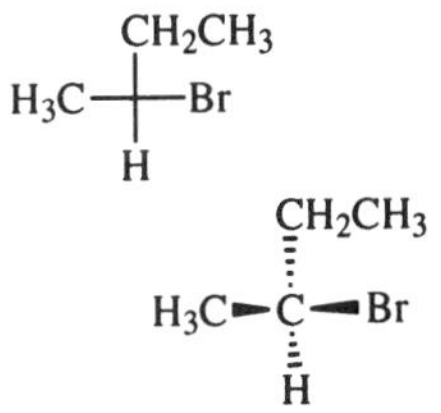

FIGURE 8.41: A Fischer projection and dash and wedge representation of 2-bromobutane.

Again, the dash and wedge figure is rotated to bring the hydrogen directly behind the stereocenter, with the priorities seen in Figure 8.40.

Flipping the Fischer projection has indeed resulted in the *R* enantiomer of the original (*S*)-2-bromobutane.

Finally, in Rule Four we see the effect of moving three of the four substituents around the stereocenter. If the ethyl group is held constant in the original projection in Figure 8.33, but each of the other three substituents is moved to a new location, Figure 8.41 results.

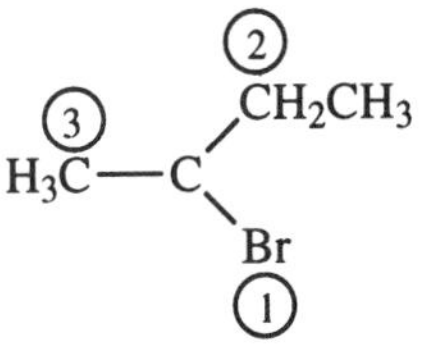

FIGURE 8.42: Assigning priorities in (*S*)-2-bromobutane.

Again, rotation of the hydrogen atom directly behind the stereocenter results in Figure 8.42.

Note that the original "*S*" configuration is retained, and Rule Four is confirmed.

Now let us consider a molecule that contains two stereocenters, (2*R*,3*S*)-2-bromo-3-chlorobutane, Figure 8.43

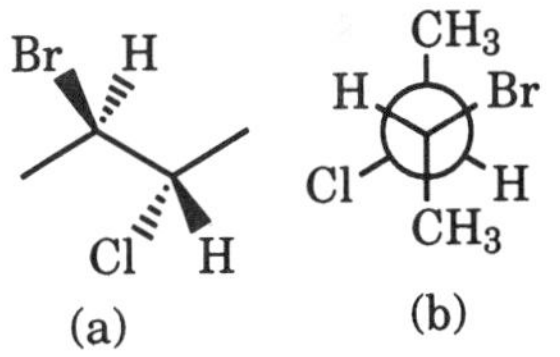

FIGURE 8.43: a) Sawhorse and b) Newman projection of (2*R*,3*S*)-2-bromo-3-chlorobutane.

In order to convert this staggered conformation of the molecule to a Fischer projection , it is helpful to redraw the molecule in the eclipsed conformation as shown in Figure 8.44. The eclipsed conformation was formed by holding the front carbon stationary and rotating the back carbon 180°. The Fischer projection is simply the drawing of what you would see if you were looking down onto the eclipsed conformation.

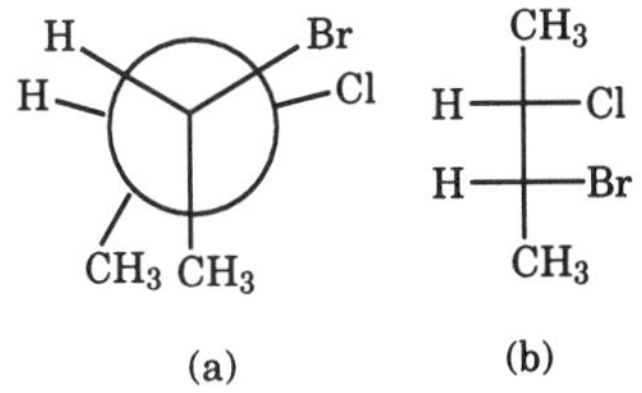

FIGURE 8.44: a) Eclipsed conformation and b) equivalent Fischer projection of (2*R*,3*S*)-2-bromo-3-chlorobutane.

Concept Check: Fischer Projections, Geometry, and Optical Activity

Problem: From the indicated Fischer projection of 2,3-dihydroxybutanal, draw the sawhorse representation and the Newman projection sighting from the C3 to the C2 carbon. Also, indicate the stereochemistry at the #2 and #3 carbons.

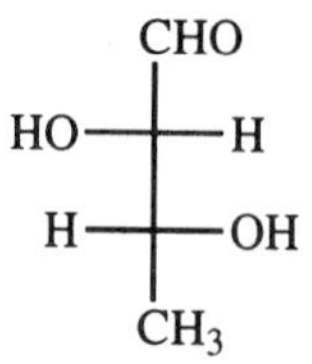

CHO
HO—+—H
H—+—OH
CH_3

Solution: For the #2 carbon, the order of priority is: OH > CHO > $CH(OH)CH_3$. This ordering appears to be in a clockwise direction. But in this representation, the H is pointed towards the viewer. Consequently, when the H is oriented away from the viewer, the rotation would be counterclockwise. Thus the #2 carbon is designated as "*S*". For the #3 carbon, the order of priority is: OH > CH(OH)CHO > CH_3. This ordering appears to be in a counterclockwise direction. Again, however, the H is pointing out towards the viewer. Consequently, when properly oriented, the rotation is clockwise, and the #3 carbon is designated "*R*".

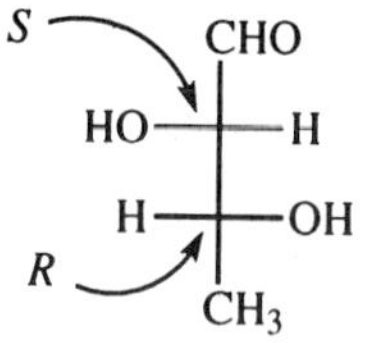

S → CHO
HO—+—H
H—+—OH
R → CH_3

To draw the sawhorse and Newman projections, remember that in a Fisher projection bonds along the vertical axis are bending away from the viewer, and that horizontal bonds are bending towards the viewer. Consequently, the CHO (#1 carbon) and the CH_3 (#4 carbon) are bending down, while the OH and H are bent up, resulting in the following eclipsed sawhorse and staggered Newman projections.

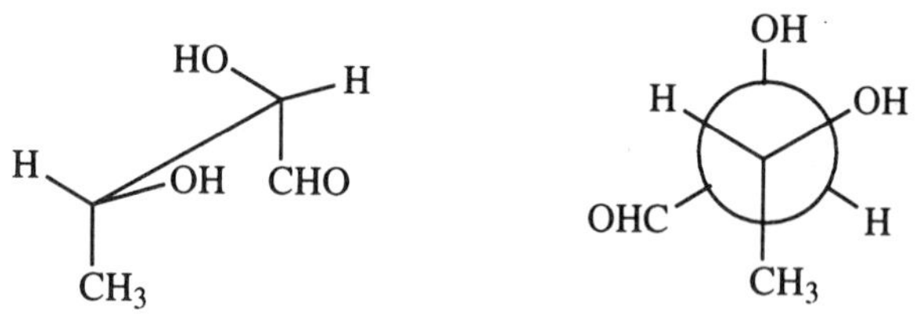

HO, H, H, OH, CHO, CH_3

OH, H, OH, OHC, H, CH_3

8.8 RESOLUTION

You may be wondering at this point how enantiomers are separated especially since many chemical reactions produce racemic mixtures instead of a pure enantiomer. If you think back to our discussion of the physical properties you will remember that enantiomers have identical physical properties and thus cannot be separated by any physical means. On the other hand, diastereomers have different physical properties and are separable. Therefore, one strategy for separating a racemic mixture is to convert the enantiomers into diastereomers, separate the diastereomers, and then regenerate the enantiomers in optically pure form. The process of separating a racemic mixture into optically pure enantiomers in known as **resolution.**

We will illustrate this process by discussing the separation of a racemic mixture of a carboxylic acid, 2-methylbutanoic acid. In order to conduct the resolution in an efficient manner we want to be able to synthesize the diastereomers and regenerate the optically pure carboxylic acids in high yield. If you recall, carboxylic acids can be converted to amides when they are reacted with amines, and amides can be hydrolyzed to reform carboxylic acids. Both of these processes are usually accomplished in high yield. Therefore, one strategy for resolving this chiral carboxylic acid would be to react the racemic mixture with an optically pure amine to generate a diastereomeric pair of amides, separate the diastereomers based on physical properties, and then hydrolyze the amide bond to regenerate the carboxylic acids. This process is illustrated next. Resolution by the process of generating diastereomers is a classical, but useful, process; other methods of resolution include chiral chromatography and enzymatic resolution.

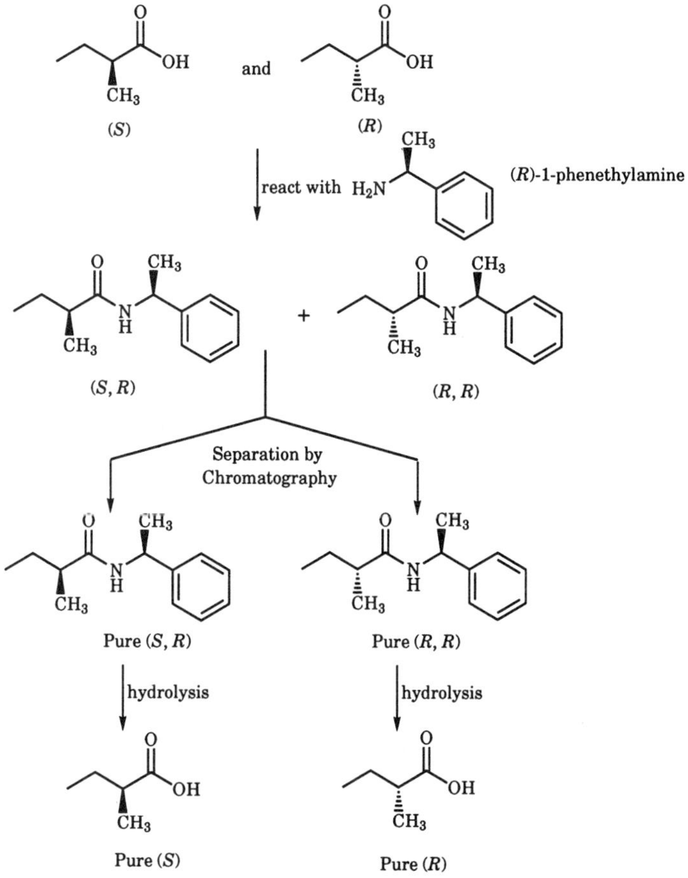

(*S*) and (*R*)

react with H_2N — (*R*)-1-phenethylamine

(*S*, *R*) + (*R*, *R*)

Separation by Chromatography

Pure (*S*, *R*) — Pure (*R*, *R*)

hydrolysis — hydrolysis

Pure (*S*) — Pure (*R*)

FIGURE 8.45: Resolution of 2-methylbutanoic acid with phenethylamine.

8.9 CHIRALITY WITHOUT A CHIRAL CARBON

Until now our discussion of chiral molecules has focused on molecules that contain a chiral carbon atom. However, molecules may be chiral even though they do not contain chiral carbons. Remember, a chiral molecule is not superimposable on its mirror image.

Consider the two molecules in Figure 8.46, both of which are chiral.

Neither compound is superimposable on its mirror image because of restricted rotation.

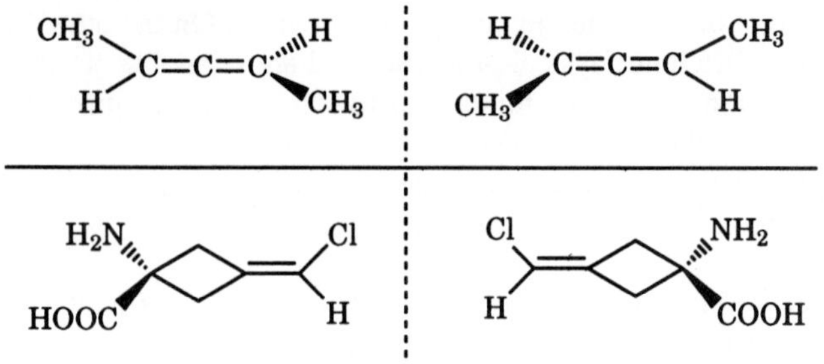

FIGURE 8.46: An allene derivative, a spirocompound, and their mirror images.

Atoms other than carbon can also be chiral. Recalling that the electronic geometry around nitrogen is tetrahedral, you should expect that a nitrogen atom surrounded by four different groups has the potential to exist as a pair of enantiomers. N-methyl-N-ethyl-N-propylbutylammonium ion is, in fact, chiral.

Phosphorus surrounded by four groups also has tetrahedral geometry and enantiomers of these kinds of compounds are isolable. In fact, one enantiomer of certain organophosphorus pesticides may actually be much more potent than the other enantiomer. Shown below in Figure 8.48 is the structure of Soman, a potent nerve gas.

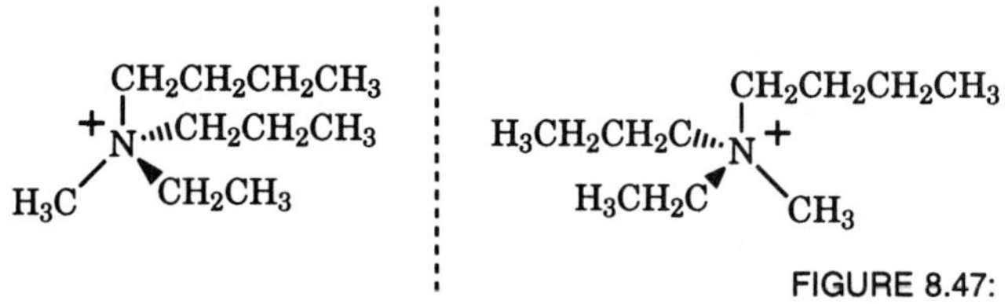

FIGURE 8.47: Mirror images of N-methyl-N-ethyl-N-propyl-butylammonium.

It is known that the isomer that has the *R* configuration at phosphorus is much more potent than the isomer that has the *S* configuration.

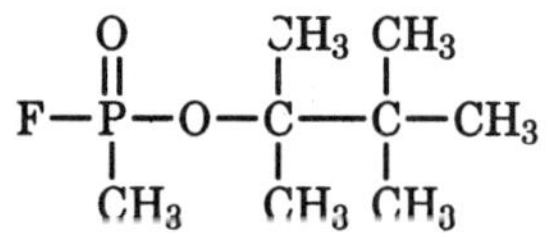

FIGURE 8.48: Structure of Soman.

Chiral Center(Stereocenter): An atom (usually carbon) that is surrounded by four different substituents or space groups.

Diastereomers: All stereoisomers that are not related as enantiomers.

Dextrorotatory: Rotation of plane polarized light to the right.

Enantiomers: Stereoisomers that are non-superimposable mirror images.

Fischer Projection: A two-dimensional representation of a three-dimensional molecule in which bonds along the vertical axis are oriented away from the observer and bonds along the horizontal axis are oriented toward the viewer.

Geometric isomers: Stereoisomers that result from restricted rotation. Geometric isomers are related as diastereomers.

Levorotatory: Rotation of plane polarized light to the left.

Meso compound: A molecule that contains chiral centers, but is achiral because of an internal plane of symmetry.

Optical rotation: Rotation of plane polarized light as it passes through a solution of an optically active compound.

Racemic mixture: A 50:50 mixture of two enantiomers. Racemic mixtures do not exhibit optical rotation.

Resolution: The process of separating optically pure enantiomers by chemical or biochemical means.

Specific rotation: A value (used for standardization) that relates the observed rotation of a given compound to the concentration of the compound and the length of the polarimeter tube.

Stereoisomers: Compounds with the same molecular formula and connectivity that differ in their orientation in space.

HOMEWORK Problems

1. Rank the following substituents in order of increasing priority using the Cahn-Ingold-Prelog rules.

(a) -Cl, -F, -I, -Br
(b) $-CH_3$, $-CH_2CH_2CH_3$, $-CH_2CH_3$, $-CH_2CH_2CH_2CH_3$
(c) -OH, $-NH_2$, $-CO_2CH_3$, $-CO_2H$
(d) $-CH_2OH$, $-CH_2Br$, $-CH{=}CH_2$, -C=O

2. Determine if the following compounds are E or Z.

a. Br, CH_3 / C=C / CH_3, NH_2

b. CH_3CH_2, CH_2OH / C=C / CH_3, CH_3

c. $CH_3O{-}C({=}O)$, CH_2COOH / C=C / CH_3, H

d. $CH_2{=}C(H)$, CH_2NH_2 / C=C / H, CH_2OH

3. Provide the IUPAC name for the following alkenes.

a. H, H / C=C / CH_3, $CH_2CH_2CH_3$

b. CH_3CH_2, H / C=C / H, $CH_2CH_2CH_2CH_3$

c. CH_3CH_2, CH_3 / C=C / $CH_3CH_2CH_2$, CH_2OH

d. CH_3, CH_2Br / C=C / H−C(=O), CH_2CH_3

e.

f. CH_3, CH_3 / C=C / H, $CH_2CH_2CH_2$ — $CH_2CH_2CH_2$, CH_2CH_3 / C=C / H, CH_3

4. Draw the structure of the following compounds.

a. *cis*-2-pentene
b. (*E*)-5-methyl-2-heptene
c. *trans*-cyclodecene
d. (*Z*)- 3-octene
e. (2*E*, 4*Z*)- octadiene

5. Squalene is the biosynthetic precursor to cholesterol. Determine the *E* or *Z* configuration at the bonds denoted in the following drawing of squalene.

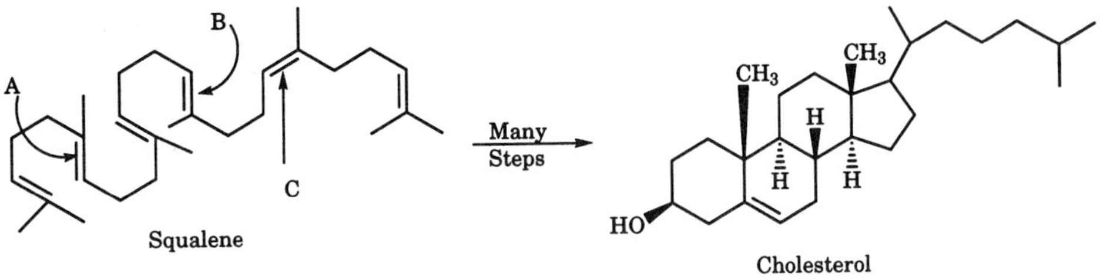

A B C

Squalene

Many Steps

CH_3 CH_3 H H H HO

Cholesterol

6. Draw arrows pointing to all stereocenters in the following molecules.

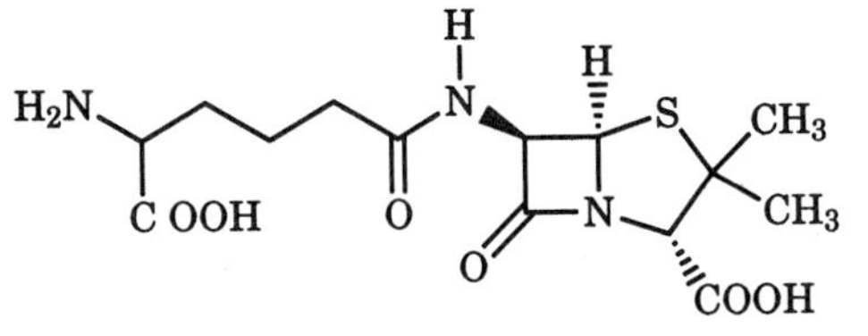

H_2N C OOH H N O H S CH_3 CH_3 N O COOH

Penicillin N

H_3C N O C O—CH_3 O—C O H

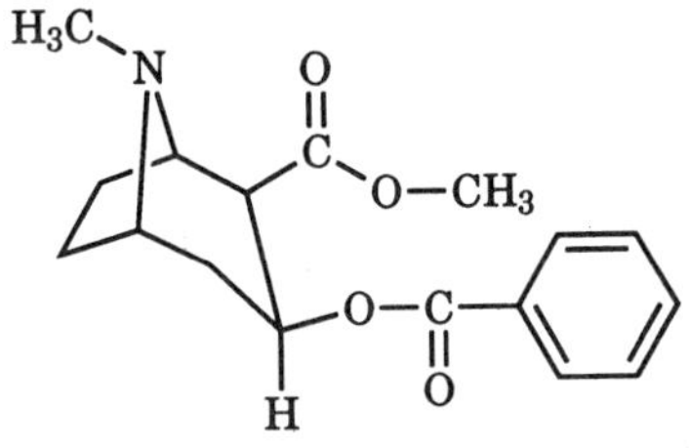

Cocaine

7. Assign *R* or *S* configuration at the stereocenters in the following molecules.

a. 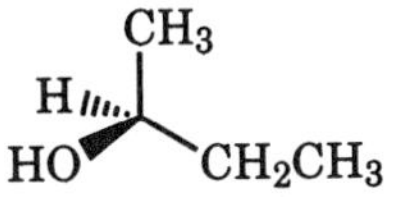 CH_3 H HO CH_2CH_3

b. 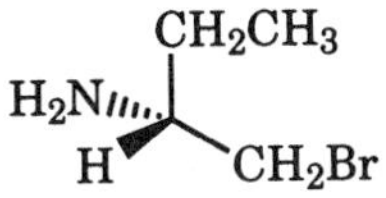 CH_2CH_3 H_2N H CH_2Br

c. 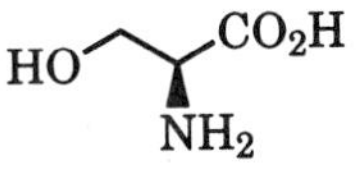 HO CO_2H NH_2

d. 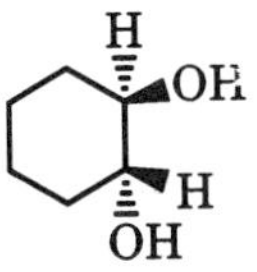 H OH H OH

e. 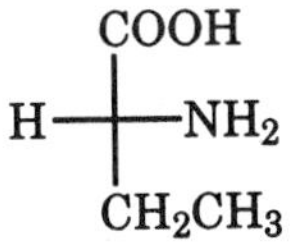 COOH H—NH$_2$ CH_2CH_3

f. 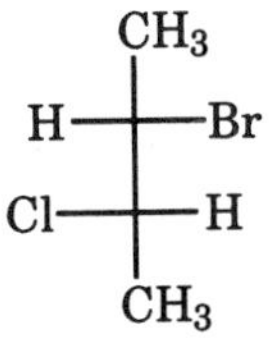 CH_3 H—Br Cl—H CH_3

8. Name the following compounds. Be sure to designate stereochemistry when necessary.

a. 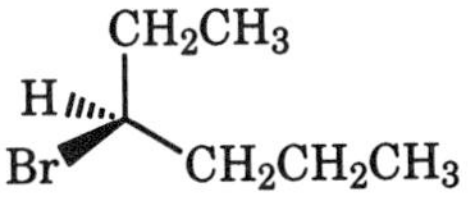 CH_2CH_3 H Br $CH_2CH_2CH_3$

b. 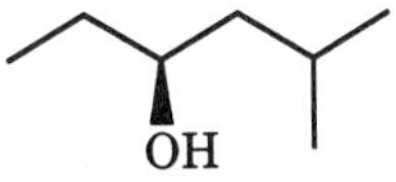 OH

c. 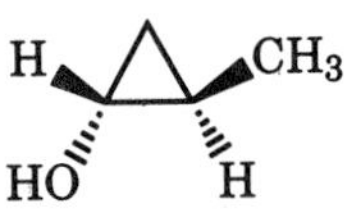 H CH_3 HO H

d. 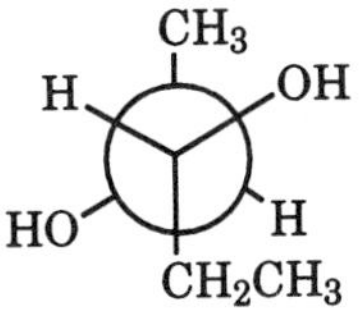 CH_3 H OH HO H CH_2CH_3

9. Draw structures of the following compounds.

a. (*S*)-2-pentanol
b. (*R*)-3-bromo-1-hexene
c. (1*S*, 2*S*)- 1-amino-2-methylcyclohexane
d. (2*R*, 3*S*)-dibromobutane

10. For each of the following pairs, decide if the stuctures are identical, structural isomers, geometric isomers, enantiomers, or diastereomers.

a.

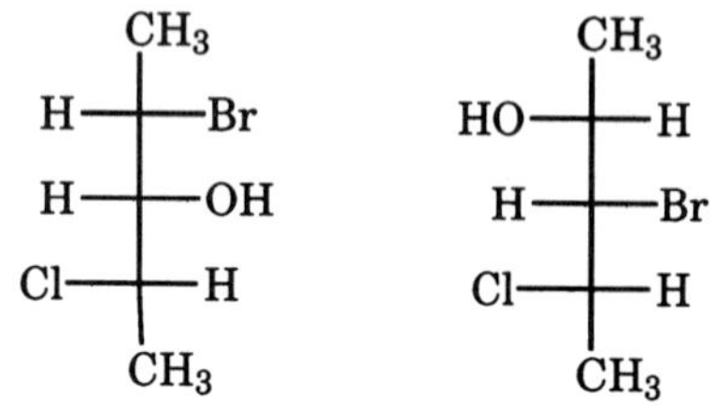

b.

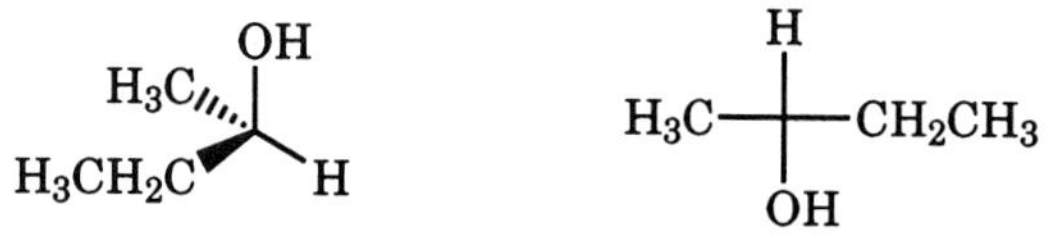

c.

d.

e.

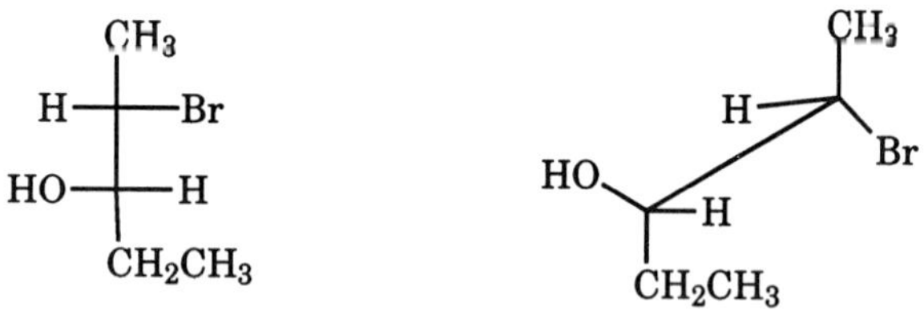

11. 1,2-butanediol and 1,3-butanediol both contain a stereocenter and are chiral. 2,3-butanediol contains two stereocenters, and may or may not be chiral. Explain.

12. Tartaric acid (2,3-dihydroxybutanedioic acid) has three unique forms (not counting a racemic mixture). For each of the following, draw the Fisher projections, label the chiral carbons as R or S, and indicate if the species is or is not optically active, and why:

a. (2*R*, 3*R*) - 2,3-dihydroxybutanedioic acid
b. (2*R*, 3*S*) - 2,3-dihydroxybutanedioic acid
c. (2*S*, 3*S*) - 2,3-dihydroxybutanedioic acid

13. 5.00 g of a pure enantiomer of 1,2-butanediol is dissolved in ethanol to form 25.0 ml of solution at 20°C. When placed in a 10.0 cm pathlength cell, the solution rotates light from a sodium lamp 2.90° to the right. Calculate $[\alpha]$ for this enantiomer.

14. 5.00 g of a pure enantiomer of 1,3-butanediol is dissolved in ethanol to form 25.0 ml of solution at 20°C. When placed in a 10.0 cm pathlength cell, the solution rotates light from a sodium lamp 3.76° to the left. Calculate $[\alpha]$ for this enantiomer.

15. 5.00 g of a known pure sample of the levorotatory entantiomer of cocaine is dissolved in chloroform to form 10.0 ml of solution at 20°C. When placed in a 10.0 cm pathlength cell, the solution rotates light from a sodium lamp 8.15°. Resolution of a racemic mixture of cocaine results in 14.0 g of product that, when dissolved in chloroform to form 25.0 ml of solution, shows an optical rotation of 8.20°. What is the enantiomeric excess of the levorotatory isomer in the partially resolved sample?

16. 15.0 g of a known pure sample of quinine harvested from cinchona bark is dissolved in ethanol to form 50.0 ml of solution at 15°C. When placed in a 10.0 cm pathlength cell, the solution rotates light from a sodium lamp -43.6°. A sample of quinine produced in the laboratory is a racemic mixture. Resolution of this mixture results in 6.50 g of product that, when dissolved in ethanol to form 25.0 ml of solution, shows an optical rotation of -22.0°. What is the enantiomeric excess of the levorotatory isomer in the partially resolved sample?

17. A sample of enantiomerically pure liquid nicotine is reported to have a specific rotation of -169°. How could one experimentally demonstrate that this is in fact the correct value, and not +191° ?

18. A 0.100 molar aqueous solution of morphine hydrochloride ($C_{17}H_{20}ClNO_3$, MW = 321.81) is put into a 25.0 cm cell at 25°C and found to have an optical rotation of - 9.13°. What is the specific rotation of morphine hydrochloride under these conditions?

19. A 0.300 molar aqueous solution of glyceraldehyde (2,3-dihydroxy-propanal) is put into a 50.0 cm cell at 25°C and found to have an optical rotation of + 13.0°.

    a. What is the specific rotation of glyceraldehyde under these condi tions?
    b. Draw the two possible Fisher projections of glyceraldehyde, and assign the configuration (*R*/*S*) at each chiral center.

20. (*S*) - 3-chloro-1-pentene is optically active. When reacted with hydrogen in the presence of a catalyst, the resulting product is no longer optically active. Write the reaction and explain the loss of optical activity.

21. An unknown compound was analyzed to be 40.00% carbon, 6.71% hydrogen, and 53.29% oxygen. Mass spectral analysis shows the highest m/z peak to be at 90. The IR spectrum shows a strong and wide absorbance between 2500 and 3500 cm^{-1} and a strong absorbance around 1700 cm^{-1}. The carbon NMR spectrum shows three distinct carbon environments. The spin coupled carbon spectrum shows a singlet, a doublet, and a quartet.

    a. What is the empirical formula of the unknown?
    b. What is the molecular formula of the unknown?
    c. What functional group is indicated by the IR spectrum?
    d. Draw the structure of the unknown compound, and explain the NMR spectrum in relation to that structure.
    e. Do you expect the unknown compound to be optically active? Explain.

22. An unknown compound was analyzed to be 40.00% carbon, 6.71% hydrogen, and 53.29% oxygen. Mass spectral analysis shows the highest m/z peak to be at 90. The IR spectrum shows a strong and wide absorbance between 2500 and 3500 cm^{-1} and a strong absorbance around 1700 cm^{-1}. The carbon NMR spectrum shows three distinct carbon environments. The spin coupled carbon spectrum shows a singlet and two triplets.

    a. What is the empirical formula of the unknown?
    b. What is the molecular formula of the unknown?
    c. What functional group is indicated by the IR spectrum?
    d. Draw the structure of the unknown compound, and explain the NMR spectrum in relation to that structure.
    e. Do you expect the unknown compound to be optically active? Explain.

23. Leucine (2-amino-4-methylpentanoic acid) has a specific rotation of - 10.4° in water at 25°C. Isoleucine (2-amino-3-methylpentanoic acid) has a specific rotation of + 11.29°. A sample known to be either leucine or isoleucine has a carbon NMR spectrum indicating 5 unique carbon environments.

a. Draw the structures for leucine and isoleucine.
b. Is the sample leucine or isoleucine? Explain by relating the NMR spectrum to the structures shown in part (a).
c. Is an aqueous solution of the sample expected to rotate light clock wise or counterclockwise?

24. Isoleucine has two chiral carbons. Naturally occuring isoleucine (2*S*,3*S*) - 2-amino-3-methylpentanoic acid) tastes bitter and has a specific rotation of + 64.2° in glacial acetic acid at 20°C. The "allo" form of isoleucine (2*S*,3*R*) - 2-amino-3-methylpentanoic acid) tastes sweet and has a specific rotation of + 55.7° in glacial acetic acid at 20°C. A vial is discovered in the lab, labelled only "purified isoleucine". A portion of the contents of this vial, weighing 2.50 g, is dissolved in glacial acetic acid to form 25.0 ml of solution, which fill a cell with a pathlength of 25.0 cm. The measured optical rotation through this solution at 20°C was + 16.0°.

a. Draw the Fisher projection of the (2*S*,3*S*) and the (2*S*, 3*R*) forms of isoleucine.
b. From the optical rotation data, determine if the material in the vial should taste bitter or sweet.

25. Squalene ($C_{30}H_{50}$, MW = 410.70) is converted into cholesterol ($C_{27}H_{46}O$, MW = 386.64) as pictured in problem #5. One eventually ends with a 50.0 ml ether solution in which the only optically active substance, the cholesterol, is dissolved. This solution has an observed rotation of - 2.5° for a pathlength of 10.0 cm at 20°C.

a. If one started with 10.0 g of squalene, what is the maximum mass of cholesterol that could be formed?
b. The literature value of the specific rotation for a 1.0 g/ml solution of cholesterol in ether solution was - 31.5° for a 10.0 cm pathlength at 20°C. Assuming 100% conversion of the squa lene into the levorotatory enantiomer of cholesterol, what is the anticipated value of optical rotation that would be mea sured for the 50.0 ml ether solution?
c. What estimate of % yield of cholesterol is provided by the mea surement of optical activity?

INSIDE THE Atom

INSIDE THE Atom

In Chapter One, we noted the philosophical origin of the concept of the atom as an indestructible and fundamental unit of matter. Inability to experimentally prove or disprove the concept of atoms kept atomism as a philosophical rather than a scientific hypothesis for over two thousand years. In Chapter Five, we saw how Dalton re-invigorated an old and largely ignored concept with indirect but compelling experimental evidence. In this chapter, we examine the development of the **atomic theory** and examine the structure within the atom.

The story begins with John Dalton, one of six children in a working class family, who grew up near the industrial city of Manchester, England. In those days a person of such humble beginnings, regardless of ability, was not able to attend Oxford or Cambridge. Instead Dalton was largely self-schooled, teaching himself Latin and Greek at night, but concentrating primarily upon math and science. Dalton became a schoolmaster, and began to attend local scientific meetings to listen to the presentation of papers by a wide variety of researchers. At one such meeting in 1800, Dalton reported his research into mixtures of gases. The result of his research is known as Dalton's law of partial pressures (which is addressed in Chapter 15). Dalton's strength, however, was as a theoretician, not as an experimentalist. His key contribution to science was his proposal of the indivisible, eternal atoms. Throughout the nineteenth century, evidence amassed to confirm Dalton's view of atoms as the fundamental unit of matter. By his death in 1844, Dalton had seen most of the scientific community adopt his theory of the atom.

Twelve years after Dalton's death, Joseph John (J.J.) Thomson was born into a middle class family near Manchester. With the early death of his father, J.J. Thomson nearly had to quit school. A scholarship fund set up by the citizens of Manchester in memory of Dalton enabled J.J. Thomson to complete his schooling and then attend Cambridge. After completing his education, J.J. Thomson continued his research at Cambridge and eventually became director of its world-famous Cavendish Laboratories. It was in these laboratories that Dalton's eternal and indivisible atoms began to show some serious cracks.

Thomson's investigations focused primarily upon radiation given off in an electrical discharge tube, known as a **Crookes tube** (after Sir William Crookes). These tubes have two metal pieces with a large voltage difference between them. If a significant amount of gas is present in the tube, a spark will eventually jump from one electrode to another, much like a lightning bolt. As the pressure of gas in the tube is reduced, the gas itself may eventually begin to glow, as in a neon sign. A further reduction in pressure eliminates that glow, and a very faint greenish luminescence appears at one end of the tube. Various experiments by Crookes and others (Figure 9.1) showed conclusively that the radiation originated at the negative electrode, and thus the term **"cathode ray"** was used to describe them. In the late 1800s, the nature of these rays was very much in question, and Thomson devoted himself to resolving the issue.

William Crookes and other English researchers held the position that cathode rays were negatively charged particles. They could be bent in a magnetic field in a way consistent with that hypothesis (Figure 9.2)

But Heinrich Hertz and other German researchers held the position that cathode rays were waves (like radio waves, the frequency of which are now denoted in mega or kilo Hertz.) Hertz showed that the cathode rays could penetrate gold foil, which was thought to be an unlikely attribute of a particle. More significantly, Hertz's attempt to deflect cathode rays by an electric field showed no deflection, leading him to oppose Crookes' idea of charged particles.

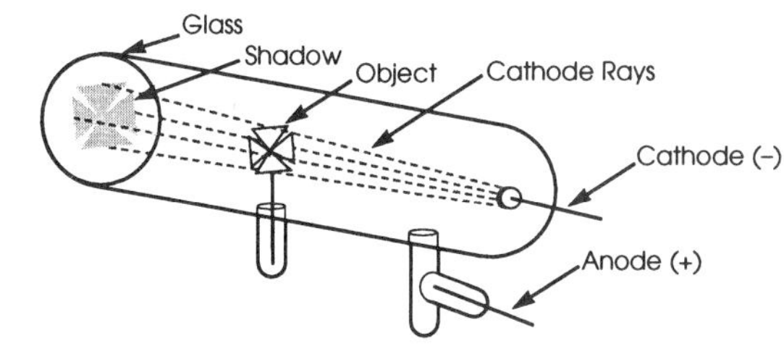

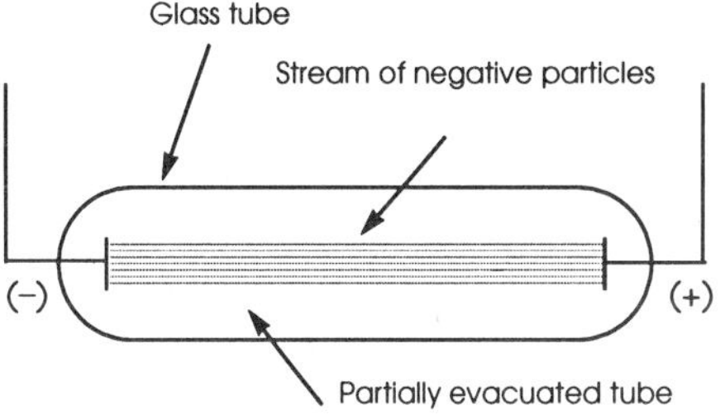

FIGURE 9.1: Various Crookes tubes.

Thomson resolved the issue by repeating Hertz's work. He noted a slight deflection of the cathode ray when his tube was first turned on, but this immediately disappeared and he could not make the cathode rays deflect after that regardless of the voltage he used. Thomson realized that the residue gas in the tube could be the problem. If the cathode rays were charged particles, they might hit the atoms of gas in the tube, which would become ions and migrate to the charged plates, neutralizing the plate's charge. Thomson repeated his measurements under very high vacuum (removing almost all the gas in the tube) and observed the deflection of the cathode ray seen in the figure below. This result was consistent with the cathode ray being a stream of negatively charged particles.

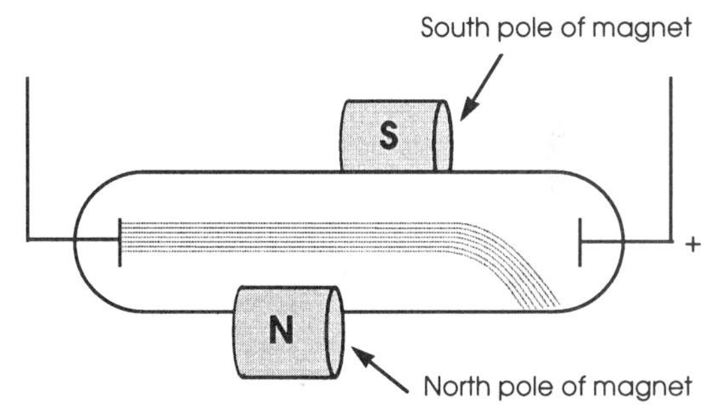

FIGURE 9.2: Cathode ray tube with magnet.

The charged nature of the cathode rays was demonstrated conclusively by Jean Perrin in 1897, when the rays were shown to form a negative charge when they hit a piece of metal.

In solving one puzzle, Thomson had created others. What was the nature and origin of these charged particles? These particles were the same regardless of the metal used in the electrodes or the gas used in the tube. In fact, Thomson later showed that these same rays were formed when ultraviolet light hit the surface of an active metal. As such, the particles would have to be common to all atoms, which would make them yet more fundamental than the atom. Could the cathode rays in fact be sub-atomic particles? To answer these questions definitively, Thomson knew he had to determine the mass of the negative particles.

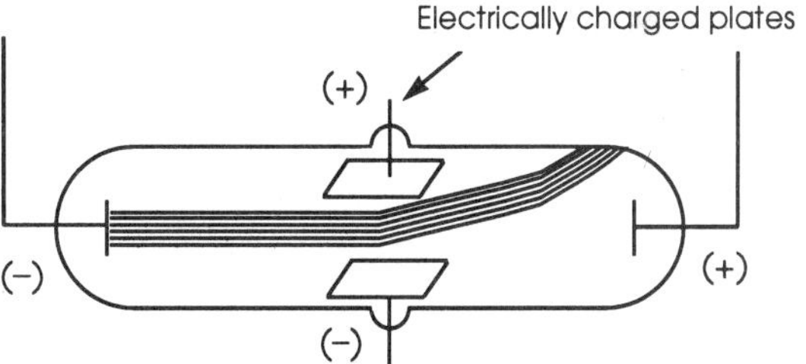

FIGURE 9.3: Cathode ray tube with electric plates.

Thomson designed a Crookes tube that imposed both a magnetic field and an electric field on the cathode rays (Figure 9.4)

By adjusting the strengths of the electric and magnetic fields and measuring the deflection in the path traveled by the cathode rays, Thomson was able to determine the charge (q) to mass (m) ratio of the cathode ray particles. His experimental value of q/m was 1.758×10^8 Coulombs per gram. (See Appendix F.1 for a more complete derivation.) This value was more than a factor of one thousand greater than the next highest measured charge to mass ratio, that of the hydrogen ion. That meant that if the charges were the same in both cases, the cathode rays particles were only one-one thousandth the mass of a hydrogen atom.

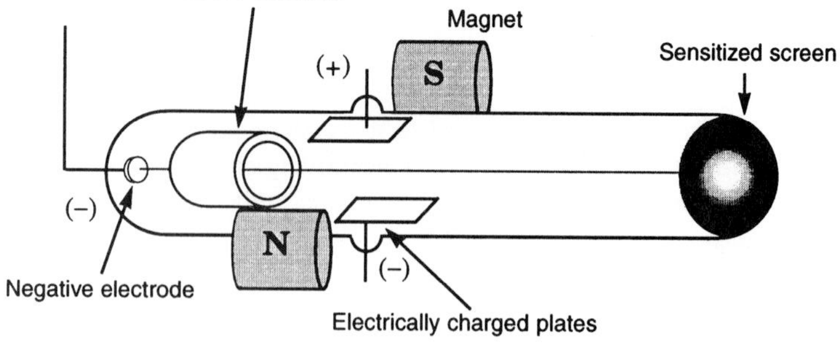

FIGURE 9.4: Thompson's cathode ray tube.

Thomson was now convinced that cathode rays were subatomic particles, and were in fact the electrons hypothesized since the time of Faraday as a fundamental unit of electricity. After 20 years of work, Thomson was prepared to announce his discovery of a sub-atomic particle. On April 30, 1897, before a meeting of the Royal Society, Thomson outlined his work and the underlying theory. The century which had begun with an Englishman fashioning immutable atoms out of a chaos of chemical data, ended with another Englishman shaking the atom apart.

9.2 THE MASS OF THE ELECTRON AND THE RECIPE FOR PLUM PUDDING

J.J. Thomson received the Nobel prize for his work in 1906. Work continued in the laboratories he directed to try to determine the charge of the electron, so that the mass could be calculated from the charge to mass ratio. One such attempt was made by C. T. R. Wilson. Wilson had used a mist of water to try to photograph the path of an electron via its interaction with water droplets. Wilson's work resulted in a device called a cloud chamber. This instrument ultimately resulted in the tracking and identifying of numerous sub-atomic particles. But water droplets were variable in mass due to their volatility, and were inconclusive in determining the charge of the electron.

The charge on the electron was ultimately quantified by the oil drop experiment conducted by Robert Millikan. Born in Morrison, Illinois, Millikan's research at the University of Chicago sought to determine the value of the charge "q" on the electron. His device, shown in the figure below, allowed a small droplet of oil sprayed from an atomizer to fall through a small hole in a brass plate.

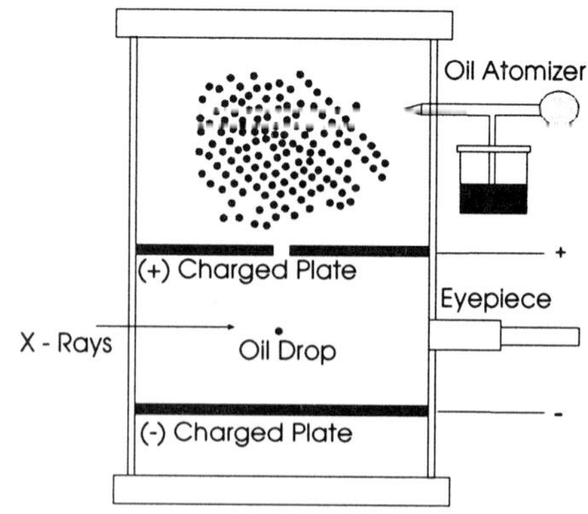

FIGURE 9.5: Millikan oil drop apparatus.

That plate, and a plate below it, were electrically charged. The oil drop falling through the hole was ionized by radiation, picking up one or more electrons. The rate at which the oil drop fell depended on the electrical charge on the oil droplet. By varying the voltage and taking numerous such measurements to account for a variable charge caused by the different numbers of electrons sticking to the droplet, Millikan was able to show the fundamental unit of charge on the electron was 1.602×10^{-19} Coulombs. Combining Thomson's mass to charge ratio with Millikan's charge on the electron, the mass of the electron was then calculated:

$$\left(1.602 \times 10^{-19}\ \text{Coulombs}\right)\left(\frac{1\ \text{gram}}{1.759 \times 10^{8}\ \text{Coulombs}}\right) = 9.109 \times 10^{-28}\ \text{grams}$$ EQ. 9.1

The result showed that the mass of the hydrogen atom, 1.673×10^{-24} g, was in fact 1,837 times heavier than the electron (mass = 9.109×10^{-28} g) in accord with Thomson's declaration.

Thomson proposed a model of the atom based upon the relative mass of the electron and the hydrogen ion. The hydrogen ion had a charge equal in magnitude but opposite in sign to that of the electron. In order to have a neutral atom, it was clear that a negatively charged subatomic particle had to have a positively charged counterpart, which was termed positive electricity. In order to picture the atom, Thomson reasoned that since 99.9% of the mass of the atom was composed of the positive electricity, 99.9% of the volume of the atom was positive electricity. This assumed that the density (mass to volume ratio) of the positive and negative sub-atomic elements were comparable. The result was Thomson's model of the atom, Figure 9.6 . This is sometimes known as the "Plum Pudding" model, since the electrons were scattered here and there like a few raisins in a plum pudding (which oddly enough traditionally contains no plums.)

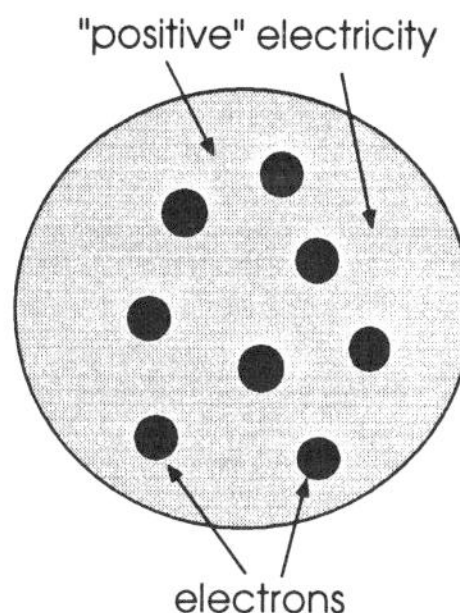

FIGURE 9.6: Plum pudding model of an atom.

Thompson's model of the structure within the atom has since proven to be faulty, but by getting the scientific community's interest in atoms aroused, Thomson initiated much of what was to follow.

9.3 THE TOOLS ARE ASSEMBLED FOR ATOMIC EXPLORATION

Dalton's models of atoms were solid billiard balls. He used a set of wooden balls color-coded for the different elements. (Dalton also had identifying marks on each, since being color-blind he could not otherwise easily distinguish between them. Dalton was the first scientist to investigate color-blindness, which is sometimes referred to as "Daltonism" in his honor.) Thomson's work made it clear that the billiard balls were not the ultimate reduction of matter, and proposed a model of the atom composed of both positive and negative electricity. The positive electricity, comprising the bulk of the atom, was uniformly distributed throughout the atom; the electrons were sparingly scattered here and there within the atom. But just as Dalton's model of the atom had to give way based upon the research of Thomson, Thomson's model of the atom was put to the experimental test by later scientists.

The years of 1895 and 1896 were a turning point in the evolving model of the atom. In this interval Wilhelm Roentgen discovered x-rays, Antoine Henri Becquerel, Marie and Pierre Curie began work on radioactivity, and Ernest Rutherford came to J.J. Thomson's labs.

In Paris, Becquerel was conducting research into phosphorescence, a phenomenon in which a chemical first absorbs light and then later emits light. Becquerel exposed various materials to sunlight, then put the material on top of a fresh photographic plate in a dark room. Any light given off caused a darkening of the photographic plate. Becquerel began to examine uranium salts for phosphorescence. At this point uranium had been known for over one hundred years, and was of no particular interest other than as an additive used to color glass. His experiment with a uranium salt was stalled when clouds covered the sky for several days. Becquerel stored the uranium sample above a pile of photographic plates still in their protective covers, and waited for the weather to change. Not wanting to waste time, he thought the bad weather gave him a chance to do some quality control work. He developed one of the unexposed photographic plates which had been stored under the uranium salt. Instead of the expected blank plate confirming that his materials and techniques were still good, he found the plate had been exposed. Concerned, he repeated the experiment and found an image on the film approximating that of the container of the uranium salt. He began a series of studies, varying the amount of uranium, and determined that the intensity of the film exposure varied directly with the amount of uranium, and did not depend upon the salt being first exposed to light. Clearly, a phenomenon other that phosphorescence was at work. Then, in a study of a sample of a uranium ore called pitchblende, Becquerel found a much greater intensity of exposure than could be attributed to the amount of uranium present. He discussed the problem with Marie Curie, who agreed to undertake the chemical analysis of the ore called pitchblende as part of her doctoral thesis.

Marie Curie, working with electrical measuring equipment invented by her husband Pierre, began her investigations. She confirmed that the reaction Becquerel first observed was an atomic property of uranium, and she gave it the name "radioactivity." She then proved that the pitchblende ore contained something far more radioactive than uranium, or any other known element. For Marie Curie, this meant that she was on the trail of a new and very unique element. But to prove it, she knew she would have to isolate the pure element. She consulted Mendeleev, whom she had met as a young girl in her native Poland. Mendeleev agreed with her, and indicated that there likely was one or more such elements.

With this encouragement, and no money, the Curies began the backbreaking work of isolating the new element from literally tons of pitchblende ore. The work was done in a leaky shed, with borrowed equipment, over the course of several years. Starting with shovels and cauldrons, and progressing to funnels and test tubes, slowly, grudgingly, the components of the ore were isolated. A very small amount of a metal salt was isolated. The metal was unlike any previously known, and was about three hundred times more radioactive than uranium. Marie named the element polonium, after her native country. But Marie Curie continued her research into the other isolated fractions. After two years of exhaustive fractional recrystallizations, a minuscule amount of another new material was being purified. On entering their shed at night, the Curies could see their glassware faintly glowing from the product they were isolating. Finally, a few crystals of a salt of another new metal was isolated, which was over a million times more radioactive than uranium. Marie Curie named it "radium."

The effects of radium turned out to be extremely potent. Its presence near the skin could cause intense burns, but could also be used to successfully treat skin cancer. Instead of capitalizing on the radium they isolated, the Curies donated it to hospitals and research centers. They refused even to patent the isolation process they took years to establish, knowing the scientific community and the world at large needed immediate access. The Curies shared the 1903 Nobel prize in physics with Becquerel for their early work in radioactivity. In 1911, Marie Curie became the first person in history to win a second Nobel prize, this time in chemistry for her work on radium.

9.4 RADIOACTIVITY AND TRANSMUTATION OF THE ELEMENTS

One of the recipients of Curie's generosity was a new research student in Thomson's lab: Ernest Rutherford. Rutherford's experimental ability was a natural compliment to Thomson theoretical mastery. Rutherford's dexterity in fabrication and expertise in the newly discovered X-rays proved crucial to Thomson's discovery of the electron. Rutherford chose radioactivity as an area ripe for further discovery. Together with Frederick Soddy, Ernest Rutherford conclusively demonstrated that the radioactivity of thorium and radium compounds was the result of atoms of one element becoming atoms of another element. Dalton's indivisible, eternal atoms had proven to be neither.

Using naturally radioactive samples as sources, Rutherford demonstrated that their radiation was influenced by the presence of an electric field. He initially isolated two different types of radiation, α and β, and a third, γ, soon followed. See Figure 9.7.

The α particles, attracted to the negative plate, had a positive charge. They could be stopped by a thick piece of paper. **Alpha particles** were later recognized to be helium atoms with both electrons removed: ${}^{4}_{2}He^{2+}$. The β particles, attracted to the positive plate had a negative charge. They could be stopped by a thin piece of aluminum. The **Beta particles** particles were soon shown to be highly energetic (fast moving) electrons. **Gamma radiation** was uncharged, extremely penetrating, and was not initially observed.

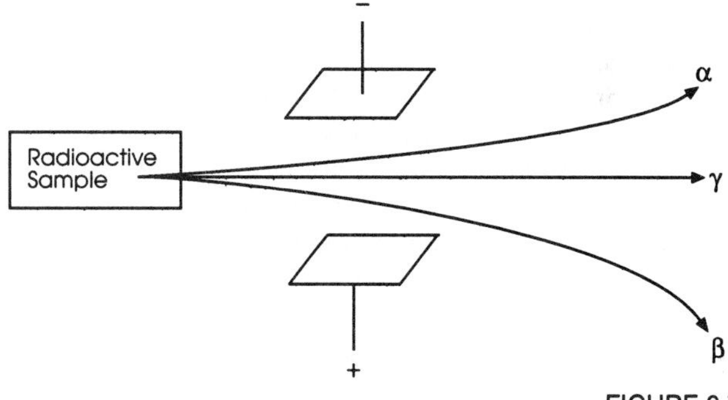

FIGURE 9.7: Alpha, beta, gamma separation.

The α particles were much more massive than electrons, and Rutherford spoke of them as "bullets" and "artillery shells" to help convey to others the amazing energy these particles carried. Rutherford bombarded different kinds of matter with his newly discovered α particles.

When hydrogen gas was bombarded with alpha particles, Rutherford's scintillation detector picked up an anomalous, particularly bright flash. Repeating the experiment with nitrogen gas showed characteristics of the same particle. Rutherford immediately suspected that he had found a new sub-atomic particle. He christened it "proton." Upon careful analysis, the sample of nitrogen bombarded with alpha particles was shown to have formed traces of hydrogen and oxygen.

SIDEBAR: WRITING NUCLEAR BOMBARDMENT REACTIONS

We could write a balanced reaction for this process:

$$^{14}_{7}N + ^{4}_{2}He \rightarrow ^{17}_{8}O + ^{1}_{1}p$$

The proton, symbolized by "p," is knocked out of the nucleus as a result of the collision. Note that the atomic masses (given as superscripts) are conserved, and that the number of protons (given as subscripts) are conserved. Such bombardment reactions in which one atom and a particle collide to give another atom and particle are sometimes written in the following fashion $^{14}_{7}N\,(\alpha,p)\,^{17}_{8}O$. In this convention the mother and daughter isotopes are separated parenthetically by the particles bombarding and being emitted.

CONCEPT CHECK: BOMBARDMENT REACTIONS

Speaking of mothers and daughters, Marie Curie's daughter, Irene Curie-Joliot, along with her husband Frederick, produced the first artificial radioisotope. They bombarded ^{27}Al with alpha particles, producing the isotope along with a neutron. Determine what artificial isotope was produced by completing the following:

$$^{27}_{13}Al\left(\alpha,n\right)\ ?$$

Solution: The key is to conserve the atomic number and mass number, as indicated by the subscripts and superscripts. Thus, it would help to write these out for all the species involved:

$$^{27}_{13}Al + ^{4}_{2}\alpha \longrightarrow ^{1}_{0}n + ?$$

The unknown isotope must have an atomic number of 15, and a mass number of 30. Thus we complete the bombardment equation as:

$$^{27}_{13}Al\left(\alpha,n\right)^{30}_{15}P$$

The spectroscopic fingerprints of hydrogen were found when alpha particles bombarded other elements, including metals and non-metals. The fundamental nature of the new particle was firmly established. The protons he created in this fashion were also shown to behave in electric and magnetic fields as expected: they exhibited a charge equal in magnitude but opposite in sign to the electron. The mass of the proton, however, was nearly two thousand times the mass of the electron. The missing piece in Thomson's atomic puzzle (positive electricity) was isolated and characterized as the proton. But how did those puzzle pieces of the atom, the electron and the proton, fit together to make the atom?

9.5 RUTHERFORD'S NUCLEAR ATOM

Rutherford used C.T.R. Wilson's cloud chamber to track the path of alpha particles as they traveled though a sample of nitrogen gas. The alpha particles typically traveled several inches, past thousands of atoms. On some occasions the path of these alpha particles seemed to be bent, as if slightly deflected. On rare occasions, they appeared to rebound off something in their path.

A more extensive set of experiments was conducted with extremely thin gold foil and a new detection system that could count individual subatomic particles. The previous tests on nitrogen were clearly substantiated: the vast majority of alpha particles went straight through, but several were deflected and a few even rebounded. Clearly, due to its much lighter mass and opposite charge, the electrons could not cause such an effect. As Rutherford himself noted: "An electron would have little more effect on an alpha particle than a fly on a rifle bullet." The deflection of the alpha particle, Rutherford reasoned, had to be due to something extremely dense and, like the alpha particle, positively charged. These crucial tests had, in fact, just begun when Rutherford received news that he had been awarded the 1908 Nobel prize in chemistry for his work in radioactivity. The tests with alpha scattering on gold foil, however, proved to be more important than his Nobel prize winning work.

In 1911, Rutherford announced a new model of the atom, consisting of a small but incredibly dense positive nucleus surrounded by a diffuse volume of electrons. Figure 9.8 is not to scale; the nucleus is shown as a fairly large part of the atom in order to illustrate the scattering phenomenon. Less than one percent of the volume of the atom is actually occupied by the positive nucleus, even though the nucleus contains more than 99.9% of the mass of the atom.

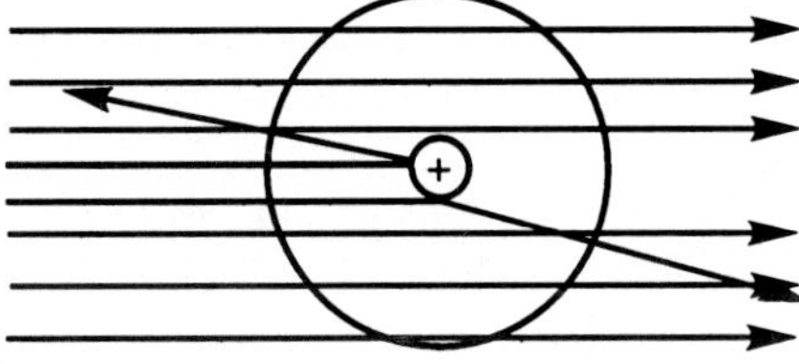

FIGURE 9.8: Alpha particles scattering off the nucleus.

As seen in the figure, the majority of incident alpha radiation passes through the volume occupied by the electrons, with essentially no effect. A close approach to the positive nucleus, however, results in a considerable deflection of the alpha particle's trajectory. The nuclear atomic model replaced the plum pudding model of Rutherford's mentor, J.J. Thomson. But as Thomson's student, Rutherford, had altered his preceptor's model, so too would three students of Rutherford alter his model of the atom: Henry Moseley, James Chadwick, and Niels Bohr.

9.6 THE PROBLEM OF ATOMIC NUMBER AND ATOMIC MASS

Rutherford's model of the atom, composed only of protons in the nucleus and electrons around it, led him to predict that the charge on the nucleus (equal to the number of protons) ought to be proportional to the mass of the nucleus. This assumed that the nucleus was composed of protons only. Rutherford assigned Henry Moseley the task of investigating this hypothesis. Roentgen had earlier discovered x-rays, and Max von Laue had recently learned that crystals were capable of isolating x-rays according to their frequency (symbolized by ν,) much like a prism can separate visible light into its various frequencies (colors.) Moseley set about photographically recording x-rays that were emitted from the anode of a Crookes tube. The wavelength of the x-rays clearly depended upon the metal used as the electrode.

Moseley conducted an exhaustive set of experiments, analyzing all readily available metals from aluminum to gold. The result is shown graphically in Figure 9.9.

The square root of the frequency of the radiation from each metal had a linear relation with a series of integers, symbolized by Z. This Moseley called the **atomic number**: 1 corresponding to hydrogen, 13 to aluminum, 79 for gold, etc. Up to this time, the periodic table had been arranged according to mass, with various inconsistencies. Although the atomic weight of cobalt is greater than nickel, Moseley's work correctly placed cobalt prior to nickel in the ordering of the periodic table. Ordering the elements by Moseley's atomic number, the number of protons in the nucleus, cleared up other problems: argon and potassium; tellurium and iodine. Moseley, at the age of 26, had established a new periodic table, including space for seven yet undiscovered elements (which, given the impetus of Moseley's work, were soon discovered.)

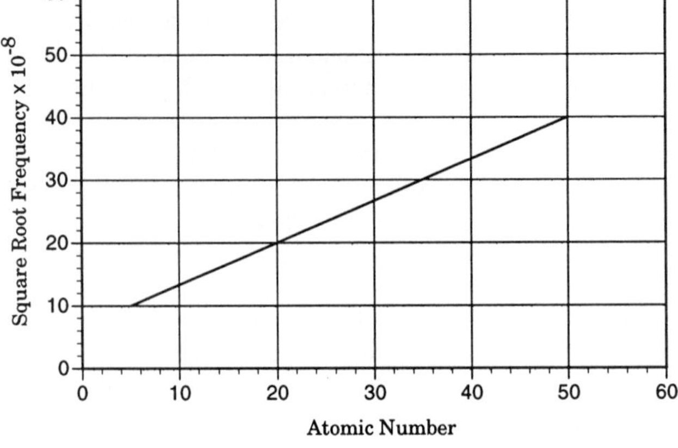

FIGURE 9.9: Plot of square root of frequency vs. Z.

The model of the atom had to be adjusted to account for Moseley's discovery. For example, the mass of a proton is roughly 1 amu (atomic mass unit). The nucleus of a typical carbon atom was shown to have a charge of +6 (corresponding to 6 protons) but a mass of 12 amu (corresponding to about 12 protons.) Rutherford first tried to reconcile these facts with theories incorporating some number of electrons into the nucleus, but abandoned that approach in favor of another, yet undiscovered subatomic particle. This particle would have to have considerable mass but no charge. Postulating such a particle was easy compared to the task of isolating and identifying it.

James Chadwick, another Englishman born in Manchester and schooled at Cambridge, is credited with the discovery of the neutron. After working with Rutherford in the discovery of the proton, Chadwick studied the interaction of alpha particles with lighter elements, including beryllium. As a result of the collision of alpha particles with beryllium atoms, a new and intense radiation was observed. The radiation was uncharged, since it was not deflected by electric fields.

We can write a balanced nuclear reaction as follows:

$$^{9}_{4}\text{Be} + ^{4}_{2}\alpha \rightarrow ^{12}_{6}\text{C} + ^{1}_{0}\text{n} \qquad \text{EQ. 9.3}$$

where n represents the neutron with atomic number 0 (no protons) and atomic mass of 1. Again, note the conservation of mass and nuclear charge.

When this new radiation impacted matter, high energy protons were knocked out of the nucleus. Chadwick was able to demonstrate that phenomenon was due to neutrons. Having no charge, the neutrons were capable of penetrating the nucleus of an atom. These high energy particles, with mass nearly that of a proton, were able to easily penetrate to the nucleus of an atom. As World War

II research into the atomic bomb showed, nuclear fission chain reactions were made possible by neutrons hitting large nuclei, like those of uranium. When fission occurred, the large uranium atom would split into two fairly large isotopes, plus some extra neutrons. These neutrons could hit more uranium atoms, causing more neutrons to be formed, which continued the chain reaction.

SIDEBAR: NUCLEAR FISSION AND THE ATOM BOMB

On December 2, 1942, beneath the University of Chicago's football stadium, the first controlled fission reaction occurred. The fission brought about by a neutron hitting uranium is a somewhat random process. In addition to producing more neutrons, the uranium atom typically splits into two daughter nuclides. Over 300 such daughter nuclides have been determined for this reaction. Yet each reaction balances in terms of mass and nuclear charge. Some possible fission products forming from a neutron hitting uranium 235 are shown:

$$^{235}_{92}U + ^{1}_{0}n \rightarrow ^{140}_{56}Ba + ^{93}_{36}Kr + 3\ ^{1}_{0}n$$ EQ. 9.4

$$^{235}_{92}U + ^{1}_{0}n \rightarrow ^{144}_{54}Xe + ^{90}_{38}Sr + 2\ ^{1}_{0}n$$ EQ. 9.5

Such nuclear reactions release tremendous amounts of energy (8 x 10^7 kJ) per gram of fuel undergoing fission. For comparison, burning a gram of traditional fuel, like gasoline, releases only about 50 kJ per gram. This led to the ^{235}U atom bomb dropped onto Hiroshima on August 6, 1945, and the ^{239}Pu atom bomb dropped on Nagasaki a few days later.

Chadwick received the 1935 Nobel prize in physics for his discovery of the neutron, which completed the list of major subatomic particles of interest in non-nuclear chemistry:

CONCEPT CHECK: BALANCING NUCLEAR REACTIONS

A radioactive isotope of iodine, ^{131}I, is used as a medical tracer in radiology. As ^{131}I decomposes, one of the products is a beta particle ($^{0}_{-1}e$). Write the balanced radioactive decay process, indicating the other product that forms. Is this other product a likely health concern?

SOLUTION: Remember that both the mass number (superscripts) and atomic number (subscripts) must be in balance.

$$^{121}_{53}I \longrightarrow ^{0}_{-1}e + ?$$

The unknown product must have an atomic number of 54 (i.e., xenon) and a mass number of 131. The balanced reaction is:

$$^{121}_{53}I \longrightarrow ^{0}_{-1}e + ^{121}_{54}Xe$$

Xenon, as a noble gas, is expected to be chemically unreactive.

TABLE 9.1: PROPERTIES OF SUBATOMIC PARTICLES

TABLE 9.1

PARTICLE	GRAMS	AMU	CHARGE	SYMBOL
electron	9.109389×10^{-28}	0.0005485799	-1	${}^{0}_{-1}e$
proton	1.672623×10^{-24}	1.007276	+1	${}^{1}_{1}p$
neutron	1.674929×10^{-24}	1.008665	0	${}^{1}_{0}n$

9.7 SIDELIGHT ON LIGHT

At the end of the nineteenth century, the nature of light was largely understood to be wave like. A wave has distinct characteristics: amplitude (the height of the wave); wavelength λ (the distance between identical points on the wave; and frequency ν (the number of wave cycles passing a point per unit time). The speed at which a wave travels equals the product of the wavelength and the frequency. This is shown in the following equation:

$$(c, \text{speed}) = (\lambda, \text{wavelength})(\nu, \text{frequency})$$

EQ 9.6

$$\left(\frac{\text{distance}}{\text{time}}\right) = \left(\frac{\text{distance}}{\text{wave}}\right)\left(\frac{\text{waves}}{\text{time}}\right)$$

CONCEPT CHECK: WORKING WITH WAVES

Sound waves travel in air at a speed of about 340 meters sec^{-1}. The lowest frequency the average human ear can hear is about 20 Hz. What is the wavelength of this sound wave?

The equation needed to relate speed, wavelength, and frequency is:

$$c = \lambda\nu$$

Substituting the known values and solving for wavelength yields:

$$\lambda = \frac{c}{\nu} = \frac{340\ \text{m/s}}{20\ \text{s}^{-1}} = 17\ \text{m}$$

Another characteristic of waves is that waves interact to form an interference pattern. Two waves of the same frequency that meet at a point where the amplitudes of both waves match (said to be in phase) have a constructive interference, seen in the figure below. Waves of the same frequency that meet where the amplitudes are opposite to one another (said to be out of phase) have destructive interference.

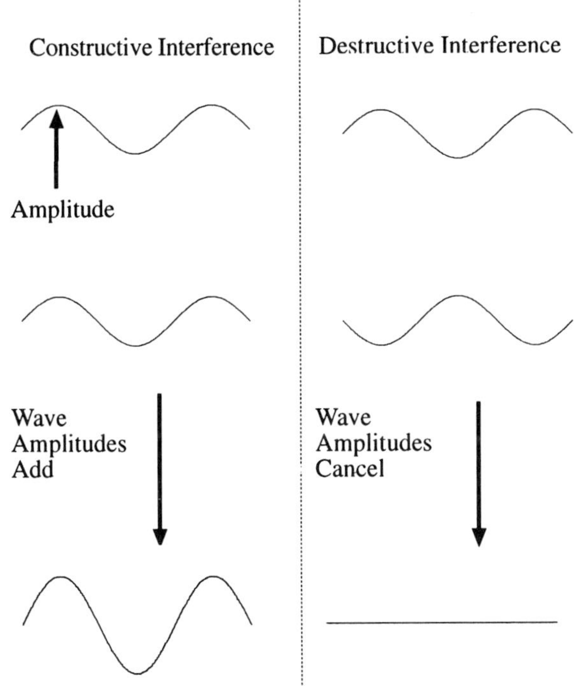

FIGURE 9.10: Interference patterns of waves.

Simple experiments with two wave sources in a pool of water readily illustrates the point. In 1803, Thomas Young conducted the classic double-slit experiment, in which light, striking two parallel narrow openings, showed an interference pattern. The case for light being a wave seemed to be conclusive.

But Max Planck, in solving one of physics long-standing problems of black-body radiation, showed that energy was quantized. The energy was proportional to a frequency ν. The equation is written as

$$E = h\nu \qquad \text{EQ. 9.7}$$

with h being Planck's constant (6.626×10^{-34} J sec) and ν being the frequency. Albert Einstein followed up Planck's quantum theory by applying it to light. Einstein showed that interpreting light as quantized bundles of energy, called photons, allowed a solution of another problem baffling physics at the time: the photoelectric effect. Einstein showed that light can be simultaneously thought of as a particle (with quantized energy) and as a wave (with a frequency ν.) The energy of the particle is related to the frequency of the wave according to Planck's equation. Light's schizophrenic manifestations are known as **wave-particle duality.**

Concept Check: Particle Nature of Light and the Photoelectric Effect

The particle nature of light was supported by Einstein's explanation of the photoelectric effect. Photons (bundles of quantized light energy) hit a metal surface and either do or do not dislodge an electron. The amount of energy a photon brings in is related to its frequency via

Planck's equation. The amount of work required to remove an electron from a metal is called the work function. The work function of magnesium is 352 kJ/mole. What is the longest wavelength of light that can free an electron from the surface of magnesium? What frequency of light is this?

SOLUTION: The work function of 352 kJ/mole means 352 kJ of energy is needed for a mole of photons to free a mole of electrons. In order for one photon to free one electron, the required energy is

$$5.84 \times 10^{-19}\ \text{J}$$

The wavelength and frequency of light associated with this energy can be calculated using Planck's constant and the speed of light:

$$E = h\nu\ ;\ \nu = \frac{E}{h} = \frac{5.84 \times 10^{-19}\ \text{J}}{6.626 \times 10^{-34}\ \text{J·s}} = 8.82 \times 10^{14}\ \text{s}^{-1}$$

$$c = \lambda\nu\ ;\ \lambda = \frac{c}{\nu} = \frac{2.9979 \times 10^{8}\ \text{m/s}}{8.82 \times 10^{14}\ \text{s}^{-1}} = 3.40 \times 10^{-7}\ \text{m}$$

Thus, only light with this wavelength (or shorter) have photons with sufficient energy to cause an electron to be emitted. Brighter , more intense light with a longer wavelength may have more total energy, but the energy per photon is too small to cause the electron to be emitted.

9.8 THE DEVELOPMENT OF THE BOHR ATOM

Rutherford's nuclear atom, with a central positively charged nucleus containing protons and neutrons surrounded by negatively charged electrons had theoreticians troubled. According to classical physics, if the electrons were stationary within the atom, the Coulombic force of attraction between opposite charges should pull the electrons into the nucleus, and the atom should essentially implode. But if the electrons were in motion around the nucleus, classical physics predicted that energy should be radiated out of the atom, with the electron spiraling down into the nucleus as well. Heads or tails: either result was untenable!

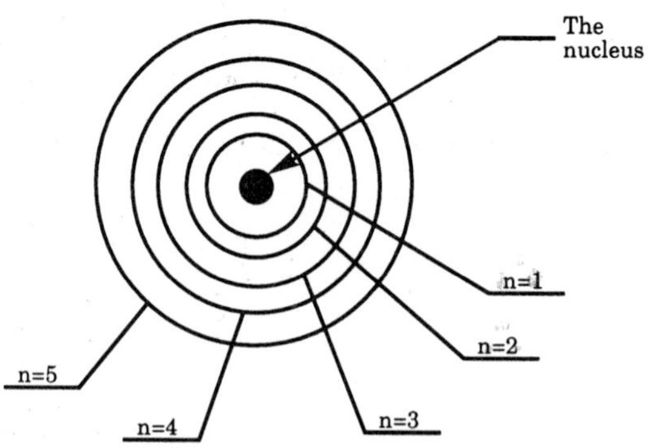

Neils Bohr, a student of Rutherford from Denmark, found an ingenious solution to the problem. He approached the problem using a planetary model. The electrons, he hypothesized, were in planetary-like orbits around the nucleus of the atom, and while in such orbits they neither gained nor lost energy. When an electron changed orbits, he proposed that an amount of energy exactly equal to the difference in the energy between the orbits was given off or absorbed (depending on whether the electron was going to a higher or lower energy orbit.)

FIGURE 9.11: Simple Bohr model.

Starting with hydrogen as his model atom, Bohr had a proton for a nucleus and a single electron in a circular orbit around the nucleus. Electrons could exist in a stable orbit, he reasoned, by moving in a way to create a centripetal force away from the nucleus which exactly offset the force of Coulombic attraction towards the nucleus. But not any old orbit was allowed: only those which met Bohr's criterion of quantized angular momentum were permitted by his theory. The electron thus could be in orbit 1, orbit 2, orbit 3, but never between those states. (A ladder analogy may help: you can be stable on the first rung, or the second, but not at the 1.4 rung.) The rationale and consequences of Bohr's approach are developed more fully in Appendix F.2 and Appendix F.3.

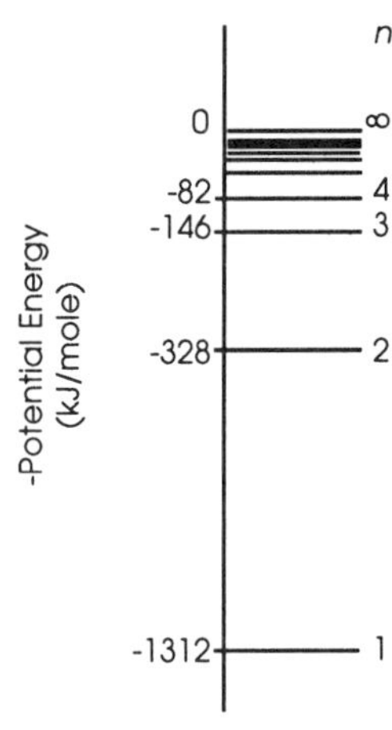

FIGURE 9.12:
n vs. energy level.

Bohr described the orbits allowed using the symbol n. This n can be any positive integer: 1, 2, 3, etc. Bohr was then able to show that as a consequence of his theory, predictions could be made that correlated with experimental results. Bohr could calculate the size of the orbit, and thus the size of the atom. For the ground state hydrogen atom, with n=1, Bohr calculated the radius of the orbit to be 0.05299 nm. The experimentally determined value was 0.053 nm, in excellent agreement.

Bohr showed that the energy of an electron in a particular orbit was related to the value n^2. The greater the value of n, the higher the energy. The relative energy spacings of the various n levels is shown next in Figure 9.12.

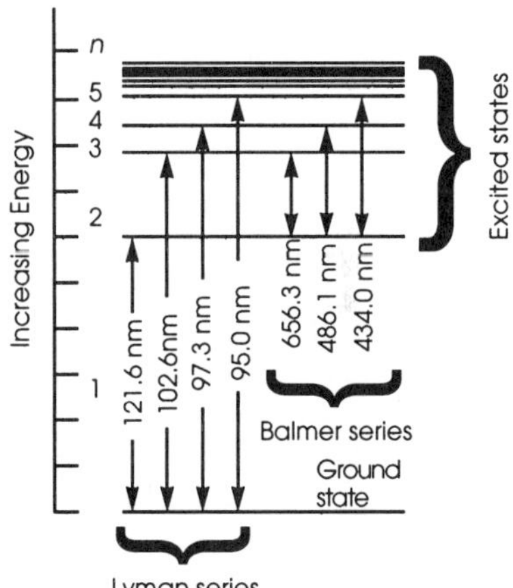

FIGURE 9.13:
Lyman, Balmer,
Paschen, etc. series.

Bohr then calculated the energy associated with all possible transitions between an initial n_i orbit and a final n_f orbit. Examination of hydrogen's spectrum showed infrared, visible and ultraviolet light of the exact frequencies which correlated (through Planck's equation) to the energies predicted by Bohr (E_f - E_i.) Such transitions are shown schematically in Figure 9.13.

Thus Bohr's planetary model for the hydrogen atom had exceptional experimental confirmation. One of Bohr's initial postulates, allowing only those orbits with specific values of angular momentum, was, however, an act of faith on his part. It was simply necessary in order to get the right answers. But Louis de Broglie showed that electrons could be treated as waves as well as particles. De Broglie's wave interpretation of the electron justified Bohr's restriction on orbital sizes. (See Appendix F.4.) Figure 9.14 shows a schematic representation of how some orbits are appropriate for a wave, while other size orbits are not. Orbits that result in constructive interference of waves correspond to orbits Bohr allowed. Orbits that resulted in destructive interference resulted in orbits that are not allowed.

De Broglie extended the wave-particle duality formerly associated only with light to any particle provided that the mass was very small. Experimental verification followed when Davisson and Germer in the US and George Thomson in the UK showed that electron beams in fact demonstrated wave like properties of interference. George Thomson received the Nobel prize in 1937 for proving that the electron was a wave. In the audience was his father, J.J. Thomson, who had won the Nobel prize in 1906 for proving that the electron was a particle. Bohr's model for the hydrogen atom was extremely successful: a variety of

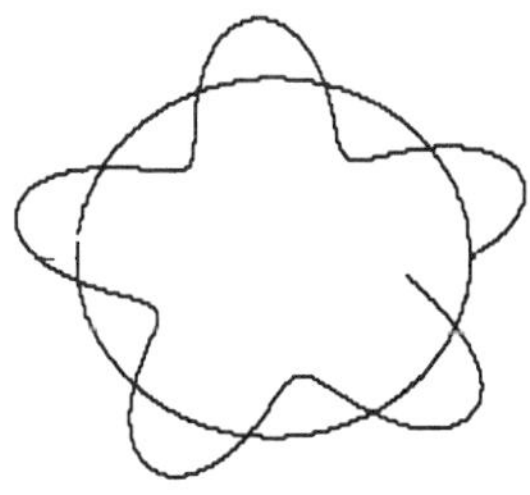
FIGURE 9.14:
de Broglie waves
around Bohr orbit.

stunningly accurate predictions were experimentally confirmed, and the postulates upon which he based the theory found additional corroboration. However, all attempts to extend the model beyond hydrogen or hydrogen like ions (e.g., He^{+1}, Li^{2+}, and other one electron systems) were doomed to failure.

9.9 ELECTRON WAVES AND PROBABILITY PICTURES

Bohr was unable to extend his model of the atom to include larger atoms with more than one electron. In his approach to calculating the Coulombic forces within the hydrogen atom, he had to be concerned only with r, the average separation between the proton and the electron. But in a two electron system, for example, there are electrostatic forces not only between the proton and electrons, but also between the electrons themselves (as seen in Figure 9.15 .)

In order to replicate Bohr's calculations for a multi-electron system, one would have to know the exact position of each electron relative to all the other electrons. Even for a one electron system, exact knowledge of position assumes the electron has only particle features. Stipulating a unique position (i.e., x, y and z coordinates) does not make sense for a wave phenomenon.

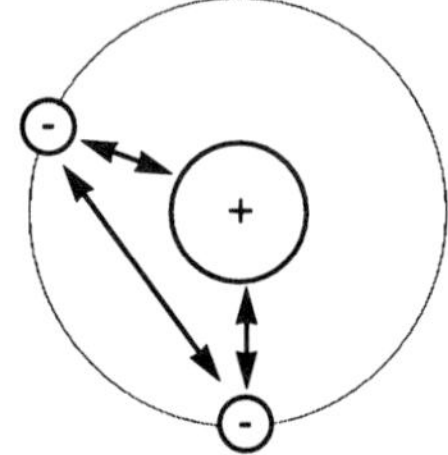

FIGURE 9.15: Electrons around a nucleus.

This difficulty is not unique to the electron. The theoretical physicist, Werner Heisenberg, showed that certain variables were always coupled in such a way that their combined uncertainties always exceeded a finite limit. Position and energy (in the form of momentum) form such a couple. According to his theory, position and momentum could not simultaneously both be determined beyond a specified limit of uncertainty. Heisenberg's **uncertainty principle**, as this was called, was the result of theoretical study, not a limit on experimental ability. For example, if more accurate measuring devices were to be invented tomorrow, Heisenberg's uncertainty principle would still remain.

Many people have tried to conduct "thought experiments" (in which the mechanics of doing the experiment are ignored) in order to imagine how the position and momentum of an electron could be simultaneously determined. The results always end up increasing the precision of the knowledge of one parameter at the expense of the accuracy of the other parameter. One detects the position of the electron, for example, by scattering some particle or radiation off of it. But that scattering event itself changes the momentum of the electron. This is like locating a small animal on a dark country road by running into it with a car: the animal's position at a point in time is determined, but the animal's momentum is altered.

In 1925, Erwin Schrödinger proposed a new approach to describing the position of the electron in an atom. It avoided Bohr's difficulty with multiple electrons by a very different approach to the problem. Schrödinger showed that he could describe electrons in terms of three dimensional **wave functions** (symbolized as Ψ.) The square of the wave function, Ψ^2, yielded a probability density. That the electron could be described only in terms of probability was in accord with the Heisenberg uncertainty principle. The solution of Schrödinger's equation for the hydrogen atom produced the same numerical results as Bohr's theory and thus were also in agreement with experiment. In solving his model for the atom, Bohr needed only one variable called n. The value of n yielded the size and energy of the Bohr orbit. In solving the mathematics of his wave equations, Schrödinger needed three variables, now symbolized as n (the principal quantum number,) l (the angular momentum quantum number,) and m (the magnetic quantum number.) Later experimental results showed that a fourth quantum number, s (the spin quantum number) was also needed.

Although the exact solution of Schrödinger's wave equations are beyond the scope of this book, the probability pictures derived from those results are quite useful. Thus, while a plot of Ψ^2 versus distance from the nucleus might be seem obtuse, a pictorial representation of the same relationship provides meaningful images of the electron distribution in an atom (figures 9.16, 9.17).

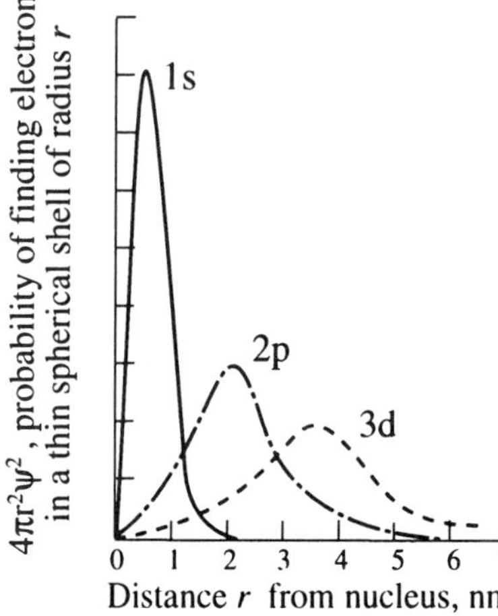

FIGURE 9.16: Plot of Ψ^2 vs. r for various energy levels.

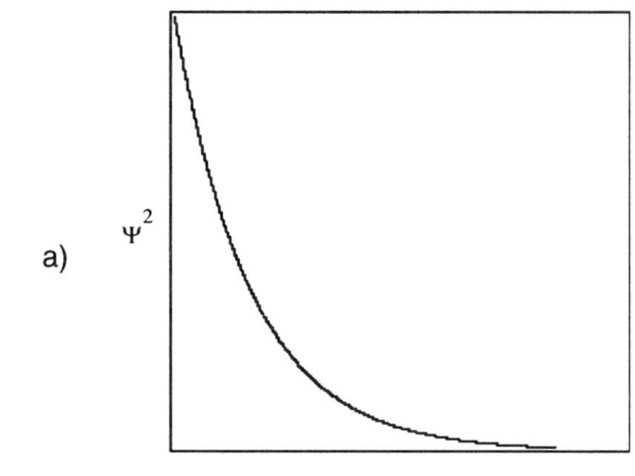

b)

c)

FIGURE 9.16: Electron in the 1s orbital as pictured by a) graph, b) probability shell, and c) probability density.

The models in Figure 9.17 can be interpreted as a snapshot picture of how the electron wave is spread out in a single moment in time, or as a multiple exposure photograph of a single particle's motion over time. Note that Ψ^2 versus r approaches the x axis asymptotically, meaning that there is some small chance of finding the electron very far away from the nucleus. Thus we select a 90% or a 99% probability shell (i.e., one can not show a finite volume which contains all the electron density.) Depending upon the actual values of the

quantum numbers, a wide variety of electron distributions in space results, as seen in Figure 9.18.

FIGURE 9.18: s, p, d orbitals.

9.10 QUANTUM NUMBERS AND ELECTRON CONFIGURATION

The specific solution of the Schrödinger equation for the hydrogen atom can be extended to other atoms. The key to understanding the energy and distribution of electrons within these atoms are Schrödinger's quantum numbers. Far from being obtuse mathematical imaginings, these quantum numbers represent very important properties of electron distribution within the atom. Since the material world is made of atoms, and 99% of the volume of atoms is electron distribution, quantum numbers in a sense represent most of reality.

The principal quantum number n: n represents the size of the electron orbital. The larger the value of n, the greater the extent of the electron distribution. For example, an electron in an n = 2 orbital is, on average, farther away from the nucleus than an electron in an n = 1 orbital. Allowed values for n are positive integers: 1, 2, 3, 4, etc. The value of n also determines the energy of the electron in a hydrogen atom.

The angular momentum quantum number l: l represents the shape of the electron distribution. The larger the value of l, the more fragmented the electron distribution becomes. Values of l range from 0 up to and including n-1. That is, for n=1, the only allowed l value is 0; for n=2, l can equal either 0 or 1; for n=3, l can equal 0, 1 or 2, and so on. When l = 0, the electron distribution is spherically symmetric. This is also denoted by an "s" designation after the principle quantum number. That is, n=1, l=0 is called a 1s orbital; n=2, l=0 is called a 2s orbital. Both are spherically symmetric, but the 2s is larger than the 1s. A 3s orbital is again bigger than the 2s orbital, and so on (Figure 9.19).

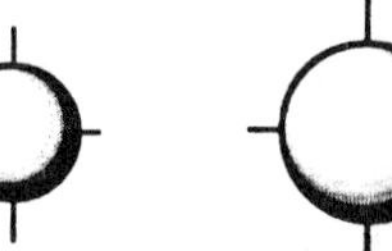

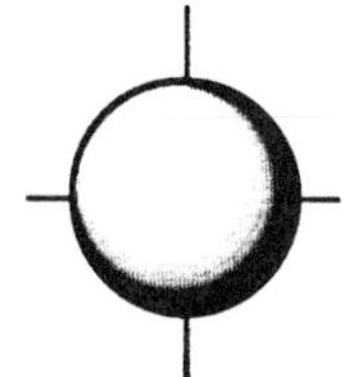

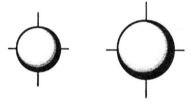

FIGURE 9.19: 1s, 2s, 3s orbitals.

When l=1, the electron distribution is dumbbell shaped, as shown in Figure 9.20. Electron distributions with l=1 have a nodal plane bisecting two equal parts. A node represents a zero probability of finding an electron in that place. Such orbitals are described as being cylindrically symmetric about one axis, and are designated by "p". There are no 1p orbitals, since the only allowed value for l is 0 when n=1. The orbital with n=2 and l=1 is described as 2p. If n=3 and l=1, the orbital is designated 3p, and so on.

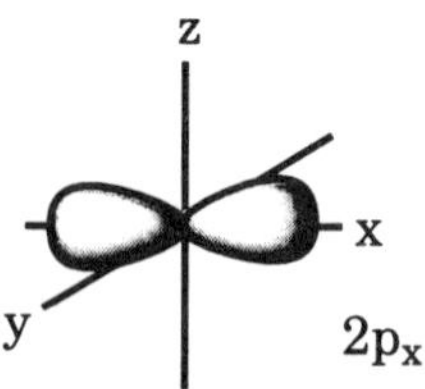

FIGURE 9.20: 2p orbital.

When l=2, the distribution becomes increasingly fragmented, with more nodal surfaces. Orbitals with l=2 are termed "d" orbitals. Since l=2 is possible only if n=3 or higher, the first such orbital is the 3d, followed by the 4d, 5d, etc.

When l=3, the orbitals become even more fragmented. The exact shapes of these orbitals will not concern us in this book. Such orbitals are termed "f". These orbitals occur for n=4 and higher: 4f, 5f, etc.

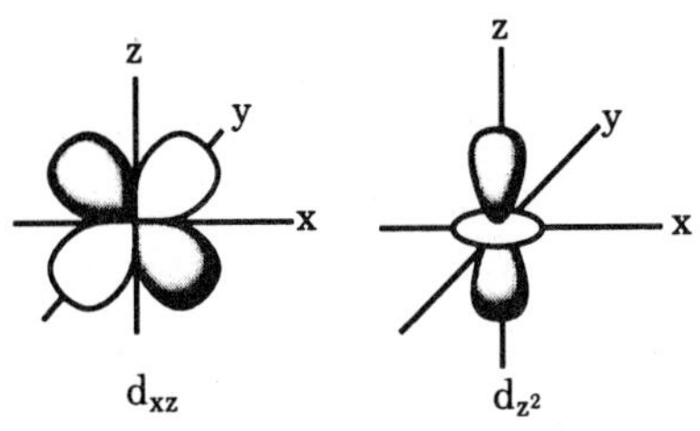

FIGURE 9.21: $3d_{xy}$ and $3d_{z^2}$.

The combination of n and l determines the overall energy of the electron orbital. The higher the sum of n + l, the higher the energy: 2s is higher than 1s; 2p is higher than 2s. When ties occur in the sum of n+l (e.g., 5s, 4p, 3d all have n + l =5,) the orbital with the lower n value is lower in energy. For the same example, the energy of the 3d orbital is lower than that of the 4p, which is lower than 5s.

The magnetic quantum number m: m (or sometimes symbolized m_l) represents the spatial orientation of the orbital (i.e., in terms of an x,y,z coordinate system.) Values of m range in integral steps from - l to + l. Thus if l=0 (an s orbital) the only m value allowed is 0. There is only one way to orient a sphere in space. When l=1 (a p orbital) the allowed values for m are -1, 0 and +1. These correspond to three orientations of the dumbbell in space: p_x, p_y and p_z. Note that the assignment of a specific m value to orientation along a specific axis is arbitrary.

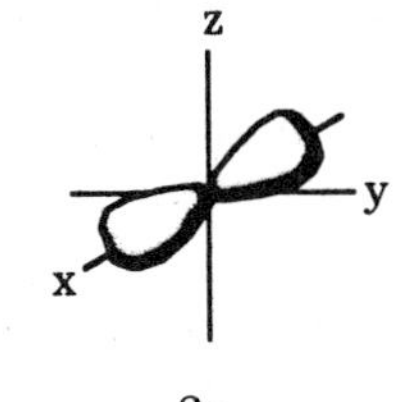

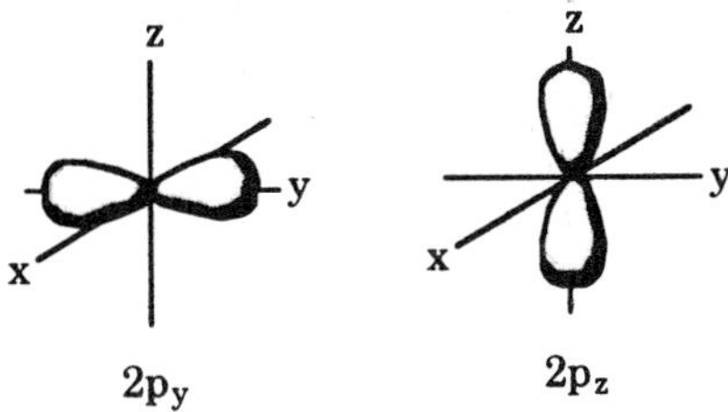

FIGURE 9.22: Px, Py, Pz.

When l=2 (d orbitals) there are 5 different m values: -2, -1, 0, +1, and +2. These correspond to five different orientations of d orbitals in space: d_{xy}, d_{yz}, d_{xz}, $d_{x^2-y^2}$, and d_{z^2}. For l=3 (f orbitals) there are seven possible m values. Again, we are not concerned with f orbital geometry.

Orbitals with the same n and l values are degenerate, meaning they have the same energy. Their energies differ only in the presence of an external magnetic field which favors one orientation over the others; hence the name "magnetic" quantum number.

The spin quantum number s: symbolized as s (or sometimes as m_s to avoid confusion with the symbol used for an l=0 orbital) the spin quantum number can be thought of as either a clockwise or counterclockwise spin. The only two allowed values for s are +1/2 or -1/2. While not a direct outcome of the solution of Schrödinger's wave equation, Wolfgang Pauli showed that fine details in atomic emission spectra could be interpreted this way.

Schrödinger's wave approach for the hydrogen atom produces a variety of electron distributions. The electron in a hydrogen atom occupies one of the possible energy levels, or orbitals, around the nucleus. In the ground state, the electron occupies the lowest energy level. As we extend the theory to bigger atoms, we follow the "Aufbau", or "building up" principle. In this case the electrons in the helium atom should occupy orbitals like those for a hydrogen atom, except room must be made for two electrons instead of one. As bigger atoms are considered, more electrons are added, and the lowest energy of all the electrons is achieved in the ground state.

The rules associated with the quantum numbers thus dictate not only the shape of electron distribution within the hydrogen atom, but electronic configurations for all the elements and the shape of the periodic table as well. As shown in Table 9.2, for n=1, there are only two allowed combinations of quantum numbers. (As we noted earlier, Pauli's exclusion principle forbids two electrons in the same atom from having all four quantum numbers identical.) Thus the first period has only two elements: hydrogen and helium. For n=2, Table 9.2 shows 8 possible combinations; two from the 2s orbitals and 6 from the triply degenerate 2p orbitals. Thus the second period has 8 elements from lithium through neon.

TABLE 9.2: n, l, m, s VALUES FOR LAST ELECTRON IN VARIOUS ELEMENTS

TABLE 9.2

n	*l*	*m*	*s*	**Max Number of Electrons In Shell**
1	0	0	+1/2	
	0	0	−1/2	2
2	0	0	+1/2	
	0	0	−1/2	
	1	-1	+1/2	
	1	-1	−1/2	
	1	0	+1/2	
	1	0	−1/2	
	1	+1	+1/2	
	1	+1	−1/2	8
3	0	0	+1/2	
	0	0	−1/2	
	1	-1	+1/2	
	1	-1	−1/2	
	1	0	+1/2	
	1	0	−1/2	
	1	+1	+1/2	
	1	+1	−1/2	
	2	-2	+1/2	
	2	-2	−1/2	
	2	-1	+1/2	
	2	-1	−1/2	
	2	0	+1/2	
	2	0	−1/2	
	2	1	+1/2	
	2	1	−1/2	
	2	2	+1/2	
	2	2	−1/2	18

For n=3, we find 18 possible combinations of quantum numbers. Yet the third row on the periodic table is only 8 elements wide: sodium through argon. This comes about as a result of the n + l rule, which puts the 3d orbital at a higher energy than the 4s orbital. In the next chapter, we examine the periodic table in greater detail.

9.11 ELECTRON ENERGY LEVELS AND HUND'S RULE

Scientists often find the use of energy level diagrams to be helpful in understanding how electrons are arranged in atoms. The calculations of quantum mechanics show that the energies of the first several energy levels in an atom are arranged as shown in Figure 9.23. Energy is plotted in the vertical direction on this diagram. Higher energy levels are plotted upward. The horizontal axis has no meaning except that, by convention, similar orbital types are arranged in columns along the horizontal axis.

4p
3d
4s
3p
3s
2p
2s
1s
Energy

FIGURE 9.23: Energy level diagram.

The lowest possible energy state (that in which it is in the most stable state) is termed the ground state. All other energy states (those in which one or more electrons are in energy levels above the lowest ones they could occupy) are termed excited states. If we place one electron in an atom, that electron will go into the lowest energy orbital, the 1s orbital. Of course, the neutral atom with one electron is hydrogen. The energy level diagram for hydrogen is shown in Figure 9.24.

Adding another proton and electron to the atom produces the element helium. The second electron in helium also goes in the 1s level, but it must go in with opposite spin for otherwise it would violate the Pauli Exclusion Principle. This energy level diagram is shown in Figure 9.25.

The third electron must enter the 2s level. The fourth electron enters the 2s level with the opposite spin, and fills that orbital. The fifth electron then enters the 2p level to produce an energy level diagram for boron shown in Figure 9.26.

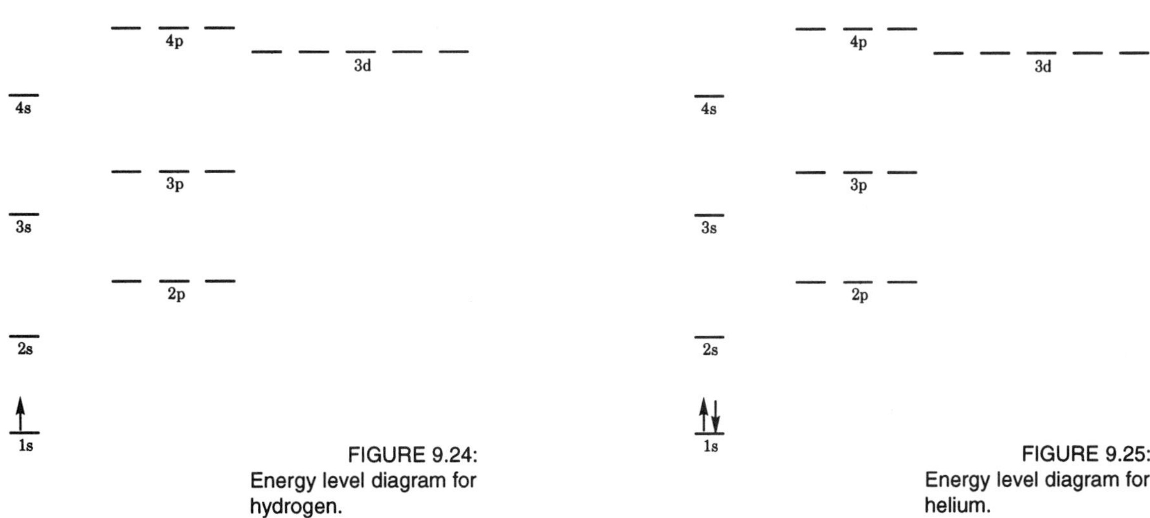

FIGURE 9.24: Energy level diagram for hydrogen.

FIGURE 9.25: Energy level diagram for helium.

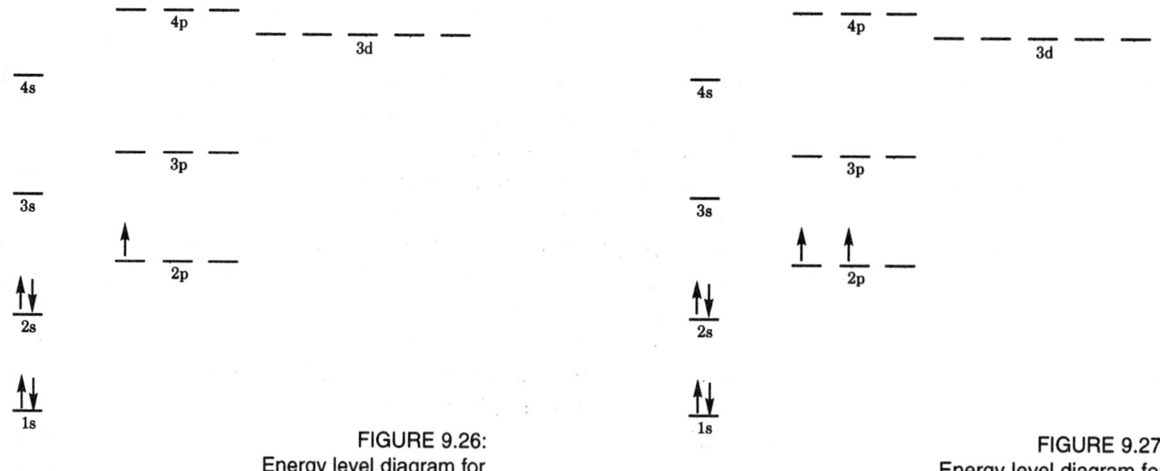

FIGURE 9.26: Energy level diagram for boron.

FIGURE 9.27: Energy level diagram for carbon.

When the element carbon is reached and the sixth electron is added, the diagram appears to indicate several different possibilities for placement of the additional electron. It could be placed in a different p orbital with the same spin (also called parallel spin), a different p orbital with opposite spin (also called spin–paired), or in the same p orbital with its spin paired. While this simple diagram suggests that all of those electron placements are equal in energy, more sophisticated calculations and experimental evidence tell us that the most energetically favorable condition is the first–named condition: the electron goes in one of the other p orbitals with its spin parallel to the first electron in that orbital. This is stated in **Hund's Rule** that, in essence, states that when electrons occupy a set of degenerate (equal energy) orbitals, they occupy those levels singly with parallel spins until there are enough electrons in the orbitals to require spin pairing. Thus the energy level diagram of carbon is pictured in Figure 9.27.

In nitrogen, the 2p levels are totally occupied, each with one electron. At oxygen, it is no longer possible to add an electron to the atom without putting it into an orbital that already contains one electron. Thus pairing of the 2p electrons begins with oxygen and is completed with neon.

You might wonder how it is that scientists know whether the electrons in an atom or molecule have their spins parallel or are all paired. This information can be determined by measuring the magnetic properties of a collection of the atoms or molecules. Atoms or molecules in which all the electrons are paired are weakly repelled by a magnetic field. Such species are termed **diamagnetic.** If an atom or molecule contains at least one unpaired electron, it is weakly attracted to a magnetic field and exhibits a property known as paramagnetism. Careful measurement of the strength of the **paramagnetic** effect can be used to determine the number of unpaired electrons in an atom, molecule, or ion. Measurements of this type were used to confirm Hund's Rule.

9.12 ELECTRON CONFIGURATION NOTATION

Energy level diagrams are good for understanding how the electrons are distributed within an atom, but drawing out an energy level diagram would be an unwieldy method of conveying this information. Scientists quickly developed a convention for expressing electron configurations that took up much less space. Soon another, even more compact system was developed to write electron configurations. The electron configurations for the ground state of the first 25 elements are presented in Table 9.3. Both the expanded and the shortened form of the notation are given for each species.

TABLE 9.3

TABLE 9.3: ELECTRON CONFIGURATIONS OF THE ATOMS

ATOM	EXPANDED NOTATION	SHORTENED NOTATION
H	$1s^1$	$1s^1$
He	$1s^2$	$1s^2$
Li	$1s^22s^1$	$[He]2s^1$
Be	$1s^22s^2$	$[He]2s^2$
B	$1s^22s^22p^1$	$[He]2s^22p^1$
C	$1s^22s^22p^2$	$[He]2s^22p^2$
N	$1s^22s^22p^3$	$[He]2s^22p^3$
O	$1s^22s^22p^4$	$[He]2s^22p^4$
F	$1s^22s^22p^5$	$[He]2s^22p^5$
Ne	$1s^22s^22p^6$	$[He]2s^22p^6$
Na	$1s^22s^22p^63s^1$	$[Ne]3s^1$
Mg	$1s^22s^22p^63s^2$	$[Ne]3s^2$
Al	$1s^22s^22p^63s^23p^1$	$[Ne]3s^23p^1$
Si	$1s^22s^22p^63s^23p^2$	$[Ne]3s^23p^2$
P	$1s^22s^22p^63s^23p^3$	$[Ne]3s^23p^3$
S	$1s^22s^22p^63s^23p^4$	$[Ne]3s^23p^4$
Cl	$1s^22s^22p^63s^23p^5$	$[Ne]3s^23p^5$
Ar	$1s^22s^22p^63s^23p^6$	$[Ne]3s^23p^6$
K	$1s^22s^22p^63s^23p^64s^1$	$[Ar]4s^1$
Ca	$1s^22s^22p^63s^23p^64s^2$	$[Ar]4s^2$
Sc	$1s^22s^22p^63s^23p^64s^23d^1$	$[Ar]4s^23d^1$
Ti	$1s^22s^22p^63s^23p^64s^23d^2$	$[Ar]4s^23d^2$
V	$1s^22s^22p^63s^23p^64s^23d3$	$[Ar]4s^23d^3$
Cr	$1s^22s^22p^63s^23p^64s^13d^5$	$[Ar]4s^13d^5$
Mn	$1s^22s^22p^63s^23p^64s^23d^5$	$[Ar]4s^23d^5$

In this notation the number in front of the orbital letter designation is the principal quantum number for that orbital. The number, written as a superscript, after the orbital letter designation is the number of electrons in that set of degenerate orbitals. It is simply understood that Hund's Rule is obeyed and that, where possible, the electrons occupy the orbitals singly with parallel spins. If this were not true because, for instance, an excited state were being described, the orbital designations would be expanded even further to indicate the population of the individual orbitals (e.g., p_x, p_y, or p_z).

In the compacted notation listed in the third column, the electron configurations are further shortened by placing brackets around the preceding inert gas element and then continuing with the electron configuration notation from that point forward. It is important to note that the bracket notation indicates the electron configuration of the preceding inert gas and cannot be used for the present inert gas. A more complete listing of electron configurations is provided in Appendix D.

Alpha particle: one of the forms of radioactivity; this particle, corresponding to a helium nucleus with two protons and two neutrons, is emitted during the decay of certain radioactive species.

Atomic number: (symbolized as Z) the number of protons in a specific nucleus, which determines the chemical identity of the species.

Atomic theory: a variety of explanations of the structure of the atom. Those covered in this chapter include Dalton's model (indestructible atom), Thompson's model (plum-pudding atom), Rutherford's model (nuclear atom), and Bohr's model (planetary system atom.)

Aufbau Principle: the assumption that electron configurations for all atoms can be built by adding electrons, one after another, into the energy levels determined for the hydrogen atom.

Beta particle: one of the forms of radioactivity; this particle, corresponding to a high energy electron, is emitted during the decay of certain radioactive species.

Cathode ray: a radiation originating from the cathode, later determined to be a stream of high velocity electrons.

Crookes tube: the early predecessor of a cathode ray tube, it consisted of a glass envelope, two metal electrodes, and a port through which gas could enter or leave.

Diamagnetic, diamagnetism: used in reference to an atom, ion or molecule in which all the electrons are spin–paired.

Gamma radiation: one of the forms of radioactivity; this short wavelength electromagnetic radiation is emitted during the decay of certain radioactive species. Unlike alpha or beta particles, gamma radiation has no mass and no charge.

Hund's Rule: a statement by Hund that electrons occupy degenerate energy levels singly with parallel spins until forced to pair. The exact statement says that the electronic state of maximum multiplicity (the one with the largest number of unpaired electrons) is the lowest energy state of those possible electronic states with the same electron distribution.

Paramagnetic, paramagnetism: used in reference to an atom, ion or molecule in which at least one of the electrons is not spin–paired with another electron.

Quantum Hypothesis: Proposal by Planck that energy is quantized, *i.e.*, it comes only in multiples of a basic unit. For electromagnetic radiation, the energy is proportional to the frequency (ν) of the light. This is stated in the equation $E = h\nu$.

Quantum numbers: values symbolized by n (principle), l (angular momentum), and m (magnetic quantum number) which are part of the solution of wave equations describing the behaviour of an electron in an atom. Quantum numbers typically also include s (spin quantum number.)

Uncertainty Principle: proposed by Heisenberg, this principle states that certain pairs of variables (*e.g.*, position and momentum, energy and time) are related in such a way that increased precision in the determination of one of the linked variables necessarily results in decreased precision in the determination of the other variable.

Wave function: mathematical description of the behaviour of an electron in an atom based upon the electron's wavelike characteristics.

Wave-particle duality: the ability of light (and small particles like electrons) to behave as either a wave or a particle, depending upon experimental circumstances.

HOMEWORK Problems

NUCLEAR REACTIONS

1. Balance the following nuclear reactions by completely identifying the species (elemental symbol, mass number and atomic number) given by "?".

(a) $^{9}_{4}Be + ? \rightarrow ^{6}_{3}Li + ^{4}_{2}He$

(b) $^{32}_{16}S + ^{1}_{0}n \rightarrow ^{1}_{1}H + ?$

(c) $^{43}_{20}Ca + ? \rightarrow ^{46}_{21}Sc + ^{1}_{1}H$

2. Complete the following nuclear reaction symbols by identifying the species (elemental symbol, mass number and atomic number) given by "?".

(a) $^{59}_{27}Co\,(^{1}_{0}n, ^{1}_{1}H)\,?$

(b) $^{7}_{3}Li\,(?, ^{1}_{0}n)\,^{7}_{4}Be$

(c) $?\,(^{1}_{0}n, ^{4}_{2}He)\,^{24}_{11}Na$

FREQUENCY, WAVELENGTH AND ENERGY

3. A certain AM radio station broadcasts at a frequency 720 kilocycles (i.e., a frequency of 7.20×10^{5} sec^{-1}.) Using the known speed of light, calculate the wavelength, in meters, of this radio signal. Calculate the energy, in Joules, of a single photon of this frequency.

4. A certain pulsed laser emits 5.00×10^{3} Joules of energy in a single burst of light with a wavelength of 400.0 nm. Calculate the number of photons in this burst.

5. Sodium street lamps emit light with a wavelength of about 590 nm. What frequency is this light?

6. Strontium is used to give some fireworks a distinctive color. If a photon emitted from an excited strontium atom has an energy of 3.00×10^{-19} Joules, what is the wavelength of the strontium emission?

7. The name "thallium" comes from the Greek meaning "green shoot", referring to the characteristic color thallium emits when excited. The color corresponds to a wavelength of 535 nanometers. Calculate the frequency of this light, and the energy of such a photon.

8. The bond dissociation energy of H_2 is 432 kJoules/mole H_2. That is, 432 kiloJoules of energy is required to break the H-H bond in a mole of hydrogen molecules. If a single photon were to cause one hydrogen molecule to dissociate, what is the maximum wavelength of light with sufficient energy?

Bohr Model

9. According to the Bohr theory, the energy of an electron in a hydrogen atom is determined by the orbital number "n". The difference in energy between the initial and final orbitals, n_i and n_f, in a hydrogen atom is given by the Rydberg equation:

$$\Delta E = \left(-2.1799 \times 10^{-18}\,J\right)\left(\frac{1}{n_f^2} - \frac{1}{n_i^2}\right)$$

Calculate the energy of the photon (equal to the difference in energy between the orbitals) emitted when an electron falls from the n=4 to the n=2 orbital in a hydrogen atom.

10. Calculate the energy, wavelength and frequency of light which is absorbed by a hydrogen atom when the electron is excited from the ground state (n=1) to the n= 6 orbital.

11. The ionization potential is the amount of energy required to remove an electron from the atom. This corresponds to taking an electron from the ground state of the atom and moving it infinitely far away, so that it becomes a free electron. Starting with the equation given in problem #9, calculate the ionization potential of the hydrogen atom in kJ/mole.

Waves and Particles

12. The de Broglie wavelength λ is given by the equation

$$\lambda = \frac{h}{m\,v}$$

Calculate the wavelength of a 150 g baseball thrown at 100 miles per hour. Hint: convert all units to metric.

13. Calculate the wavelength of an electron traveling at 1.0×10^8 m sec^{-1} (about one-third the speed of light.)

14. One piece of evidence supporting the particle nature of light was the photoelectric effect, in which a metal exposed to light of a minimum frequency will eject electrons. Light of lower frequency, regardless of intensity, will not cause the effect. Each metal has a characteristic "work function", the amount of light energy required to eject electrons from the metal's surface. Calculate the "work function" of a piece of sodium metal, in kiloJoules per mole, if the minimum frequency of light which causes the effect is 5.51×10^{14} sec^{-1}.

15. The work function of gallium is about 400 kJoules per mole. Calculate the maximum wavelength of light which can cause the photoelectric effect with gallium.

Quantum Numbers and Electron Orbitals

16. Rank the following orbitals in order of increasing energy (lowest energy first, highest last.)

(a) 4p, 5f, 5s, 4d, 5p

(b) 5d, 4f, 6p, 7s, 6d

17. List each of the allowed combinations of n, l and m when the principle quantum number is 4.

CHAPTER NINE

OVERVIEW Problems

18. Earlier we saw that the stretching of a carbonyl bond is associated with the absorbance of light with a wavenumber of about 1700 cm^{-1}. (Remember, the wavenumber is the reciprocal wavelength.)
   (a) Calculate the wavelength, in meters, associated with that wavenumber.
   (b) Calculate the frequency associated with that wavenumber.
   (c) Calculate the energy of such a photon.

19. The stretching of a carbon to carbon double bond corresponds to a wavenumber of 1620 cm^{-1}, while a carbon to carbon triple bond corresponds to a wavenumber of 2100 cm^{-1}. Which type of bond requires more energy to stretch? Show this quantitatively from the IR absorbance data.

20. Some chemical reactions are initiated by light. Consider the reaction between gaseous hydrogen and gaseous bromine to form hydrogen bromide, which can be initiated by light dissociating the bond in the bromine molecule. The strength of the bromine to bromine bond is 192 kJ/mole.
   (a) Write the balanced chemical reaction
   (b) Calculate the longest wavelength of light which could initiate the reaction.

21. In a certain strength magnetic field, a proton will undergo nuclear magnetic resonance by absorbing a radiofrequency of 270 MHz (see Figure 4.1). What is the difference in energy (in Joules) between the lower and higher energy spin states of the proton under these conditions?

22. In Chapter Five, we noted that the Law of Conservation of Mass applied to normal chemical reactions. Nuclear reactions, on the other hand, often occur with a measureable loss of mass. The loss of mass is associated with a liberation of energy, where the energy released, ΔE, equals the product of the change in mass, Δm, and the square of the speed of light, c^2. Consider the fusion reaction of ^1H (1.007825 amu) with ^3H (3.01605 amu) to form ^4He (4.00260 amu). Remember: a Joule is a kg m^2 s^{-2}. 1 amu = 1.6602×10^{-27} kg
$c = 2.9979 \times 10^8$ m s^{-1}.
   (a) Calculate the energy, in Joules, when one atom of helium is formed.

(b) Calculate the energy, in Joules, when one gram of tritium (^3H) reacts.

23. Cesium is a very reactive metal. It reacts with light, with a work function of 206 kJ/mole, and it reacts with water to form hydrogen gas.

(a) Write the balanced chemical reaction of cesium with water.

(b) What is the longest wavelength of light capable of causing the photoelectric effect with cesium?

24. Barium is a reactive alkaline earth metal. It reacts with dilute aqueous acids to form hydrogen gas. Light with a wavelength of 459 nm is sufficient to cause the photoelectric effect in barium.

(a) Write the balanced net ionic reaction of barium with hydrochloric acid

(b) Calculate the work function, in kJ/mole, for barium.

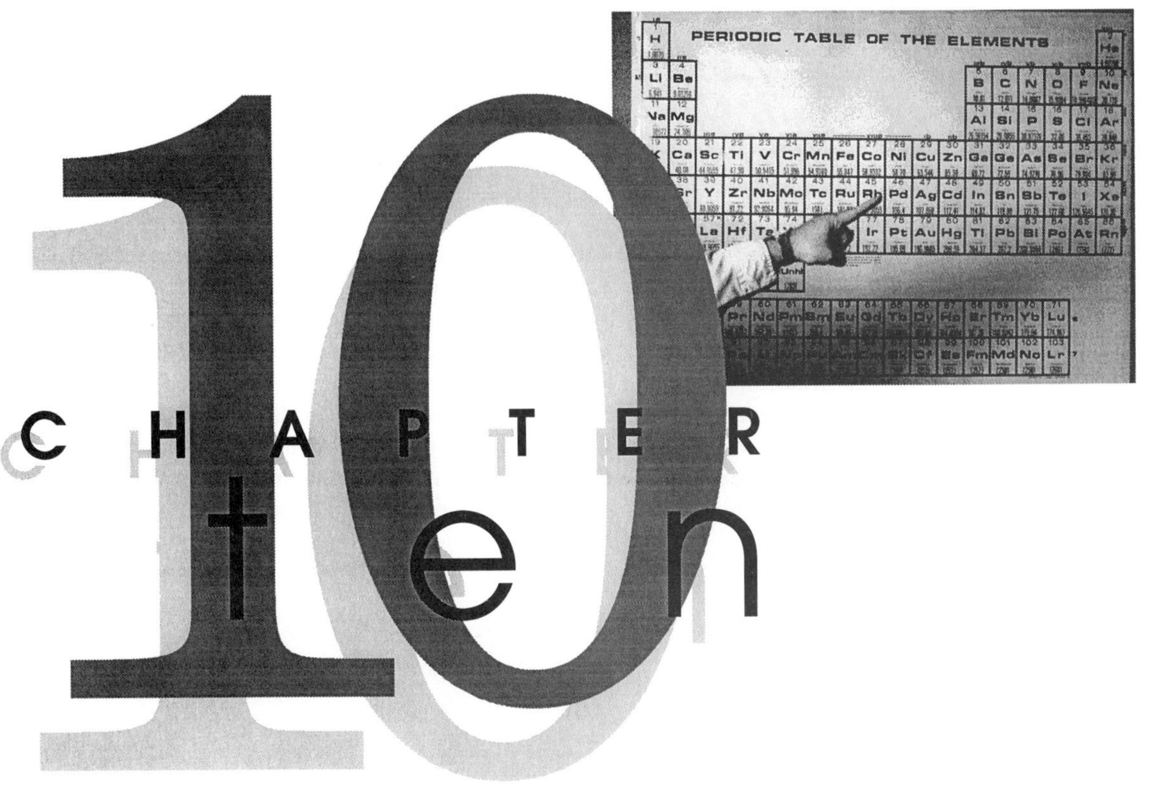

CHAPTER ten

The Table

THE Table

The history of chemistry, as the history of all the natural sciences, is best told as a continuous search for an underlying principle about which knowledge in that branch of science can be organized. The search for this principle usually begins early in the life of the science but is not commonly able to make much headway until a "critical mass" of the facts of the particular science have been accumulated. When this has happened one or two or three persons will step forth to provide the insights that will lead to the statement of this organizing principle and then, usually after a short time, to quite rapid exploitation of this principle in the uncovering of new knowledge. The science of chemistry had reached this point in the mid 1800's.

10.1 EARLY STUDIES AND INSIGHTS

The German chemist, inventor and pharmacist Johann Wolfgang Dobereiner (1780–1849) was the earliest to note trends in the properties of the elements. In his studies, begun around 1817, he noted that there were a number of instances in which triads of elements had quite similar properties. **Dobereiner's Triads** consisted of the elements calcium, strontium, and barium; sulfur, selenium, and tellurium; and chlorine, bromine, and iodine.

In the mid 1860's, the English chemist John Alexander Reina Newlands (1837–1898), after arranging the 62 elements known at the time in order of increasing atomic weight, noted that an element with similar properties occurred every eight elements in the list. He, thus, formulated his Law of Octaves which was not well–received at the time.

In the late 1860's a German chemist, Lothar **Meyer** (1830–1895), and a Russian chemist, Dmitri Ivanovich **Mendeleev** (1834–1907), independently found the insight that led directly to the periodic table as we know it today. Both worked from a list of known elements arranged in order of increasing atomic weight. Unlike other workers, however, both recognized, and left blank spaces, where they believed undiscovered elements should be fit in. Having done that, both found a much tighter and neater arrangement of physical and chemical properties than had been seen before. Although Meyer's work actually pre–dates that of Mendeleev by about two years, Mendeleev is more commonly given the credit for the periodic table because his work was the more complete being based on a discussion of the chemical properties of the elements rather than, as Meyer's, on the physical properties of the elements. Both had undertaken to predict the properties of a number of yet undiscovered elements. Mendeleev's predictions spanned the elements gallium, scandium, and germanium (he called them ekaaluminum, ekaboron, and ekasilicon). It was the discovery of these elements, within about 10 years of their prediction and description by Mendeleev, that turned the skepticism of the chemical community to widespread genuine belief in the new table of the elements proposed by Mendeleev.

10.2 EVIDENCE OF PERIODICITY

Figure 10.1 shows a modern periodic table with the symbols of those elements known by the year 1865.

1 H																1 H	
3 Li	4 Be											5 B	6 C	7 N	8 O		
11 Na	12 Mg											13 Al	14 Si	15 P	16 S	17 Cl	
19 K	20 Ca		22 Ti	23 V	24 Cr	25 Mn	26 Fe	27 Co	28 Ni	29 Cu	30 Zn			33 As	34 Se	35 Br	
37 Rb	38 Sr	39 Y	40 Zr	41 Nb	42 Mo		44 Ru	45 Rh	46 Pd	47 Ag	48 Cd	49 In	50 Sn	51 Sb	52 Te	53 I	
55 Cs	56 Ba	57 La		73 Ta	74 W		76 Os	77 Ir	78 Pt	79 Au	80 Hg	81 Tl	82 Pb	83 Bi			

58 Ce							65 Tb			68 Er			
90 Th		92 U											

FIGURE 10.1: Periodic table of elements known in 1865.

Let us examine some of the evidence that led Mendeleev and Meyer to their insight. First let us look at data for two complete triads known in 1865: the sodium, potassium, and rubidium grouping and the chlorine, bromine, and iodine grouping.

TABLE 10.1: CHEMICAL REACTIVITY DATA FOR ALKALI METAL TRIAD

TABLE 10.1

ELEMENT	Na	K	Rb
Atomic Weight	22.99	39.1	85.47
Density/g cm^{-3}	0.968	0.856	1.532
M.p./K	371	336.5	312.1
B.p./K	1154.6	1038.7	961
State	solid	solid	solid
Compound With			
oxygen	Na_2O	K_2O	Rb_2O
chlorine	NaCl	KCl	RbCl
Reactivity			
with H_2O	vigorous	vigorous	vigorous
with $HCl_{(aq)}$	vigorous	vigorous	vigorous
with $NaOH_{(aq)}$	vigorous	vigorous	vigorous

TABLE 10.2 CHEMICAL REACTIVITY DATA FOR ALKALI HALOGEN TRIAD

TABLE 10.2

ELEMENT	Cl	Br	I
Atomic Weight	35.45	79.9	126.91
Density/g cm^{-3}	0.0032	3.1	4.94
M.p./K	172.2	266	386.7
B.p./K	239.2	332.7	458.4
State	gas	liquid	solid
Compound with			
sodium	NaCl	NaBr	NaI
aluminum	$AlCl_3$	$AlBr_3$	AlI_3
Reactivity			
with H_2O	mild	NR	NR
with HCl(aq)	mild	mild	NR
with $NaOH_{(aq)}$	mild	mild	mild

As seen in these tables, members of a triad either share characteristics or show trends in their properties. For instance, all the elements in Table 10.1 are low melting, silvery metals that react vigorously with water and aqueous solutions, especially acids. The vigor of the reactions increases as the atomic weight increases. All form compounds in which the element exists as ions with a positive one charge. All form oxides and chlorides with identical stoichiometry. When the elements known in 1865 are arranged in atomic weight order, potassium immediately follows chlorine and rubidium immediately follows bromine. The pattern of consistent chemical properties and regular trends in physical properties holds true for the elements in Table 10.2. Chlorine, bromine, and iodine all form ionic compounds of identical stoichiometry with sodium, all show similar reactivity and stoichiometry toward aluminum, and all show similar reactivity toward water and aqueous solutions.

Other groups of elements showed similarities as striking as these. In many cases it was easy to see a regular pattern in the chemical properties. For instance, a series of elements M arranged in order of increasing atomic mass formed chlorides with the formulas of MCl, MCl_2, MCl_3, and MCl_4. There was also a regular pattern in the properties of the chlorides beginning with high melting solids (MCl) and ending with low melting solids or liquids (MCl_4). The pattern would then begin again. Occasionally the pattern was broken and, for instance, the chloride formulas went from a high–melting solid MCl_2 to a low melting liquid MCl_5. Mendeleev reasoned that this change from the regular pattern seen earlier was due to the presence of some, yet undiscovered, elements. An additional piece of evidence supporting this argument was that the increase in atomic weight between adjacent elements was larger than would be expected. Representative data for the elements carbon, silicon, and tin are presented in Table 10.3.

TABLE 10.3: CHEMICAL REACTIVITY DATA FOR CARBON, SILICON, AND TIN

TABLE 10.3

ELEMENT	C	Si	Sn
Atomic Weight	12.01	28.09	118.71
Density/g cm^{-3}	2.266	2.336	7.26
M.p./K	3823.2	1683.2	505.2
B.p./K	no data	3553	2896
State	solid	solid	solid
Compound With			
oxygen	CO_2, CO	SiO_2	SnO, SnO_2
chlorine	CCl_4	$SiCl_4$	$SnCl_2$, $SnCl_4$
Reactivity			
with H_2O	NR	NR	NR
with $HCl_{(aq)}$	NR	NR	NR
with $NaOH_{(aq)}$	NR	mild	mild

These data (e.g., the large jump in atomic weight between Si and Sn) along with data for the elements on both sides of silicon and tin, make a case for there being a missing element in this series located between silicon and tin. Mendeleev named this element *eka*–silicon. *Eka* is from the Sanskrit and means first. *Eka*–silicon means, literally, “first comes silicon”. Looking at the properties of silicon, tin, and the known elements on either side of *eka*–silicon, Mendeleev predicted the properties of *eka*–silicon. The predicted properties of *eka*–silicon and those of the real element, Germanium (discovered in 1886, at least partially because of Mendeleev’s predictions) are presented in Table 10.4.

TABLE 10.4

TABLE 10.4: PROPERTIES OF *EKA*-SILICON AND GERMANIUM

Property	*Eka*–silicon (1871)	Germanium (1886)
atomic weight	72	72.6
density (g/cm^3)	5.5	5.35
color	gray	gray–white
formula of oxide	EsO_2	GeO_2
density of oxide (g/cm^3)	4.7	4.703
formula of chloride	$EsCl_4$	$GeCl_4$
boiling point of chloride	<100 °C	86 °C
density of chloride (g/cm^3)	1.9	1.887

In the course of his work, Mendeleev discovered and corrected a number of problems with the chemical information then known. For instance, the then accepted value for the atomic weight of indium was approximately 76 — based on an assumption that its oxide had the formula InO because of its association with zinc oxide ores. This placed it between arsenic and selenium, two obviously non–metals. Mendeleev proposed that the assumed formula for indium oxide was wrong and that it was really In_2O_3. Recalculation of its atomic weight based upon this formula provided an atomic weight of approximately 113. This placed it below aluminum and in the space between cadmium and tin. This made it fit properly into the new table proposed by Mendeleev and was used to support his hypothesis of its formula. He solved another "problem", the inverted ordering of tellurium and iodine, by assuming that the measurements of atomic weight were incorrect.

10.3 MODERN PERIODIC TABLE ARRANGEMENTS

Although Mendeleev did predict the correct ordering for the elements tellurium and iodine based on their chemical properties, his assumption that their atomic weights had been incorrectly determined was wrong. It was the work of Henry **Moseley** (1887–1915) that established the ordering of the elements according to what he called atomic number rather than atomic weight. Thus ordered, the atomic weight inversions that occur with element pairs 18 and 19, 27 and 28, and 52 and 53 became of no significance to the periodic table. In the last chapter we saw that the atomic number of an element is the chief determinate of its electron configuration. The position of an element in the periodic table is determined by its atomic number and is an indication of its electron configuration.

The modern periodic table consists of vertical columns of elements with similar chemical properties known as **Groups**. . Many slightly different versions of the periodic table exist. The internationally accepted one numbers groups from 1 to 18. Older, more useful, numbering schemes assigned two sets of numbers from 1 to 8 in each row and used A's and B's to keep the two groups separate. In the table in this book, we shall use one of the older schemes with the suffix A indicating an element belonging to one of the Main Groups, and the suffix B indicating an element belonging to the **Transition Metals.** Thus Group IA consists of the elements Li, Na, K, Rb, Cs, and Fr and Group VIIA consists of F, Cl, Br, I, and At. Several of the groups have names associated with them. Group IA is often referred to as the ***Alkali Metals***, group IIA is the **Alkaline**

Earth Metals, and group VIIA is referred to as the **Halogen** group. Several other groups are also named, but their names are used much less frequently and will not be presented here.

Horizontal rows are called **Periods** and consist of elements whose properties change regularly as we move across the table. Because of the rules governing the assignment of quantum numbers and the filling of the orbitals as electrons are added, the number of elements in a period varies from 2 in the first period, to 8 in periods 2 and 3, to 18 in periods 4 and 5, and finally to 32 in the fifth and sixth periods.

Groups IA and IIA, the Alkali metals and the Alkaline earth metals, are those elements whose outermost electrons are in s orbitals. The elements in groups IIIA through VIIIA are elements whose outermost electrons are in p orbitals. These elements make up the list of elements known as the **Main Group Elements**. The transition metals, groups IIIB through IIB, reading from left to right on the periodic table, have outer electron configurations such as $ns^2(n-1)d^y$, while the **rare earths** (sometimes called the **lanthanides** and **actinides** or the **inner transition metals**), gathered at the very bottom of the table, are those elements whose last electron(s) was (were) added to the f sub–shell. All s and p electrons added since the last inert gas plus any d or f electrons in unfilled shells are referred to as **valence electrons**. These outermost, or valence, electrons disproportionately affect the chemical behavior of an element and, therefore, the valence electrons are of greatest importance to most chemists.

10.4 MAJOR SECTIONS OF THE PERIODIC TABLE

Figure 10.2 is a representation of the periodic table with the electron blocks shaded to indicate the sub–shell currently filling. The s block elements have the lightest shading, the p block the darkest. In–between shadings indicate the d block (the darker of the middle two) and the f block. The broad white gap running diagonally between boron and polonium separates elements with primarily metallic character from elements which are typically non–metallic. Because there is not a sharp transition from metallic to non–metallic behavior, elements close to the line have characteristics of both metals and non–metals. Hydrogen is a unique element, having some characteristics that would place it with the alkali metals and some that would place it with the halogens.

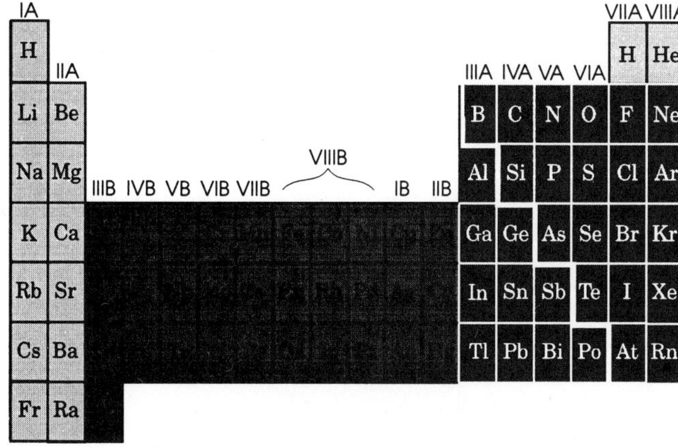

FIGURE 10.2: Periodic table of elements.

Thus, in this periodic table, hydrogen appears twice: once as a member of Group IA and once as a member of Group VIIA. Other periodic tables solve the dilemma of where to put hydrogen by putting it alone above the center of the table.

Let us now see what variations in properties to expect as we move about the periodic table. We should always keep in mind that we are using these variations to help us learn more about the properties of materials so we become able to predict the chemical behavior of new elements and compounds.

10.5 PERIODIC TRENDS IN ATOMIC AND IONIC RADIUS

Figure 10.3 shows a plot of atomic radius vs. atomic number. It is noteworthy that we see a repeating pattern of large values for alkali metals followed by a rather steep decline until the next alkali metal.

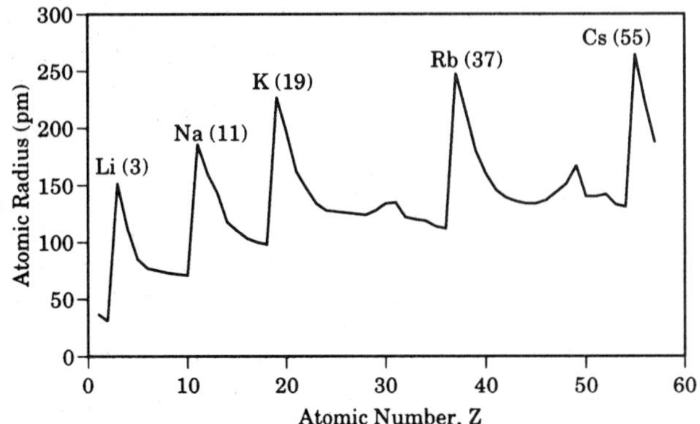

FIGURE 10.3: Radius as function of atomic number.

It is, perhaps, more informative to examine what happens to the radius of an atom as we move down the periodic table in the same group and as we move across the table in the same period. Figures 10.4 and 10.5 present typical data of this type. In figure 10.4 we see the radii of the alkali metals. Note that these radii increase as we go down the table. This is easily explained since moving down the periodic table results in increasing the value of the n quantum number for the outermost electron. Since the n quantum number is the one that primarily determines the distance from the nucleus, we would expect the atoms to become larger.

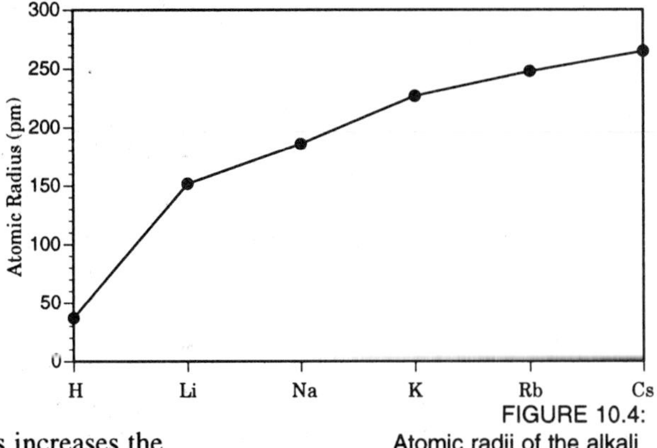

FIGURE 10.4: Atomic radii of the alkali metals.

We might expect that moving across a period would also result in larger atoms since each succeeding atom has one more proton, one more electron, and one or more neutrons than the previous element. Figure 10.5 shows clearly that this is not the case and, therefore, we must look more closely to find the explanation. If we really think about the situation, we realize that all the electrons being added as we move across a short period are being added to orbitals with the same value of the principal quantum number, n. This means that all of these electrons should be approximately the same distance from the nucleus. However, one additional proton is being added to the nucleus for each electron. This increases the nuclear charge and, as a result, the force of attraction between the nucleus and the electron. This increased attractive force is at its highest for any given period at the point that the inert gas configuration is reached and, at that point, closing the electron shell seems to exert a shielding effect on additional electrons added past the inert gas configuration. Thus, when the next alkali metal is reached by adding one more proton and electron, that electron appears to feel a nuclear charge approximately equivalent to just one proton.

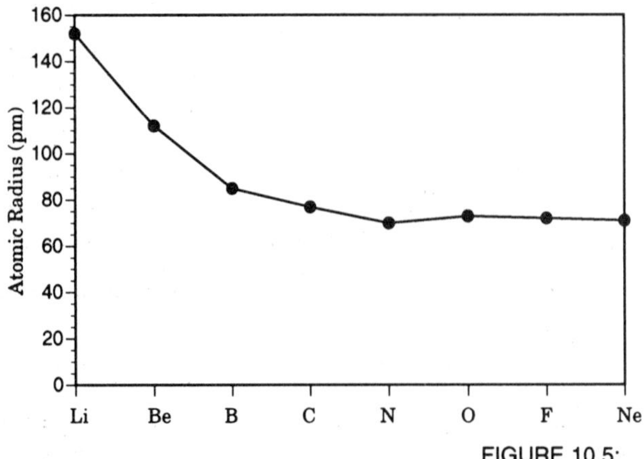

FIGURE 10.5: Atomic radii of the second periodic elements.

The trend of decreasing atomic radius with increasing atomic number is repeated for main group elements in the third period, Figure 10.6(a), and main group elements in the fourth period, Figure 10.6(b).

Figures 10.6(a) and (b) show the influence on atomic size when electrons are placed into s or p orbitals. There is a much less pronounced decrease in size, and in some cases actually a slight increase in size, as d or f orbitals are filled. Figure 10.6(c) shows the influence of 3d electrons on the atomic radius. This is explained by remembering that the electrons being added are actually going into orbitals with a smaller principal quantum number than that of the s and p valence electrons. This means that the d electrons being added are actually closer to the nucleus than the s and p electrons in the valence shell. It is the distance of the outermost electrons from the nucleus, of course, that accounts for the overall size of the atom. Adding d electrons to an inner shell approximately cancels the effect of increased nuclear charge on the outermost s and p electrons, and the atomic size remains fairly constant.

With few exceptions, however, the general trend remains that atomic size increases going down the table and decreases going across the table.

At this point it is also instructive to consider the sizes of cations and anions of closely related atoms. Clearly we expect that a cation will be smaller and an anion will be larger than a neutral atom of the same element since the number of protons is the same and the number of electrons varies. However, what would we expect for an **isoelectronic** series? For instance, consider the following series: O^{2-}, F^-, Na^+, Mg^{2+}, and Al^{3+}. Figure 10.7 plots the size of these species and shows a definite trend. Since all of these species have the same number of electrons, the trend is easily explained as being due to the increased nuclear charge of those atoms with the higher atomic number pulling on the same number of electrons.

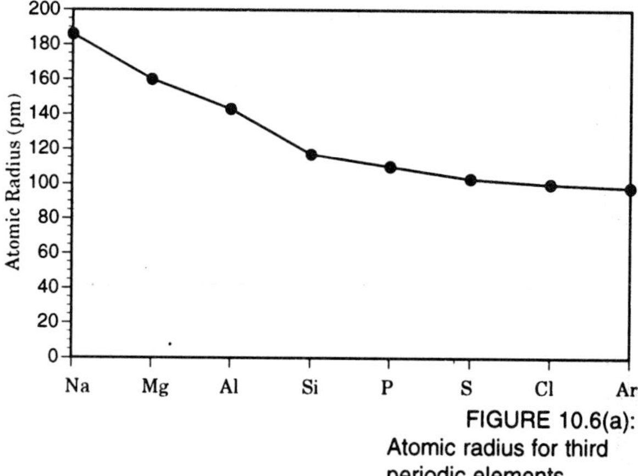

FIGURE 10.6(a): Atomic radius for third periodic elements.

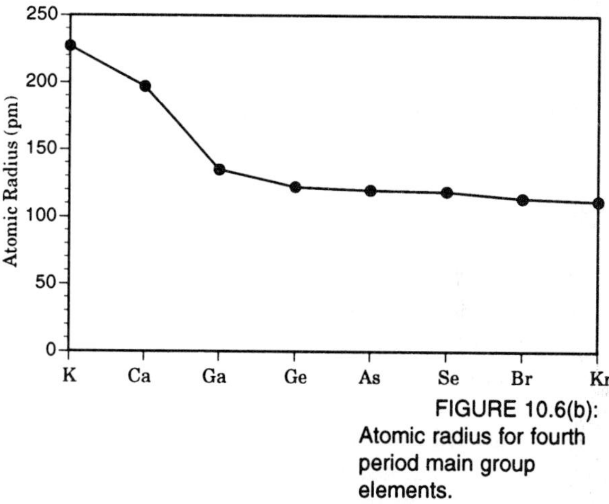

FIGURE 10.6(b): Atomic radius for fourth period main group elements.

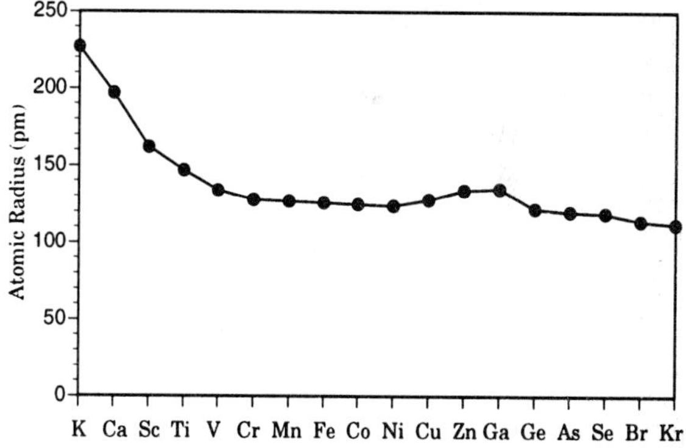

FIGURE 10.6(c): Atomic radius for fourth period elements.

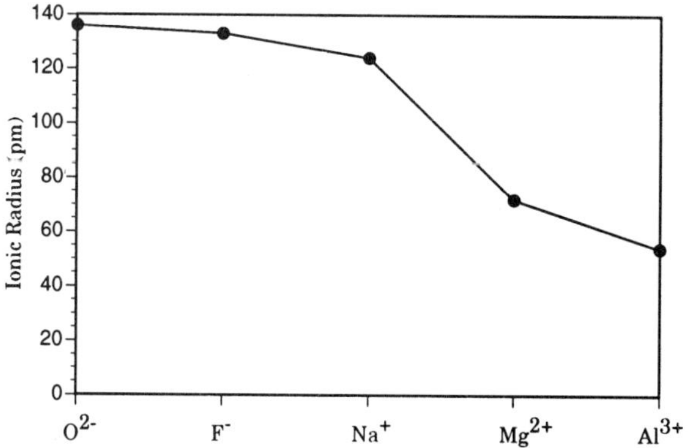

FIGURE 10.7: Radii of an isoelectronic series.

10.6 PERIODIC TRENDS IN IONIZATION ENERGY

In this section we shall determine how the energy required to ionize an atom changes with its position in the period table. The energy change for the reaction:

$$E(g) \longrightarrow E^{+}(g) + e^{-}$$

is termed the first **ionization energy**. In addition to the first ionization energy, second, third, and subsequent ionization energies are tabulated for some atoms. The second ionization energy for the element E is the energy required in the reaction:

$$E^{+}(g) \longrightarrow E^{2+}(g) + e^{-}$$

Since the removal of an electron from a neutral atom or a positively charged species always requires energy, all ionization energies for such species are positive. We would expect that ionization energies will be lowest for those elements that tend to form cations, i.e., the metals, and higher for the non–metals.

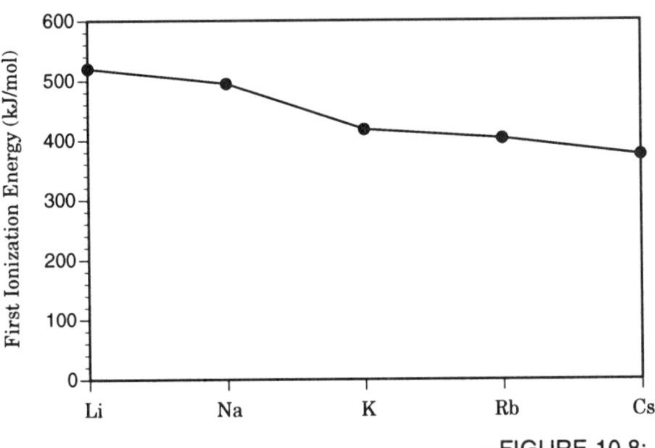

FIGURE 10.8: First ionization energies of the alkali metals.

Figure 10.8 is a plot of the first ionization energies of the alkali metals. As expected, the ionization energy decreases as we move toward the bottom of the periodic table because the electron being removed is farther from the nucleus. The situation as we move across a period is somewhat more interesting, however, as noted in Figure 10.9. While the general trend as we move from left to right on the periodic table is upward because of the increased pull of the nucleus with its extra protons, there are some spots where there are reversals. Between Be and B and again between N and O there are decreases in the ionization energy. These are exceptions to the general trend. Both of these anomalies can be explained by considering the electrons that are being removed. In the case of Be, the electron being removed is a 2s electron. Hence the electron is being taken out of a completed sub–shell. In the case of B, the electron being removed is the single electron in the 2p sub–shell. There is a general rule that says that it is easier to remove a single electron in a sub–shell than to remove an electron from a completed sub shell. Similarly, there is another general rule that says that half–filled and completely filled sub–shells have extra stability. Removal of an electron from an oxygen atom would form a half–filled shell, while removal of the electron from N would break up the half–filled shell.

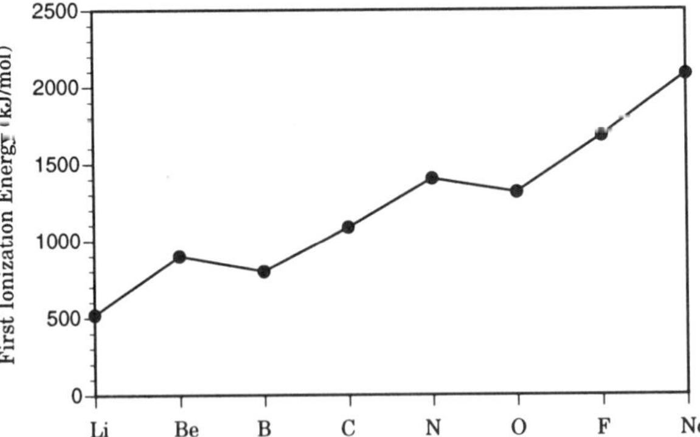

FIGURE 10.9: First ionization energies of the second periodic elements.

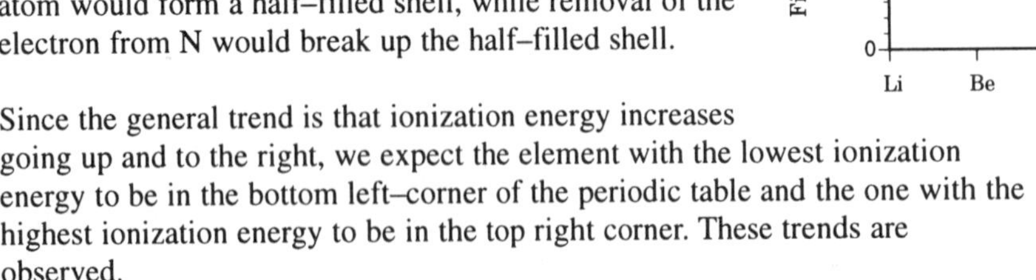

Since the general trend is that ionization energy increases going up and to the right, we expect the element with the lowest ionization energy to be in the bottom left–corner of the periodic table and the one with the highest ionization energy to be in the top right corner. These trends are observed.

Similar trends occur when looking at the second **ionization potentials.** Figure 10.10 presents the second ionization potentials of the second period elements. Note the especially high second ionization potential for lithium and the decrease in potential between B and C, and between O and F. These anomalies can be explained by looking at the electron being removed. In lithium the second electron to be removed would be a 1s closed shell electron; in B, the removal of a second electron breaks the 2s shell while in carbon the electron being removed is one singly occupying the 2p level. The situation between O and F is analogous to the N — O situation discussed above.

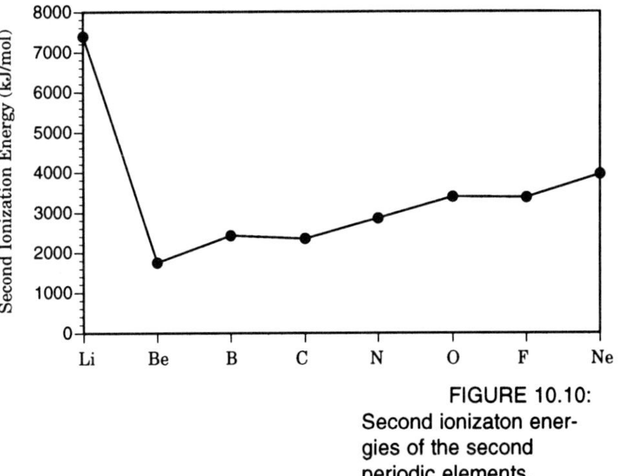

FIGURE 10.10: Second ionizaton energies of the second periodic elements.

10.7 PERIODIC TRENDS IN ELECTRON AFFINITY

The term **electron affinity** refers to the energy associated with the reaction:

$$E^{-}(g) \longrightarrow E(g) + e^{-}$$

Since, for most anions, energy is required to remove an electron from the gaseous ion, electron affinities are normally positive. The major exceptions to this rule occur with the **inert gases** where the ions are less stable than the neutral molecule. Ignoring the inert gases for the moment, we would expect that electron affinities would be largest for those elements that are most likely to form anions and smallest for those elements that normally form cations, i.e., highest for non–metals and lowest for metals. Figures 10.11 and 10.12 show the electron affinities of the alkali metals and the second period elements. The general trend, with some exceptions, is as expected: electron affinities increase going up and toward the right on the periodic table. The exceptions that we see in Figure 10.12 are readily explained. Beryllium, nitrogen, and neon have lower than anticipated electron affinities due to the extra stability of the neutral atom's electron configuration: beryllium having a filled s shell, nitrogen having a half-filled p shell, and neon having a completed shell.

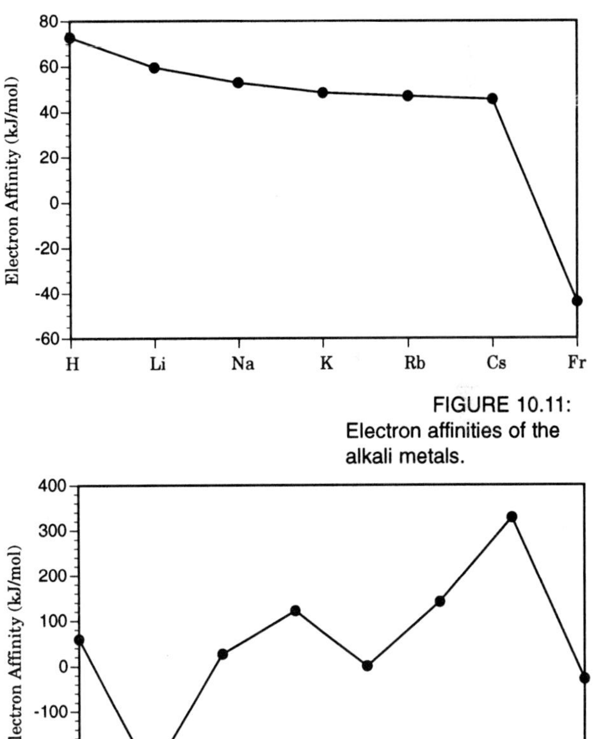

FIGURE 10.11: Electron affinities of the alkali metals.

FIGURE 10.12: Electron affinities of the second periodic elements.

10.8 PERIODIC TRENDS IN ELECTRONEGATIVITY

We can use the elemental properties of ionization energy and electron affinity to distinguish between metals and non–metals; metals have low ionization potentials and low electron affinities, and non–metals the reverse. For other considerations, especially for deciding what type of bond is likely to form between two atoms, it is useful to have a single characteristic to describe an atom. Several different chemists have proposed such a scheme for classifying elements, among them the American Noble Prize winning chemist Linus Pauling (1901–1994). Each scheme for such classification uses a formula that takes into account both the ionization energy and the electron affinity of the element and arrives at a single number termed the **electronegativity**. Although there are slight differences between the actual values calculated by the several equations, all the values for electronegativity range from approximately 1 for the most metallic elements to approximately 4 for the most non–metallic element. Figures 10.13 and 10.14 show the electronegativities for the alkali metals and the second period elements. The trend these figures show is exactly the trend we would expect with the most electronegative elements being in the upper right corner of the table (ignoring the inert gases) and the least electronegative (most electropositive) elements being in the lower left corner of the table.

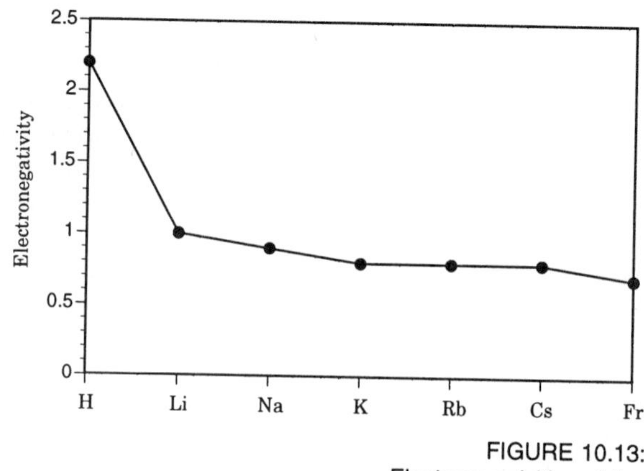

FIGURE 10.13: Electronegativities of the group IA elements.

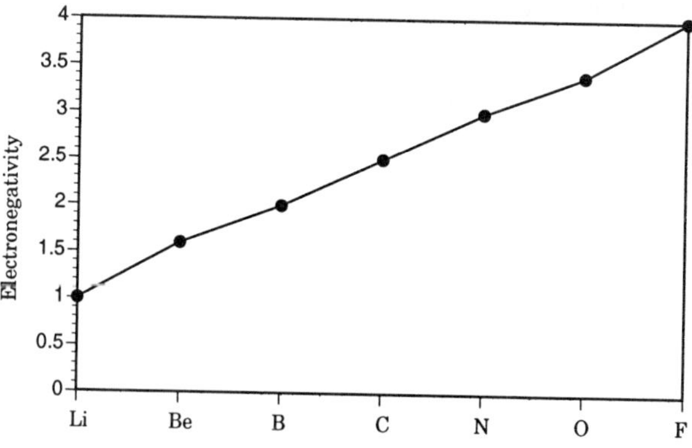

FIGURE 10.14: Electronegativity of the second periodic elements.

10.9 ELECTRON CONFIGURATIONS AND IONS

From the aufbau principle introduced in Chapter 9, and a periodic table indicating the s, p, d, and f blocks (such as Figure 10.2) the electron configuration of atoms can be readily determined. While there are rare exceptions, nearly all the electron configurations of neutral atoms are determined by filling the lowest energy levels available, following the (n + l) rule and Hund's rule.

The situation for monatomic anions is consistent with the procedure for neutral atoms. The distinction is simply that additional electrons are added to the neutral species. As before, the lowest energy levels will be filled. For example, oxygen atom (with 8 total electrons) has an electron configuration of $1s^22s^22p^4$. Its position in the "p block" in Figure 10.2 assures us that the highest energy electrons are in a p orbital. The oxide ion, O^{2-}, will have 10 total electrons. It is isoelectronic with neon (which also has 10 total electrons). Like neutral neon, oxide anion will have an electron configuration of $1s^22s^22p^6$.

For cations, the situation may be more complicated. In every case, however, the electrons removed come from the outermost orbitals. For representative elements (A groups), the highest energy electrons are also the electrons most distant from the nucleus. For elements in IA and IIA, the ns electrons are both the highest energy and the farthest distance from the nucleus. For elements in IIIA through VIIIA, the np electrons are both the highest energy and the farthest distance from the nucleus. For these elements, the outermost electrons that are removed are also the highest energy electrons. In terms of the aufbau for electron configuration, ionization means last in, first out.

Let us consider an example of a representative metal. The neutral Pb atom has an electron configuration of $[Xe]6s^24f^{14}5d^{10}6p^2$. The Pb^{2+} cation has lost its two outermost electrons. The highest n value orbitals are farthest from the nucleus. In case of a tie between orbitals with the same n value, the higher l value orbital is a farther distance from the nucleus. The electrons that are ionized (i.e., the electrons farthest from the nucleus) are the 6p electrons. Thus the electron configuration of Pb^{2+} is $[Xe]6s^24f^{14}5d^{10}6p^0$. If we wish to write the electron configuration for Pb^{4+}, we must determine which two electrons are next removed. Although both the 5d and 4f energy levels are higher than the 6s energy level, the 6s orbital is outermost from the nucleus. The 6s electrons are lost next, resulting in an electron configuration of Pb^{4+} of $[Xe]6s^04f^{14}5d^{10}6p^0$.

The situation is the same for transition metals. Zinc has an electron configuration of $[Ar]4s^23d^{10}$. The Zn^{2+} cation forms by the loss of the two outermost electrons. Again, these are not the highest energy electrons, but those electrons whose orbital is farthest removed from the nucleus. The electron configuration of Zn^{2+} is $[Ar]4s^03d^{10}$. That Zinc is stable with a 2+ oxidation state is reflected in its electron configuration that shows only filled shells and subshells. A similar situation results for iron. Neutral iron has an electron configuration of $[Ar]4s^23d^6$. We know that both the ferrous ion and the ferric ion are common. Ferrous ion (Fe^{2+}) has a configuration of $[Ar]4s^03d^6$, while ferric ion (Fe^{3+}) has a configuration of $[Ar]4s^03d^5$. The +2 oxidation state common in many transition metals results from the loss of electrons in the outermost ns orbital.

10.10 A SUMMARY OF PERIODIC TRENDS, ELECTRIC CONFIGURATIONS, AND OXIDATION STATES

Throughout this chapter, we have seen how the periodic table summarizes much of our knowledge of the chemistry of the elements. Although the periodic table was originally only a visual summary of selected empirical data, the current periodic table also reflects the atomic structure and quantum mechanics developed in Chapter 9. Specifically, the periodic trends for atomic size, ionization energy, electron affinity, and electronegativity explain much of observed chemical reactivity of the elements.

Metals, found at the left side of each period and towards the bottom of most groups, are characterized by their readiness to lose electrons. This can be interpreted on the basis of atomic size and on the basis of electron configuration. Since atomic size increases going left across a period and going down a group, metals tend to be larger atoms than non-metals in the same period or family. Thus the valence electrons of a metal are, in general, farther removed from the nucleus than the valence electrons of a comparable non-metal, and consequently are more easily removed. Alternatively, the electron configuration of metals, especially the Group IA, Group IIA, transition, and rare earth metals, have few electrons in the outermost shell. To achieve the energetically more stable electron configuration similar to that of the noble gases, metals have to lose only a few electrons in the outermost orbitals. Representative metals, like Na, Mg, and Al, will readily become Na^+, Mg^{2+}, and Al^{3+} to become isoelectronic with neon. Transition metals and inner transition metals typically form cations with a 2+ or 3+ charge. While this typically does not make these metals isoelectronic with a noble gas (because of additional d and/or f electrons) it does remove the outermost valence electrons (ns^2) and leave the outermost (n - 1) p shell filled, as it is for noble gases. Depending on their group number, metals may have oxidation states higher than +3, but only in combination with other atoms, not as simple cations.

Non-metals, conversely, have a smaller atomic size than comparable metal atoms. Since the negatively charged electrons are closer to the positively charged nucleus, Coulombic forces make it more difficult to remove an electron from a non-metal. Non-metals tend to gain electrons to form monatomic anions. Since non-metals are on the right side of the periodic table, they achieve noble gas electron configurations by gaining electrons. The most common charge on a simple non-metallic anion is (the group number - 8.) Thus halides (formed from group VIIA halogens) have a charge of 7 - 8 = -1. Chalconides (formed from the group VIA **chalcogens**) have a charge of 6 - 8 = -2. While non-metals typically form monatomic anions with a charge predictable from their group number, non-metals frequently have higher oxidation states, ranging up to positive the group number, but only in combination with other non-metals.

As we previously saw, and as we will examine in more detail next chapter, simply knowing the common charges of the elements is sufficient for us to picture how ionic compounds are formed, and to predict the formulas for simple ionic compounds (e.g., NaCl, $MgCl_2$, Al_2O_3, etc.) To understand more complicated ionic compounds (i.e., with polyatomic anions like CO_3^{2-}, SCN^-, $S_2O_3^{2-}$) and to understand covalent compounds, formal oxidation states prove useful.

As noted previously in Chapter Six, oxidation numbers for elements follow a pattern based upon the element's position in the periodic table. For monatomic atoms or ions, the formal charge and the oxidation state are the same. Metals and non-metals in their elemental states [Na(s), O_2(g)] have a zero oxidation state. Monatomic ions of both metals and non-metals will have an oxidation number equal to the charge: Na^+ has sodium in a +1 oxidation state; O^{2-} has oxygen in a -2 oxidation state. In compounds, Group IA and IIA metals will always have a +1 and +2 oxidation state respectively. Any metal in a compound involving non-metals will always have a positive oxidation state. Thus metals can vary in oxidation number from 0 (for the metal itself) to positive the group number (when combined with a more electronegative element).

Next to fluorine, oxygen is the most electronegative element. Oxygen also forms the most compounds with other elements. In combination with metals, either as a neutral oxide salt (e.g., Na_2O) or as a polyatomic ion (e.g., CrO_4^{2-})

oxygen will nearly always be in a -2 oxidation state. The rare exceptions to this are compounds of oxygen with Group IA or Group IIA elements, and these are easily recognized since those elements are invariably in a +1 or +2 oxidation state respectively. Oxygen can form a peroxide (e.g., H_2O_2, with oxygen in a -1 oxidation state) or a superoxide (e.g., CsO_2, with oxygen in a -1/2 oxidation state).

Elemental metals are good reducing agents, since they readily give up electrons to form metallic cations. Transition metals, however, can form species in which the metal is in a high oxidation state. For example, Cr in $Cr_2O_7^{2-}$ is in a +6 oxidation state; Mn in MnO_4^- is in a +7 oxidation state. These species are good oxidizing agents, since the metal is in an abnormally high oxidation state (equal to the group number in the two examples previously cited.)

The non-metals with the highest electron affinities (e.g., F_2, O_2, Cl_2) readily gain electrons, and thus are good oxidizing agents. When a non-metal is combined with other non-metals to reach a positive oxidation state, that species also can function as an oxidizing agent (e.g., N in NO_3^-, I in IO_4^-).

In the next chapter, we will see how these periodic properties of the elements help to explain the different types of chemical bonding.

Actinides: rare earth elements in which the 5f subshell is filling.

Alkali metals: those elements in group IA.

Alkaline earth metals: those elements in group IIA.

Chalcogens: those elements in group VIA.

Dobereiner triads: groups of three elements with similar chemical properties that eventual led to the periodic table concept.

Electron affinity: the energy associated with the removal of an electron from a -1 charged ion in the gas phase to produce the neutral atom in the gas phase.

Electronegativity: the relative ability of an atom in a molecule to draw electrons to itself.

Groups: vertical areas of the periodic table holding elements with similar chemical properties.

Halogens: those elements in group VIIA.

Inert gases: those elements in group VIIIA whose outer electron configuration is ns^2np^6 (except He which is just $1s^2$) Also called the inert gases.

Ionization energy: the amount of energy necessary to remove an electron from a gaseous monatomic species to produce an electron and an ion with a charge greater than the previous ion by one unit (commonly given in joules or KJ/mole).

Ionization potential: see ionization energy (the units are commonly electron volts instead of joules for ionization potential).

Isoelectronic: species that have the same electron configuration.

Lanthanides: rare earth elements in which the 4f subshell is filling.

Main group elements: elements in which the s or p sub–shells are filling.

Mendeleev: Russian chemist who did early work leading to the periodic table concept.

Meyer: German chemist who did early work leading to the periodic table concept.

Moseley: English physicist whose work with X–Ray spectra led to the current arrangement of the periodic table by atomic number instead of atomic weight.

Periods: horizontal areas of the periodic table holding elements whose outer electrons are all in the same shell.

Rare earth elements: elements whose last added electron was in the f sub–shell; sometimes called the inner transition elements.

Transition elements: elements whose last electron added was in the d sub–shell.

Valence electrons: the outermost electrons in an atom — those electrons added since the last inert gas.

HOMEWORK Problems

1. The elements known in 1862 are shown in the figure below. From the properties of the known elements at that time, predict the atomic weight, density, physical state, formula(s) of the chloride(s) and oxide(s) for the element directly above Thallium. Compare your predicted properties to those of the element Indium which was discovered in 1863.

1 H																1 H	
3 Li	4 Be											5 B	6 C	7 N	8 O		
11 Na	12 Mg											13 Al	14 Si	15 P	16 S	17 Cl	
19 K	20 Ca		22 Ti	23 V	24 Cr	25 Mn	26 Fe	27 Co	28 Ni	29 Cu	30 Zn			33 As	34 Se	35 Br	
37 Rb	38 Sr	39 Y	40 Zr	41 Nb	42 Mo		44 Ru	45 Rh	46 Pd	47 Ag	48 Cd		50 Sn	51 Sb	52 Te	53 I	
55 Cs	56 Ba	57 La		73 Ta	74 W		76 Os	77 Ir	78 Pt	79 Au	80 Hg	81 Tl	82 Pb	83 Bi			

58 Ce							65 Tb			68 Er			
90 Th		92 U											

Elements Known in 1862

2. The elements known in 1885 are shown in the figure below. From the properties of the known elements at that time, predict the atomic weight, density, physical state, and the formula(s) of the compounds formed with sodium and calcium for the element above Chlorine. Compare your predicted properties to those of the element Fluorine which was discovered in 1886.

1 H																1 H	
3 Li	4 Be											5 B	6 C	7 N	8 O		
11 Na	12 Mg											13 Al	14 Si	15 P	16 S	17 Cl	
19 K	20 Ca	21 Sc	22 Ti	23 V	24 Cr	25 Mn	26 Fe	27 Co	28 Ni	29 Cu	30 Zn	31 Ga		33 As	34 Se	35 Br	
37 Rb	38 Sr	39 Y	40 Zr	41 Nb	42 Mo		44 Ru	45 Rh	46 Pd	47 Ag	48 Cd	49 In	50 Sn	51 Sb	52 Te	53 I	
55 Cs	56 Ba	57 La		73 Ta	74 W		76 Os	77 Ir	78 Pt	79 Au	80 Hg	81 Tl	82 Pb	83 Bi			

58 Ce	59 Pr	60 Nd		62 Sm		64 Gd	65 Tb		67 Ho	68 Er	69 Tm	70 Yb	
90 Th		92 U											

Elements Known in 1885

3. Some of the melting points of period 2 fluorides are listed below. Fill in the table with estimates of those missing.

COMPOUND	M.P. (°C)
LiF	845
BeF_2	800 (sub)
BF_3	-126.7
CF_4	
NF_3	-206
OF_2	

4. The solubilities of some of the alkali metal hydroxides in cold water are given below. Fill in the table with estimates of those missing.

COMPOUND	SOLUBILITY (g/100 ml cold water)
LiOH	12.8
NaOH	42
KOH	
RbOH	180
CsOH	

5. Name the main group element which is chemically related to nitrogen and is in the period containing the inert gas Xenon.

6. Which main group element has the smallest atomic weight and has a fluoride whose formula is EF_3?

7. There are some similarities in the properties of the main group elements and those of the transition metals which bear the same numeric group designation (i.e., IIB vs. IIA). Which transition metals do you expect are most like carbon and silicon? After looking up the properties of these elements, cite a few such similarities.

8. Arrange the following elements in increasing order of atomic radius: Na, B, N, Li.

9. Arrange the following species in order of increasing radius: S^{2-}, Ar, K^+, Cl^-.

10. Arrange the following elements in increasing order of electronegativity: As, O, F, Se, S.

11. Of the following species: He, Be^{2+}, Li^+, B^{3+}, which would you expect to require the most energy for the removal of one (except for He, additional) electron?

12. Arrange the following in increasing order of electron affinity: C, Ne, Na, F.

13. Refering only to a periodic table, write the electron configuration for the sulfur atom and the sulfide anion. Briefly explain (using Lewis dot diagrams) why sulfur analogues of functional groups involving oxygen are common. [e.g., thiols, R-S-H, resembling alcohols, R-O-H.]

14. Write the electron configuration of the mercury atom, and the electron configuration of mercurous in the +1 oxidation state. Briefly explain why mercurous ion is found in a diatomic state (Hg_2^{2+}) in a fashion similar to H_2.

15. Write the Lewis dot diagram of F_2O. Which is the more electronegative element? As such, what oxidation state would you predict for F and O in this compound?

16. Write the electron configuration of the silver atom and silver in the +1 oxidation state. (Hint: remember Hund's rule.) In Mendeleev's original table, silver was grouped along with lithium, sodium, potassium, rubidium and cesium. Explain how this made sense to Mendeleev based on chemical similarities. Why is silver called a Group IB element?

17. Write the electron configuration of the cadmium atom and cadmium in the +2 oxidation state. In Mendeleev's original table, cadmium was grouped along with beryllium, magnesium, calcium, and strontium. Explain how this made sense to Mendeleev based on chemical similarities. Why is cadmium called a Group IIB element?

18. In Mendeleev's original table, vanadium was grouped along with nitrogen, phosphorus, arsenic, and antimony. Explain how this made sense to Mendeleev based on chemical similarities. Why is vanadium called a Group VB element?

19. In Mendeleev's original table, molybdenum was grouped along with oxygen, sulfur, and selenium. Explain how this made sense to Mendeleev based on chemical similarities. Why is molybdenum called a Group VIB element?

20. Cerium is a lanthanide, or rare earth metal.
   (a) Write the electron configuration of cerium.
   (b) Explain why cerium frequently occurs as the +3 ion.
   (c) Cerium (IV) can be a powerful oxidizing agent when reacted with organic compounds. Write a balanced redox reaction in which cerium (IV) reacts with formic (methanoic) acid to form carbon dioxide and cerium (III).
   (d) Cerium (IV) can also be a selective oxidizing agent when reacted with organic compounds. Write a balanced redox reaction in which cerium (IV) reacts with 1,2-butanedione to form formic acid, propionic acid, and cerium (III).

21. Element "X" forms a compound with the formula XF_5, which has a boiling point of 7°C. X forms another compound with the formula K_3X, with a melting point of 812°C. There are several oxides of X, with varying % X by mass. The smallest %X observed is 75.27%.

(a). Is the fluoride of X an ionic or covalent compound?
(b) Is K_3X ionic or covalent?
(c) What is the anticipated formula of the oxide of X that has X in its highest oxidation state?
(d) Calculate the atomic mass of X.

22. Element "Q" forms a compound with hydrogen H_2Q that is a gas at room temperature. Q forms compounds with flourine including QF_6, another gas. Q forms three different oxides, with 88.9% Q, 79.9% Q, and 72.6% Q by mass.
(a) Is H_2Q ionic or covalent?
(b) Is QF_6 ionic or covalent?
(c) To what Group does Q belong?
(d) Specify the formulas of each of the three oxides (i.e., identify the values of a and b in Q_aO_b)
(e) Calculate the atomic mass of Q.

23. Element "Z" forms compounds with nitrogen in both a 3:2 and 3:4 "Z:N" ratio. Element Z also forms ZI_2 and ZI_4 compounds.
(a) To what main Group does Z belong?
(b) Draw the Lewis dot diagram of ZI_4.
(c) What two oxides of Z would be expected? (i.e., identify the values of a and b in Z_aO_b)
(d) These oxides of Z are 81.9% and 69.4% Z by mass. Calculate the atomic mass of Z.

24. Element "J" reacts with arsenic in a 1:1 mole ratio. When reacted with excess F_2, the compound JF_3 results. The mono-, di-, and tri-bromides of J are the only bromides observed.
(a) To what main Group would J belong?
(b) Draw the Lewis dot diagram of JF_3. (Remember, fluorine never multiply bonds.)
(c) Three oxides of J are known. Indicate their likely formulas. (i.e., identify the values of a and b in J_aO_b)

CHAPTER eleven

BONDING: New Orbitals from Old

BONDING: New Orbitals from Old

CHAPTER ELEVEN

11.1 WHEN AN ATOM MEETS AN ATOM

Atoms can combine chemically to form bonds. There are three classic types of chemical bonds: metallic, covalent and ionic. The type of bond formed has much to do with the relative electronegativities and ionization potentials of the atoms involved. In short, the type of bonding depends upon whether the atoms involved are both metals, both non-metals, or one of each.

Earlier we characterized metals as elements that were malleable, ductile, lustrous, and good conductors of electricity and heat. Metals are found to the lower left of the periodic table. Non-metals do not have the properties associated with metals: non-metals are poor conductors; solid non-metals tend to be brittle or waxy, and can not be hammered into sheets or pulled into wires; and non-metals generally lack the sheen characteristic of metals.

When a metal atom bonds with other metal atoms, the bonding is described as **metallic**. When a non-metal atom bonds with another non-metal atom, the bonding is described as **covalent**. When a metal atom bonds with a non-metal atom, the bonding is described as **ionic**.

Substances formed from metallic bonds tend to have metallic properties. Such substances are usually dense solids that are lustrous, malleable, ductile and good conductors of heat and electricity. These substances are soluble in other metals, often have high melting points, and very often have high boiling points.

Substances formed from covalent bonds typically have non-metallic character. These substances can be solid, liquid or gas. The solids tend to be brittle or waxy, and are poor conductors. Covalent molecules tend to have low densities, low melting points and low boiling points. These substances are often soluble in hydrocarbons.

Substances formed from **ionic bonds** are hard, brittle crystalline solids. They are dense, with very high melting points and boiling points. As solids they are poor conductors, but are good conductors when melted. Many ionic substances are soluble in water.

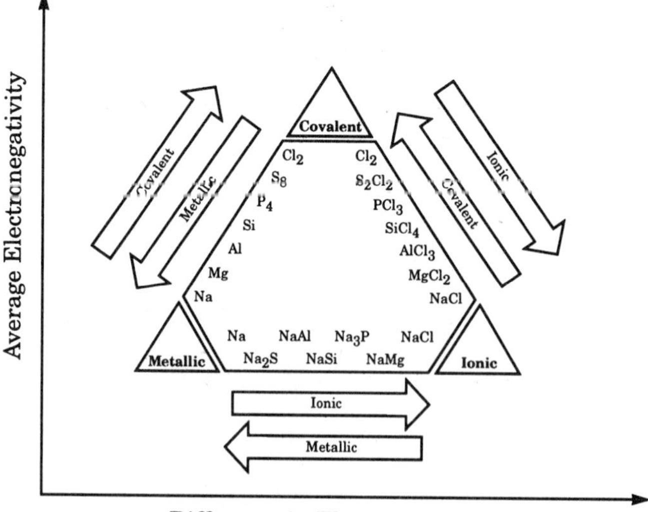

FIGURE 11.1: Pyramid of bond types.

Figure 11.1 shows a continuum of bond types from metallic to ionic to covalent. The distinction in types of bonds can be no more clear cut than the distinctions between metals and non-metals. The descriptions of the substances formed from the three types of bonding have many exceptions. We have also glossed over some distinctions that will be made later. But for now, let us focus on an example of each type of bond.

11.2 SODIUM, CHLORINE, AND SODIUM CHLORIDE

Table 11.1 shows some characteristic properties of sodium, chlorine and sodium chloride.

TABLE 11.1: PROPERTIES OF SODIUM, CHLORINE, AND SODIUM CHLORIDE

TABLE 11.1

	Na	Cl_2	NaCl
appearance	soft, silvery metal	yellow-green gas	colorless crystals
m.p. (°C)	98	-101	801
b.p. (°C)	883	-35	1413
conductance (as solid)	good	poor	poor
(as liquid)	good	poor	good

Sodium atoms interact with one another to form metallic bonds. We picture the interaction in Figure 11.2.

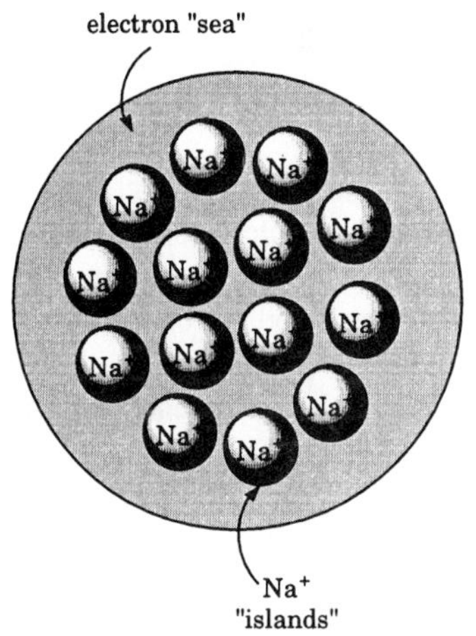

FIGURE 11.2: Metallic bonding in Na.

Sodium has a low ionization potential and low electronegativity. Thus we picture sodium metal as a sea of electrons with isolated islands of sodium ions. There is an attraction between the sodium cations and the electrons in the sea, but there is no clear identification of one electron with one cation. The result is that the electrons are free to move, and are termed "delocalized". That metals like sodium are malleable and ductile implies that the individual atoms are also relatively free to move. Unlike the other two types of bonding, there is no bond breakage when one atom moves relative to another. The change in position of the sodium atoms is accommodated by the movement of the electron sea. The free motion of the delocalized electrons within the metal also explains other physical properties of the substance. The high electrical conductivity of metals is explained by picturing the passage of electric current as one electron jumping into the east side of the "electron-sea", and having a different electron jumping out of the west side. The high heat conductivity of metals is also due to the free movement of the electrons, which readily transmit kinetic energy and thus heat from one end of the metal to the other. The luster associated with metals is also due to the delocalized electrons. Electrons on the surface of the metal absorb and re-radiate light at all angles, giving rise to the metal's shininess. Finally, we note that sodium is an especially soft, low melting metal partially because of the size of the sodium atom and the low charge of the sodium ion. The attraction of the cation island for the electron sea becomes stronger as the charge on the cation increases from M^+ to M^{2+} to M^{3+}, *etc.*; this increased attraction usually leads to harder, higher melting solids. This simple model of metallic bonding will suffice for our current concerns. Later, properties of metals based upon band theory and molecular orbitals will be examined.

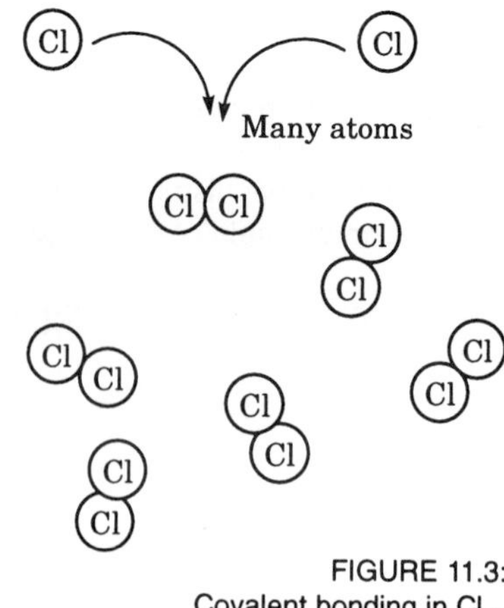

FIGURE 11.3:
Covalent bonding in Cl_2.

Two chlorine atoms interact to form a covalent bond and the molecule Cl_2. Since both atoms have high electronegativities, both chlorine atoms attempt to control the electron pair they share. The covalent bond formed by the two chlorine atoms is highly localized between the chlorine nuclei. Both atoms complete their octet by cooperating in this fashion. As a result, individual diatomic molecules of chlorine have little attraction for one another. The consequence of that weak attraction is that chlorine is a gas at room temperature. Chlorine gas can be condensed to a liquid and finally solidified by lowering the temperature considerably. The melting and boiling points are both low due to the weakness of the attraction between chlorine molecules. The localization of the electrons within a chlorine molecule is responsible for its lack of metallic characteristics: its solid phase can not be deformed into sheets or wires; chlorine is a poor conductor; and instead of reflecting light in a lustrous fashion, solid chlorine absorbs light and is greenish yellow. How atoms share electrons and the resulting consequences in terms of structure and properties will be a major topic in this chapter.

A sodium atom and a chlorine atom combine in yet a third way. The low ionization potential of sodium and high electron affinity of chlorine act to completely transfer an electron from the sodium atom to the chlorine atom, forming a sodium cation and a chloride anion. These ions combine to form a regular crystalline lattice, with each sodium cation having 6 chloride nearest neighbors, and each chloride anion having six sodium ion nearest neighbors. The coulombic forces holding the ions together in the lattice are very strong. The result is very high melting and boiling points. The electrons are localized on the ions, so the solid (with the ions held in place) does not conduct electricity. But when melted, the sodium and chloride ions can move freely, and an electrical current can be carried by the ions (as opposed to moving free electrons in the case of metallic conduction). The strong localized bonds make for a brittle solid, which shatters rather than deform to make sheets or wires. Sodium chloride crystals are colorless and thus do not absorb visible light. Salt crystals reflect light, but only at small angles; thus they do not have a metallic luster.

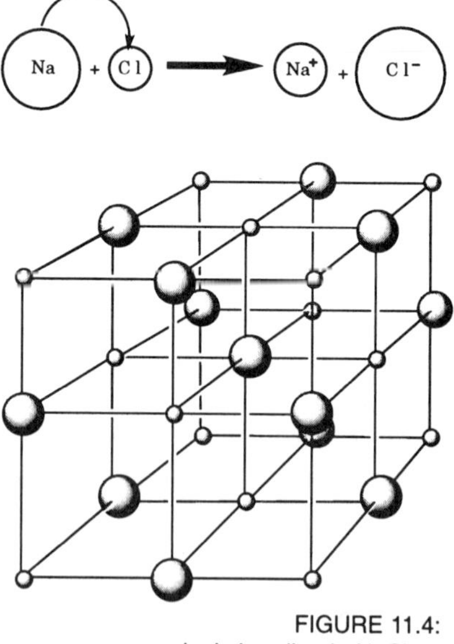

FIGURE 11.4:
Ionic bonding in NaCl.

11.3 BOND STRENGTH

A chemical bond forms when atoms combine. Such a combination is always the result of favorable energetics: the combination of atoms is at a lower energy than the separate atoms. Fig 11.5 shows a plot of energy versus distance of separation between two atoms; in one case a bond forms, in another case no bond forms.

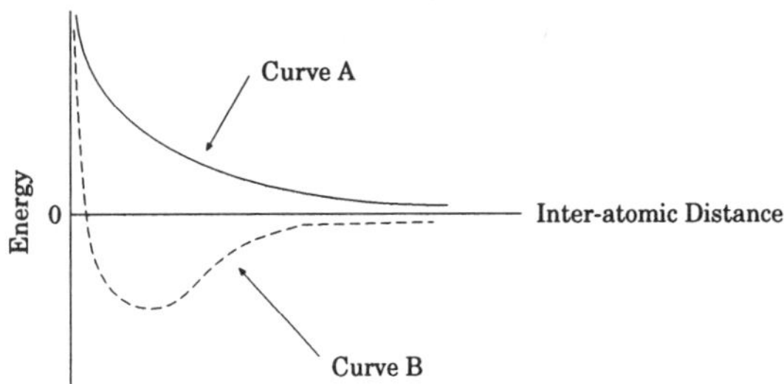

FIGURE 11.5: Energy vs. bond distance.

At the extreme right of the plot, the atoms are so far apart that they do not interact energetically. As the atoms approach one another, moving left along the curves, the interaction either increases or decreases their energy. If energy increases, the atoms repel each other and no bond is formed (curve A). If the energy decreases, a bond is formed (curve B). Note that, even in the case when a bond is formed, if the atoms are pressed too close together repulsion occurs. There is an optimum distance of separation between the atoms (*i.e.*, a minimum energy) which corresponds to the average bond length. From the principle of conservation of energy it is clear that if the energy of the system is lowered by forming a bond, that extra energy does not disappear. It is likely that the energy released by bond formation will be given off as heat.

Starting with an existing bond (in the bottom of the energy well on curve B, Fig 11.5) the atoms can be separated only by going uphill energetically. This means that energy has to be input in order to break the bonds. The relation between energy and bond making or breaking can be written in the following equations:

$$\text{Atom} + \text{Atom} \longrightarrow \text{Compound} + \text{Energy} \quad \text{(heat released)} \qquad \text{EQ. 11.1}$$

$$\text{Compound} + \text{Energy} \longrightarrow \text{Atom} + \text{Atom} \quad \text{(heat consumed)} \qquad \text{EQ. 11.2}$$

The reaction that forms a bond releases heat and is called **exothermic**. The value of heat for this process is symbolized by ΔH; since heat is given off, the value of ΔH for an exothermic reaction is negative. The reaction that breaks a bond consumes heat and is called **endothermic.** Endothermic reactions consume heat and have a positive value for ΔH. You may have previously studied reactions that were described as "breaking bonds to release energy." Burning gasoline or oxidizing carbohydrates can be viewed as breaking apart a large molecule. Both of those reactions liberate heat. Simply breaking apart the large molecules, however, consumes and does not produce heat. Heat is produced when large hydrocarbons or carbohydrates are oxidized because more energy is released in making the bonds in carbon dioxide and water than consumed in breaking the bonds of the reactants.

There are generally two types of covalent **bond strength** values. One is "bond dissociation energy", a measured value of the energy required to break a specific bond between two atoms in a binary compound. This value is readily determined for diatomic species, and becomes increasingly difficult to measure accurately as more atoms are included in the molecule. The other type is "average bond energy", which is an averaging of bond dissociation energies for several compounds containing the bond in question.

TABLE 11.2: AVERAGE BOND ENERGIES, KJ/MOLE

TABLE 11.2

	H	C	N	O	F
H	436				
C	412	348			
N	388	305	163		
O	463	360	157	146	
F	565	484	270	185	155

Table 11.2 shows average strengths for single bonds between various non-metal atoms. By convention, these values are listed as positive. The value given is the average energy required to break the specified single bond.

11.4 COVALENT BOND STRENGTH AND ELECTRONEGATIVITY

Fig 11.6 shows the electronegativity value of the elements. Bonds formed between atoms that differ by more than 1.7 electronegativity units are considered ionic. The 1.7 ΔEN value is arbitrary, and as seen in Fig 11.7, there is a continuum of bonding in terms of % ionic character. Comparing data from Table 11.2 with that in Fig 11.7 shows an important trend: covalent bonds increase in strength as the ionic character of the bond increases. Consider the bonds formed between chlorine and other third period elements. Sodium and magnesium both differ in electronegativity from chlorine by more than the 1.7 benchmark. These chlorides are considered ionic, and are treated later. The remaining third period elements that combine with chlorine (aluminum through chlorine) have bond strengths and ΔEN values shown in Table 11.3.

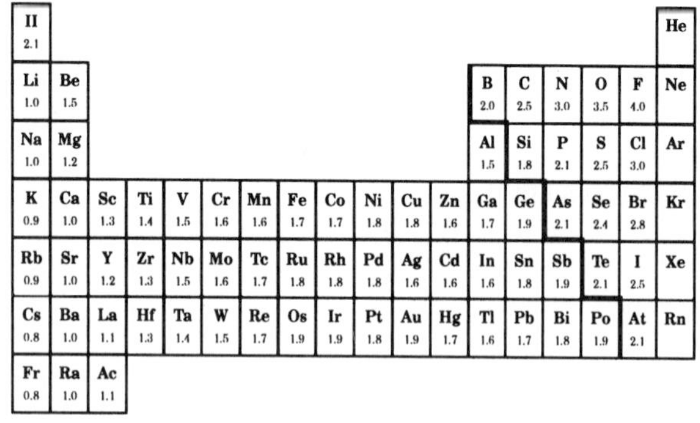

FIGURE 11.6:
Periodic table with EN values.

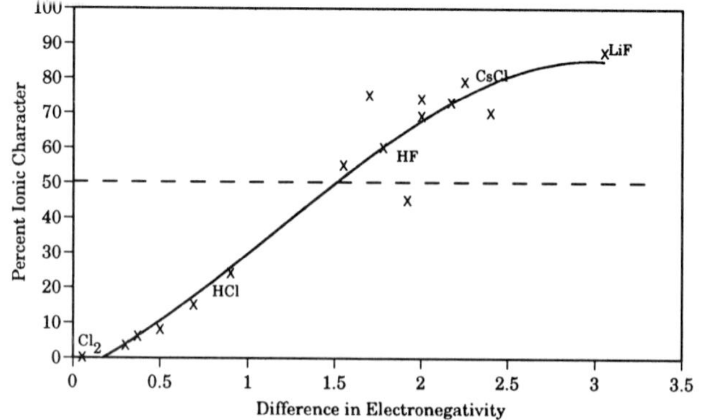

FIGURE 11.7:
ΔEN vs. % ionic character.

TABLE 11.3

TABLE 11.3: BOND STRENGTH AND ELECTRONEGATIVITY

BOND TYPE	BOND STRENGTH (KJ/MOL)	ΔEN
Al-Cl	494	1.5
Si-Cl	456	1.2
P-Cl	289	0.9
S-Cl	250	0.5
Cl-Cl	243	0.0

The chlorine molecule, with a ΔEN value exactly 0, has a perfectly covalent bond, meaning an absolutely equal sharing of the electron pair between the two chlorine atoms. Those bonds formed from elements that differ in electronegativity, but not enough to be termed ionic, are called **"polar covalent."** The more polar the covalent bond is, the greater the % ionic character and the stronger the bond.

The distinction between ionic and covalent bonds is made clear by noting the properties of the third period chlorine compounds. Table 11.4 shows the melting points for the specified compounds.

TABLE 11.4

TABLE 11.4: THIRD PERIOD BINARY CHLORIDES

COMPOUND	MELTING POINT, ^{o}C
NaCl	801
$MgCl_2$	714
$AlCl_3$	190
$SiCl_4$	-70
PCl_3	-112
S_2Cl_2	-80
Cl_2	-101

Sodium chloride and magnesium chloride clearly differ from the other compounds in terms of their melting point and, by inference, the bonds that hold the compound together. As shown in Table 11.3, the bonds in aluminum chloride are polar covalent. Aluminum falls near the break between metals and nonmetals, and it is not surprising to find that the melting point of aluminum chloride is intermediate between the clearly ionic sodium and magnesium chloride and the clearly covalent silicon tetrachloride, phosphorous trichloride, disulfur dichloride and chlorine. The gradation is confirmed when conductivity of the molten compounds in Table 11.4 is determined. Molten sodium chloride and magnesium chloride are good conductors, molten aluminum chloride is a weak conductor, and the rest are non-conductors.

11.5 STRENGTH OF BONDS IN IONIC SOLIDS

From our model of ionic solids like NaCl, Fig 11.4, we see that there is no molecule formed between the sodium and chloride ions; instead a large three-dimensional array of regularly spaced ions, called a lattice, is formed. Unlike the case in covalent molecules, specific bond strengths can not be measured for ionic compounds. Instead, the energy associated with changing one mole of the solid salt into separate gas phase ions is determined. For sodium chloride, the reaction is:

$$\text{Heat} + \text{NaCl (s)} \longrightarrow \text{Na}^+\text{(g)} + \text{Cl}^-\text{(g)}$$ EQ. 11.3

Note that all such reactions require the input of heat; these reactions are therefore termed endothermic, and have a positive value for ΔH. Table 11.5 shows values for the **"lattice energy"**, the quantity of heat associated with the reaction, for various ionic solids.

TABLE 11.5: LATTICE ENERGY OF SELECTED SALTS

TABLE 11.5

IONIC SOLID	LATTICE ENERGY (KJ/MOLE)	TYPE SALT CATION: ANION CHARGE
NaF	926	1:1
NaCl	787	1:1
KCl	717	1:1
MgO	3850	2:2
MgS	3406	2:2
$MgCl_2$	2326	2:1
Na_2O	2481	1:2

The lattice energy, the amount of energy required to break apart a mole of salt, is the means by which we estimate the strength of the bonds formed in an ionic solid. While many energy factors are involved in the actual lattice energy value, note that one feature, the charge on the ions, plays a crucial role. The lattice energies for salts with singly charged cations and anions (1:1 salts) are much less than the corresponding values for doubly charged cations and anions (2:2 salts). Lattice energies for 2:1 salts (doubly charged cation with singly charged anion, or vice versa) are between those for 1:1 and 2:2 salts. Remember that the coulombic forces holding the solid together increase with increasing charge. Thus the dependence of the lattice energies on those same charges is consistent with our use of lattice energies as ionic bond strengths.

11.6 STRENGTH OF METAL BONDS

As was the case in ionic solids, metal solids do not have specific molecular formulas. Fig 11.8 shows that solid metals form three-dimensional arrays, or lattices like those formed by ionic solids.

FIGURE 11.8:
Solid metal lattice.

Depending on the metal, an individual atom in a lattice may have 6, 8 or 12 nearest neighbors. Individual lattice sites in metals are occupied by atoms, not ions. Since metallic solids are found in these large arrays, the strength of individual bonds can not be directly measured. Instead, the energy required to change a mole of metal from the solid phase to gas phase atoms is determined. For sodium metal, the reaction is:

$$\text{Heat} + \text{Na(s)} \longrightarrow \text{Na(g)} \quad \text{EQ. 11.4}$$

Table 11.6 gives the value of the heat required for this process for a variety of metals. This value is also known as the **enthalpy of atomization**, and is equivalent to the bond strength of the metal. The melting points of the metals are also included in the table.

TABLE 11.6: ENTHALPY OF ATOMIZATION FOR SELECTED METALS

TABLE 11.6

	METAL	ΔH ATOMIZATION KJ/MOLE	MELTING POINT K
3rd	Na	107.3	371
Period	Mg	147.3	923
	Al	326.4	932
4th	K	89.2	336
Period	Ca	177.7	1123
	Sc	377.8	1670
	Ti	469.9	2000
	V	514.2	200
	Cr	396.6	2173
	Mn	280.7	1517
	Fe	416.3	1808
	Co	424.7	1766
	Ni	429.7	1728
	Cu	338.3	1356
	Zn	130.7	693
	Ga	277.0	303
	Ge	376.6	1211

Trends of bond strengths for metals are not immediately obvious. For sodium, magnesium and aluminum, the trend is as expected: increasing the number of valence electrons increases the attraction between the cationic "islands" and the electron "sea". This results in an increase in both the enthalpy of atomization (bond strength) and the melting point going from Na to Mg to Al. The metals from potassium to germanium do not increase smoothly; instead, both the bond strength and melting point maximize around vanadium and chromium.

As we stated earlier, generalizations will have exceptions, and categories of bonding tend to overlap. Chromium, for example, is thought to use some of the d-orbital electrons to covalently bond with chromium atoms on adjacent lattice sites. This may account not only for chromium's high bond strength and high melting point, but for its exceptional hardness as well. Later courses will examine band theory, a more advanced view of metal bonding, which will help explain some of the observed trends in metal properties.

11.7 A CLOSER LOOK AT THE PROPERTIES OF COVALENTLY BONDED SUBSTANCES

Substances that have metal bonds or ionic bonds are almost always solids at room temperature. Solid metals and ionic compounds all have extended regular structures, or lattices, with the metal bonds (or ionic bonds) responsible for holding the atoms (or ions) together in the lattice. Covalent compounds have a much wider range of typical properties, with solid, liquid and gaseous covalent compounds all common. While covalently bonded substances can have a lattice structure, the forces holding this extended array of atoms/molecules together may or may not be covalent bonding.

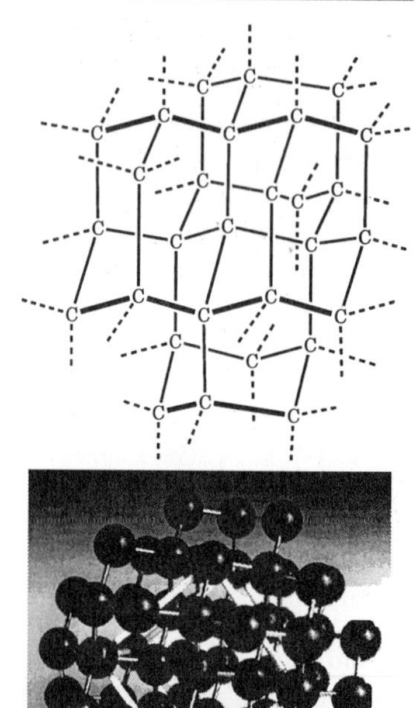

FIGURE 11.9: Diamond structure.

The prototype covalent substance we discussed earlier was chlorine. This is a perfectly covalent molecule. We referred to "weak forces" that held chlorine molecules together in the liquid or solid state. Covalent compounds that are gases or liquids at room temperature can be assumed to have only such "weak forces" holding them together in the solid phase, usually in a lattice arrangement with molecules occupying the lattice sites. Such solids are sometimes termed **"molecular solids"** to emphasize that fact. These molecular solids are soft, low melting, and non-conducting. The forces holding the molecules together in a lattice are much weaker than the covalent bonds holding the atoms together in the separate molecules. The nature of these weak forces will be examined in Chapter 16.

Other covalent substances have a regular three-dimensional array, where the lattice sites are individual atoms, and the forces holding the atoms together are **covalent bonds.** These are often called **"covalent network solids"**, and diamond is a commonly used example. Diamond is one of the allotropes of carbon, and has each carbon atom covalently bonded to four other carbon atoms. A small segment of diamond structure is shown in Figure 11.9.

In one sense diamond, silicates and other network solids can be thought of as a single giant molecule. Such solids will all have extraordinarily high melting points. Other properties will depend upon the specific structure of the solid. The rigid, three-dimensional structure resulting from the bonding for diamond results in its extraordinary hardness. But graphite (another covalent network solid with a structure shown in Figure 11.10) although high melting, is exceptionally soft. The difference in physical properties of the allotropes is the result of the difference in bonds used.

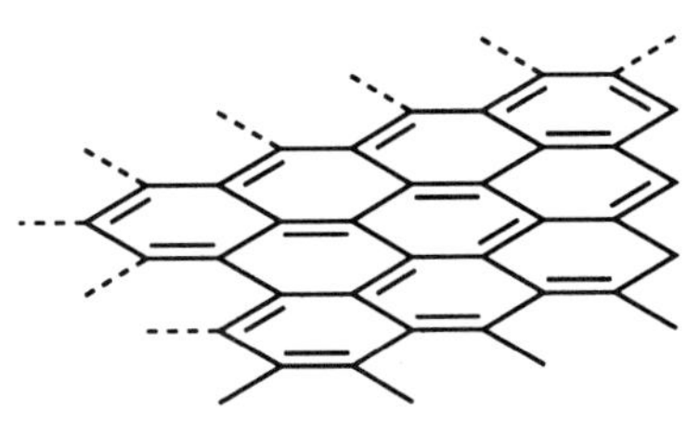

FIGURE 11.10:
Graphite structure.

Graphite has carbon atoms bonded in planes of hexagons with shared sides. Some of the electrons are shared among the rings and are delocalized throughout the plane. Individual layers of graphite are held together by forces much weaker than covalent bonds, and this results in graphite's softness. The delocalized electrons means graphite can conduct electricity along (but not perpendicular to) the planes. Graphite's delocalized electrons absorb visible light and graphite appears black (in contrast to the colorless diamond).

11.8 ATOMIC ORBITALS AND COVALENT BOND FORMATION

So far, we have considered covalent bonds to be formed by the sharing of electrons between atoms. Electrons initially in atomic orbitals (s, p, d, f, described in Chapter 9) are thought to overlap, causing electron pairs to be localized between the atoms sharing them. This is pictured in Fig 11.11, showing the overlap of a single electron in a 1s orbital in a hydrogen atom with a single electron in a 1s orbital of a second hydrogen atom.

The resulting distribution in space tends to concentrate the electron probability between the two positive nuclei. The combination of the single electrons in the separate hydrogen atoms to form a shared pair is thus the "glue" that holds the hydrogen molecule together. Overlapping two helium atoms, both with an electron configuration of $1s^2$, would mean putting more than two electrons in the same region of space, which violates Pauli's exclusion principle; consequently we neither expect nor obtain He_2 molecules.

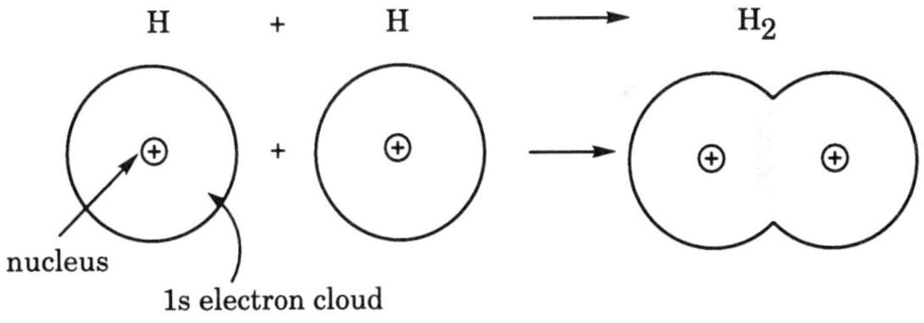

H + H → H_2

nucleus

1s electron cloud

FIGURE 11.11:
Bonding in H_2.

From our earlier study of VSEPR theory and from experimental evidence, we anticipate linear electronic geometry (and linear molecular geometry) for hydrogen. This is in accord with our picture of atomic orbital overlap in Fig 11.11, and so far, so good. The case of two electron pairs around a central atom, e.g., BeF_2, is quite different. Each fluorine atom, with an electron configuration of $1s^2$ $2s^2$ $2p^5$, has an unpaired electron to share with beryllium. The electron configuration of beryllium, however, is $1s^2$ $2s^2$, and all electrons are already paired. An additional problem is that no two atomic orbitals available to beryllium are oriented 180° apart, which is the predicted and experimental

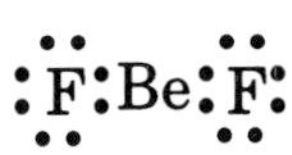

:F:Be:F:

FIGURE 11.12:
Lewis dot diagram of BeF_2.

geometry for BeF_2. One might consider a p atomic orbital on beryllium, which has the requisite geometry. However, as shown in Fig 11.12, two pairs of electrons (4 total electrons) must be situated around beryllium. One pair of electrons fills an individual p orbital. If both a p_x and p_y orbital were used, the electronic geometry would be 90°, not 180°.

The theoretical picture for BF_3 and CF_4 becomes even more difficult. Boron ($1s^2\ 2s^2\ 2p^1$) would be predicted on the basis of electronic configuration of atomic orbitals to form one bond, not three. Carbon ($1s^2\ 2s^2\ 2p^2$) would be predicted to form two bonds, not four. There is the final insurmountable problem of finding means by which s and p atomic orbitals can provide the observed 120° and 109° bond angles.

11.9 HYBRID ATOMIC ORBITALS AND COVALENT BOND FORMATION

Inconsistencies between theory and fact must be resolved by modifying the theory to accommodate the facts. In this case, the theoretical assumption that atomic orbitals (s, p, d, f) are the orbitals involved in all covalent bonding must be discarded. In their place we substitute hybrid atomic orbitals. A hybrid, in a biological sense, is the offspring from parents of widely differing genetic constitution (*e.g.*, a mule is the offspring of a donkey and a horse.) By combining two or more atomic orbitals, a variety of hybrid atomic orbitals result. We begin our examination of hybrid orbitals with Lewis dot diagrams.

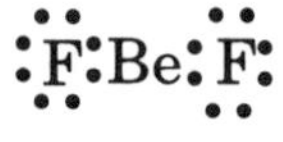

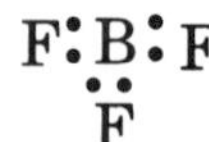

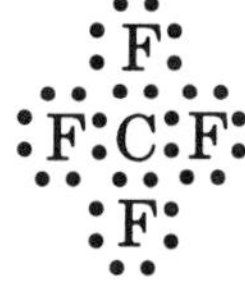

FIGURE 11.13: Lewis dot diagrams of BeF_2, BF_3, and CF_4.

BeF_2, BF_3 and CF_4 have two, three and four regions of electron densities respectively. In these molecules, beryllium is said to have a **"steric number"** of 2, boron a steric number of 3, and carbon a steric number of 4. The steric number is the key feature in predicting what types of hybrid orbitals form.

sp HYBRIDIZATION

To explain the bonding in BeF_2, we return to the electronic configuration of beryllium: $1s^2\ 2s^2\ 2p^0$. The valence shell of beryllium has 2s, $2p_x$, $2p_y$ and $2p_z$ orbitals. In the ground state of the beryllium atom, only the 2s orbitals are occupied. This is shown schematically in Fig 11.14.

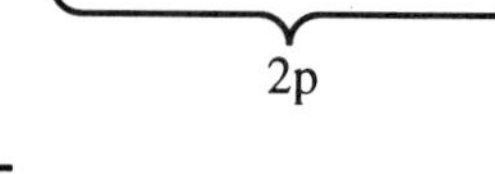

FIGURE 11.14: 2s and 2p energy levels for Be.

Hybridization is thought to occur by combining the 2s and one of the 2p atomic orbitals of beryllium. This hybridization is outlined in Fig 11.15, which also shows the resulting geometry.

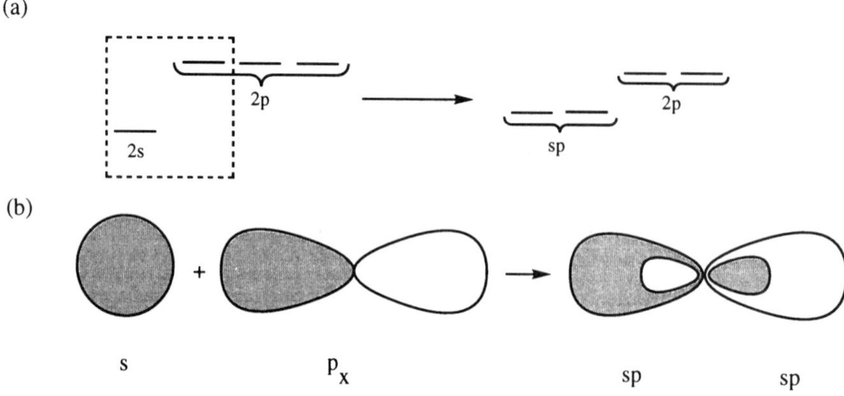

FIGURE 11.15: a) sp hybridization and b) orbital geometry.

The hybrid orbitals pictured are the result of wave functions generated by the mathematical combination of the wave functions for the 2s and $2p_x$ orbitals. A key feature of hybridization is that the number of orbitals is always conserved. Since two atomic orbitals were mixed (s and p_x), two hybrid orbitals are formed. Hybrid orbitals are named from the parent atomic orbitals. Thus the two hybrid orbitals in Fig 11.15 are termed "**sp**" hybrid orbitals. These hybrid orbitals are degenerate, meaning they have the same energy value. The energy level of hybrid orbitals will be an average of the energies of the combined atomic orbitals.

Note that the hybrid orbitals in Fig 11.15(b) share some characteristics of a p atomic orbital, especially the 180° orientation and general dumbbell shape. A key difference is that there are two orbitals pictured in Fig 11.15(b); these orbitals can hold two pairs of electrons, while a single p_x orbital can contain a maximum of one pair of electrons.

Beryllium initially had 2 valence electrons, both in a 2s orbital. If hybrid orbitals are used, the s and the p_x atomic orbitals disappear, and in their place two sp hybrid orbitals appear. (The unhybridized p_y and p_z orbitals are unaffected by this process.) The number of electrons will also be conserved. The electron configuration of beryllium using these hybridized orbitals is shown in Fig 11.16.

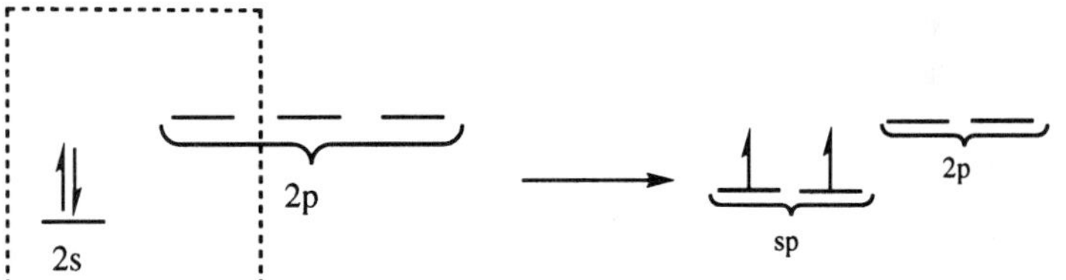

Note that Hund's rule is obeyed in hybrid orbitals also: the two valence electrons enter a degenerate energy level with parallel spin until forced to pair (when equivalent orbitals are all singly occupied.) The total energy of the two electrons in the hybrid orbitals is higher than the two electrons initially in the unhybridized 2s atomic orbital. That such hybrid orbitals would be used is counterintuitive, because we expect nature to operate in a fashion which seeks a minimum energy. But in fact, nature is doing that, because by unpairing the two 2s electrons into two singly occupied sp hybrid orbitals, beryllium can form two bonds. And as we have previously seen, bond forming is an energy lowering process. It is as if beryllium has struck a quid pro quo bargain, investing some energy by promoting two electrons into a hybrid orbital in return for the energy payback of two covalent bonds being formed.

FIGURE 11.16: Be electron configuration using sp hybrids.

As we develop hybrid orbitals further, we will see that the steric number (two in the case of BeF_2) must equal the number of degenerate orbitals formed, and thus the number of atomic orbitals combined. It is also worth emphasizing that the two sp hybrid orbitals on beryllium are oriented 180° apart, which is predicted by VSEPR and which is experimentally confirmed.

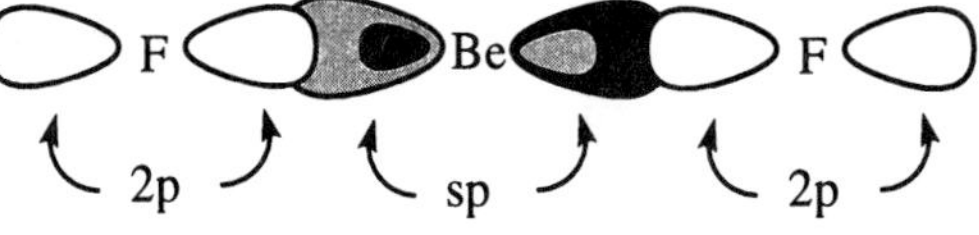

FIGURE 11.17: Bonding in BeF_2 using hybrid orbitals.

sp² Hybridization

The bonding involved in the molecule BF_3 is also rationalized using hybrid orbitals. Boron's ground state electron configuration ($1s^2\ 2s^2\ 2p^1$) has only one

FIGURE 11.18:
sp^2 hybrid scheme for BF_3.

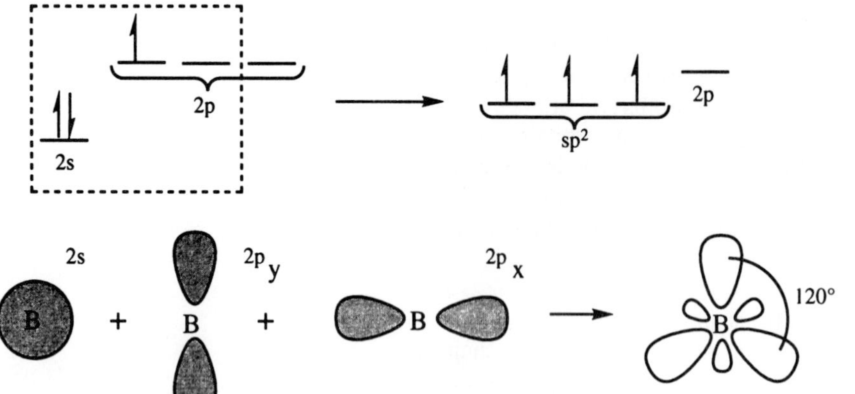

unpaired electron, and thus is expected to form only one bond. In order to form the three regions of electron density observed in the Lewis dot diagram for BF_3, and to form them with 120° bond angles as predicted by VSEPR, atomic orbitals have to be abandoned in favor of hybrid orbitals.

As shown in Fig 11.18, the s, p_x and p_y atomic orbitals are mathematically mixed to form three new hybrid orbitals. Again, the name of the hybrid is based on the atomic orbitals combined in making the hybrid. Since one s and two p atomic orbitals were combined, each of the three hybrid orbitals is termed $\mathbf{sp^2}$. Again, some of the parental heritage comes through in the hybrid: since p_x and p_y were involved in making the hybrid orbitals, the orbitals are oriented in the xy plane. The combination of atomic orbital wave functions also results in the **sp^2 hybrid orbitals** being oriented 120° apart, the result predicted by VSEPR and observed experimentally.

Note that the steric number of three in BF_3 meant three degenerate hybrid orbitals had to be formed, and conservation of orbitals required that three atomic orbitals had to be mixed in the recipe to make the hybrid orbitals. The electrons in the 2s atomic orbital had to be unpaired, and the three valence electrons entered the degenerate energy levels with parallel spins. The energy spent by boron in promoting electrons into a higher total energy configuration using hybrid orbitals is "paid back" when two additional bonds are formed.

SP^3 HYBRIDIZATION

The bonding involved in CF_4 similarly must utilize hybrid orbitals. Carbon's electron configuration ($1s^2\ 2s^2\ 2p_x{}^1\ 2p_y{}^1$) means that a maximum of two bonds can form from the two unpaired electrons in the atomic orbitals; these bonds would have 90° bond angles. In order to accommodate a steric number of 4, all four valence energy levels have to be hybridized, to form four degenerate hybrid orbitals, as shown in Figure 11.19.

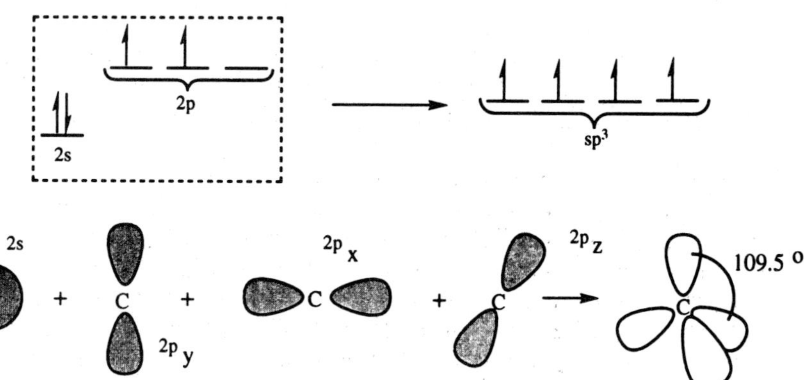

FIGURE 11.19: sp^3 hybrid orbitals for CF_4.

The mathematical combination of the s, p_x, p_y and p_z atomic orbitals yields four degenerate hybrid orbitals with a tetrahedral orientation. This is predicted by VSEPR and experimentally confirmed by the 109° bond angles.

Again, the number of regions of electron density, or steric number, is the key in determining the hybridization used. Whether the regions of electron density around the atom are lone pairs or bonding pairs makes no difference. Ammonia, water, and hydrogen fluoride, with the Lewis dot diagrams pictured in Fig 11.20, all have central atoms with four regions of electron density. In each case, the central atom is sp^3 hybridized, as shown in Fig 11.21. Unshared electron pairs occupy their own sp^3 hybrid orbital.

H:N:H H:O:H H:F:
H

FIGURE 11.20: Lewis dot diagram for NH_3, H_2O, and HF.

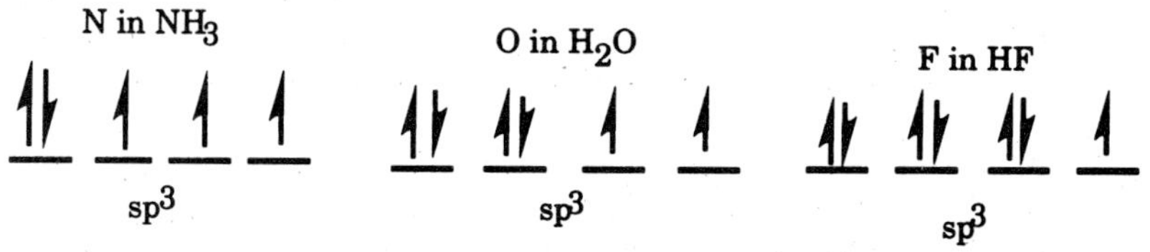

N in NH_3 — O in H_2O — F in HF

sp^3 — sp^3 — sp^3

FIGURE 11.21: sp^3 hybrid orbital schemes by N, O, and F.

11.10 MULTIPLE BONDS AND HYBRID ORBITALS

The examples for hybrid orbitals previously examined have had only single bonds and unshared pairs as regions of electron density. As we pointed out earlier, a multiple bond between two atoms is counted as a single region of electron density for purposes of determining a steric number and making VSEPR predictions. That is also true in determining the hybrid orbitals used. Fig 11.22, for example, shows ethene with a double bond between the carbon atoms.

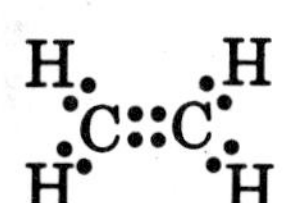

FIGURE 11.22: Lewis dot diagram of ethene.

Each carbon in ethene forms four bonds, implying that hybrid orbitals must be used. Each carbon has three regions of electron density, implying that three atomic orbitals must be combined to make three degenerate hybrid orbitals. The recipe calls for mixing s, p_x and p_y to make three sp^2 hybrid orbitals. The hybridization scheme starts with s, p_x, p_y and p_z, and ends with three sp^2 orbitals, and an unhybridized p_z atomic orbital. The four valence electrons occupy each of the final orbitals with parallel spin (Fig 11.23).

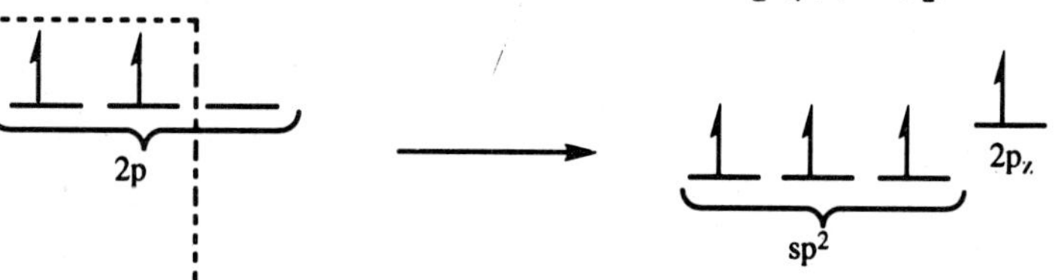

FIGURE 11.23: Hybrid orbitals for C in C_2H_4 (sp^2 + p_z).

Again, there is an energy cost for promoting the electrons above their ground state. The payback is in the form of extra bond formation. Instead of two unpaired p electrons forming two bonds, the sp^2 hybridization of carbon allows it to form a bond with each of the singly occupied sp^2 orbitals, and a bond with the singly occupied p_z orbital. Both carbon atoms in C_2H_4 are sp^2 hybridized, and the overlap of the sp^2 hybrid orbitals is pictured in Fig 11.24.

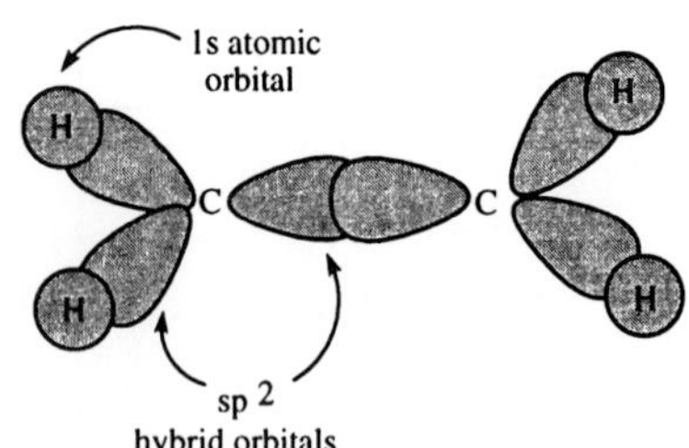

FIGURE 11.24: Sigma bonds in ethene.

The bonds formed by the overlap of hybrid orbitals directly along the internuclear axis are termed **"sigma bonds"**. Each carbon forms sigma bonds with the two hydrogen atoms as well as with the other carbon. The sigma bonds are formed from sp^2 hybrids made from s, p_x and p_y. Thus these three orbitals lie in the xy plane, and the two carbon atoms and four hydrogen atoms are all in the same plane. Each carbon atom has, in addition, an unhybridized, singly occupied p_z atomic orbital. The second bond seen in the Lewis dot diagram for ethene is now explained by the side-to-side overlap of these two p_z atomic orbitals. See Fig 11.25.

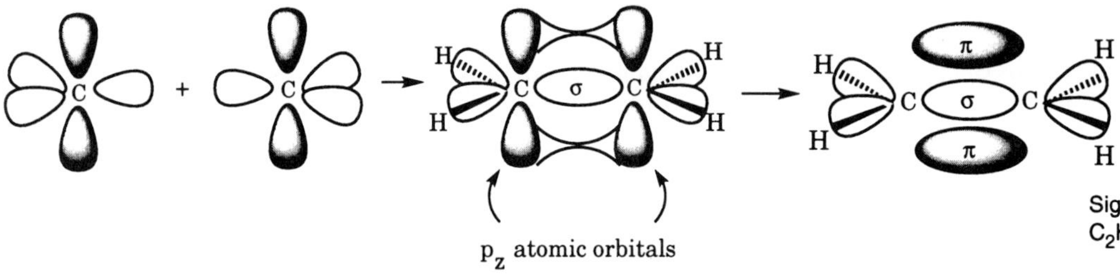

FIGURE 11.25: Sigma and pi bonding in C_2H_4.

This is a much different type bond than the sigma bonds seen earlier. As seen in Fig 11.24, sigma bonds are symmetric along the internuclear axis; rotation around this axis by any amount yields an unchanged electron distribution. But **pi (p) bonds** have their electron distribution above and below (not along) the inter nuclear axis (Fig 11.25). As a result, a pi bond is symmetric when rotated around the internuclear axis only when rotated by an integral multiple of 180°, or π radians.

Now consider the nitrogen molecule. The triple bond, seen in Fig 11.26, counts as a single region of electron density. Combined with the lone pair, each nitrogen atom then has two regions of electron density, a steric number of two, and two degenerate hybrid orbitals. The result is sp hybridization. Nitrogen goes from having valence electrons of $2s^2\ 2p^3$ to two sp orbitals and two unhybridized, singly occupied p atomic orbitals (Fig 11.27.)

:N:::N:

FIGURE 11.26: Lewis dot diagram of N_2.

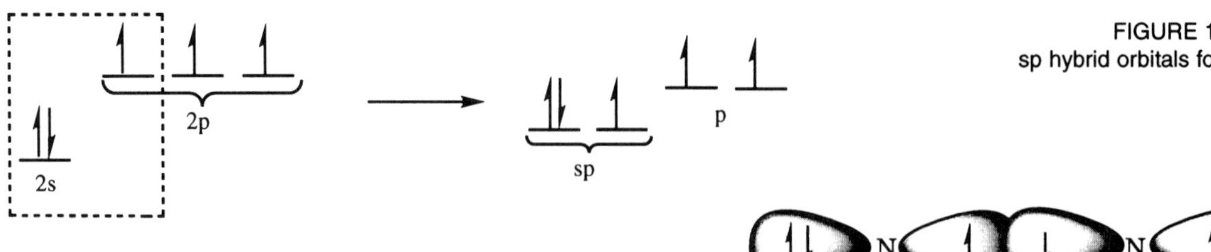

FIGURE 11.27: sp hybrid orbitals for N_2.

The doubly occupied sp orbital represents the lone pair in the Lewis dot diagram. The singly occupied sp orbital leads to the formation of the sigma bond directly between the nitrogen atoms. See Fig 11.28.

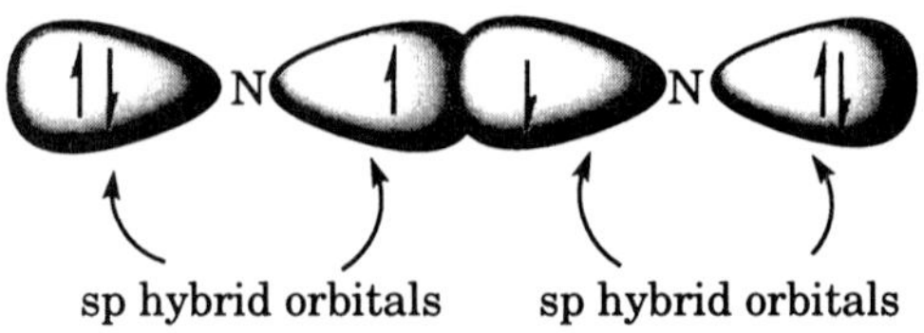

FIGURE 11.28: sp overlap for N_2.

With the p_x orbital used in making the sp hybrid orbitals, the p_y and p_z orbitals remain unhybridized. Again, a side-to-side overlap of p atomic orbitals produces pi bonds. One of these is along the y axis, the other is along the z axis. The triple bond in nitrogen is thus seen as a sigma bond, resulting from the overlap of sp hybrid orbitals, and two pi bonds, resulting from the p_y-p_y and p_z-p_z atomic orbital overlap. See Fig 11.29.

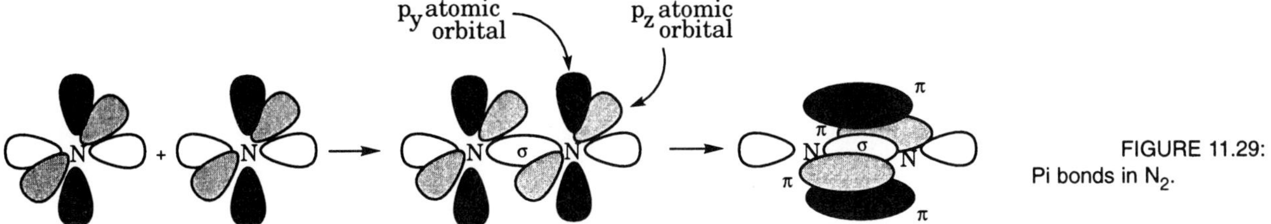

FIGURE 11.29: Pi bonds in N_2.

11.11 BOND ORDER, BOND LENGTH, AND BOND STRENGTH

As we saw in Chapter 3, atoms in a molecule continuously move relative to one another, vibrating around an average distance of separation called the bond length. The length of the bond, or the distance separating the two nuclei, is related to the size of the atoms involved. It is also related to the strength of the bond formed. Table 11.7 summarizes some data for bonds involving carbon, nitrogen and oxygen. The data indicate that increasing the number of bonds **(bond order)** between the atoms increases the bond strength and decreases the bond length.

TABLE 11.7: BOND STRENGTH AND BOND LENGTH

TABLE 11.7

BOND TYPE	BOND STRENGTH (kJ/mol)	BOND LENGTH (pm)
C-C	356	154
C=C	598	134
C≡C	813	120
N-N	160	145
N=N	418	123
N≡N	946	110
O-O	146	145
O=O	498	121
C-N	285	147
C=N	616	128
C≡N	866	116
C-O	336	143
C=O	695	120

In general, a single bond between two atoms is most often a sigma bond; the second and third bonds between the atoms are most often pi bonds. While increasing the number of bonds increases the bond strength, it does not necessarily do so proportionately; double bonds are not necessarily twice as strong as single bonds, and triple bonds are not three times the strength of single bonds.

The length and strength relationship holds as well for single bonds in families of binary compounds. Table 11.8 summarizes this for the hydrogen halides.

TABLE 11.8: BOND STRENGTH OF THE HYDROGEN HALIDES

TABLE 11.8

BOND TYPE	BOND STRENGTH (kJ/mol)	BOND LENGTH (pm)
H-F	565	91.7
H-Cl	431	127.4
H-Br	364	141.4
H-I	297	160.9

Here we see two trends at work. One is increasing atomic size going down a group; iodine is larger than bromine, which is larger than chlorine, which is larger than fluorine. The second trend mirrors the first: fluorine is more electronegative than chlorine, which is more electronegative than bromine, which is more electronegative than iodine. Increasing electronegativity and decreasing atomic size both act to make shorter and stronger covalent bonds. Later we will examine several chemical consequences of these trends in bond length and bond strength.

Bond order: the number of electron pairs shared between two atoms.

Bond strength: the energy required to break a specific bond in a molecule.

Covalent bond: the type of chemical bond formed between non-metals wherein electron density is shared between the atoms.

Covalent network solid: high melting crystalline solids in which atoms occupy the lattice sites. The atoms are all connected by covalent bonds into a single rigid molecule.

Endothermic: describes a reaction that consumes heat; ΔH for the reaction is positive.

Enthalpy of atomization: the amount of energy required to transform a mole of substance into the separate gas phase atoms. It is roughly analogous to the strength of the metallic bond.

Exothermic: describes a reaction that produces heat; ΔH for the reaction is negative.

Ionic bond: the type of chemical bond formed between a metal and a non-metal wherein the metal assumes a positive charge and the non-metal assumes a negative charge.

Lattice energy: the amount of energy required to transform a mole of solid salt into the separate gas phase ions. It is roughly analogous to the strength of the ionic bond in the solid.

Metallic bond: the type of chemical bond formed between metal atoms wherein electrons are freely shared within an array of metal cations.

Molecular solids: soft, low melting solids with molecules occupying lattice sites. The attraction between the molecules in a molecular solid is weak.

Pi bond: a bond formed by sharing an electron pair whose density is greatest above and below the internuclear axis (a line connecting the two atoms which are bonded) but whose density falls to zero along the internuclear axis. The side-to-side overlap of two singly occupied p atomic orbitals results in a pi bond.

Polar covalent bond: a bond formed by an unequal sharing of electron density between atoms of differing electronegativity.

Sigma bond: a bond formed by sharing an electron pair whose density is greatest directly along the internuclear axis.

sp hybrid orbitals: 2 degenerate electron orbitals around an atom which are formed by combining one s and one p atomic orbital. The two resulting sp hybrid orbitals are linearly arranged (180°).

sp^2 hybrid orbitals: 3 degenerate electron orbitals around an atom that are formed by combining one s and two p atomic orbitals. The three resulting sp^2 hybrid orbitals are oriented towards the corners of an equilateral triangle (120°).

sp^3 hybrid orbitals: 4 degenerate electron orbitals around an atom that are formed by combining one s and three p atomic orbitals. The four resulting sp^3 hybrid orbitals are oriented towards the corners of a tetrahedron (109°).

Steric number: the number of regions of electron density around an atom shown in a Lewis dot diagram. The steric number is the sum of sigma bonds.formed and unshared electron pairs.

HOMEWORK Problems

BOND TYPES AND PROPERTIES

1. In a few sentences, clearly distinguish covalent, ionic and metallic bonding.

2. Specify the type bonding used for each of the following:
   (a) NaK, an alloy of sodium and potassium
   (b) cesium iodide (CsI)
   (c) iodine (I_2)
   (d) acetic acid ($HC_2H_3O_2$)
   (e) sodium hydroxide (NaOH)

3. Which of the substances in problem #2 would readily conduct electricity
   (a) as a solid?
   (b) as a liquid?
   (c) when added to water?

4. Without reference to data tables, indicate which of the following solids has the higher melting point.

   (a) HCl (a molecular solid); BN (a covalent network solid)
   (b) IF (a molecular solid); FrF (an ionic solid)
   (c) NaCl (an ionic solid); $TiCl_4$ (a molecular solid)

5. The bond energy of F_2(g) is 159 kJ/mole. What amount of energy is involved in making 10.0 g of fluorine atoms from fluorine molecules? Is this energy consumed or released during the reaction?

6. The bond energy of O_2(g) is 498 kJ/mole, which is significantly larger than the value for F_2 in the previous problem. Discuss why this is so.

HYBRID ORBITALS AND GEOMETRY

7. Draw Lewis dot diagrams for each of the following and indicate the steric number around the underlined atom.

   (a) $\underline{N}_2$
   (b) $\underline{Se}O_2$
   (c) $\underline{Si}Cl_4$
   (d) $\underline{N}I_3$

8. For each of the compounds in problem 7, indicate the type of hybrid orbital the underlined atom would likely use (e.g., sp, sp^2, sp^3). Name the geometric distribution of the hybrid orbitals (e.g., linear, trigonal planar, tetrahedral).

9. For each of the compounds in problem 7, indicate the actual geometry of the molecule as a whole. If this does not match the geometry given in your answer to problem 8, explain why not.

10. Draw Lewis dot diagrams for each of the following and indicate the steric number around the underlined atom, and the type of hybrid orbitals that atom uses in its bonding.

(a) $\underline{C}OCl_2$ (carbon is central atom, bonded to other three atoms)
(b) $\underline{C}HCl_3$
(c) $HC\underline{C}CH_3$

11. List the number and symmetry of each bond made by the underlined atom in each part of problem 10. (*e.g.*, 2 sigma bonds and 3 pi bonds)

12. Draw Lewis dot diagrams for each of the following and indicate the steric number around the underlined atom, and the type of hybrid orbitals that atom uses in its bonding.

(a) $CH_3\underline{O}H$
(b) $CH_3C\underline{N}$
(c) $CH_3\underline{N}H_2$

13. List the number and symmetry of each bond made by the underlined atom in each part of problem 12. (*e.g.*, 2 sigma bonds and 3 pi bonds)

CHAPTER ELEVEN OVERVIEW Problems

14. Draw the structure of *E*-3-bromo-2-butene. Indicate the hybrid orbitals used by each carbon atom.

15. Draw the structure of isopropyl acetate. Indicate the hybrid orbitals used by each of the oxygen atoms. How many unique carbon environments would be determined by CNMR?

16. Draw the structure of 4-cyanobutanamide. Indicate the hybrid orbitals used by each nitrogen atom.

17. A compound is 74.19% sodium and 25.81% oxygen by mass.
(a) Determine the empirical formula of the compound.
(b) Is the compound metallic, ionic, or covalent?
(c) What is the oxidation state of each element in the compound?

18. A compound is 35.36% chromium, 26.58% potassium, and 38.07% oxygen.
(a) Determine the empirical formula of the compound.
(b) Is the compound metallic, ionic, or covalent?
(c) What is the oxidation state of each element in the compound?

19. A compound is 74.41% mercury and 25.59% sodium.
(a). Determine the empirical formula of the compound.
(b) Is the compound metallic, ionic, or covalent?

20. A compound is 52.54% chlorine, 23.76% sulfur, and 23.71% oxygen.

(a) Determine the empirical formula of the compound.

(b) Is the compound metallic, ionic, or covalent?

21. A compound with the formula C_3H_6O has a CNMR spectrum with two signals: one about 30 ppm, another at 208 ppm. The HNMR spectrum is a singlet at 2.2 ppm. The IR spectrum shows a strong absorbance at 1710 cm^{-1}.

(a) Draw the structure and name the compound.

(b) Interpret the IR data.

(c) Interpret the NMR data by relating spectral peaks to specific atoms shown in (a).

(d) Specify the type hybrid orbitals used for each carbon in (a) .

22. A compound is 64.81% carbon, 13.60% hydrogen, and 21.59% oxygen. The HNMR spectrum shows a triplet at 1.2 ppm, and a quartet at 3.5 ppm. The integrated signals for the triplet and quartet are in a 3:2 ratio.

(a) What is the empirical formula of the compound?

(b) Assuming that this is the molecular formula, what is the IHD?

(c) Draw the structure of the compound.

(d) Interpret the NMR data by relating spectral peaks to specific atoms shown in (c).

(e) What hybrid orbitals are used the oxygen atom.

(f) What will be the approximate bond angle around the oxygen?

23. N_2O_5 reacts with water to form a molecular compound. Note that the oxidation state of nitrogen remains unchanged in this reaction.

(a) What is the constant oxidation state of nitrogen?

(b) Write the balanced combination reaction.

(c) Draw a Lewis dot diagram of the product molecule.

(d) What hybrid orbitals are used by the nitrogen atom?

24. Cl_2O reacts with water to form a molecular compound. Note that the oxidation state of chlorine remains unchanged in this reaction.

(a) What is the constant oxidation state of chlorine?

(b) Write the balanced combination reaction.

(c) Draw a Lewis dot diagram of the product molecule.

(d) What approximate bond angle is expected in the molecule?

25. An aqueous solution of 2-butanol is reacted with potassium dichromate in the presence of sulfuric acid.

(a) Write the structure of 2-butanol

(b) What hybrid orbitals are used by the oxygen atom in (a) ?

(c) Draw the structure of the predicted organic product.

(d) What hybrid orbitals are used by the oxygen atom(s) in (c) ?

26. Propanitrile is reacted with lithium aluminum hydride.

(a) Write the structure of propanitrile

(b) What hybrid orbitals are used by the nitrogen atom in (a) ?

(c) Draw the structure of the predicted organic product.

(d) What hybrid orbitals are used by the nitrogen atom(s) in (c) ?

ACIDS AND Bases

ACIDS AND Bases

In this chapter we begin the study of two of the most important classes of compounds in the chemical world: acids and bases. Natural philosophers in ancient Egypt, Greece and elsewhere sought to explain the properties of compounds, including those we now refer to as acids and bases. In today's terms we would call such an explanation a theory of acid base behavior.

The process used by the natural philosophers to devise these theories makes use of the first several steps of the modern scientific method: observation and hypothesis. Unfortunately this is where the process stopped, partially because the learned peoples of the time did not believe in doing the manual labor involved in experimentation, and partially because, given the nature of chemistry at that time, it would have been difficult to devise experiments that could have tested the hypothesis.

12.1 EARLY ACID/ BASE THEORIES

The natural philosophers made the observation that substances now known as **acids** all had a very sour taste. Your own observations will confirm this: lemon juice is sour because of its citric acid content; vinegar is sour because of its acetic acid content, etc. Combining these observations with the atomic theory of **Democritus**, the ancient theory said that acids were sour because the "atoms" of acids were shaped like little needles. These needles pierced the tongue to cause the sharp sensation associated with sourness. At the same time they noted that solutions of **bases** produced a slippery sensation when spilled on the skin. Perhaps you have had the misfortune to spill some sodium hydroxide solution on your fingers and have noted that they slide over one another quite easily. This was explained by hypothesizing that the "atoms" of such bases were shaped like spheres — little ball bearings. They then went on to explain neutralization, the process by which an acid and a base react to form a solution having neither the sour taste of the acid nor the slippery feel of the base, by suggesting that the base's "atoms" had holes in them into which the sharp points of the acids could fit. A pictorial view of a neutralization reaction is given in Figure 12.1.

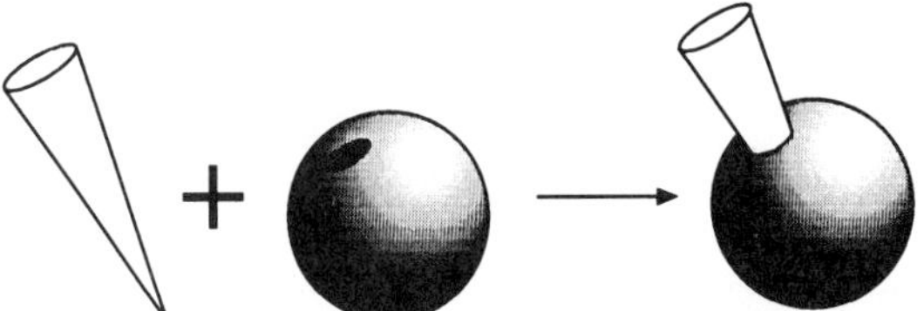

FIGURE 12.1: Neutralization reaction picture.

Despite the naiveté associated with these pictures of acid and base particles, this "theory" did accomplish the task of a theory in science: it explained the observations. What the proposers of the theory did not do, and could not do given the equipment of the time, was to subject their theory to experimental test for verification.

As science progressed, later ideas of acids and bases were essentially means of classifying compounds into one category or the other. During the Middle Ages acids were thought of as compounds that turned certain plant dyes one color and bases were compounds that turned these plant dyes another color. These cannot truly be called theories because they did not attempt to explain what it was about the compound that caused the plant dyes to turn color.

12.2 ARRHENIUS ACID/ BASE THEORY

The first of the "modern" theories of **acids** and **bases** was proposed by Svante **Arrhenius** (1859–1927) who defined an acid as a compound that ionized in water to produce hydrogen ions (H^+) and a base as a compound that ionized in water to produce **hydroxide ions** (OH^-). Neutralization was then defined as the reaction between the hydrogen and hydroxide ions to produce water.

Thus hydrogen chloride is an acid because when it dissolves in water:

$$HCl(g) \longrightarrow H^+(aq) + Cl^-(aq)$$ EQ. 12.1

and NaOH is a base because:

$$NaOH(s) \longrightarrow Na^+(aq) + OH^-(aq)$$ EQ. 12.2

When aqueous solutions of hydrochloric acid and sodium hydroxide are mixed, the following net ionic neutralization reaction occurs:

$$H^+(aq) + OH^-(aq) \longrightarrow H_2O(l)$$ EQ. 12.3

This theory worked (and still works) well for substances dissolved in water, but it does have a few problems. One of the immediate problems with this theory was how to explain the basic properties of ammonia, NH_3. All the properties of ammonia clearly classified it as a base, yet ammonia does not even have oxygen, let alone an ionizable hydroxide ion. Arrhenius solved this problem by saying that ammonia dissolves in water with the formation of a new compound that he called ammonium hydroxide (NH_4OH). There was much to support this conclusion: the ammonium ion was already well known in many other compounds, and aqueous solutions of ammonia were definitely basic. Consequently, Arrhenius' explanation was readily accepted. Other, more difficult to solve, problems arose with Arrhenius' explanations. We next examine several theories that have subsequently supplanted Arrhenius' theory.

12.3 THE SOLVENT THEORY OF ACIDS AND BASES

One of the most important problems with Arrhenius Acid–Base Theory is that it is tied to water as a solvent. There are two problems with this: first, many chemists, especially organic chemists, do most of their work in non–aqueous solvents, and second, some acid–base reactions occur in the gas phase. The first of these problems was solved by a slight broadening of the Arrhenius theory to include other solvents. In this theory, often called the Solvent System Definition of Acids and Bases, an acid is defined as a compound that dissolves in a solvent with the formation of the **lyonium ion** (the positive ion of the solvent) and a base is defined as a compound that dissolves in a solvent with the formation of the **lyate ion** (the negative ion of the solvent). This is most easily seen if we consider an aprotic solvent (one not containing ionizable hydrogen atoms) such as liquid phosgene, $COCl_2$. In this solvent, the autoionization reaction:

$$COCl_2(l) \longrightarrow COCl^+(sol) + Cl^-(sol)$$ EQ. 12.4

occurs to a small extent. Thus any compound, such as NaCl that might dissolve with the formation of chloride ions would be a base in this solvent. Notice that this definition of acids and bases does not preclude water or protic solvents. In fact, Arrhenius Theory is simply a special case of Solvent System Theory. Solvent System Theory has never been widely accepted probably because it places too much emphasis on solution reactions and because, shortly after it was put forth, two other theories of acids and bases were proposed which made it unnecessary.

12.4 BRØNSTED-LOWRY ACID/ BASE THEORY

In 1923, a Danish chemist Johannes Nicolaus **Brønsted** (1879–1947) and an English chemist Thomas Martin **Lowry** (1874–1936) independently arrived at a new explanation of acid-base properties. These two chemists proposed that acids and bases should be classified on the basis of a chemical reaction involving a proton transfer between two species. An acid is defined to be a proton donor and a base to be a proton acceptor in Brønsted–Lowry theory. The products of the reaction are, themselves, a new acid and a new base since, in essentially all acid–base reactions, the products can react to reform the reactants. Hence, a reaction between the acid, HA, and the base, B, could be written:

$$HA + B \rightleftharpoons A^- + HB^+$$ EQ. 12.5

In this reaction, the acid HA has a corresponding base A^-, and the base B has a corresponding acid HB^+. HA and A^- are referred to as a **conjugate acid–base pair**, as are HB^+ and B. Note that the acid and base in each pair differ by the presence or absence of a proton (H^+); the acid has the proton, the base does not have the proton.

One of the important consequences of Brønsted–Lowry theory is illustrated in the following reaction which is a very important reaction in aqueous chemistry:

$$H_2O + H_2O \rightleftharpoons H_3O^+ + OH^-$$ EQ. 12.6

Note that one of the water molecules on the left side of the reaction is functioning as an acid and the other is functioning as a base in this reaction. Thus a molecule is not intrinsically an acid or a base but may function both ways. Species that can react as either a proton donor or a proton acceptor are referred to as **amphiprotic**. Many of the commonly used hydroxylic solvents are amphiprotic. Water is the most common example, but the list also contains the alcohols as well as carboxylic acids in their pure state. The amphiprotic properties of many species, including the bicarbonate ion, HCO_3^-, and all amino acids, also play a key role in biochemistry.

The Brønsted–Lowry theory provides a convenient way of talking about relative strengths of acids and bases. Before we consider the topic, however, let us discuss what we mean when we say that acid A is stronger than acid B. Consider the following reactions:

$$HCl + H_2O \rightleftharpoons H_3O^+ + Cl^-$$ EQ. 12.7

$$CH_3COOH + H_2O \rightleftharpoons H_3O^+ + CH_3COO^-$$ EQ. 12.8

The first of these two reactions (Eq. 12.7) essentially goes to completion for all concentrations of HCl dissolved in water. The second of these two reactions (Eq. 12.8) goes only approximately 1% of the way to the right under ordinary laboratory conditions. As we have noted earlier, acids that are essentially completely ionized in a solvent are termed **strong acids**, and acids that are only slightly ionized are termed **weak acids**. Thus hydrogen chloride is a strong acid and acetic acid is a weak acid when dissolved in water. The last part of that sentence is, however, a very important part, for hydrogen chloride does react as a weak acid in some solvents, viz.:

$$HCl + H_2SO_4 \rightleftharpoons H_3SO_4^+ + Cl^-$$ EQ. 12.9

That is, when dissolved in liquid sulfuric acid, hydrogen chloride is only slightly ionized, and thus would be categorized as a weak acid. Alternatively, acetic acid completely ionizes in some solvents, such as liquid ammonia, and reacts as a strong acid:

$$CH_3COOH + NH_3(l) \rightleftharpoons NH_4^+ + CH_3COO^-$$ EQ. 12.10

It should be readily apparent, then, that the base to which the proton is to be transferred must be specified if one is to classify a particular acid as weak or strong. This is another way of saying that it is the relative acid strength of the old and new acid/base pairs that determines whether or not a particular reaction goes strongly toward the right or stays mostly on the left. In the case of HCl, the hydrogen chloride molecule is a stronger acid than the **hydronium ion** (H_3O^+) and the water molecule is a stronger base than the chloride ion. Hence, Eq. 12.7 goes strongly toward the right (products). On the other hand, the hydronium ion is a stronger acid than is the acetic acid molecule and Eq. 12.8 goes so slightly toward the right that acetic acid is considered a weak acid in water.

Any acid that is intrinsically stronger than the hydronium ion will behave as a strong acid in water since the transfer of the proton from the stronger acid to the water molecule will be essentially complete. Thus the hydronium ion is the strongest acid that can exist in water solution. Acids stronger than the hydronium ion are said to be *leveled* by dissolving them in water — that is to say that differences in intrinsic strength go unnoticed when the acids are dissolved in water. In general, for any solvent: *the lyonium ion is the strongest acid that can exist in a solvent.* A similar argument can be made concerning base strengths with the generalization that the *lyate ion is the strongest base that can exist in a solvent.* As a result, bases stronger than the hydroxide ion are leveled to the strength of the hydroxide ion when dissolved in water.

12.5 FACTORS AFFECTING THE STRENGTH OF PROTONIC ACIDS

We have just discussed the relative strengths of acids in terms of the degree to which the acids ionize in a given medium. We have not yet discussed why one acid should ionize to a greater extent than another acid. To answer this question, we must look at what happens when an acid ionizes. This is most easily done by considering the reaction between a hypothetical binary acid, HX, and water:

$$H—X + H_2O \longrightarrow H_3O^+ + X^-$$ EQ. 12.11

For this reaction to occur, the H–X bond must be broken and both electrons forming the bond must remain with the X atom or group, while the proton is transferred to the water molecule. In order to predict the relative acid strengths within an analogous series of compounds, numerous factors must be considered. Key among the relevant features are bond length, bond strength, and electronegativity differences. In other cases, the strength of the interaction between the X^- ion and water molecules may also be important. Table 12.1 summarizes the relevant data for some example compounds.

Table 12.1: DATA FOR SELECTED BINARY HYDROGEN COMPOUNDS

TABLE 12.1

Compound	Bond Length (in pm)	Bond Dissociation Energy (kJ/mole)	Δ Electronegativity
H-CH_3	110	435	0.4
H-NH_2	98	460	0.9
H-OH	94	492	1.4
H-F	92	565	1.9
H-Cl	127	431	0.9
H-Br	142	366	0.7
H-I	161	299	0.4

(100pm = 1 Angstrom) (4.184 kJ/mol = 1 kCal/mol)

Note that in the series methane, ammonia, water, and hydrogen fluoride, we are moving left to right across the second period for the non-hydrogen species. The length of the bond decreases slightly as atomic size decreases in this series. The H-X bond becomes significantly stronger, however, due to the rapidly increasing polar character of the covalent bond. In the series hydrogen fluoride, hydrogen chloride, hydrogen bromide, and hydrogen iodide, the "X" species is moving down the halogen family. The bond length increases significantly, with a resulting significant decrease in bond strength. When the compounds listed in Table 12.1 are dissolved in water, HCl, HBr and HI are strong acids, HF is a weak acid, and the rest either have no effect on hydronium ion concentration or are weakly basic.

One simplified explanation of the strength of binary acids focuses on two trends: the likelihood of X attracting both bonding pair electrons to itself (bond polarity), and the likelihood of the HX bond breaking (bond strength). If we move across a period comparing acid strengths of binary hydrides in the same period, *viz.*, CH_4, NH_3, H_2O, and HF, acid strengths increase. Since the electronegativity of the X atom, and hence bond polarity, increases as we move to the right, clearly the bond polarity is the chief determinate of the acid strength in this series. However, as we move down the periodic table, *viz.*, HF, HCl, HBr, and HI, we find that, although bond polarities decrease going down the periodic table, acid strengths actually increase going down the table. This is not what we would expect using only the bond polarity argument. In this case, the size difference of the X atoms is such that the H–X bonds vary significantly in bond energy. This means that the compound with the weakest H–X bond, HI, is the strongest acid.

Ternary acids have three constituent elements: hydrogen, oxygen, and another element. In ternary acids, hydrogen is almost always bonded through an oxygen atom to the rest of the molecule. Sulfuric acid and phosphoric acid, shown in Figure 12.2, are common examples of ternary acids:

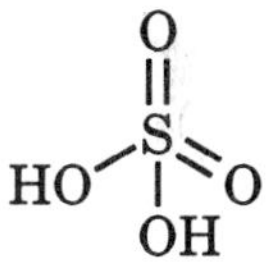

Sulfuric Acid

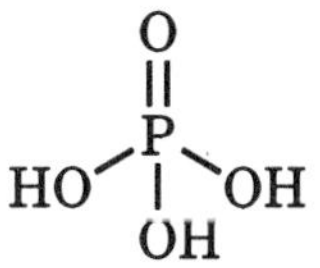

Phosphoric Acid

FIGURE 12.2: Structure of two ternary acids.

There are also two different trends that we can examine when dealing with ternary acids in which the ionizable hydrogen atom is bonded directly to an oxygen atom. In essentially all such acids, the structure of the acid, H_mXO_n, is usually best represented by the formula $(HO)_mXO_{(n-m)}$. If we look at a series of related acids in which the atom X is varied within the same family, we find that increasing the electronegativity of X increases the strength of the acid. This is easily explained because increasing the electronegativity of X causes the electrons in the X—O bond to be held more tightly to the X atom. This causes a corresponding increase in the "pull" of the oxygen atoms for the electrons in the O—H bond. This increases the polarity of the O—H bond and increases the ease with which the hydrogen atom can ionize. Thus sulfuric acid (H_2SO_4) is a stronger acid than is selenic acid (H_2SeO_4) which is, in turn, stronger than telluric acid (H_2TeO_4).

Another trend can be observed if we look at series of ternary acids in which the X atom remains the same but the values of m and/or n (usually only n) change. In these cases increasing the value of n causes an increase in the acid strength. This occurs because increasing the number of oxygen atoms surrounding the X atom draws electrons away from the X atom. This causes X indirectly to "pull" harder on the electrons in the O—H bond, thus increasing the polarity of that bond. Thus perchloric acid ($HClO_4$) is a stronger acid than is chloric acid ($HClO_3$) than is chlorous acid ($HClO_2$) than is hypochlorous acid (HClO). Note that in such series of ternary acids, a higher oxidation number for X results in a stronger acid.

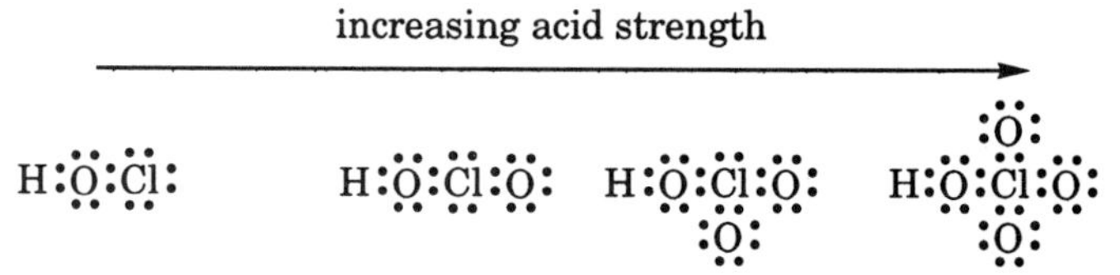

increasing acid strength →

H:O:Cl: — Hypochlorous Acid
H:O:Cl:O: — Chlorous Acid
H:O:Cl:O: with :O: — Chloric Acid
H:O:Cl:O: with :O: above and below — Perchloric Acid

FIGURE 12.3: Influence of oxidation number on the strength of related ternary acids.

Carboxylic acid strength also correlates to the structure of the "X" group and related bond polarity. We can examine this by comparing two closely related carboxylic acids: acetic acid and chloroacetic acid.

A comparison of the acidities of these two compounds shows that chloroacetic acid is a stronger acid than acetic acid. In this case, the chlorine atom is pulling on the electrons in the O-H bond, making the bond more polar. This shifting of electron density in a bond due to the electronegativity of a nearby atom is referred to as the **inductive effect**. By increasing the polarity of the O-H bond, chlorine has made it easier for the proton to be removed. As we will see in the next chapter, the chlorine atom also helps stabilize the carboxylate anion that forms when the proton is removed.

H—C(H)(H)—C(=O)—O—H

Acetic Acid

Cl—C(H)(H)—C(=O)—O—H

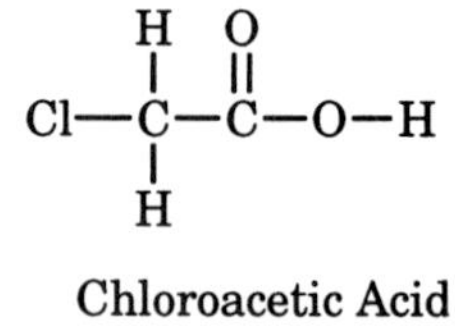

Chloroacetic Acid

FIGURE 12.4: Acetic acid and chloroacetic acid.

As the number of electron withdrawing groups on "X" or the electronegativity of the group(s) on "X" increases, acid strength will also increase. Thus, dichloroacetic acid is a stronger acid than chloroacetic acid, and fluoroacetic acid is stronger than chloroacetic acid.

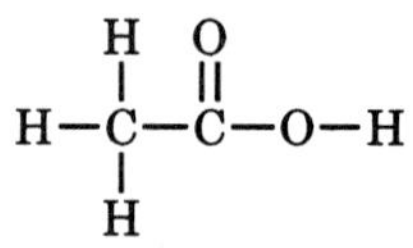

$$^{\delta^-}\text{Cl} \leftarrow \text{CH}_2 \leftarrow \text{C}(=\text{O}) \leftarrow \text{O}-\text{H} \longrightarrow {}^{\delta^-}\text{Cl} \leftarrow \text{CH}_2 \leftarrow \text{C}(=\text{O}) \leftarrow \text{O}^- + \text{H}^+$$

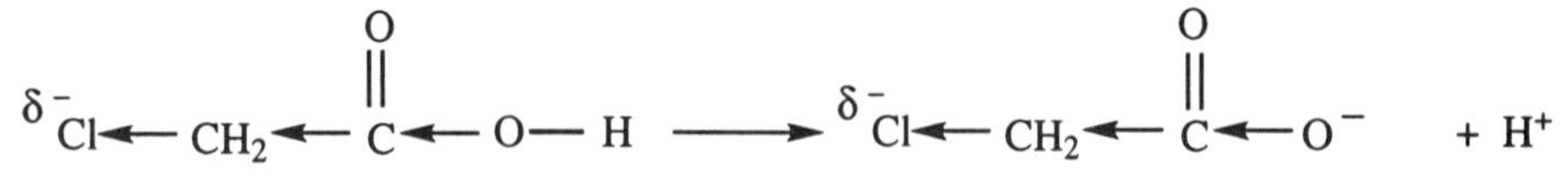

FIGURE 12.5: Inductive effect of chlorine on carboxylic acid strength.

12.6 LEWIS ACID/ BASE THEORY

Brønsted–Lowry theory is extremely useful when talking about the vast majority of acids encountered in everyday life. It also helped to clarify the role of the solvent in determining acid–base behavior. However, for some types of reaction, chiefly those of interest to organic chemists, Brønsted–Lowry theory's reliance of the proton as the chief defining characteristic of an acid or base makes it an unacceptable means of defining an acid or a base. For instance, organic chemists recognize aluminum chloride ($AlCl_3$) as a very acidic material although it is incapable of donating a proton. The American chemist G.N. **Lewis** (1875–1946) put forth a theory of acids and bases that is the most encompassing theory of all. This theory, first proposed in 1923 at about the same time as the Brønsted–Lowry theory, defines an acid as an electron–pair acceptor and a base as an electron–pair donor. An acid–base reaction then occurs when an acid accepts an electron pair from a base and forms a covalent (sometimes called coordinate covalent) bond.

To illustrate the use of Lewis theory to discuss acid–base reactions, let us first consider a number of reactions that we have already seen to be acid–base reactions on the basis of other theories. For instance, consider the reaction between hydrogen ion and hydroxide that forms the basis for Arrhenius theory:

$$H^+ + \overset{-}{:\ddot{\underset{\cdot\cdot}{O}}:}H \longrightarrow H\diagup\ddot{\underset{}{O}}\!\!:\diagdown H$$ EQ. 12.12

From the standpoint of Lewis theory the hydrogen ion is electron deficient, and acts as an electron pair acceptor (*i.e.*, an acid.) The oxygen in the hydroxide ion has several unshared pairs of electrons, which it donates (*i.e*, a base.) The reaction between the hydroxide ion and the proton forms a covalent bond in water, the product.

In the reaction between hydrogen chloride and ammonia, the proton on the hydrogen chloride is electron–deficient because of the strong attraction of the electron pair in the hydrogen – chlorine bond by the very electronegative chlorine. This proton then forms a bond with the nitrogen atom on the ammonia. This nitrogen atom is somewhat electron–rich because of the presence of an unshared pair of electrons. A covalent N-H bond forms as a result of the acid-base reaction.

$$\begin{matrix} H \\ H:\ddot{\underset{\cdot\cdot}{N}}:H \end{matrix} + H:\ddot{\underset{\cdot\cdot}{Cl}}: \longrightarrow \left[\begin{matrix} H \\ H:\ddot{\underset{\cdot\cdot}{N}}:H \\ H \end{matrix}\right]^+ + \left[:\ddot{\underset{\cdot\cdot}{Cl}}:\right]^-$$ EQ. 12.13

Both of the previous examples could just as easily have been understood using other acid–base theories. The real advantage of Lewis theory comes in the consideration of non–protonic acids and bases. For instance, many metal atoms, either as ions or as central atoms in other species, react as acids toward species containing easily accessible electron pairs. Simple examples of this behavior are seen in the reactions of the classic electron–deficient compounds of Group III metals with halide ions, *viz.*:

$$\begin{matrix} :\ddot{Cl}: \\ :\ddot{\underset{\cdot\cdot}{Cl}}:\ddot{Al}:\ddot{\underset{\cdot\cdot}{Cl}}: \end{matrix} + \left[:\ddot{\underset{\cdot\cdot}{Cl}}:\right]^- \longrightarrow \left[\begin{matrix} :\ddot{Cl}: \\ :\ddot{\underset{\cdot\cdot}{Cl}}:\ddot{\underset{\cdot\cdot}{Al}}:\ddot{\underset{\cdot\cdot}{Cl}}: \\ :\underset{\cdot\cdot}{Cl}: \end{matrix}\right]^-$$ EQ. 12.14

Less obvious examples include the interactions of metal ions such as silver (I) ion with the pi electrons of unsaturated molecules such as ethylene, acetylene, and benzene.

Much of the chemistry of organic compounds can be rationalized by applying the Lewis theory. Remember that in some cases, other theories may also be applicable, but the Lewis theory is typically the most general. We will begin to use Lewis acid-base theory for organic compounds here, and we will continue to apply this method throughout the text. As you have probably noticed, many organic compounds (*e.g.*, alcohols, carbonyl containing compounds, amines, thiols, and ethers) contain unshared pairs of electrons that are easily donated. These compounds, then, act as **Lewis bases** in many types of reactions. Many of them, for example, can react with a proton-containing acid to become protonated.

$$C_5H_{10}\ddot{N}H + H^+ \longrightarrow C_5H_{10}\overset{+}{N}H_2 \qquad \text{EQ. 12.15}$$

$$CH_3\ddot{S}H + H^+ \longrightarrow CH_3\overset{+}{S}H_2 \qquad \text{EQ. 12.16}$$

It is probably convenient, at this point, to introduce the "curvy arrow" convention for indicating the movement of electrons. When trying to understand a reaction, it is often helpful to look at bond formation and breakage, and to try to determine where electrons have moved. The "curvy arrow" convention is the chemist's way of keeping track of electron movement. Consider the reaction between ethanol and hydrochloric acid. The unshared pairs of electrons on the oxygen atom allow ethanol to act as a Lewis base. A bond is formed with the proton of hydrochloric acid. The products of this reaction are the oxonium cation and the chloride anion.

$$H_3CH_2C{-}OH + H{-}Cl \longrightarrow Cl^- + H_3CH_2C{-}\overset{+}{O}H_2 \qquad \text{EQ. 12.17}$$

To show this schematically, a "curvy arrow" is drawn showing the unshared pair of electrons moving toward the proton and the electrons that make-up the H-Cl bond moving toward the chlorine. **Notice that a full-headed curvy arrow is drawn to indicate the direction of movement of a pair of electrons.** (Later in the text, we will use half-headed curvy arrows for the motion of a single electron.)

$$H_3CH_2C{-}\ddot{O}H + H{-}Cl \longrightarrow Cl^- + H_3CH_2C{-}\overset{+}{O}H_2 \qquad \text{EQ. 12.18}$$

Compounds containing functional groups with unshared pairs of electrons can also react with non-protonic **Lewis acids**. Consider, for example, the reaction between cyclohexanone and boron trifluoride. The boron atom contains an empty orbital that is capable of accepting a pair of electrons. One of the unshared pairs on oxygen forms a complex with the boron trifluoride.

$$F_3B + O{=}C_6H_{10} \longrightarrow F_3B^{-}{-}O^{+}{=}C_6H_{10}$$ EQ. 12.19

Other types of compounds that have pairs of electrons that are relatively available can also act as Lewis bases. Consider the reaction between 2-butene and hydrogen bromide. As you already know, the H-Br adds across the double bond to give 2-bromobutane as the product. If we look more closely at this reaction, though, we might predict that the first step in the reaction could be an acid/base reaction between the alkene and the H-Br. One of the pairs of electrons that make up the double bond forms a new bond with the proton from H-Br. The first step in this reaction sequence forms the intermediate carbocation and the bromide ion, shown in the equation below.

$$H_3C{-}C(H){=}CHCH_3 + H{-}Br \longrightarrow H_3C{-}C^{+}(H)(H){-}CHCH_3 + Br^-$$ EQ. 12.20

Reaction of the bromide ion with the carbocation results in formation of the final product.

Another example of such a reaction is the hydration of alkenes with acid and water. As you recall, the hydration of alkenes with acid and water gives the Markovnikov addition product. So, for example, the product of the hydration of 2-methyl-2-butene with sulfuric acid and water is 2-methyl-2-butanol (as seen in Eq. 12.21.) Again, the first step is a Lewis acid/base reaction between the alkene and the acid catalyst to form a carbocation. You may have noticed that it is possible to form two different carbocations when 2-methyl-2-butene is protonated. It turns out that the more substituted carbocation (*i.e.*, the positive carbon atom with more R groups attached) is more stable, as shown in Eq 12.22.

$$H_3C{-}C(CH_3){=}CHCH_3 \xrightarrow{H_2SO_4,\ H_2O} H_3C{-}C(CH_3)(OH){-}CH_2CH_3$$ EQ. 12.21

$$H_3C{-}C(CH_3){=}CHCH_3 + H^+ \longrightarrow \underset{\text{more stable}}{H_3C{-}C^{+}(CH_3){-}CH_2CH_3} + \underset{\text{less stable}}{H_3C{-}C(CH_3)(H){-}C^{+}HCH_3}$$ EQ. 12.22

The tertiary carbocation (the one with three R groups around it) is more stable than that secondary carbocation (the one with two R groups around it.) There are several ways of understanding the ordering of stability of carbocations shown in figure 12.6. One is simply the inductive effect. Based upon electronegativity, we know that an alkyl group is better able to donate electron density than a hydrogen atom. A tertiary carbocation can draw electron density from those three carbon atoms to which it is bonded, which in effect helps to distribute and stabilize the positive charge. A carbocation bonded to three other carbons is very able to spread out the charge on the adjacent atoms. The carbocation bonded to only two other carbons is less able to do so. A primary carbocation has only one carbon which may donate electron density, and thus is even less stable. The methyl carbocation has no other carbon atoms bonded to the charged carbon, and thus is least stable. Thus we rationalize the product formed in Eq. 12.21 based upon the relative stability of the two possible carbocations formed in Eq. 12.22. The tertiary carbocation is formed in preference to the secondary carbocation. Ultimately, then, the hydroxyl group ends up attached to the tertiary carbon atom, forming 2-methyl-2-butanol. Markovnikov's rule is, therefore, a consequence of the stability of the reaction intermediates.

$$\oplus CH_2\text{—}H \;>\; \oplus CH_2\text{—}CH_3 \;>\; \oplus CH(CH_3)\text{—}CH_3 \;>\; \oplus C(CH_3)_2\text{—}CH_3$$

FIGURE 12.6: Relative Stability of Carbocations

Now let us consider a new reaction, hydroboration. Hydroboration followed by oxidation is a method of hydrating alkenes with the result being anti-Markovnikov addition. This reaction was discovered in 1956 by H.C. Brown at Purdue University. The first step in this process is, again, a Lewis acid/base reaction. Consider the reaction between 2-methyl-2-butene and borane (BH_3). An incomplete reaction pathway is shown below.

$$CH_3\text{—}C(CH_3)\text{=}CHCH_3 + H\text{—}BH_2 \longrightarrow CH_3\text{—}C(CH_3)(H)\text{—}CH(BH_2)CH_3 \qquad \text{EQ. 12.23}$$

$$CH_3\text{—}C(CH_3)(H)\text{—}CH(BH_2)CH_3 \xrightarrow{[O]} CH_3\text{—}C(CH_3)(H)\text{—}CH(OH)CH_3 \qquad \text{EQ. 12.24}$$

The first step in the process, Eq. 12.23, is the formation of a complex between the alkene and the borane. Remember, the boron atom has an empty p orbital, and is electron deficient. The alkene has a relatively available pair of pi electrons. This pair of electrons then forms a bond with the boron. This results in the formation of a partial positive charge on the tertiary carbon atom. The borane, then, transfers a hydride (H^-) to this tertiary carbon. The end result is a neutral complex between the organic compound and the borane, an alkyl borane. As you can see, removal of the boron by oxidation in Eq. 12.24 results in the formation of 3-methyl-2-butanol. (Note that the reactions in Eq. 12.21 and 12.23 both started with the same alkene. The acid catalyzed hydration via Eq 12.21 resulted in 2-methyl-2-butanol, as predicted by Markovnikov's Rule, while the hydroboration via Eq. 12.23 resulted in 3-methyl-2-butanol, the anti-Markovnikov product.)

Let us finally consider the very strong **organometallic** bases. These compounds consist of an organic portion bonded to some metal atom. Recall that in Chapter 5 we already encountered one type of these compounds, the **Grignard reagents**, under different circumstances. We discussed Grignard reagents as compounds that add to a carbon-oxygen double bonds. These compounds react as they do because the electron density in the carbon-metal bond is not at all symmetrical. In fact, most of the electron density lies toward the carbon atom. It is, therefore, often useful to think of these types of compounds as existing almost in the ionic state. For example we might think of the carbon atom in methyl magnesium bromide as bearing a negative charge and the magnesium atom bearing a positive charge.

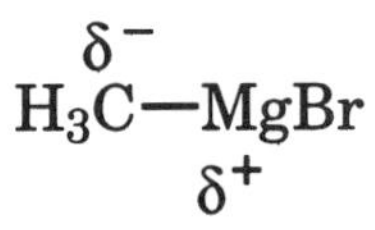

$$\overset{\delta^-}{H_3C}\text{—}\underset{\delta^+}{MgBr}$$

FIGURE 12.7: "Ionic" methyl magnesium bromide.

The "negative" methyl group, in the form of CH_3^-, is certainly capable of donating a pair of electrons and can, therefore, be thought of as a Lewis base. As we will see, these compounds are capable of deprotonating species that the more common bases (*e.g.*, sodium hydroxide) could not. Other examples of such compounds include butyllithium and lithium diisopropylamide.

Li^+ $^-$:N(CH(CH$_3$)$_2$)$_2$ — CH_3, CH, CH_3, CH, H_3C, CH_3

a.

CH_2^- Li^+

b.

FIGURE 12.8: a) lithium diisopropylamide, b) butyllithium.

12.7 SUMMARY

At this point in the discussion, it might be useful to step back to ask the question, "Which of these acid-base theories is the right one?" The best way to answer that question is to say that they are all right. The theory that should be used is the one that most easily explains — sheds the most light on — the particular phenomenon currently under discussion. Organic chemists and some inorganic chemists tend to use Lewis theory most commonly because they often work with substances that can be considered only with the use of this theory. Most of the rest of the chemical world deals more commonly with protonic acids and bases and uses the somewhat simpler Brønsted–Lowry theory because it answers all of their questions without the added complications of thinking of electron–deficient and electron–rich compounds. Even Arrhenius theory still explains many things about acids and bases and is still used in some discussions of acid–base behavior.

KEY WORDS & Concepts

Acid: a class of chemical compounds characterized by their ability to donate protons, especially in aqueous solution.

Amphiprotic: a chemical species capable of either donating or accepting a proton.

Arrhenius: a Swedish chemist who contributed much to our knowledge of electrolytic dissociation and the nature of acids and bases.

Base: a class of chemical compounds characterized by their ability to accept protons, especially in aqueous solution.

Brønsted: a Danish chemist who contributed a theory of acid–base behavior.

Conjugate acid–base pairs: two compounds that differ by a proton.

Democritus: a Greek philosopher credited with devising an early atomic theory of matter.

Grignard reagent: an organomagnesium halide reagent important in organic synthesis.

Hydronium ion: the H_3O^+ ion.

Hydroxide ion: the OH^- ion.

Lewis: an American chemist who contributed a theory of acid–base behavior.

Lewis Acid: a compound that behaves as an electron-pair acceptor.

Lewis Base: a compound that behaves as an electron-pair donor.

Lowry: an English chemist who contributed a theory of acid–base behavior.

Lyate ion: the negative ion formed by the solvent.

Lyonium ion: the positive ion formed by the solvent.

Organometallic: a word referring to a compound in which there is a carbon–metal bond.

Strong acid: an acid which is totally ionized in solution.

Weak acid: an acid which is only partially ionized in solution.

HOMEWORK Problems

1. Write a chemical equation showing the ionization of each of the following as an Arrhenius acid:

(a) HNO_3
(b) $HC_2H_3O_2$
(c) $HClO_4$
(d) HSO_4^-

2. Write a chemical equation showing the ionization of each of the following as an Arrhenius base:

(a) KOH
(b) $Ca(OH)_2$
(c) CH_3NH_2
(d) Na_2O

3. For each of the following chemical equations, identify the lyonium and lyate ions:

(a) $2\ H_2O \rightarrow H_3O^+ + OH^-$
(b) $2\ NH_3 \rightarrow NH_4^+ + NH_2^-$
(c) $N_2O_5 \rightarrow NO_2^+ + NO_3^-$

4. For each of the reactions in problem 3, give the formula and name of a chemical which would act as a base when dissolved in that solvent (assuming that the species did dissolve).

5. Write a chemical equation showing the reaction of each of the following as a Lowry–Brønsted acid:

(a) H_2SO_4
(b) NH_4^+
(c) $HC_2H_3O_2$
(d) OH^-

6. Write a chemical equation showing the reaction of each of the following as a Lowry–Brønsted base:

(a) HSO_4^-
(b) NH_3
(c) $C_2H_3O_2^-$
(d) OH^-

7. Provide the name and formula of the conjugate acid of each of the following:

(a) Cl^-
(b) HSO_4^-
(c) $HC_2H_3O_2$
(d) NH_3

8. Provide the name and formula of the conjugate base of each of the following:

(a) $HClO_4$
(b) HSO_4^-
(c) $HC_2H_3O_2$
(d) NH_3

9. Identify the conjugate acid–base pairs in each chemical equation:

(a) $HNO_3 + H_2O \longrightarrow H_3O^+ + NO_3^-$
(b) $HNO_3 + NH_3 \longrightarrow NH_4^+ + NO_3^-$
(c) $H_3O^+ + C_2H_5O^- \longrightarrow C_2H_5OH + H_2O$
(d) $H_3O^+ + NH_2^- \longrightarrow NH_3 + H_2O$

10. Arrange the following sets of acids from strongest to weakest:

(a) H_2Se, H_2Te, H_2S, H_2O
(b) $HClO_2$, $HClO$, $HClO_4$, $HClO_3$
(c) NH_3, H_2S, HBr
(d) CH_2FCOOH, CH_3COOH, CF_3COOH, CHF_2COOH

11. Classify the following as Lewis acids or Lewis bases:

(a) CH_3COCH_3
(b) GaH_3
(c) Fe^{3+}
(d) $CH_3COCH_2COCH_3$

12. Write a Lewis Dot depiction of the reaction between boron trifluoride and diethyl ether that makes clear the reaction is a Lewis Acid–Base reaction.

13. Using curvy arrows, draw the reaction you would expect to occur between the following pairs of molecules. For each reaction, indicate which starting material behaves as a Lewis Acid and which behaves as a Lewis Base.

(a) Methanol and HBr
(b) Acetone and HCl
(c) Acetone and BF_3
(d) Cyclohexene and HBr

OVERVIEW Problems

14. When 1.50 g of chlorine heptoxide (Cl_2O_7, MW = 182.90) reacts with sufficient water, 25.0 ml of aqueous solution of an acid are formed.

(a) Write the balanced chemical reaction between chlorine heptoxide and water.

(b) What concentration acid (in moles of H^+ per liter of solution) is formed?

(c) If 0.713 g of Cl_2O (MW = 86.905) were reacted to form 25.0 ml of aqueous solution, would its hydrogen ion concentration be greater than or less than that of the solution described earlier in this problem? Explain.

15. When 5.00 g of cesium oxide (Cs_2O, FW = 281.81) reacts with sufficient water, 50.0 ml of aqueous solution of base are formed.

(a) Write the balanced chemical reaction between cesium oxide and water.

(b) What concentration base (in moles of OH^- per liter of solution) is formed?

(c) If 5.00 g of Cs_2O_2 were reacted to form 50.0 ml of aqueous solu tion, would its hydroxide ion concentration be greater than or less than that of the solution described earlier in this problem? Explain.

16. Acid anhydrides have the general formula $R{-}\overset{O}{\overset{\|}{C}}{-}O{-}\overset{O}{\overset{\|}{C}}{-}R'$. Acid anhydrides react with water as shown below. When 10.0 g of acetic anhydride ($(CH_3CO)_2O$, MW = 102.09) react with excess water, 100.0 ml of aqueous solution result.

$$(R{-}CO)(R'{-}CO)O + H_2O \longrightarrow R{-}COOH + R'{-}COOH$$

(a) Write the balanced reaction between acetic anhydride and water.

(b) What concentration, in terms of moles of acetic acid per liter of solution is formed?

(c) If the acetic acid thus formed were to be completely reacted with solid NaOH, what mass of sodium hydroxide solid would be needed?

17. Acid chlorides have the general formula $R{-}C({=}O){-}Cl$

Acid chlorides react with water as shown below. When 5.00 g of acetyl chloride (CH_3COCl, MW = 78.50) react with excess water, 50.0 ml of aqueous solution result.

$$R{-}C({=}O){-}Cl + H_2O \longrightarrow R{-}C({=}O){-}OH + HCl$$

(a) Write the balanced reaction between acetyl chloride and water.
(b) What concentration, in terms of moles of acetic acid per liter of solution is formed?
(c) What mass of sodium hydroxide solid would be needed to neutral ize the solution formed in this reaction?

18. Hydoiodic acid (HI) and periodic acid (HIO_4) are both strong acids. Of the two, only periodic acid oxidizes organic molecules, such as vicinal dialcohols.

(a) Explain the relative oxidizing strengths of HI and HIO_4.
(b) Periodic acid reacts with 2,3-butanediol to form iodic acid and acetaldehyde.
Write a balanced redox reaction, using the half-reaction method.
(c) Periodic acid will also react with a carbonyl compound that has an alcohol group immediately adjacent to form an aldehyde, a carboxylic acid, and iodic acid. Write the balanced reaction between periodic acid and 3-hydroxy-2-butanone.

19. Sulfuric acid (H_2SO_4) has two ionizable protons, and is classified as a strong acid. Is HSO_4- a strong or a weak acid? *i.e.*, what is the difference, if any, in removing the second proton compared to the first proton?

20. Just as inorganic acids can have more than one ionizable proton, there are organic molecules that have more than one ionizable proton (e.g., di- or tri-carboxylic acids.) Oxalic acid (ethanedioc acid) is one such example.

(a) Write the structure of oxalic acid.
(b) Write the Lewis dot diagram of the conjugate base of oxalic acid.
(c) Oxalic acid undergoes a decomposition reaction when heated mildly. One product is formic acid. Write the balanced decomposi tion reaction.

21. The first proton ionizes from oxalic acid (ethanedioc acid) far more readily than the second ionizable proton ($pK_1 = 1.2$, $pK_2 = 4.2$.) In adipic acid (1,6-hexanedioc acid) the first and second protons ionize approximately equally ($pK_1 = 4.4$, $pK_2 = 4.4$.) Explain.

22. A popular medication is known for its "plop-plop, fizz-fizz" when added to water. This is mostly the result of an acid-base reaction between citric acid (2 hydroxy-1,2,3-propanetricarboxylic acid, FW = 192.12) and sodium bicarbonate ($NaHCO_3$, FW = 84.00).

(a) Draw the structure of citric acid.
(b) How many unique proton environments would be seen in an HNMR spectrum of citric acid?
(c) How many unique carbon environments would be seen in a CNMR spectrum of citric acid?
(d) Write the balanced reaction between citric acid and sodium bicar bonate to produce water and citrate anion.
(e) What mass of sodium bicarbonate is required to exactly neutralize 2.50 g of citric acid according to the reaction in (d)?

23. One method of synthesizing carboxylic acids involves the hydrolysis of nitriles.

(a) Write the balanced reaction that hydrolyzes acetonitrile to form acetic acid and ammonia.
(b) What hybrid orbitals were used by the carbon atom attached to nitrogen in acetonitrile?
(c) What hybrid orbitals are used by this same carbon after hydrolysis?
(d) What hybrid orbitals are used by nitrogen in acetonitrile?
(e) What hybrid orbitals are used by nitrogen in ammonia?

24. Another method for synthesizing carboxylic acids is by oxidation of the aldehyde. This can be done by silver ion (as in the Tollen's test, which yields a silver mirror in a positive test for aldehydes.) First the aldehyde is oxidized to the carboxylate anion, while the Ag^+ is reduced to the metal under basic conditions. Subsequent treatment with strong acid (HCl) converts the carboxylate anion into the carboxylic acid.

(a) Write the structure of propanal.
(b) Write the structure of propanoate, the conjugate base of propanoic acid.
(c) Write the balanced redox reaction by which silver ion and propanal are converted to propanoate anion and silver metal in basic solution.
(d) Draw the Lewis dot structure of propanoate anion.
(e) Using the curvy arrow convention, show the reaction between propanoate and hydrogen chloride.

25. Yet another method to synthesize carboxylic acids involves oxidation of the alkyl side chain on an aromatic ring. Toluene (methylbenzene, $C_6H_5CH_3$, MW = 92.13) reacts with MnO_4^- in acidic solution to form benzoic acid (C_6H_5COOH, MW = 122.12) and Mn^{2+}.

(a) Balance the redox reaction between MnO_4^- and $C_6H_5CH_3$ in acid solution.
(b) If 10.0 g of $KMnO_4$ (FW = 158.03) react with 5.00 g of toluene, what is the maximum amount of benzoic acid that could form? Note: you must do a limiting reagent calculation.

Resonance and Aromaticity

RESONANCE AND Aromaticity

13.1 RESONANCE: WHEN THEORY AND EXPERIMENT DISAGREE

FORMAL CHARGE

Let us begin our discussion of resonance by looking again at the structure of the nitrate anion. You saw two representations of the nitrate anion in Chapter 7, and these are repeated in Figure 13.1.

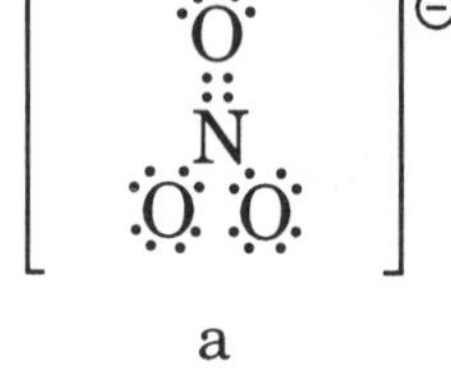

a

b

FIGURE 13.1: a) electron dot diagram of NO_3^- and b) molecular structure of nitrate ion.

These two representations both show an overall -1 charge on the entire species, but neither indicates where the charge(s) are localized within the nitrate anion. It is often useful to look more carefully at the origin of charge in chemical species because such knowledge will allow us to predict how certain reactions will occur. In order to see where the overall charge on the nitrate ion originates, we need to calculate the formal charge for each atom in this species. Formal charge is defined as the charge on each atom in a molecule or polyatomic ion assuming that bonding pairs of electrons are shared equally between atoms. The sum of the formal charges of each atom in the species will equal the overall charge. The formal charge of an individual atom can be calculated by the following formula.

Formal Charge = (number of valence electrons) – (number of bonds) – (number of unshared electrons) EQ. 13.1

Let us calculate the formal charge on each atom in the nitrate anion. First, the central nitrogen atom has 5 valence electrons, 4 bonds and no unshared electrons. Application of the formula shows that nitrogen has a +1 formal charge.

Formal Charge on Nitrogen = 5-4-0 = +1 EQ. 13.2

According to Figure 13.1, nitrate contains two different types of oxygen atoms; an oxygen atom that forms a double bond with nitrogen and two oxygen atoms that form single bonds with nitrogen. The formal charge on the doubly bonded oxygen atoms is 0, while the formal charge on the singly bonded oxygen atoms is -1. The formal charge for the oxygen doubly bonded to nitrogen is calculated as follows.

Formal Charge on Double Bonded Oxygen = 6-2-4 = 0 EQ. 13.3

The formal charge on the singly bonded oxygen atoms is calculated in the same manner.

Formal Charge on Single Bonded Oxygen Atoms = 6-1-6 = -1 EQ. 13.4

Thus the overall -1 charge is determined by the combination of one nitrogen atom with a +1 charge and two oxygen atoms that have -1 charges. A more complete representation of the nitrate ion is provided in Figure 13.2

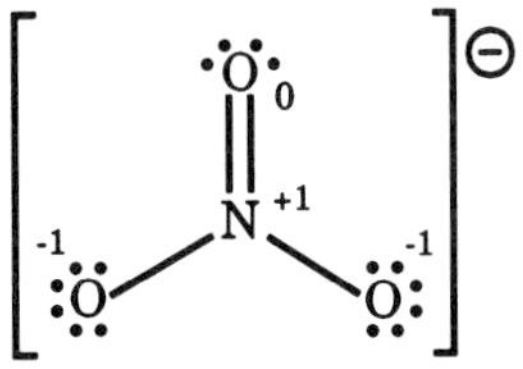

FIGURE 13.2: The nitrate ion including formal charges.

RESONANCE

According to Figure 13.2, we should expect the structure of the nitrate anion to contain two nitrogen-oxygen single bonds and one nitrogen-oxygen double bond. A double bond should be somewhat shorter than a single bond. Experimental measurement of the N—O bond lengths in the nitrate ion show that all three oxygen atoms are 122 pm from the central nitrogen atom. Thus we know that Figure 13.2 is not an accurate representation of the structure of the nitrate anion.

For each of the bonds to be the same length, each bond must be a hybrid between a single bond and a double bond. This phenomenon is referred to as **resonance** and is typically represented in the following fashion:

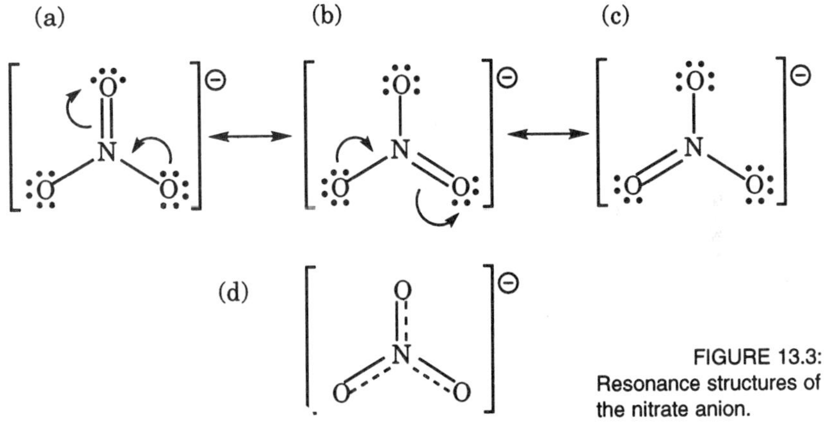

FIGURE 13.3: Resonance structures of the nitrate anion.

Figure 13.3 (a), (b) and (c) show three equivalent Lewis structures, differing in which oxygen is doubly bonded to the nitrogen. Note that in structures (a), (b) and (c), the atoms do not move, but pairs of electrons move. The double-headed single arrow between structures (a) and (b), and between (b) and (c) is used exclusively to show a correlation between equivalent Lewis structures. Such structures are known as **resonance forms**. At this point it is important to emphasize that none of the three resonance forms represent a real structure of the nitrate anion. Specifically, it is *not* the case that nitrate has structure (a) part of the time, structure (b) part of the time, and structure (c) the rest of the time. Rather, the real structure of nitrate is a hybrid of all of the resonance forms. Plant or animal hybrids, for example, have their own unique, consistent characteristics; they do not show characteristics of one parent part of the time, the other parent the rest of the time. The chemist's attempt to represent this phenomenon is shown in Figure 13.3(d), the overall resonance hybrid, where all the N—O bonds are intermediate in length between single and double bonds. It is as if the extra bond is evenly distributed, simultaneously among all three oxygen atoms. In this case, each dashed bond is essentially a third of a bond. Applying Eq. 13.1 to the structure shown in Figure 13.3(d) shows that the formal charge on each of the three oxygen atoms is -1/3. It is noteworthy that resonance results in the delocalization of charge throughout the species. Also note that the overall resonance hybrid is of lower energy (and thus is more stable) than any individual resonance form.

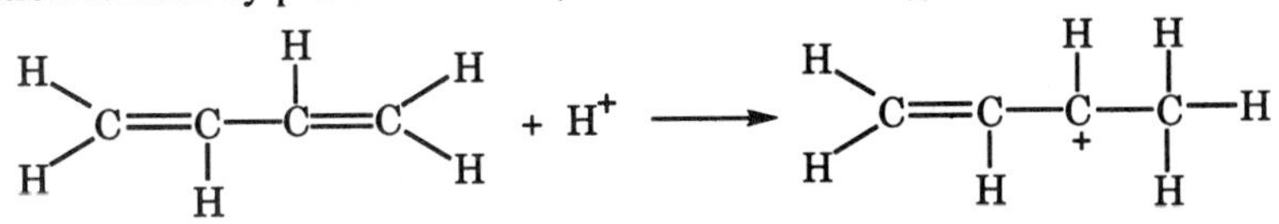

FIGURE 13.4: Formation of the allylic cation by protonation of 1,3-butadiene.

Let us examine another example, the carbocation formed by protonation of 1,3-butadiene.

This species is known as a carbocation because the positive charge resides on a carbon atom. This carbocation is termed allylic because the carbocation is one removed from the double bond, the so-called allylic position. As you might expect, and as experiment shows, this is not the only representation of this species. The following resonance structures can be drawn for this allylic carbocation.

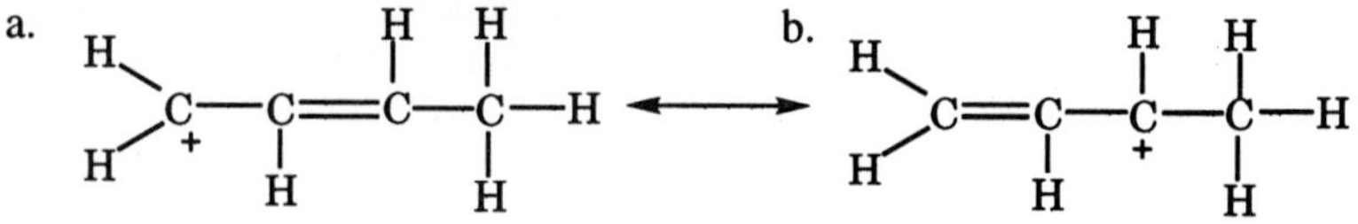

According to the two resonance forms in Figure 13.5 (a) and (b), the positive charge can occur on either the first or the third carbon atom. The resonance hybrid in Figure 13.6 is drawn to highlight this delocalization of charge. As we will see later, the concept of resonance is often very useful when attempting to rationalize reaction pathways.

FIGURE 13.5: Resonance forms of the allylic carbocation.

Resonance and Rules

As you probably noticed in the two examples above, there are systematic ways of going about drawing alternate structures that contribute to resonance. Our first two examples highlight a few of the rules to keep in mind when drawing resonance structures.

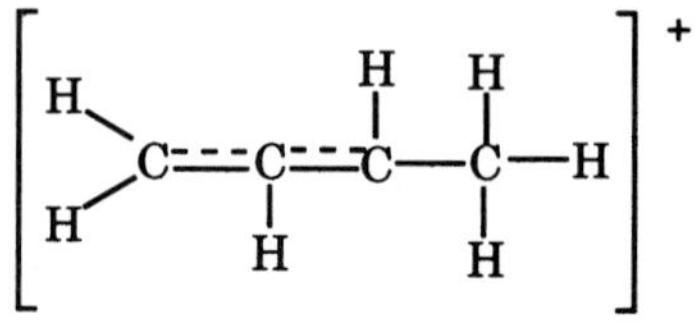

FIGURE 13.6: Resonance hybrid of the allylic carbocation.

1. Do not move atomic nuclei when drawing resonance forms; all resonance forms have the same connectivity.

2. Move only electrons involved in multiple bonds or nonbonding electrons. Electrons that form sigma bonds cannot move.

Let us look at two more examples to illustrate some other rules. First, let us consider the resonance forms of the acetate ion that are shown in Figure 13.7.

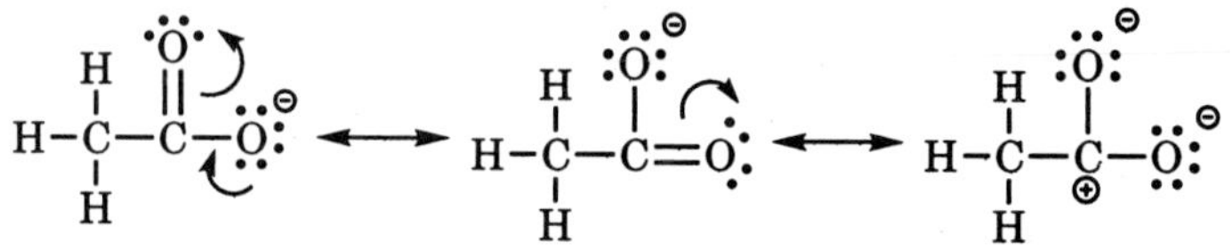

As you can see, two of the forms are identical, but the third form has only six electrons around the carboxyl carbon. In addition, this form also introduces additional charge into the species. Both of these conditions make the third form "less stable" that the other two forms. So we can add the following rules to our list.

FIGURE 13.7: Resonance forms of the acetate ion.

3. Resonance forms that contain a separation of charge contribute less to the overall resonance hybrid than those forms that do not contain a separation of charge.

4. Resonance forms in which all atoms are not surrounded by an octet of electrons contribute less to the overall resonance hybrid than those forms in which all atoms are surrounded by an octet of electrons.

FIGURE 13.8: Resonance forms of acetone.

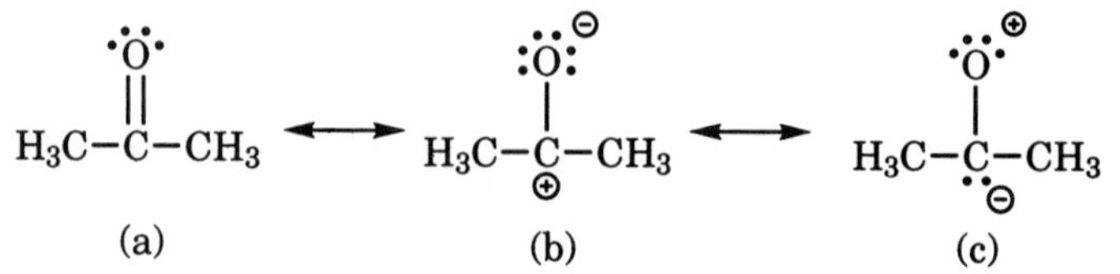

For our last example, let us consider the following resonance forms of acetone in Figure 13.8.

The neutral form (a) with no separation of charge contributes most to the overall resonance hybrid. In order to understand the chemical reactivity of ketones, however, we must decide which of the two other forms (b or c) contributes more to the overall resonance hybrid than the other. Note that the form (b) puts the negative charge on oxygen and the positive charge on carbon. Since oxygen is more electronegative than carbon, we hope that you would guess that form (b) contributes more than form (c) to the overall resonance hybrid. This leads us to the final rule.

> **5.** In cases where a negative charge can be placed on two different atoms, the resonance form that contributes more to the overall resonance hybrid will be the one in which the negative charge resides on the more electronegative atom.

13.2 RESONANCE EFFECTS ON ACIDITY

Consider the following three compounds.

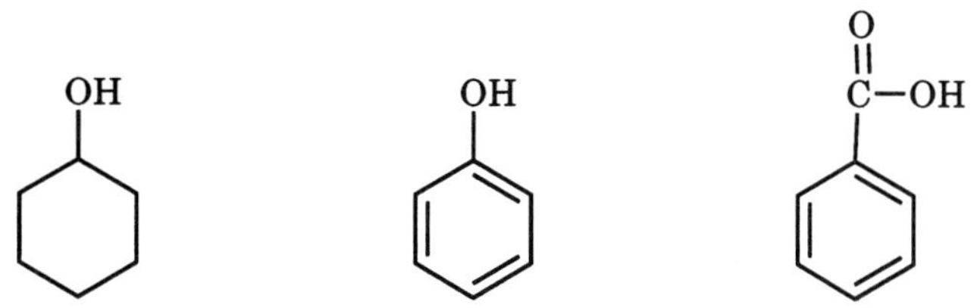

OH | OH | O ‖ C—OH

cyclohexanol $pK_a = 16$ | phenol $pK_a = 9.99$ | benzoic acid $pK_a = 4.20$

FIGURE 13.9: pK_a values of cyclohexanol, phenol and benzoic acid.

As you saw in Chapter 12, acids typically differ in their strength. The strength of an acid can be quantified in terms of its pKa value; the smaller the pKa, the stronger the acid. This allows us to compare the acidity of a series of compounds. So, for the three compounds shown in Figure 13.9, we see that benzoic acid (pKa = 4.20) is the strongest acid, phenol (pKa = 9.99) lies in the middle and cyclohexanol (pKa ≈ 16) is the weakest acid. We can rationalize this ranking by invoking resonance arguments.

Consider the alkoxide ion formed by the dissociation of cyclohexanol.

OH ⇌ O^- + H^+

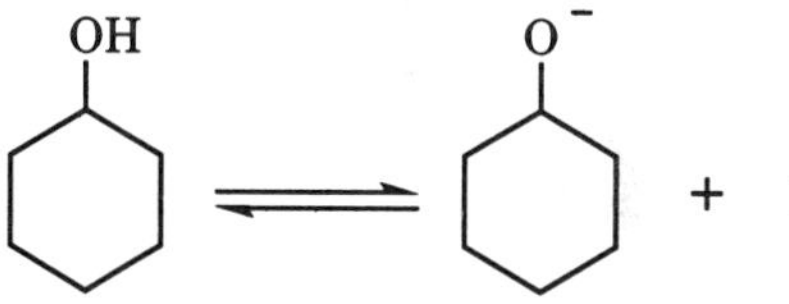

Alcohol Alkoxide

FIGURE 13.10: The dissociation of cyclohexanol.

Note that the double arrows shown above are used to indicate an interconversion between two distinct species, both of which are physically present in a definite proportion. This differs significantly from the meaning of the double headed arrows first introduced in Figure 13.3, to show a relationship between resonance forms. There are no resonance forms that can be drawn for the alkoxide ion in Figure 13.10, and we say that the ion is not resonance stabilized.

In contrast, consider the phenoxide ion formed by the dissociation of phenol in Figure 13.11.

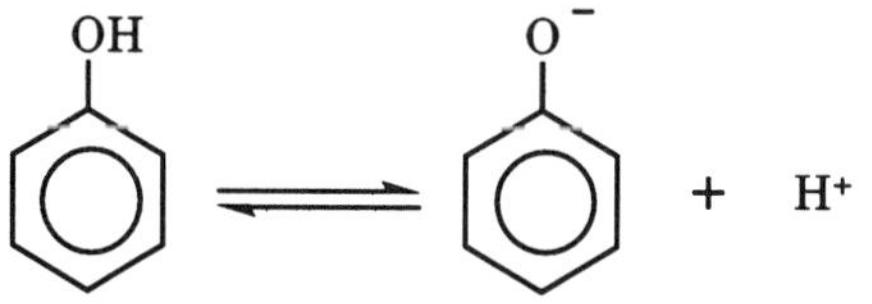

OH ⇌ O^- + H^+

Phenol Phenoxide

FIGURE 13.11: The dissociation of phenol.

The phenoxide ion has resonance stabilization, as demonstrated by the resonance forms seen in Figure 13.12. Several of these resonance forms, however, place a negative charge on carbon. These forms contribute less than those that place the negative charge on oxygen. Thus the two major contributors are only the first two resonance forms, and there is little delocalization of charge.

Finally, consider the dissociation of benzoic acid, which yields the carboxylate ion seen in Figure 13.13.

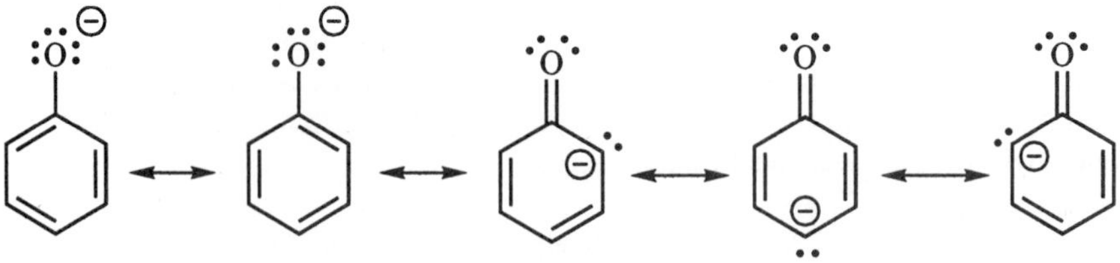

FIGURE 13.12: Resonance forms of the phenoxide ion.

In this case the resulting carboxylate ion is more resonance stabilized than the phenoxide ion because the negative charge can be delocalized between two oxygen atoms (Figure 13.14)

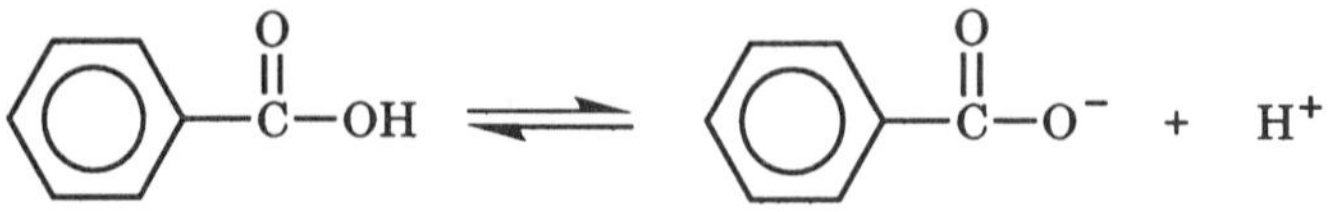

The extent to which a reaction proceeds towards products is determined by the relative energy levels of reactants and products. A reaction that has products at a lower energy level is generally favored over a reaction that has products at a higher energy level. For example, the ionization reaction of benzoic acid seen in Figure 13.13 forms an anion that is at a low energy due to resonance stabilization. By forming an anion (the conjugate base) that is more resonance stabilized than the comparable anions in Figures 13.10 and 13.11, benzoic acid dissociates to a greater extent than phenol or cyclohexanol. Thus one way to rationalize the order of acidity for a series of weak acids is to invoke resonance stabilization of the anion that is the conjugate base formed by the ionization reaction.

Benzoic Acid ⇌ Carboxylate Ion + H^+

FIGURE 13.13: The dissociation of benzoic acid.

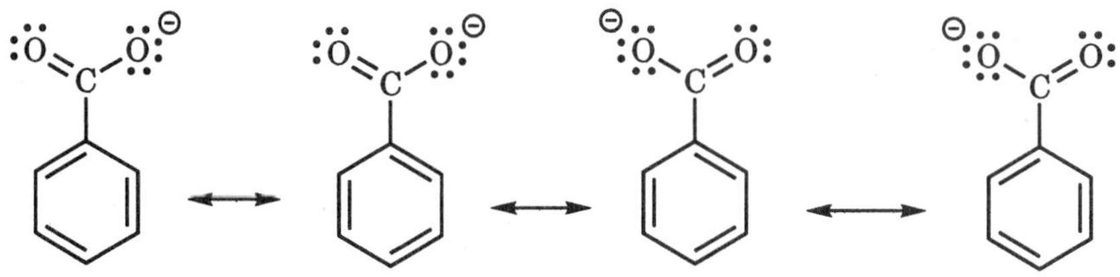

FIGURE 13.14: Resonance forms of the benzoate ion.

13.3 BENZENE

HISTORICAL BACKGROUND

Benzene was first isolated in 1825, by Michael Faraday (1791-1867), and was determined to have the formula C_6H_6. The index of hydrogen deficiency of the compound is 4 and possible structures that satisfy the formula are provided in Figure 13.15.

FIGURE 13.15: Possible structures of C_6H_6.

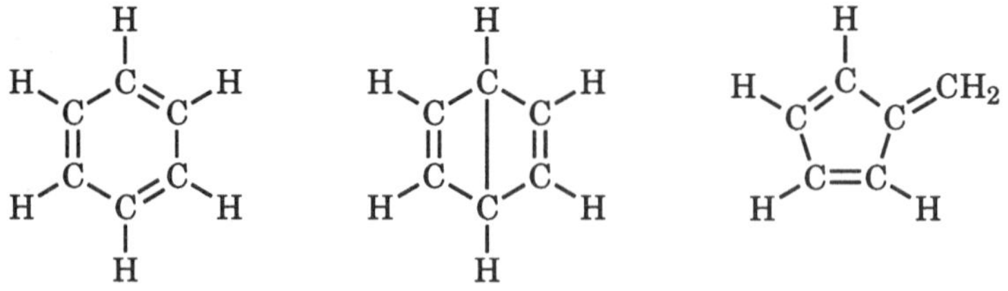

(a) Kekulé formula

(b) "Dewar" formula

(c)

$H_3C-C{\equiv}C-C{\equiv}C-CH_3$

(d)

$H_2C{=}CH\text{-}C{\equiv}C\text{-}CH{=}CH_2$

(e)

The actual structure of benzene eluded chemists for many years. It was in 1865 that August Kekulé (1829-1896) proposed the structure of benzene that is closest to the structure we accept today (Figure 13.16) after dreaming by his fireplace.

> *There I sat and wrote for my textbook; but things did not go well; my mind was occupied with other matters. I turned the chair toward the fireplace and began*

to doze. Once again the atoms danced before my eyes. This time smaller groups modestly remained in the background. My mental eye, sharpened by repeated apparition of similar kind, now distinguished larger units of various shapes. Long rows, frequently joined more densely; everything in motion, twisting and turning like snakes. And behold, what was that? One of the snakes caught hold of its own tail and mockingly whirled round before my eyes. I awoke, as if by lightening; this time, too. I spent the rest of the night working out the consequences of this hypothesis. (From K. Hafner, Angew. Chem. Internat. ed. Engl. 18, 641-651, 1979)

Kekulé proposed two co-existing forms of cyclohexatriene, shown in Figure 13.16.

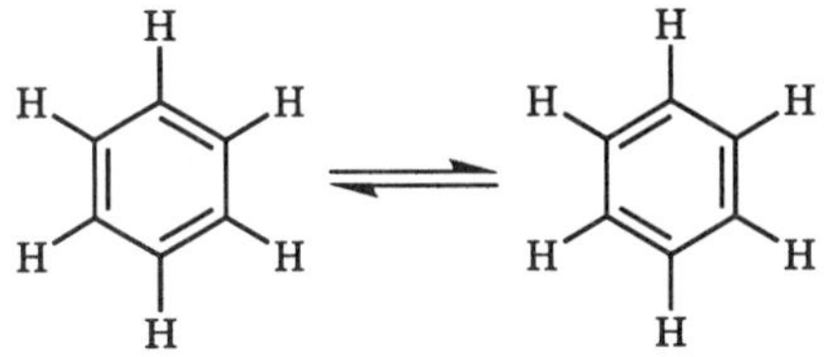

FIGURE 13.16: The Kekulé structures of benzene.

As we shall soon see, the structures of benzene shown in Figure 13.16 have to be modified to account for additional experimental results.

A More Modern View

Since Kekulé first postulated that benzene had the structure of a cyclohexatriene, additional information has been determined. First, we find that monobromination of benzene affords only one compound. That tells us that all of the hydrogen atoms in benzene are equivalent. The Kekulé structures satisfy this requirement. But we also know that addition of another bromine forms exactly three isomeric products that are shown in Figure 13.17.

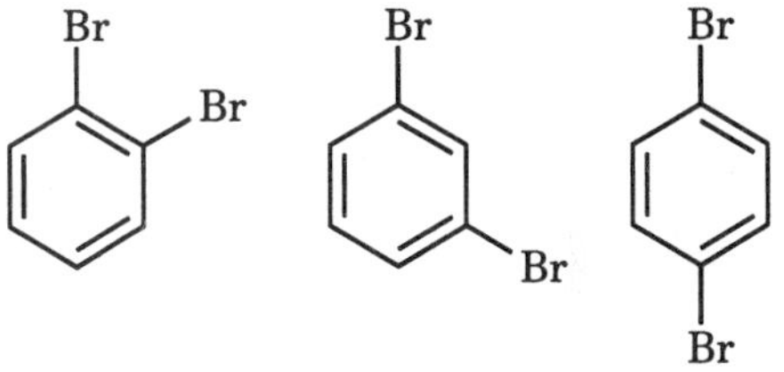

FIGURE 13.17: The three dibromobenzenes.

If the Kekulé structures were correct, that is if benzene was cyclohexatriene, then the 1,2-isomer would be slightly different than the 1,6-isomer. These two forms are shown in Figure 13.18. In one case there would be a double bond between the two carbons that are also bonded to bromine atoms, and in the other case there would be a single bond between these two carbons.

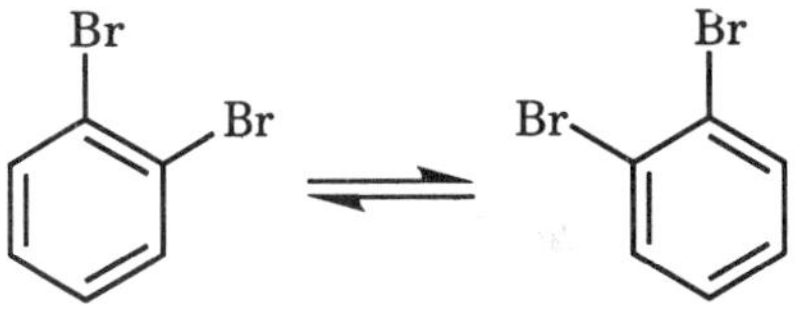

The Kekulé structures for benzene predict four, not three, isomers of dibromobenzene. Since we know that only three dibromobenzene isomers are formed, benzene must differ from the Kekulé structures. As you have probably guessed, resonance is the answer. Benzene is composed of carbon-carbon bonds that are neither single bonds nor double bonds, but are a hybrid between the two. Bond length data are additional evidence for this hypothesis. The carbon-carbon bond length in benzene is 139 pm, a hybrid between the typical carbon-carbon single bond (154 pm) and the typical carbon-carbon double bond (134 pm).

FIGURE 13.18: Kekulé structures for 1,2- and 1,6-dibromobenzene.

Thus 1,2-dibromobenzene may be drawn in any of the three ways shown in Figure 13.13. The first two representations are resonance forms using Kekulé structures. The third is a representation of a hybrid between the resonance forms; the circle inside the hexagon represents the delocalization of electrons that is characteristic of resonance.

FIGURE 13.19: Various representations of 1,2-dibromobenzene.

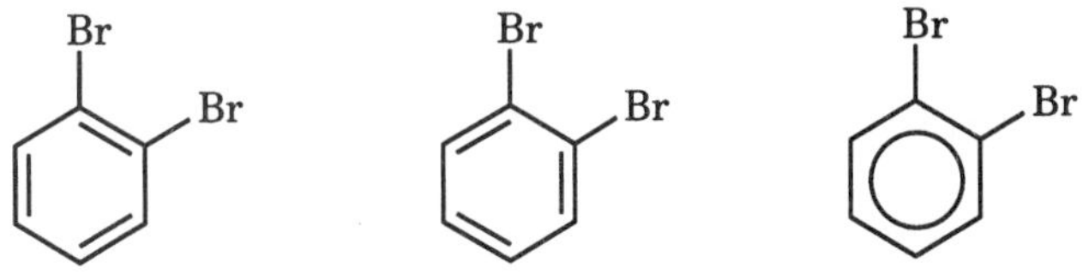

Benzene: An Orbital Picture

Benzene is a planar six-membered ring composed of six sp^2 hybridized carbon atoms with carbon-carbon bond angles of 120°. Each carbon forms a sigma bond to hydrogen by overlap of one of the sp^2 orbitals and an s orbital from the hydrogen. Carbon-carbon sigma bonds are formed by the overlap of two sp^2 orbitals from adjacent carbons. The remaining unhybridized p orbital on each carbon extends above and below the plane of the ring. This creates a pi-cloud above and below the plane of the ring. Thus the pi electrons, shown as localized in the Kekulé structures of Figure 13.16, can circulate freely around the ring and do not reside on any one carbon atom; we say the pi electrons are delocalized.

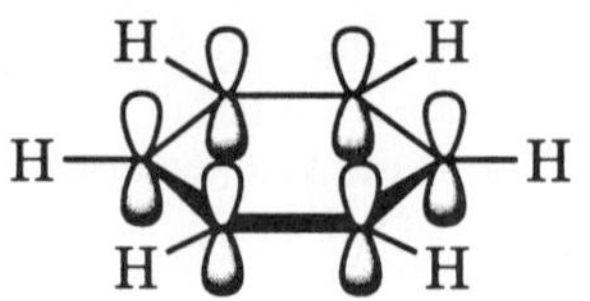

pi electron density

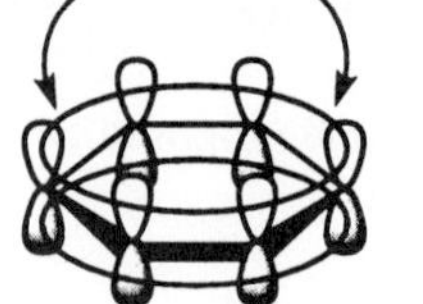

FIGURE 13.20: Orbital representations of benzene.

Benzene: A Bit More on Nomenclature

You have already seen the common names for several benzene derivatives in Chapter 3, and others are provided below. You should commit these names to memory since they are typically not named using IUPAC rules.

TABLE 13.1

Table 13.1: COMMON NAMES OF SOME MONOSUBSTITUTED BENZENES

Substituent	Name
NO_2	Nitrobenzene
CH_2CH_3	Ethylbenzene
$CH(CH_3)CH_2CH_3$	*sec* - Butylbenzene
$C(CH_3)_3$	*tert* -Butylbenzene
OCH_3	Anisole
$C(=O)H$	Benzaldehyde
$HC{=}CH_2$	Styrene
$CH(CH_3)_2$	Isopropylbenzene

As we have seen in other chapters, IUPAC and common nomenclature are often combined and it is useful to be familiar with both systems of nomenclature for a variety of compounds. Compounds derived from benzene are no exception; here again, we encounter this same mix. Earlier we saw the three isomeric dibromobenzenes: 1,2-dibromobenzene, 1,3-dibromobenzne and 1,4-dibromobenzene. We named them according to IUPAC nomenclature rules. These same compounds can also be named using a common system of nomenclature. The 1,2-isomer can also be called **ortho**-dibromobenzene (*o*-dibromobenzne), the 1,3-isomer can also be called **meta**-dibromobenzene (*m*-dibromobenzene) and the 1,4-isomer can also be called **para**-dibromobenzene (*p*-dibromobenzene).

In cases where the two substituents are different, they are arranged in alphabetical order. As with alkane nomenclature, the substituents are numbered so that the smallest possible numbers are used. It is also permissible to use common names and common prefixes when naming these types of compounds. A few examples are provided in Figure 13.22.

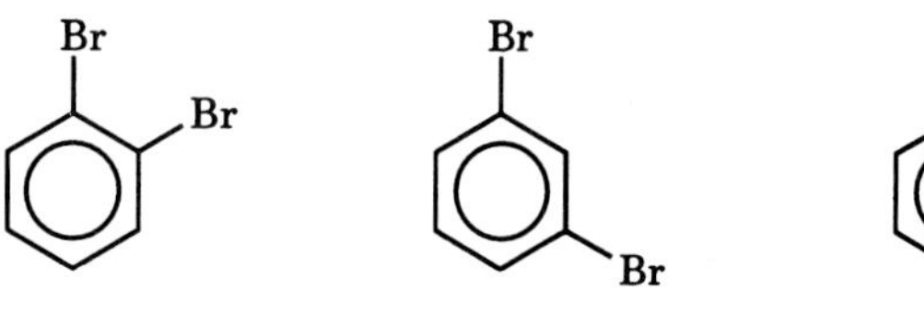

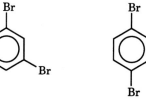

1,2-dibromobenzene — *ortho* - dibromobenzene

1,3-dibromobenzene — *meta* - dibromobenzene

1,4-dibromobenzene — *para* - dibromobenzene

FIGURE 13.21: The three isomeric dibromobenzenes.

para - xylene | 2-nitrophenol | *ortho* - cresol | *meta*-toluidine | 3-nitroaniline

FIGURE 13.22: Names of some disubstituted benzenes.

13.4 AROMATICITY AND AROMATIC COMPOUNDS

What Is Aromaticity?

Benzene and many related compounds are classified as aromatic compounds. While many aromatic compounds smell good, this is not where the term aromatic originates. A compound is considered aromatic only if it satisfies certain criteria. These criteria are as follows.

1. The compound must be cyclic.
2. The compound must be planar.
3. The compound must contain an uninterrupted cloud of pi-electrons above and below the plane of the ring.
4. The atoms in the ring must be sp^2 (or rarely, sp) hybridized.
5. The compound must contain 4n+2 pi-electrons, where n = any integer

Let us go back and look at benzene in light of these criteria. We have established that benzene is a planar, cyclic compound formed by six sp^2 hybridized carbon atoms. The orbital picture of benzene, shown in Figure 13.20, shows an uninterrupted cloud of pi-electrons. The only criterion that remains is that benzene must contain 4n+2 pi-electrons. Benzene contains six pi-electrons and so this criterion is also satisfied; in this case n = 1.

Monocyclic Aromatic Species

Certain **heterocyclic** compounds are also aromatic. Heterocyclic compounds are compounds that are cyclic, but also contain an atom or atoms other than carbon in the ring. Two examples of aromatic heterocyles are provided in Figure 13.23.

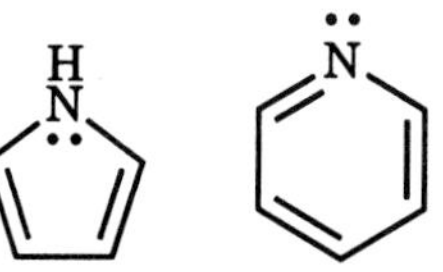

Pyrrole   Pyridine

FIGURE 13.23: Two heterocyclic aromatic compounds.

Pyridine looks very much like benzene, except that the ring now contains a nitrogen atom. Pyridine is planar and all the atoms in the ring, including nitrogen, are sp^2 hydridized. Finally, if we count the pi-electrons in pyridine we see that there are six; thus, the 4n+2 rule is met, with n = 1. The lone pair of electrons on nitrogen occupies an sp^2 orbital outside of the ring and is thus not included in the count of pi-electrons. Pyrrole also contains six pi-electrons. Four of the pi-electrons reside in the p-orbitals that comprise the two double bonds and the remaining two reside in a p-orbital on the nitrogen atom.

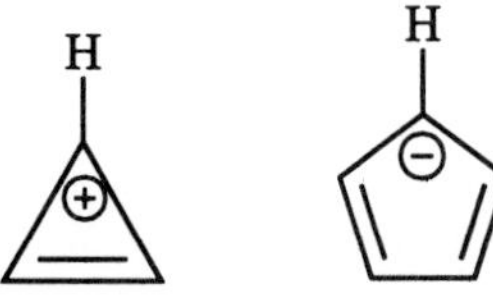

FIGURE 13.24: Charged hydrocarbon based aromatic species.

Up to this point we have looked at aromatic species that are neutral. Species that are charged, however, can also be aromatic. Two examples, the cyclopropenyl cation and the cyclopentadienyl anion, are shown in Figure 13.24.

The cyclopropenyl cation contains two pi-electons and an empty p-orbital that can participate in delocalization of these electrons. This species meets the 4n+2 rule, where n= 0. In the case of the cyclopentadienyl anion, we see that the species contains six pi-electrons and again satisfies the 4n+2 rule, where n=1. The pi-electrons in this system are also delocalized. You may want to draw resonance forms for both of these species to convince yourself of this.

It is interesting to note that some inorganic compounds also behave as aromatic compounds. One of these compounds is borazine. This compound displays bond lengths that are intermediate between boron-nitrogen single bonds and boron-nitrogen double bonds. Many of borazine's physical properties are similar to those of benzene, and it has been called "inorganic benzene". More recently, the compound hexachlorotriphosphazine was prepared, and it too exhibited aromatic properties.

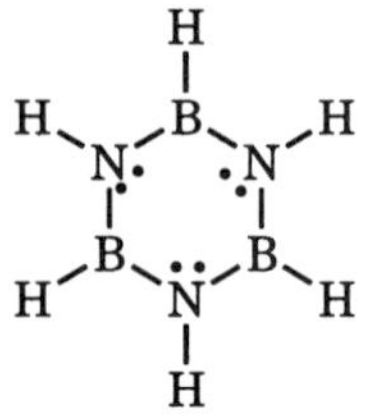

(a) borazine

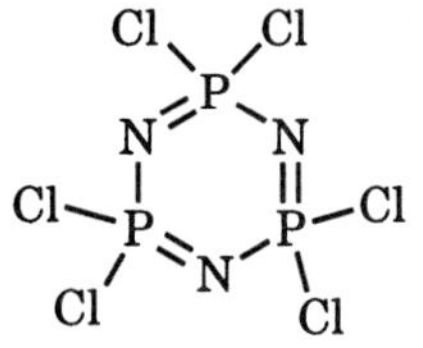

(b) a phosphazine

FIGURE 13.25: Two aromatic inorganic compounds.

Rings much larger than six atoms can also be aromatic. One hydrocarbon example is [18]annulene, pictured in Figure 13.26. The nine conjugated pi bonds along the perimeter of this macrocycle equal 18 pi electrons, a number that meets the 4n + 2 criterion, with n = 4. This same number of pi electrons is seen in the conjugated carbon chain forming the perimeter of cytochrome in Figure 13.27.

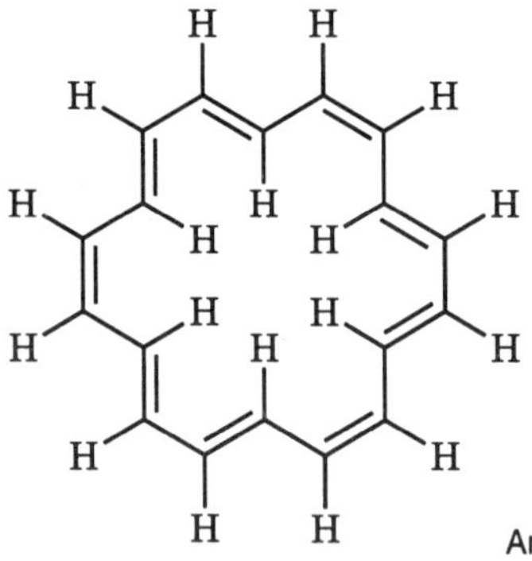

FIGURE 13.26: Annulene.

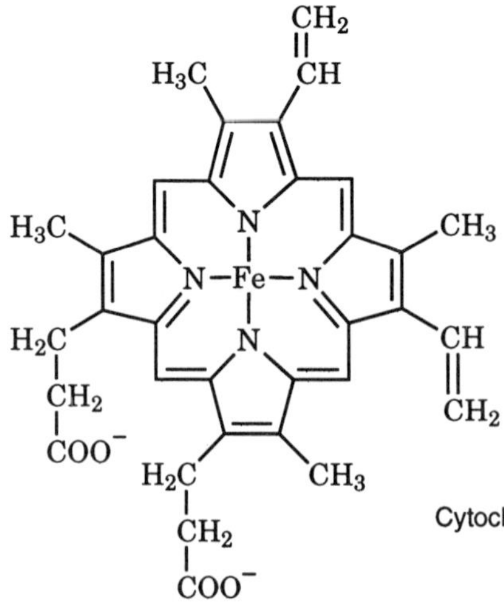

FIGURE 13.27: Cytochrome.

POLYCYCLIC AROMATIC SPECIES

We have examined a variety of monocyclic aromatic species at this point, but it is also important to note that polycyclic compounds may also have aromatic properties. The 4n+2 rule does not directly apply to polycyclic compounds and the rules for determining aromaticity of polycyclic aromatic compounds are complex. Therefore, let us look at a few examples of polycyclic aromatic compounds, with particular emaphasis on those that are biologically important.

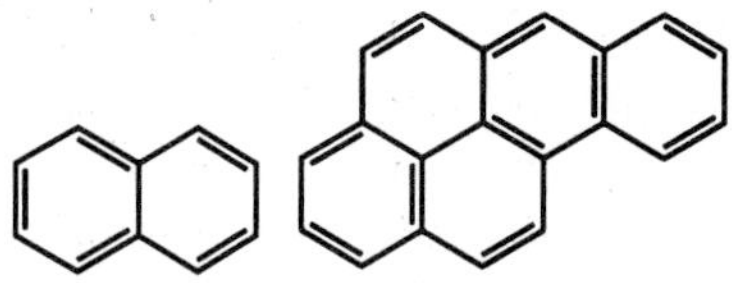

Naphthalene Benzo[a]pyrene

FIGURE 13.28: Polycyclic aromatic hydrocarbons.

Polycyclic aromatic hydrocarbons include compounds such as naphthalene, a major component of mothballs, and benzo[a]pyrene, a potent carcinogen (Figure 13.28).

Perhaps one of the most important classes of polycyclic aromatic heterocycles are the nucleic acid bases. These compounds are classified according to the parent heterocycle skeleton and are divided into two groups: the purines and the pyrimidines.

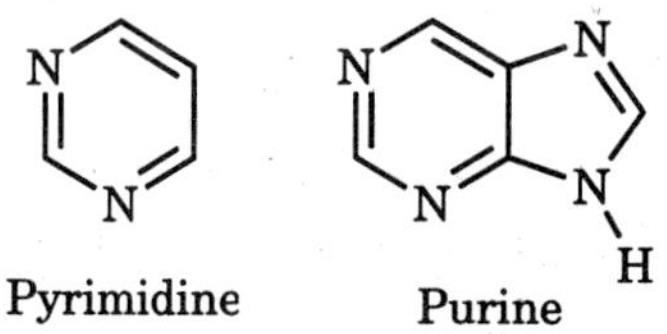

The purine bases, adenine and guanine, are of course derived from purine, while the pyrimidine bases, cytosine, thymine and uracil, are derived from pyrimidine.

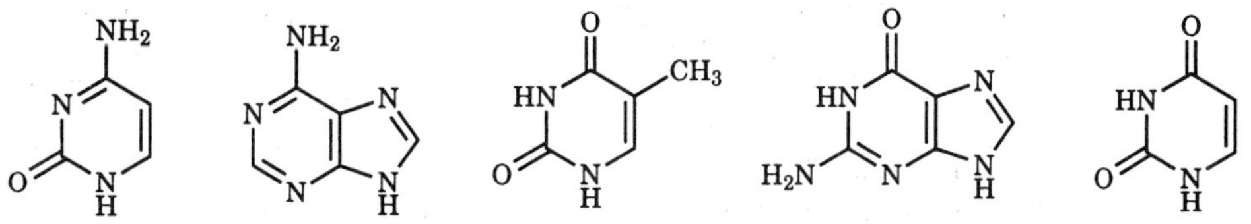

Pyrimidine Purine

FIGURE 13.29: Pyrimidine and purine skeletons.

Cytosine Adenine Thymine Guanine Uracil

FIGURE 13.30: The five nucleic acid bases.

13.5 AROMATICITY AND NMR

You may recall from Chapter 4 that the protons on an aromatic ring typically resonate between 6-8 ppm. In our initial discussion of NMR we did not address the reason that aromatic protons are shifted so far down field, but now that we have discussed the structure of aromatic compounds it is appropriate to rationalize these chemical shifts.

The value (chemical shift) at which a particular nucleus resonates is dependent upon the environment of that nucleus and is a function of how much the nucleus "sees" the applied magnetic field. When the nucleus is somehow hidden from an applied magnetic field we say that it is shielded and the resulting resonance appears upfield. Since aromatic protons resonate downfield, they must be more exposed, or deshielded, to the applied magnetic field. The question is, what causes this deshielding?

The answer lies in the structure of the aromatic ring. The electrons circulating around the pi-cloud above and below the plane of the aromatic ring induce a secondary magnetic field shown in Figure 13.31. Note that these protons on the typical aromatic ring experience an increased magnetic field, since the induced field is in the same direction as the applied field. Consequently these protons resonate downfield.

As seen in Figure 13.31, the induced magnetic field acts to increase the effective magnetic field experienced by protons outside the aromatic ring. The induced magnetic field inside the ring, however, is in the opposite direction (i.e., opposed to the applied field.) Thus protons inside the ring would experience a weaker field, and appear to be more shielded (as opposed to the deshielding effect outside the ring.) Annulene, pictured in Figure 13.26, has a proton NMR spectrum that has a resonance at about 9.3 ppm downfield from the standard TMS. This signal results from the twelve protons outside the ring. The six hydrogens inside the ring, however, show a resonance 3 ppm upfield from the standard TMS due to the magnetic field induced by the aromatic ring current.

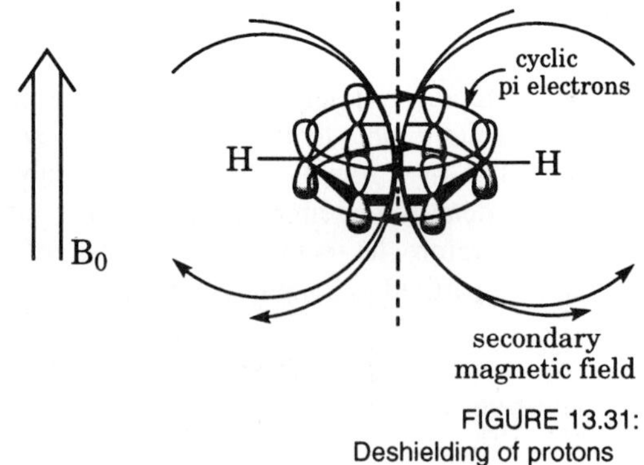

FIGURE 13.31: Deshielding of protons on benzene.

13.6 AROMATICITY AND IR

You may recall from Chapter 3 that the stretching of a carbon-hydrogen bond results in IR absorbance around 3000 cm^{-1}. In alkanes, this absorbance is strong, and occurs between 2960 and 2850 cm^{-1}. In alkenes, the absorbance was of medium intensity, between 3080 and 3020 cm^{-1}. Most often, aromatic carbon-hydrogen stretches fall between those two regions, with a common absorbance maximum at 3030 cm^{-1}. The strength of absorbances associated with aromatic carbon-hydrogen stretching is variable.

Carbon-carbon bond stretching bands also reflect the intermediate character of the bonding in aromatic compounds. As we saw earlier (Table 3.14,) isolated double bonds absorb between 1680 and 1640 cm^{-1}. Conjugation (resulting in a bond strength between that of a single and double bond) was reflected in absorbances between 1620 and 1640 cm^{-1}. The absorbance of aromatic compounds reflect this conjugation, occurring between 1600 and 1450 cm^{-1}.

Finally, substitution of other elements or groups for the hydrogen on an aromatic ring alters the IR absorbance in the fingerprint region. The region between 900 and 680 cm^{-1} characterizes the out-of-plane bending of hydrogens on an aromatic ring. The position and strength of bands in this region, although not conclusive in the absence of other data, can help assign the number and location of substituents on a benzene ring. Table 13.2 summarizes some of these characteristic absorbances.

TABLE 13.2: CHARACTERISTIC IR ABSORBANCES FOR SUBSTITUTED AROMATIC COMPOUNDS

TYPE SUBSTITUTION	POSITION	TYPE PEAK	WAVENUMBER RANGE
mono-		strong	760-730
		strong	720-680
di-	1,2 (ortho)	strong	770-740
di-	1,3 (meta)	medium	900-850
		strong	810-750
		strong	735-680
di-	1,4 (para)	strong	860-800
tri-	1,2,3	strong	800-750
		medium	745-690
tri-	1,2,4	medium	900-850
		strong	860-800
tri-	1,3,5	medium	900-860
		strong	865-810
		strong	725-675

KEY WORDS & Concepts

Allylic Position: The position on a chain or in a ring that is one carbon away from a carbon-carbon double bond.

Formal Charge: The charge on each atom in a molecule or polyatomic ion.

Heterocyclic: A ring that contains one or more atoms within the ring that are not carbon atoms.

Meta: Refers to substituents on a benzene ring that are on carbons 1 and 3.

Ortho: Refers to substituents on a bezene ring that are on carbons 1 and 2.

Para: Refers to substituents on a benzene ring that are on carbons 1 and 4.

Resonance: A phenomenon wherein a hybrid (of two or more resonance forms) delocalizes electron density. This results in a species with a lower energy than any comparable individual Lewis structure.

Resonance Form: One of several Lewis structures that are the same in terms of the position of atoms, but differ in the location of electrons.

HOMEWORK Problems

1. Draw all possible resonance structures of the following molecules or ions.
   (a) SO_2
   (b) SCN^-
   (c) NO_3^-
   (d) HNO_3

2. Draw three resonance forms of cyanic acid (HOCN) and calculate the formal charge on C, H, O and N in each form.

3. Draw the major resonance contributors of the following species.

a. 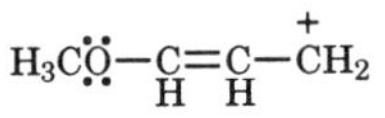 $H_3C\ddot{O}{-}CH{=}CH{-}\overset{+}{C}H_2$

b. H, +, CH_3

c. O, H

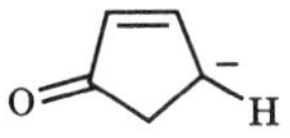

d.  +

e. 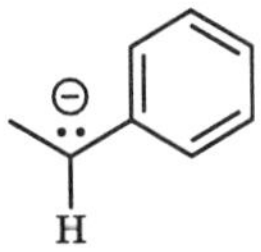 H

f.

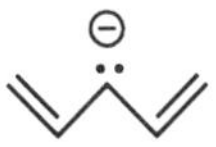

g.  O

h.

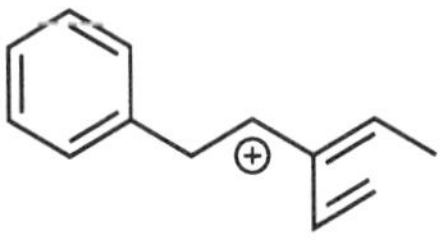

i.

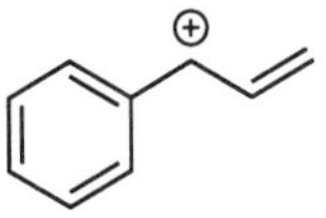

4. Provide an acceptable name for each of the following compounds.

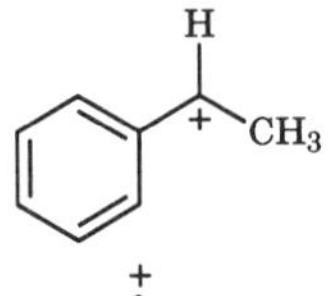

a.

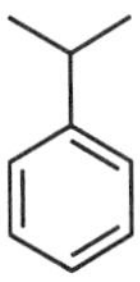

b. 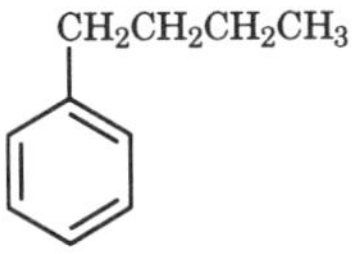 $CH_2CH_2CH_2CH_3$

c.

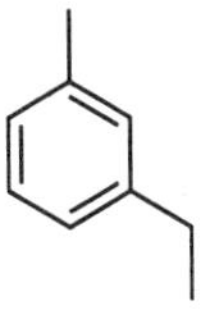

d. 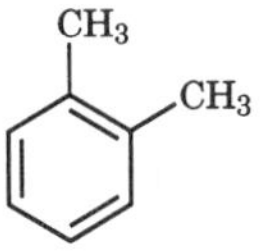 CH_3, CH_3

e. 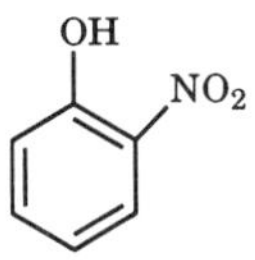 OH, NO_2

5. Draw structures that correspond to the following names.

(a) meta-cresol (b) 4-nitroaniline
(c) meta-xylene (d) 1-ethyl-3-isopropylbenzene
(e) ortho-dichlorobenzene (f) 3-nitrophenol

6. Determine which of the following compounds are aromatic.

a. 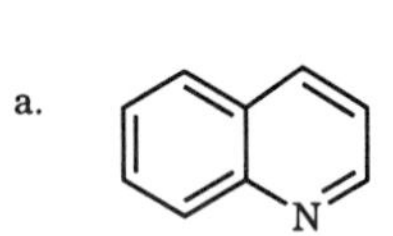

b. 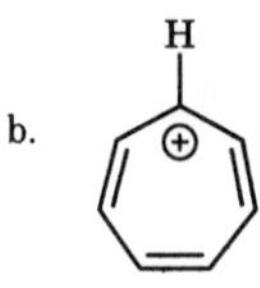

c. 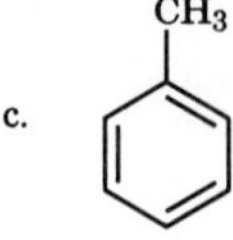

d. 

e. H

f.

7. Using resonance, explain why para-nitrophenol is more acidic than phenol.

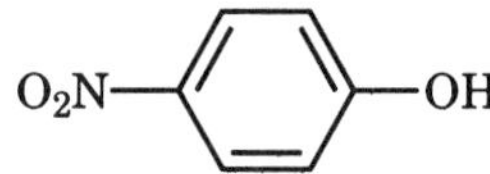

para - nitrophenol

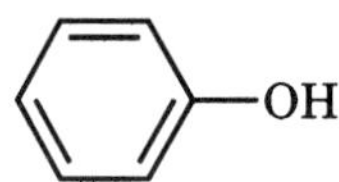

phenol

8. Earlier in the text, we mentioned that there are only three dibromobenzene isomers. This helped chemists understand the structure of benzene (C_6H_6). Show the structure of each dibromo derivative that would occur if benzene had the structure:

(a) $HC{\equiv}C{-}CH_2{-}C{\equiv}C{-}CH_3$

(b) $HC{\equiv}C{-}CH_2{-}CH_2{-}C{\equiv}CH$

9. The three dibromobenzene isomers were shown earlier this chapter (Figure 13.17). The three dibromobenzenes are known to have melting points of -7°C, 6°C, and 87°C, but which of these corresponds to ortho, meta and para? The answer was initially provided by Korner method of absolute orientation, wherein each dibromobenzene was nitrated (another aromatic hydrogen replaced by a NO_2 group.)

(a) The dibromobenzene that melts at 87°C forms only one dibromonitrobenzene derivative. Draw and name the dibromobenzene that melts at 87°C and show the structure of its one nitro derviative.
(b) The dibromobenzene that melts at 6°C forms only two dibromonitrobenzene derivatives. Draw and name the dibromobenzene that melts at 6°C and show the structures of its two nitro derviatives.
(c) The dibromobenzene that melts at -7°C forms three dibromonitrobenzene derivatives. Draw and name the dibromobenzene that melts at -7°C and show the structures of its three nitro derviatives.

10. A compound with a molecular formula of $C_8H_8O_2$ has the following NMR spectrum. The IR spectrum shows strong absorbances at 1700 and 820 cm^{-1}.

(a) Draw the structure of the compound
(b) How many unique carbon environments should there be in the CNMR spectrum of (a)?
(c) Indicate which protons in (a) are associated with which peaks in the NMR.
(d) Interpret the IR spectral information.
(e) Name the compound.

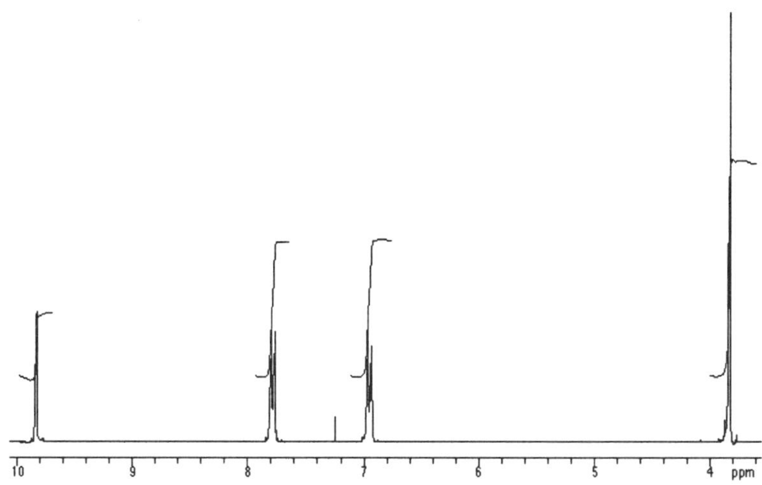

11. Another compound with a molecular formula of $C_8H_8O_2$ has the following NMR spectrum. The IR spectrum shows strong absorbances at 1700 and 750 cm^{-1}.

(a) Draw the structure of the compound
(b) How many unique carbon environments should there be in the CNMR spectrum of (a)?
(c) Label the protons in the structural formula in (a) in relation to the peaks in the proton NMR spectrum.
(d) Interpret the IR spectral information.
(e) Name the compound.

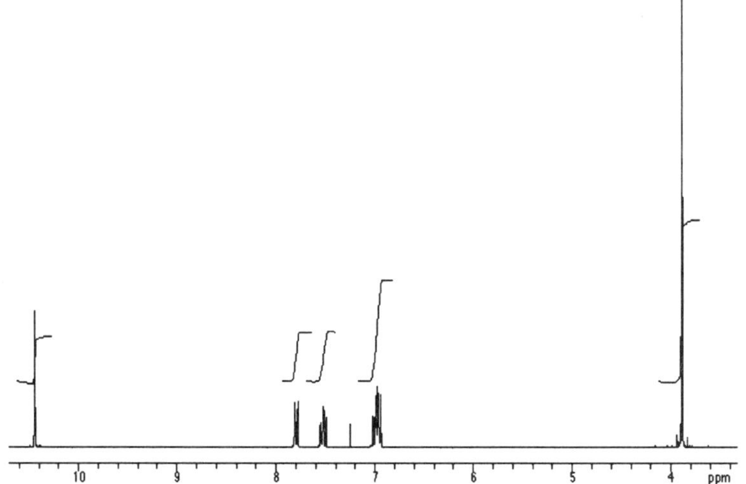

12. A compound contains 90.50% carbon and 9.50% hydrogen. The molecular ion determined by mass spectrometry indicates a mass of about 106. The HNMR spectrum shows a tight cluster of peaks at 7.0 ppm, and a singlet at 2.3 ppm. Integration of these signals at 7.0 and 2.3 ppm shows that the protons are in a 2:3 ratio respectively. The IR spectrum shows an absorbance at about 3030 cm^{-1}, and has only one absorbance between 1000 and 600 cm^{-1}: a sharp peak at 800 cm^{-1}.

(a) What is the empirical formula of this compound?
(b) What is the molecular formula of this compound?
(c) Draw the structure of this compound.
(d) Explain the significance of the IR spectral information.
(e) Label the protons in the structural formula in (c) in relation to the peaks in the proton NMR spectrum.

13. An isomer of the compound in #12 was isolated. The NMR spectrum showed that there were 5 unique carbon environments. The IR spectrum showed a medium peak at 890 cm^{-1}, and strong peaks at 690 and 780 cm^{-1}.

(a) Draw the structure of this compound.
(b) Show the 5 carbon environments in the structure in (a).
(c) Explain the significance of the IR spectral information?

14. A compound with the formula $C_8H_{10}O_2$ shows a broad IR absorbance between 3600 and 3100 cm^{-1}, and a strong absorbance at 820 cm^{-1}. The proton NMR spectrum shows a doublet at 7.3 (2H), a doublet at 6.9 (2H), a singlet at 4.6 (2H), a singlet at 3.8 (3H), and a singlet at 2.1 (1H). The carbon NMR shows 6 carbon environments.

(a) What is the IHD of the compound?
(b) What functional group is indicated by the broad IR absorbance between 3600 and 3100 cm^{-1}?
(c) Draw the structure of the compound.
(d) Label the protons in the structural formula in (c) in relation to the peaks in the proton NMR spectrum.

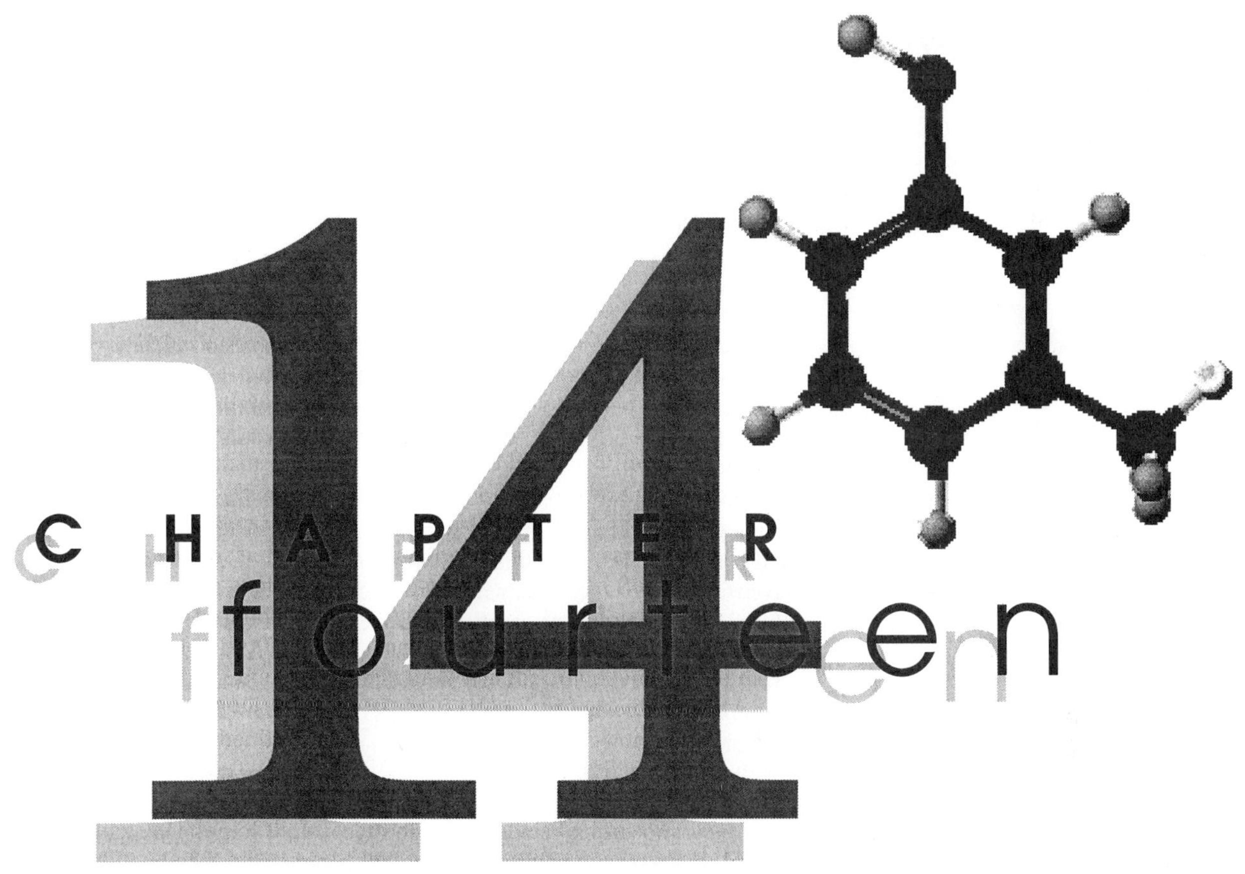

AROMATIC Reactions

AROMATIC Reactions

Benzene is often treated as the prototypical aromatic compound. In this chapter we will study the reactions of aromatic compounds, and benzene will typically be the reactant. You should, however, understand that aromatic compounds other than benzene will often undergo reactions in the same fashion as that shown for benzene.

In this chapter we begin our study of reaction mechanisms. A **mechanism** is a description of what happens between the time that reactants first come together and the time that a new compound has been formed. In this interval some bonds have broken and other new bonds have formed. For example, when one reacts benzene with nitric acid (in the presence of sulfuric acid) the products are nitrobenzene and water, shown below:

$$C_6H_6 + HNO_3 \xrightarrow{H_2SO_4} C_6H_5NO_2 + H_2O$$ EQ. 14.1

Sulfuric acid is shown over the arrow, indicating that it must be present for the reaction to proceed. The amount of sulfuric acid in the reaction mixture does not decrease with time (it is not a reactant,) nor does it increase with time (it is not a product.) Sulfuric acid must be included for the reaction to proceed, however, because it functions as a **catalyst**. A catalyst is a substance that increases the rate of a reaction, but is not consumed by the reaction. The action of a catalyst is one example of the many types of clues chemists can use to solve the puzzle of how reactions occur.

In your study of aromatic chemistry, you will see numerous reactions; each reaction may occur via a series of individual steps. You will also be asked to make predictions about reactions that you have never previously seen. Each such reaction superficially may appear to be unique, and the task of predicting products thus may appear to be daunting. By understanding a few common types of reaction mechanisms, you will soon find that there are far more similarities among reactions involving aromatic compounds than there are differences.

14.2 ELECTROPHILIC AROMATIC SUBSTITUTION

As we saw last chapter, there are numerous ways of representing benzene. Figure 14.1 emphasizes the pi cloud of electrons above and below the benzene ring. This picture of benzene helps us to understand the type of species that may react with aromatic compounds. Although benzene has no charge, the concentration of electrons above and below the ring indicates that electron-deficient species may be strongly attracted to the benzene ring (Figure 14.1). These electron-deficient species are termed **electrophiles**. Lewis acids, presented in Chapter 12, are examples of electrophiles.

FIGURE 14.1: Pi electrons in benzene.

The reaction between electrophiles and benzene is easier to understandand if we use the Kekulé representations of benzene (Figure 14.2) because these allow us to draw simple pictures using resonance forms to explain results.

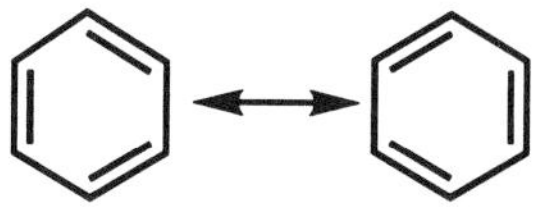

FIGURE 14.2: Kekulé structures of benzene.

Three important **electrophilic aromatic substitution** reactions are shown below: **halogenation, alkylation** and **nitration**.

chlorination (substitution of H by Cl)

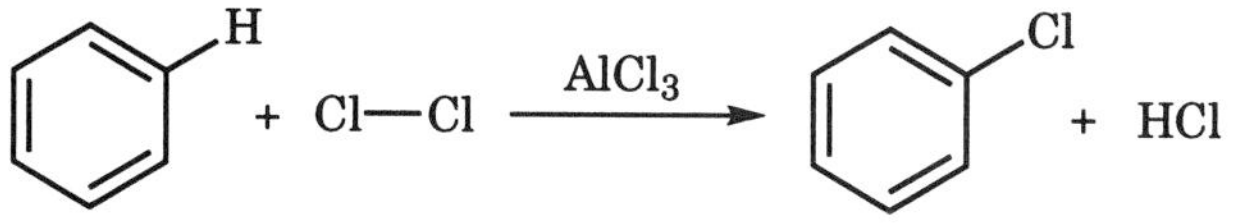

$$C_6H_5\text{-}H + Cl\text{—}Cl \xrightarrow{AlCl_3} C_6H_5\text{-}Cl + HCl \quad \text{EQ. 14.2}$$

alkylation (substitution of H by isopropyl)

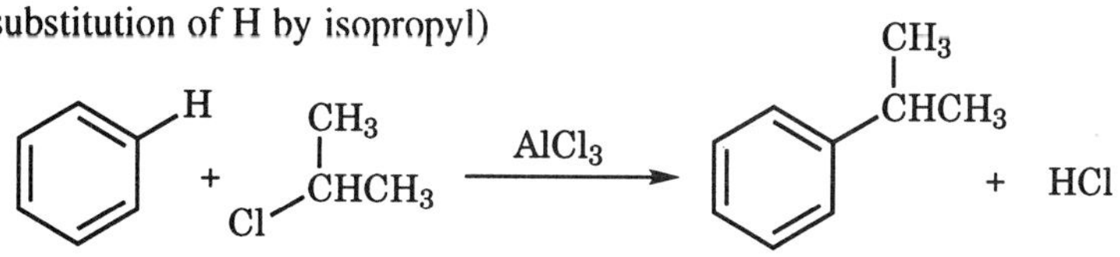

$$C_6H_5\text{-}H + Cl\text{-}CH(CH_3)CH_3 \xrightarrow{AlCl_3} C_6H_5\text{-}CH(CH_3)CH_3 + HCl \quad \text{EQ. 14.3}$$

nitration (substitution of H by NO_2)

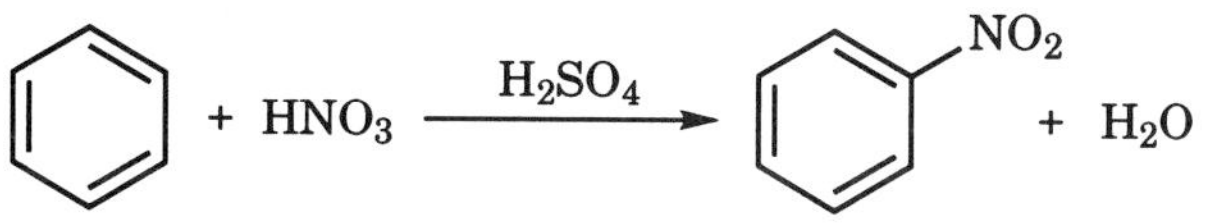

$$C_6H_6 + HNO_3 \xrightarrow{H_2SO_4} C_6H_5\text{-}NO_2 + H_2O \quad \text{EQ. 14.4}$$

These balanced equations show reactants, products, and catalyst, but offer no direct information concerning the mechanisms by which the reactions take place. From our previous discussion of the pi electron cloud of benzene, however, we would guess that the attacking species are generic electrophiles, symbolized by E^+. When we look at the products of each reaction, we would guess correctly that the specific electrophiles would be Cl^+, $(CH_3)_2CH^+$, and NO_2^+ respectively. It is easiest to think of each of these reactions as occurring via a three part mechanism: (1) formation of the electrophile; (2) reaction of the electrophile with electrons from the aromatic ring, forming a positively charged, non-aromatic intermediate; and (3) loss of H^+ from this intermediate to form the substituted aromatic ring. Thus in the chlorination example above, first the electrophile Cl^+ is formed; Cl^+ then reacts with benzene ring electrons to form a cationic intermediate; finally, the positively charged intermediate loses H^+ to form chlorobenzene. In the next three sections we will discuss the mechanisms of each of these three reactions in detail, noting their common features.

14.3 CHLORINATION (OR MORE GENERALLY–HALOGENATION)

Part 1 in the halogenation of an aromatic compound is the formation of the electrophile. In this case the electrophile Cl^+ is generated by the reaction of a Lewis base, chlorine, with a Lewis acid, aluminum chloride, as shown in the equation below. The Lewis acid and Lewis base react to form a complex. The chlorine-chlorine bond in the complex breaks asymmetrically; both shared electrons go with the chlorine atom that bonds to the aluminum atom. The other chlorine becomes positively charged, and is the electrophile (E^+).

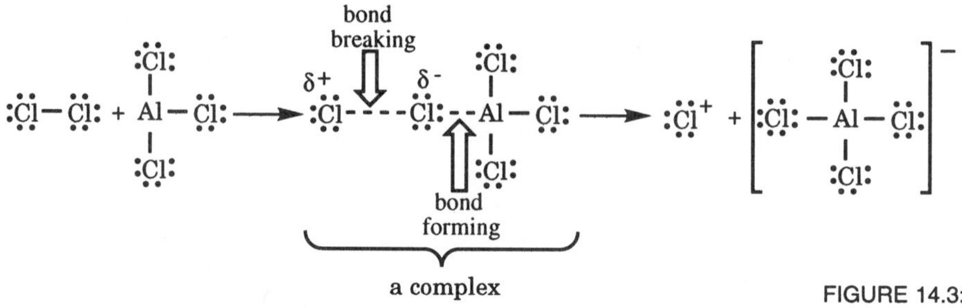

Once formed, the electrophile (Cl^+) attacks the ring as part 2 of the reaction shown in the figure below. A pair of pi electrons is donated by benzene to the Cl^+ to form a sigma bond between carbon and chlorine. This results in a positively charged carbocation **intermediate** in the benzene ring. An intermediate is a species that is formed and then rapidly destroyed during a reaction. (Note the difference between an intermediate and a catalyst.) Note that ring A in this intermediate, in addition to being positively charged, is no longer aromatic. The ring, however, still contains two double bonds, and the cation in each of the resonance forms is allylic with respect to one or both of these double bonds. As we saw in Chapter 13, an allylic carbocation is especially stable. Thus all three resonance forms add significant stability to the carbocation intermediate, and thus favor the overall reaction.

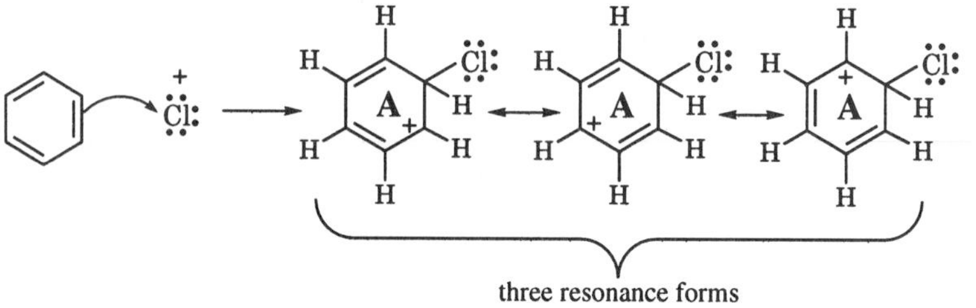

FIGURE 14.3: Part one: formation of the electrophile Cl^+.

The conversion of the resonance stabilized intermediate into final product is shown in Figure 14.5. The base, $AlCl_4^-$, removes the proton attached to the same carbon to which the chlorine is attached. This acid-base reaction forms an intermediate, $HAlCl_4$, that in turn dissociates into HCl (a final product of the overall reaction, (Eq. 14.2) and $AlCl_3$ (regenerating the catalyst that was used initially.) Note that chlorobenzene, the other product, is aromatic. The lower energy of an aromatic product helps drive the conversion of the non-aromatic, positively charged intermediate into final product.

FIGURE 14.4: Part two: electrophile Cl^+ reacts with the aromatic ring.

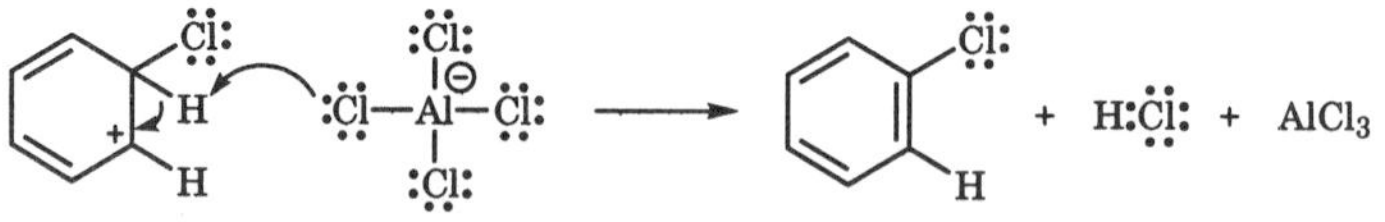

FIGURE 14.5: Part three: loss of H^+ to form chlorobenzene.

14.4 FRIEDEL- CRAFTS ALKYLATION

The reaction shown in Eq. 14.3 resembles the chlorination reaction except that the electrophile changes from Cl^+ to the isopropyl cation, $(CH_3)_2CH^+$. The first step of the mechanism is similar to that for the chlorination mechanism; the Lewis acid, aluminum chloride, complexes with the Lewis base, isopropyl chloride, to produce an isopropyl cation, as shown in the Figure 14.6.

After formation of the electrophile, the electrons in the benzene ring bond to the isopropyl cation, again producing a nonaromatic, resonance stabilized carbocation.

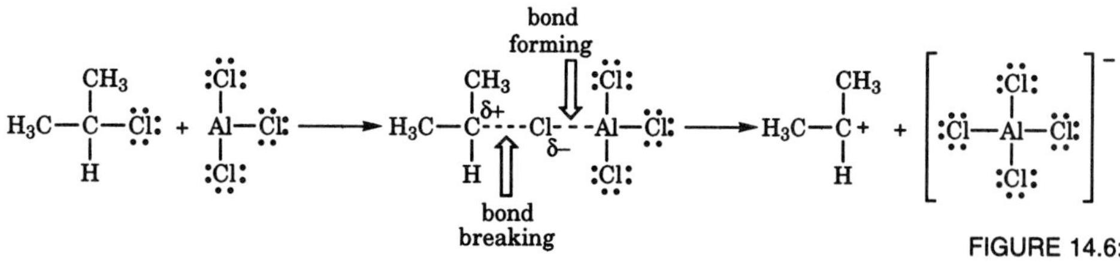

In Figure 14.8, the tetrachloroaluminum anion removes the proton, aromaticity is restored, the catalyst is regenerated, and final products are formed.

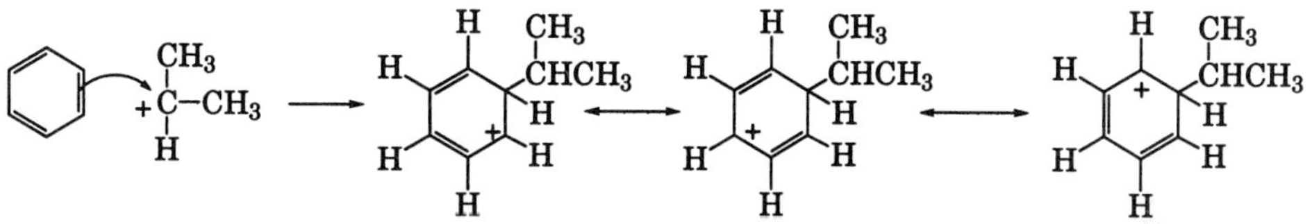

FIGURE 14.6: Part one: formation of the electrophile $(CH_3)_2CH^+$.

FIGURE 14.7: Part two: Electrophile $(CH_3)_2CH^+$ reacts with the aromatic ring.

In the reaction mechanisms of chlorination and alkylation, the same Lewis acid catalyst was used: $AlCl_3$. In the next electrophilic aromatic substitution reaction, we examine a somewhat different route to form the electrophile.

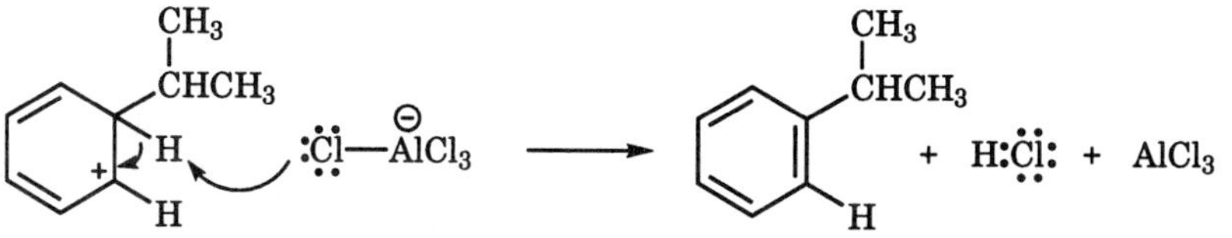

FIGURE 14.8: Part two: loss of H^+ to form isopropylbenzene.

14.5 NITRATION

Nitration is a classic example of electrophilic aromatic substitution because it has been studied so much over many years. It is also a reaction used industrially to prepare huge amounts of explosives, *e.g.*, TNT, and to prepare intemediates for the synthesis of medicines. The nitro group is important for drug manufacturing because it is easily reduced to the amino group (NH_2), that has widespread biological activity.

The electrophile in the case of aromatic nitration is nitronium ion, NO_2^+. The experimental procedure by which nitronium ion is formed uses sulfuric acid as the solvent, with lesser amounts of nitric acid. Sulfuric acid is a stronger acid than nitric acid; thus nitric acid acts as a Lewis base and accepts a proton from sulfuric acid, the Lewis acid. (You might be confused at this point as to why we wish to treat nitric acid as a base. It might help to consider that if HNO_3 were to act as an Arrhenius base, it would produce OH^- and the desired electrophile, NO_2^+.) The acid-base reaction (Figure 14.9) between sulfuric and nitric acids produces an intermediate cation, $H_2NO_3^+$, which subsequently decomposes to the nitronium cation. Note also that the water produced in this step is a final product in the overall reaction, seen in Eq. 14.4.

The reaction of the nitroniun ion with the benzene follows the mechanisms previously outlined for both the halogenation and alkylation reactions, and is seen in Figure 14.10. Again, note the formation of a resonance stabilized carbocation intermediate.

nitronium cation

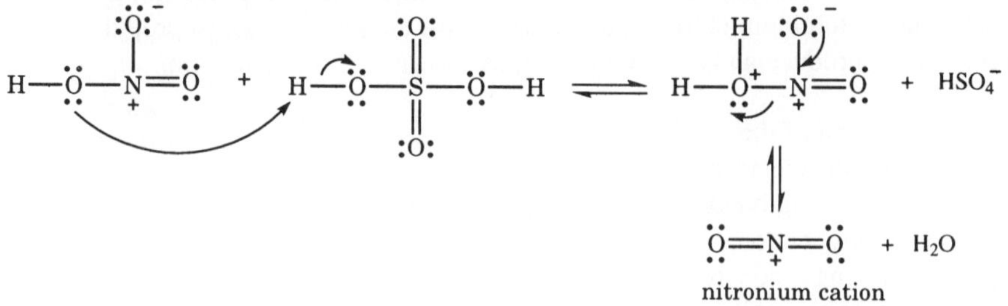

FIGURE 14.9: Part one: formation of the electrophile NO_2^+.

The nitration mechanism is completed when the carbocation reacts with bisulfate to regenerate the catalyst sulfuric acid.

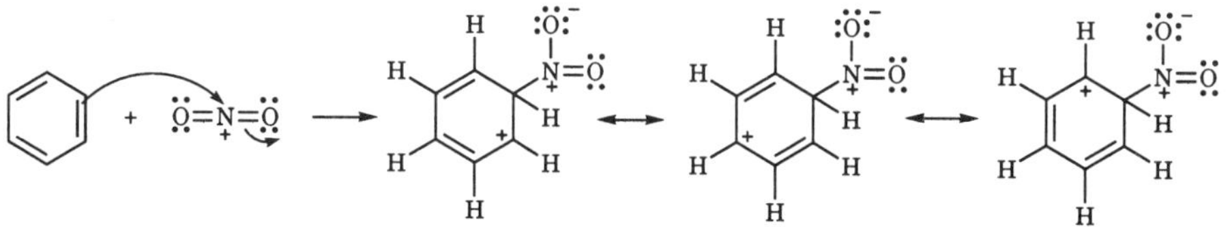

FIGURE 14.10: Part two: electrophile NO_2^+ reacts with the aromatic ring.

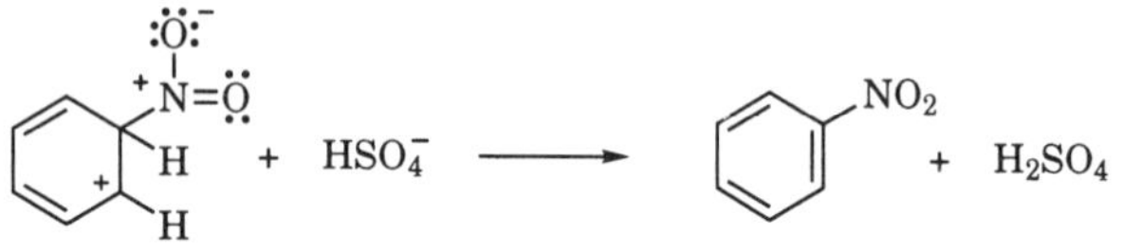

FIGURE 14.11: Part three: loss of H^+ to form nitrobenzene.

SIDEBAR

The appearance of undissociated H_2SO_4 as a product in the last step of the nitration mechanism might seem odd. Sulfuric acid appears in Figure 14.9 in its molecular form, as opposed to its common dissociation products H^+ and HSO_4^-. Furthermore, sulfuric acid is shown as a product of a Bronsted-Lowry acid-base reaction, a position traditionally held by the weaker of the two acids involved in the reaction. The first issue is resolved by noting that this reaction is taking place in an organic solvent, not water. As such, sulfuric acid is far more likely to exist in its molecular form. The second issue is resolved by noting that the carbocation intermediate is a stronger acid than sulfuric acid, just as the bisulfate ion is a stronger base than nitrobenzene.

14.6 REVIEW OF THE GENERAL ELECTROPHILIC AROMATIC SUBSTITUTION REACTION MECHANISM

As you have seen, the mechanisms for the overall reactions given in Eq. 14.2, 14.3, and 14.4, all follow the same path. First, an electrophile (E^+) is formed. Second, E^+ attacks the ring and breaks the aromatic system to give a cationic intermediate with several resonance forms that give it stability. Finally, a basic species removes an acidic proton from the ring to re-establish the aromatic nature of the ring.

1. formation of E^+ (unless already present; *e.g.*, SO_3 in fuming sulfuric acid.)

2. reaction of electrons from ring with electrophile

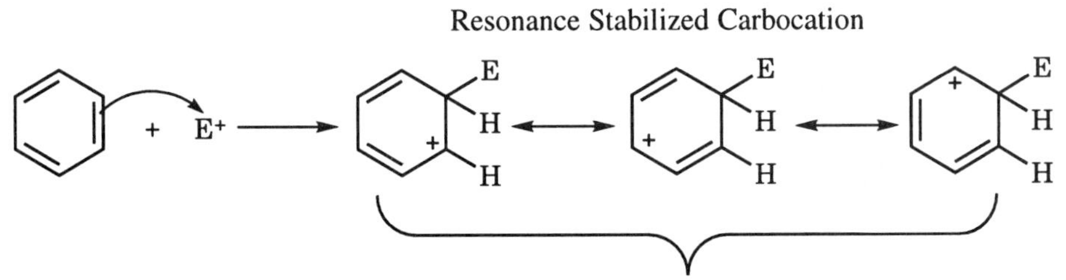

Resonance Stabilized Carbocation

+ E^+ ⟶ (E, H, +) ⟷ (E, H, +) ⟷ (E, H, +)

3. reaction of carbocation with base

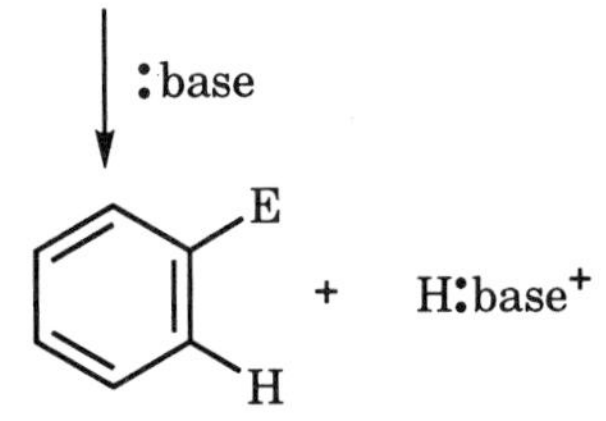

:base

(E, H) + H:base$^+$

FIGURE 14.12: General mechanism of electrophilic aromatic substitution.

So far, we have seen mechanisms for halogenation, alkylation and nitration of aromatic compounds. From these, we have formed a general mechanism of electrophilic aromatic substitution. Next we will study two other important reactions for substitution onto aromatic rings. The difference in these mechanisms is primarily in the method by which the electrophile is formed.

14.7 SULFONATION

Sulfuric acid is cheap and widely available. It is the starting material in the synthesis of many substances, including sulfonic acids. Sulfonic acids in turn are used in the production of many important substances, such as sulfa drugs. The largest use of sulfonic acids, however, is in the manufacture of household detergents. Detergents are prepared by the reaction of a sulfonic acid with sodium hydroxide. One such reaction is shown below.

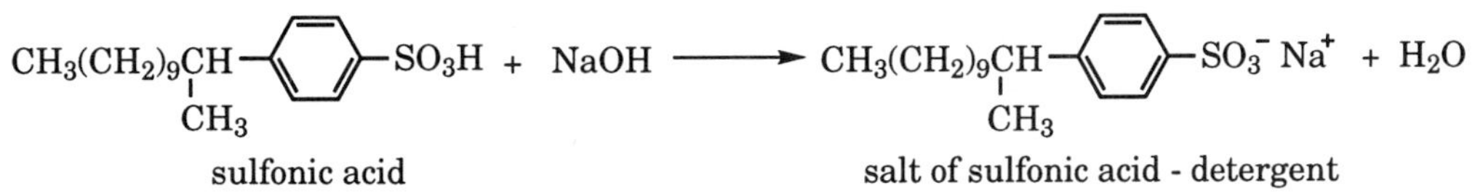

$$CH_3(CH_2)_9CH(CH_3){-}C_6H_4{-}SO_3H + NaOH \longrightarrow CH_3(CH_2)_9CH(CH_3){-}C_6H_4{-}SO_3^-\,Na^+ + H_2O$$

sulfonic acid — salt of sulfonic acid - detergent

EQ. 14.5

Sulfonic acids like that shown as a reagent in the previous equation are made by the addition of the sulfonic acid group (-SO_3H) to an aromatic ring. One of the effects of the sulfonic acid group is to increase water solubility of the aromatic ring system; this is important for detergents.

The experimental procedure for **sulfonation** uses fuming sulfuric acid, a mixture of H_2SO_4 and SO_3. The net reaction converting benzene into benzenesulfonic acid is shown below:

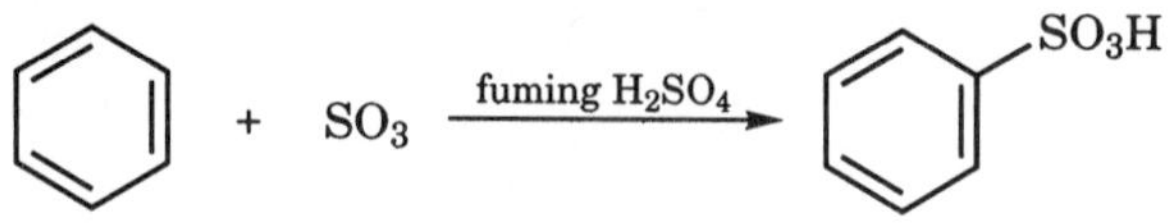

$$C_6H_6 + SO_3 \xrightarrow{\text{fuming } H_2SO_4} C_6H_5SO_3H$$

EQ. 14.6

The electrophile for sulfonation can be either the neutral compound sulfur trioxide, SO_3, or the protonated species, SO_3H^+. The reaction with SO_3H^+ as the electrophile follows the mechanisms previously developed, and is left as an exercise for the reader. The sulfonation with sulfur trioxide is somewhat different in that the electrophile is a neutral species. Several resonance forms of sulfur trioxide, shown in the figure below, show sulfur with a positive formal charge.

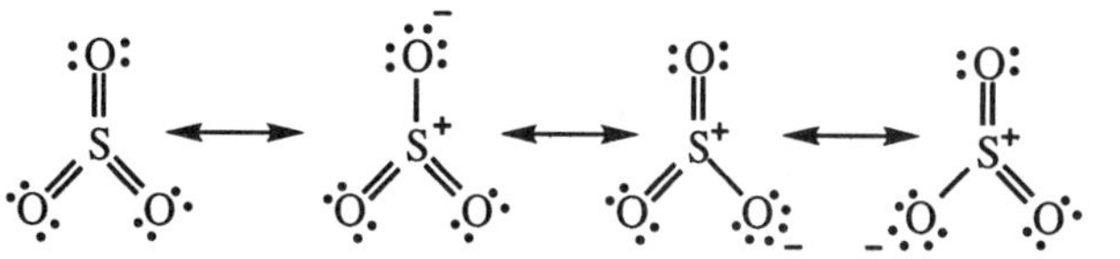

FIGURE 14.13: Resonance forms of the electrophile SO_3.

A pair of pi electrons from the aromatic ring can bond to the positive sulfur atom as shown below:

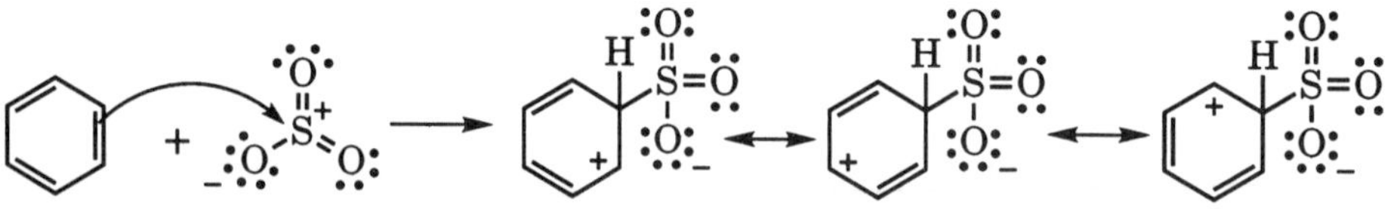

FIGURE 14.14: Electrophile SO_3 reacts with the aromatic ring.

Note that the intermediate formed contains both an oxygen with a negative charge and a resonance stabilized carbocation. Reaction of the intermediate with bisulfate ion and protonation of the sulfonate group yields the final product.

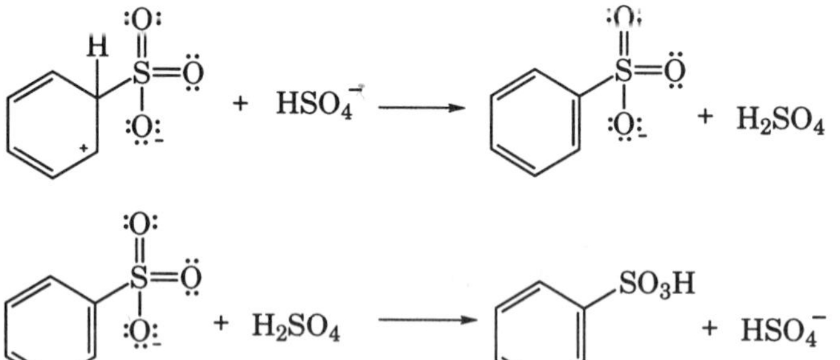

$$\text{intermediate} + HSO_4^- \longrightarrow C_6H_5SO_3^- + H_2SO_4$$

$$C_6H_5SO_3^- + H_2SO_4 \longrightarrow C_6H_5SO_3H + HSO_4^-$$

FIGURE 14.15: Formation of benzenesulfonic acid.

14.8 FRIEDEL–CRAFTS ACYLATION

The last electrophilic aromatic substitution reaction we will study is **acylation**, also named Friedel-Crafts acylation. This reaction is similar to alkylation. An acyl group (R-CO-) is added to the ring by formation of an acyl electrophile from either an acid chloride or acid anhydride.

The electrophile is formed by the reaction of an acid chloride or an acid anhydride with a Lewis acid such as aluminum chloride. (Note that this is quite similar to the reaction in Figure 14.6, when an alkyl halide reacts with a Lewis acid as the start of a Friedel Crafts alkylation reaction.)

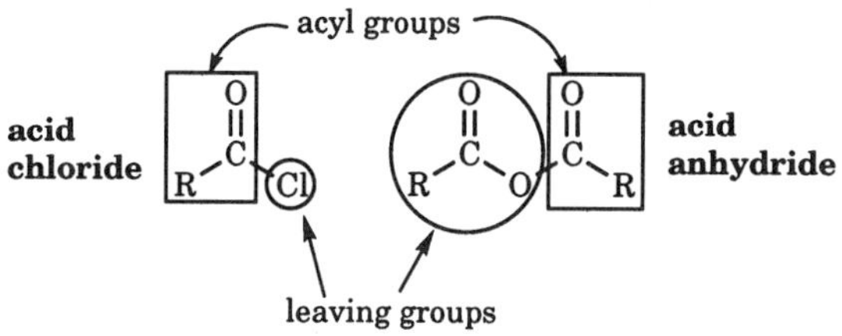

acyl groups

acid chloride

acid anhydride

leaving groups

FIGURE 14.16:
Two common acylating agents.

Also note the stabilization predicted by the resonance forms of the acyl group shown in the figure below. The mechanism for the reaction of the acylium ion with benzene proceeds in a typical manner, and is left as an exercise for the reader.

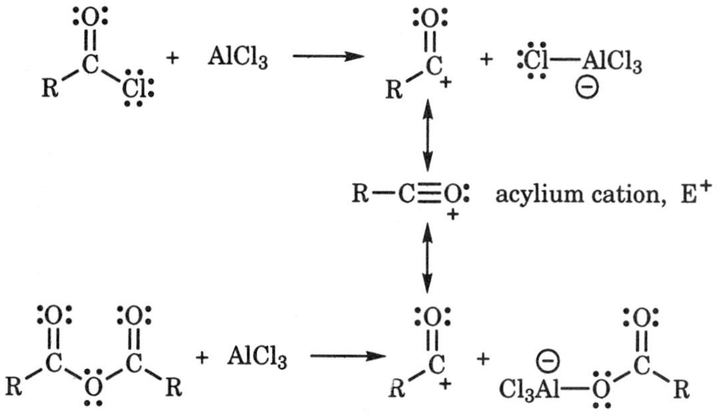

R—C≡O: acylium cation, E^+

FIGURE 14.17:
Formation of the acylium ion.

14.9 EFFECT OF GROUPS ON REACTIVITY OF THE AROMATIC RING

So far, benzene has been the only aromatic species for which substitution reactions have been presented. If a functional group is already present on the aromatic ring that is undergoing substitution, the product of this reaction is a disubstitued benzene. The overall yield of this reaction and the relative placement of one group to another on the ring both depend upon the identity of the group initially on the ring. An example of the effect that a functional group initially present on an aromatic ring has on the yield of a subsequent electrophilic addition reaction is given in Figure 14.18. This figure compares the relative amounts of monochlorination products that are produced by reacting an equimolar mixture of three aromatic compounds with chlorine and aluminum chloride.

In Figure 14.18, the final products show chlorine bonded through the side of the hexagonal ring representing benzene. This is conventionally used to signify a collection of all types of substitution products: e.g., 2-chloroanisole, 3-chloroanisole, and 4-chloroanisole.

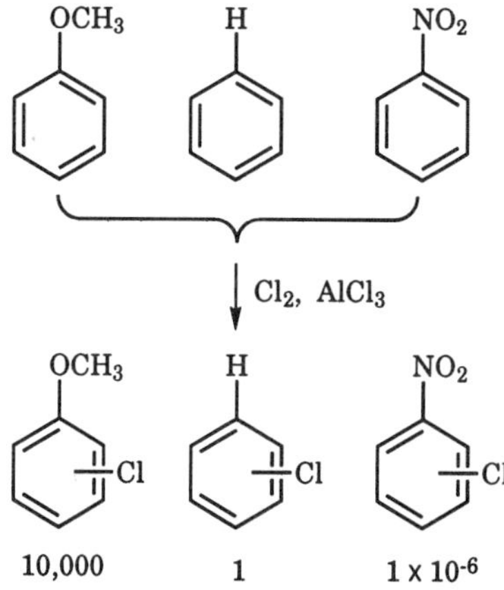

OCH_3 H NO_2

Cl_2, $AlCl_3$

10,000 1 1×10^{-6}

FIGURE 14.18:
Relative reactivity of anisole, benzene, and nitrobenzene towards chlorination.

Notice the huge difference in the relative amounts of chloroanisole, chlorobenzene, and chloronitrobenzene formed! Since benzene is unsubstituted, the relative ratio is usually based upon the substitution product for benzene being one. When this adjustment is made, the relative ratios are 10,000 to 1 to 0.000001. Thus it is clear that the presence of either a methoxy group or a nitro group on the ring has a great effect on the relative ease of electrophilic aromatic substitution.

Functional groups that are substituted onto the aromatic ring influence the electron density of the ring. Such groups can be classified either as activating groups or deactivating groups. An activating group, like methoxy, increases the electron density of the ring, making it more reactive towards electrophilic substitution. A deactivating group, like nitro, decreases the electron density of the ring, making it less reactive towards electrophilic substitution. This influence is illustrated in the figure below.

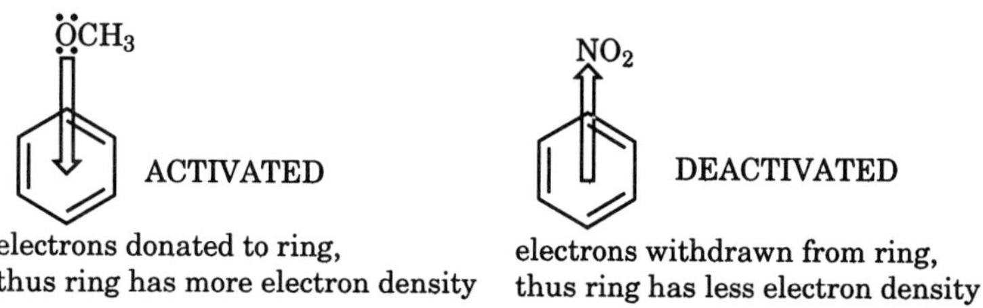

FIGURE 14.19: Activation-deactivation of the ring by different functional groups.

Activators like the methoxy group add electron density to the ring and usually have the characteristic of a pair of electrons adjacent to the ring. Table 14.1 lists some common groups that are activators.

TABLE 14.1: ACTIVATING GROUPS FOR ELECTROPHILIC AROMATIC SUBSTITUTION

EQ. 14.1

MOST ACTIVATING

$-\ddot{N}H_2$	amine
$-\ddot{O}H$	hydroxyl
$-\ddot{O}R$	alkoxy
$-\ddot{N}(H)-C(=O)-R$	amide
$-R$	alkyl

LEAST ACTIVATING

Note that although the alkyl group has no unshared pair of electrons, it donates electrons to the ring by an inductive effect.

Deactivators like the nitro group withdraw electron density from the ring. Such groups usually have an electron deficient atom adjacent to the ring. The figure at right shows resonance forms wherein the nitrogen assumes a positive charge, explaining the nitro group's ability to withdraw electron density from the ring.

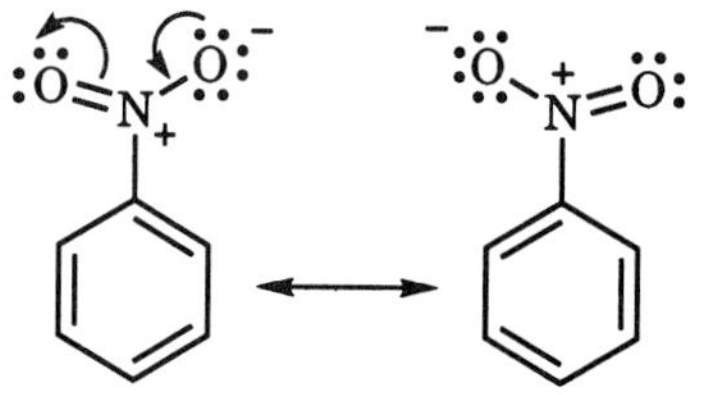

FIGURE 14.20: Resonance forms of nitrobenzene, showing nitrogen's electron deficiency.

Table 14.2 lists some common groups that are deactivators.

TABLE 14.2: DEACTIVATING GROUPS FOR ELECTROPHILIC AROMATIC SUBSTITUTION

MOST DEACTIVATING

$—\overset{+}{N}R_3$	quarternary amine
$—NO_2$	nitro
$—CN$	nitrile
$—SO_3H$	sulfonic acid
$—\overset{O}{\overset{\|\|}{C}}—R$	carbonyls
$—X$	halides

LEAST DEACTIVATING

TABLE 14.2

14.10 EFFECT OF GROUPS ON ORIENTATION OF SUBSTITUTION

The presence of a functional group on the aromatic ring undergoing substitution will not only increase or decrease the relative amount of product, it influences where on the ring the electrophile will add. For example, Eq. 14.7 shows that chlorination of anisole results almost exclusively in ortho or para chloroanisole. The meta isomer is formed only in trace amounts.

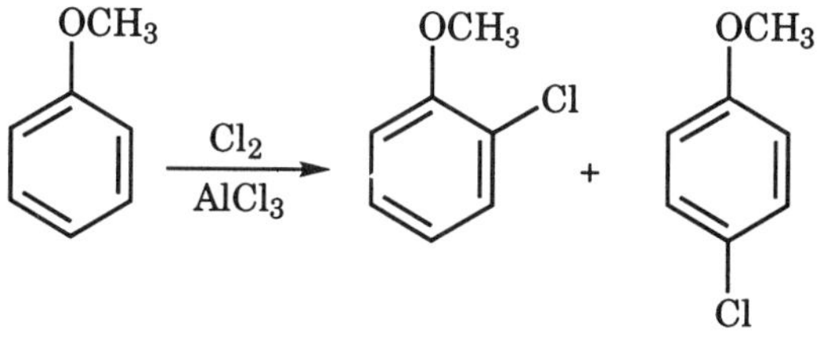

EQ. 14.7

OCH_3 $\xrightarrow[AlCl_3]{Cl_2}$ OCH_3, Cl + OCH_3, Cl

ortho para

Equation 14.8, on the other hand, shows that chlorination of nitrobenzene results almost exclusively in the meta isomer of chloronitrobenzene. The ortho and para isomers are formed only in trace amounts.

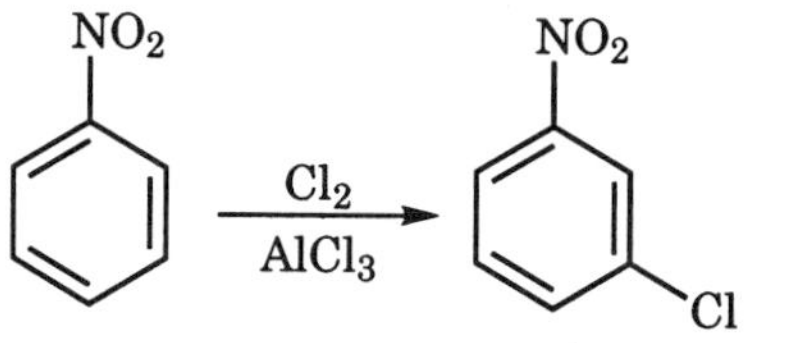

EQ. 14.8

NO_2 $\xrightarrow[AlCl_3]{Cl_2}$ NO_2, Cl

meta

Orientation effects are related to the stability of the carbocation intermediate formed when the electrophile has added to the ring. In Figure 14.21, E^+ adds to the ring, to give A, B and C as carbocation intermediates. Note that the electrophile E^+ is added at the bottom carbon in the picture and the positive charge appears on only three carbons, alternating positions around the ring, as shown in the resonance forms A, B, and C.

loss of proton, H^+

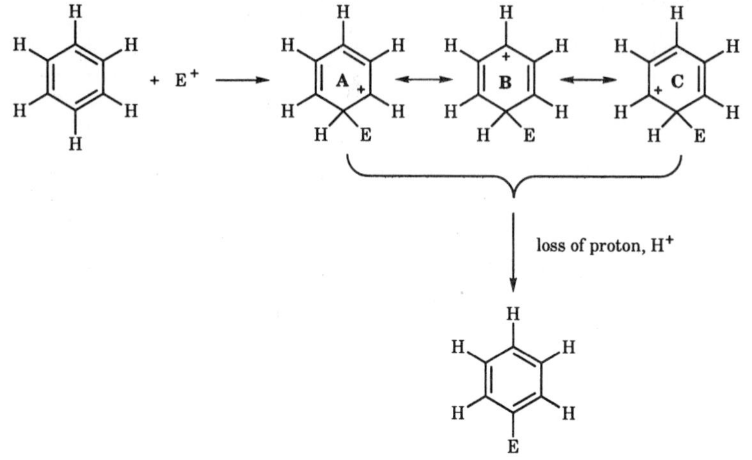

FIGURE 14.21: Resonance forms of the carbocation intermediate in electrophilic aromatic substitution.

When groups are put on the ring, the stability of at least one of the resonance forms A, B and C is affected. Activating groups will stabilize certain resonance forms and deactivating groups destabilize the same resonance forms. Since we have already used methoxy and nitro groups, let us use them to illustrate what is meant. If the group G is ortho to the electrophile, cation intermediates O1, O2 and O3 are formed. Similarly, para and meta substitutions produce cation intermediates P1, P2, P3 and M1, M2, M3 respectively.

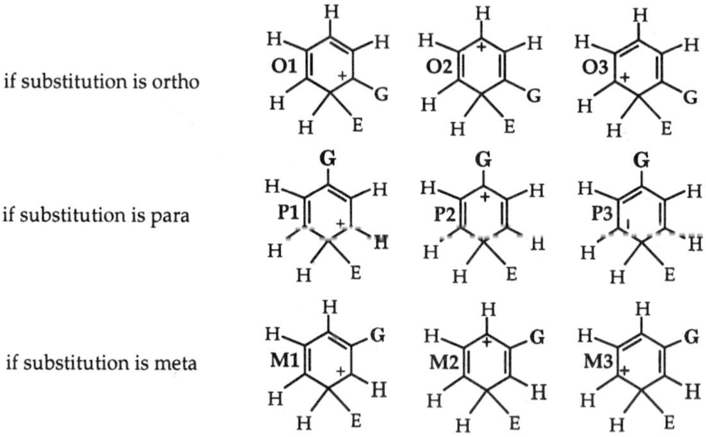

if substitution is ortho

if substitution is para

if substitution is meta

FIGURE 14.22: Resonance forms of the cationic intermediate for ortho, meta, and para substituted aromatics.

Now consider the effect of group G on these nine carbocation intermediates. If G is methoxy, with its ability to release electrons into the ring, Forms O1 and P2 are stabilized. In addition, there are other resonance forms, shown in Figure 14.23, that help to delocalize the charge. This delocalization adds to the relative stability of substitution ortho and para to the methoxy group.

Alternatively, if G is nitro, with its ability to withdraw electrons from the ring, forms O1 and P2 are destabilized because adjacent electron deficient atoms are unfavorable. Note that no forms are stabilized by the nitro group.

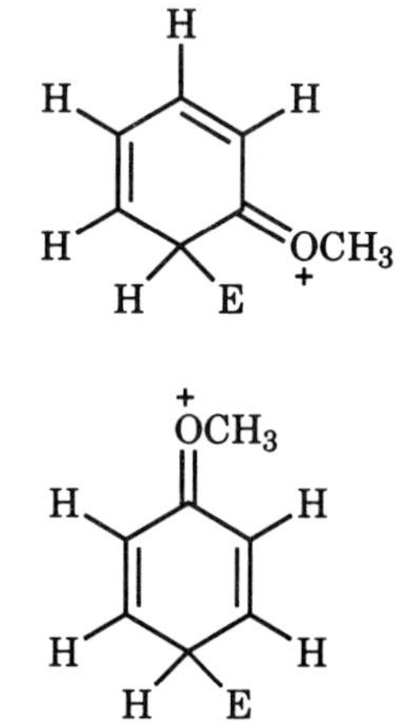

Figure 14.24 summarizes the resonance effects on the positional preference of electrophilic substitution when G is an electron donating (activating) group. Note that an activating group G especially stabilizes electrophilic substitution in the ortho and para position (indicated as "NICE!"). Thus activating groups are ortho-para directors.

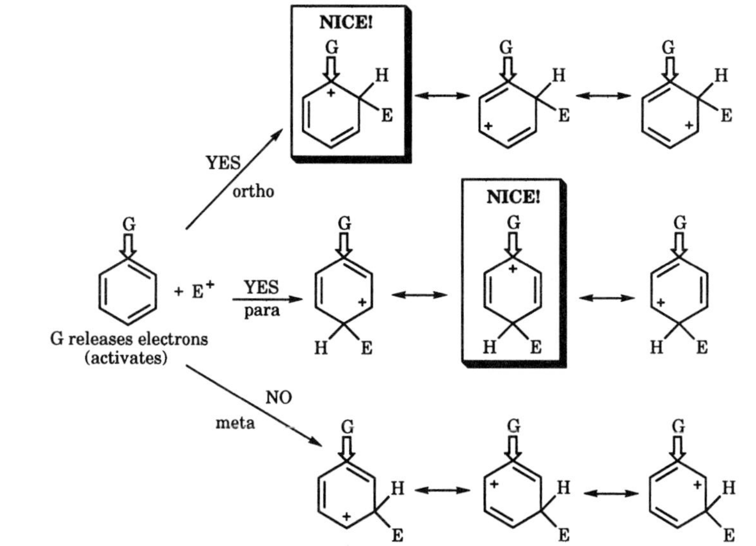

FIGURE 14.23: Additional resonance forms for ortho and para substitution.

FIGURE 14.24: Ortho para directing influence of a group that activates electrophilic aromatic substitution.

Conversely, Figure 14.25 summarizes the resonance effects on the positional preference of electrophilic substitution when G is an electron withdrawing (deactivating) group. Note that a deactivating group G especially destabilizes electrophilic substitution in the ortho and para position (indicated as "UGH!"). Thus deactivating groups not only inhibit overall electrophilic substitution, they especially inhibit ortho and para substitution. Thus the substitution that does occur on a ring with a deactivating group will predominantly occur at the meta position.

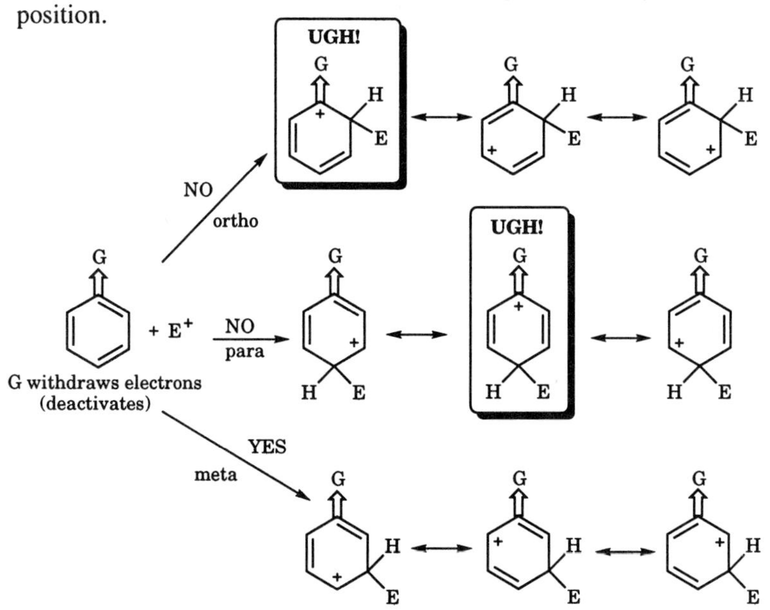

FIGURE 14.25: Meta directing influence of a group that deactivates electrophilic aromatic substitution.

The halides prove to be the exception to the general rule that deactivators are meta directing. Fluoride, chloride, bromide, and iodide are all deactivating, yet direct electrophiles to positions ortho or para to the halide. This seeming contradiction can be explained by a combination of two influences: the electron withdrawing nature of the halides (deactivating), and the resonance stabilization by halides of the carbocation intermediate formed upon substitution at the ortho and para positions. As can be expected due to the high electronegativity of the halides, the dominant influence is electron withdrawal, which means halides deactivate the ring towards electrophilic substitution. Yet the substitution that does occur is more likely at the ortho and para positions because of the resonance at those positions, shown in the figure below.

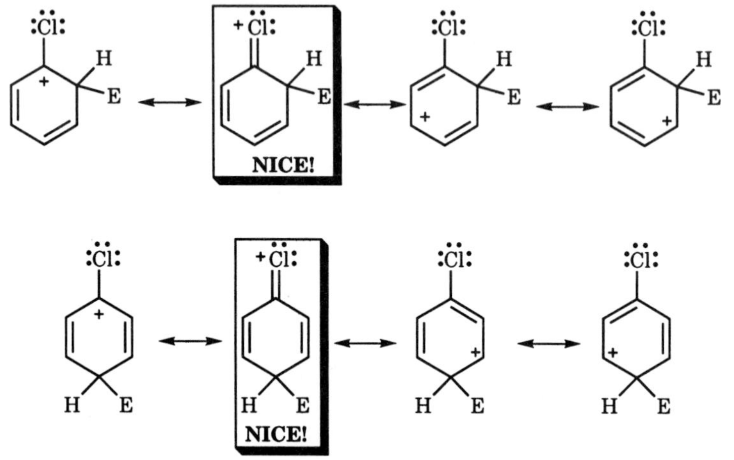

FIGURE 14.25: Resonance forms of the carbocation intermediate formed upon nitration of chlorobenzene.

In summary, deactivating groups (shown in Table 14.2) are meta directors, with the exception of the ortho/para directing halides.

14.11 TRI-SUBSTITUTED AROMATICS

The presence of one functional group on the aromatic ring has been shown to influence how readily and where a second electrophilic substitution will occur. The presence of two functional groups on the ring will also influence the amount and position of additional electrophilic substitution. For example, chlorination of *p*-nitrotoluene results in 2-chloro-4-nitrotoluene as the major product, as seen in the equation below.

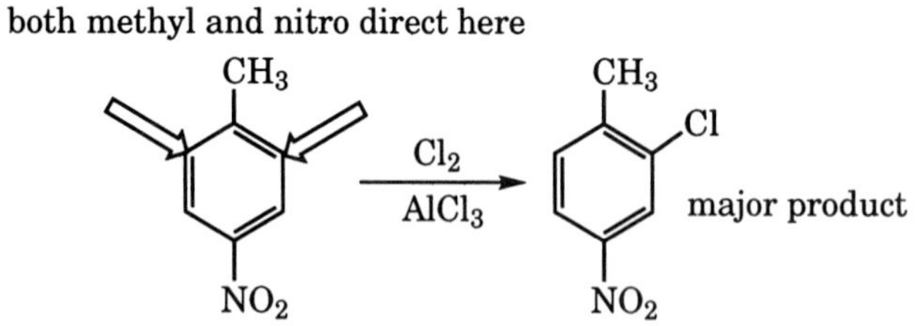

EQ. 14.9

Rule One: if the two groups currently on the aromatic ring each direct an electrophile to a common position, substitution at that position will predominate.

Since the methyl group is an activating group, it directs ortho and para. Since the nitro group is deactivating, it is a meta director. Substitution ortho to the methyl group coincides with substitution meta to the nitro group. Thus the two groups direct the electrophile to a common position.

> **Rule Two:** if the two groups currently on the ring do not direct an electrophile to a common position, the more powerful activating group has the greater effect. Often, however, mixtures of products result.

For example, consider the chlorination of *p*-cresol (4-hydroxytoluene), seen below. Both the methyl group and the hydroxyl group are activators, and thus both are ortho/para directors. The hydroxyl group is the stronger activator, as seen in Table 14.1, so the major chlorination product occurs ortho to the hydroxyl group.

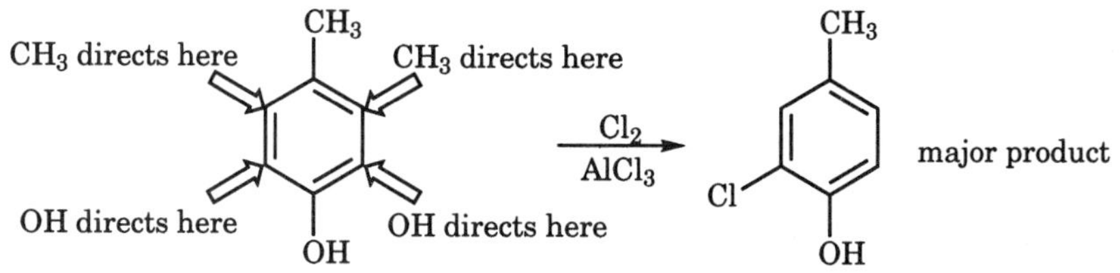

CH_3 directs here; CH_3 directs here; OH directs here; OH directs here; $\xrightarrow[AlCl_3]{Cl_2}$ major product

EQ. 14.10

> **Rule Three:** substitution rarely happens between two groups meta to each other.

The nitration of 3-chlorotoluene is shown in the equation below. Although chlorine is deactivating and methyl is activating, both are ortho/para directors. As such, the two groups direct to the three common sites indicated by the arrows. The site that is ortho to both the methyl and chlorine is thought to be too congested for the electrophilic substitution reaction to occur. Consequently the major products are 3-chloro-4-nitrotoluene and 5-chloro-2-nitrotoluene.

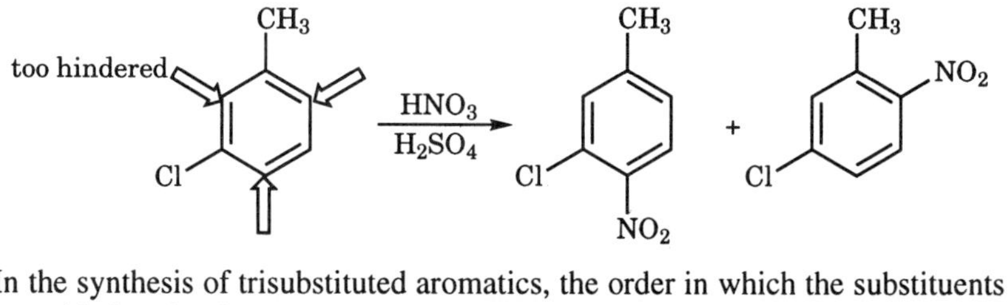

too hindered; $\xrightarrow[H_2SO_4]{HNO_3}$ +

EQ. 14.11

In the synthesis of trisubstituted aromatics, the order in which the substituents are added to the ring can matter greatly. Sometimes, however, the position of earlier substitutions make no difference in the synthesis of the final product. In the following synthesis of a suspected sex attractant of ticks, the starting material, phenol, is subjected to two sequential chorination reactions. It does not matter whether the first chlorine goes ortho or para to the hydroxyl group; the second chlorination yields the same product in both cases.

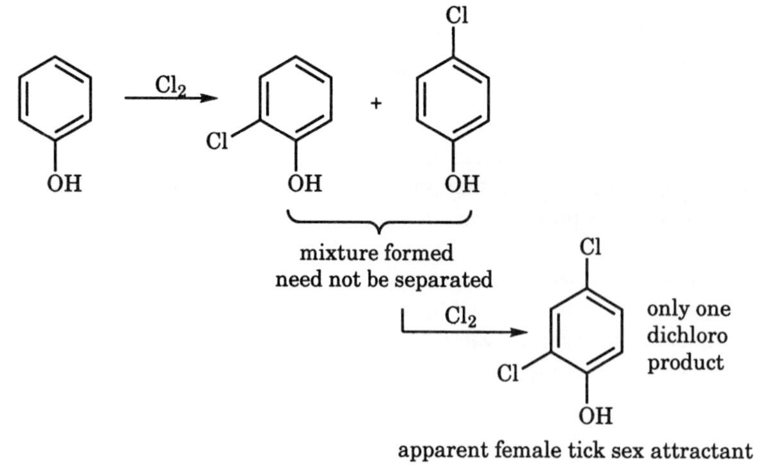
Cl
Cl₂
+
Cl
OH
OH
OH
mixture formed
need not be separated
Cl
Cl₂
only one
dichloro
product
Cl
OH
apparent female tick sex attractant

FIGURE 14.27: Sequential synthesis of a trisubstituted aromatic compound.

14.12 NUCLEOPHILIC AROMATION SUBSTITUTION

In the preceding sections, reactions were shown in which an electrophilic species substituted for a proton on an aromatic ring. Under some conditions, however, an electron rich species, a **nucleophile**, can attach to an aromatic ring. **Nucleophilic aromatic substitution** reactions require an anion as a leaving group, most commonly a halide.

There are two mechanisms for nucleophilic aromatic substitution, illustrated in Figure 14.28 and 14.29; both have the same two steps, but in opposite order. The first mechanism has an addition step followed by an elimination step; the second mechanism has an elimination step followed by addition to a **benzyne intermediate**. We shall examine both mechanisms.

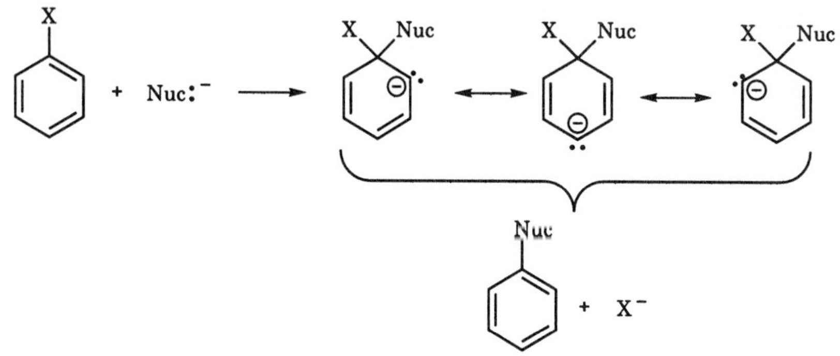
X
+ Nuc:⁻
X Nuc
X Nuc
X Nuc
Nuc
+ X⁻

FIGURE 14.28: Nucleophilic substitutions by addition-elimination.

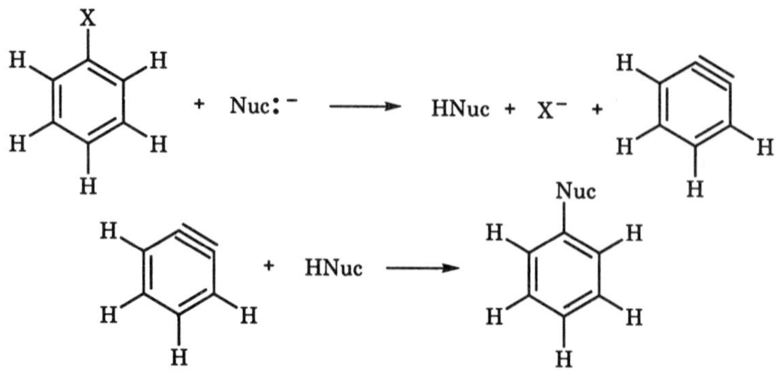
X
H
H
H
H
H
+ Nuc:⁻
HNuc + X⁻ +
H
H
H
H
+ HNuc
Nuc
H
H
H
H
H

FIGURE 14.29: Nucleophilic substitutions by elimination-addition.

Addition-Elimination Nucleophilic Aromatic Substitution

In order for a nucleophile to be attracted to a benzene ring, the electron density of the ring must be reduced. The most common group that facilitates this type of reaction is the nitro group, which we have seen is a strong ring deactivator. Nitro groups draw electron density away from aromatic rings and make the rings more susceptible to attack by nucleophiles. One of the most common types of aromatic molecules that undergo nucleophilic substitution are ortho and para nitro substituted halides. As you can see in the following example, increasing the number of nitro groups in the ortho and para positions enable the substitution to occur under milder conditions.

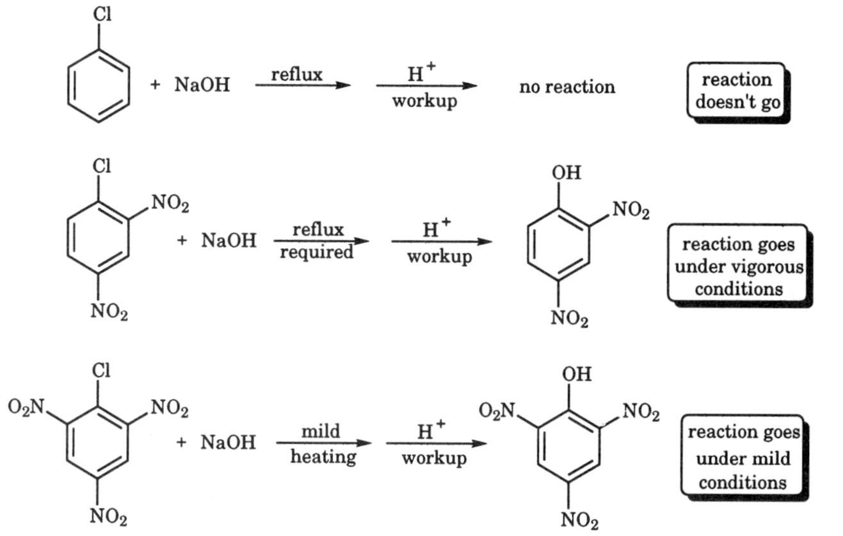

FIGURE 14.30: Influence of the nitro group on reactivity via nucleophilic aromatic substitution.

The mechanism has two steps: addition of the nucleophile, hydroxide ion, followed by elimination of the halide. As seen before, the delocalization of charge through various resonance forms plays an important role in the mechanism.

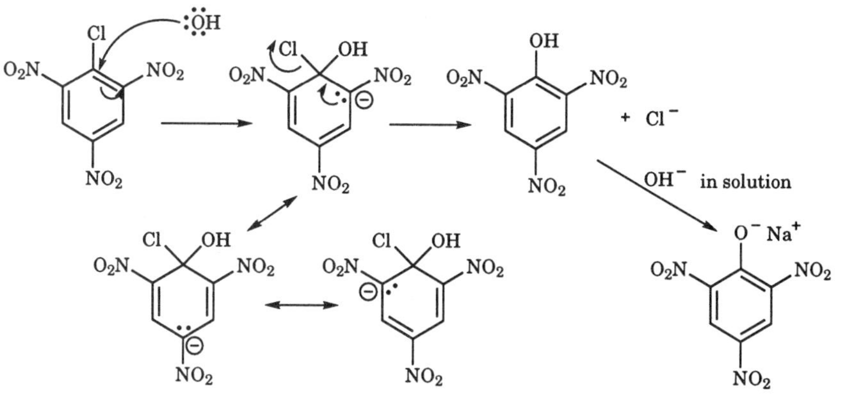

FIGURE 14.31: Reaction of 1-chloro-2,4,6-trinitrobenzene with hydroxide.

Note that resonance forms with **carbanions** adjacent to the electron withdrawing nitro group are stabilized, as shown in Figure 14.32. Increasing the number of appropriately positioned nitro groups further stabilizes the intermediate anion, and thus facilitates the nucleophilic substitution.

When the nitro groups are meta to the chloride, as in 3,5-dinitrochlorobenzene, substitution of Cl by OH does not occur. Nitro groups in the meta position do not stabilize the carbanion, as shown in Figure 14.33.

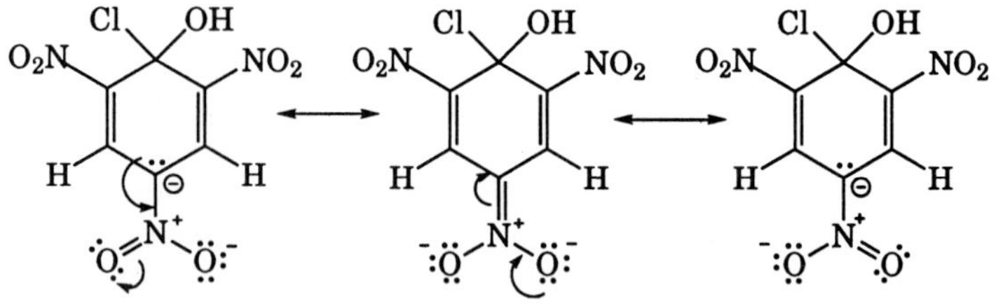

FIGURE 14.32: Anion stabilization for a nitro group para to the substitution site.

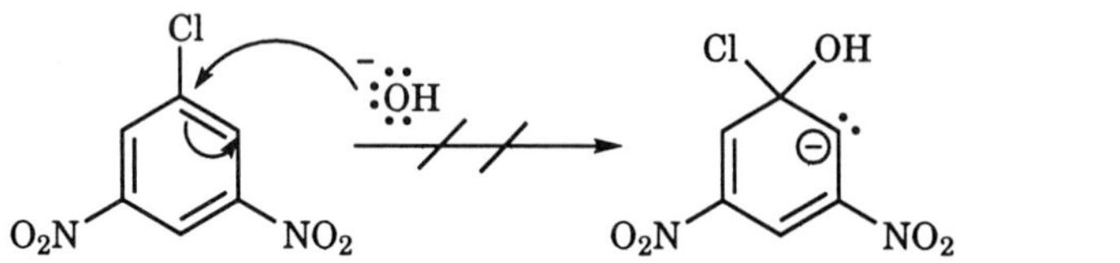

charge **not** stabilized by resonance through a nitro group

FIGURE 14.33: Mechanism showing lack of anion stabilization for a nitro group meta to the substitution site.

ELIMINATION-ADDITION NUCLEOPHILIC AROMATIC SUBSTITUTION

In the presence of a very strong base, nucleophilic aromatic substitution can begin with elimination of HX from the aromatic compound to form benzyne. The protonated nucleophile can then add to the highly reactive benzyne. For example, when $NaNH_2$ reacts with para-bromotoluene (C_7H_7-Br), toluidine (C_7H_7-NH_2) is obtained in a good yield. The toluidine synthesized in this fashion, however, contains about equal amounts of para and meta isomers!

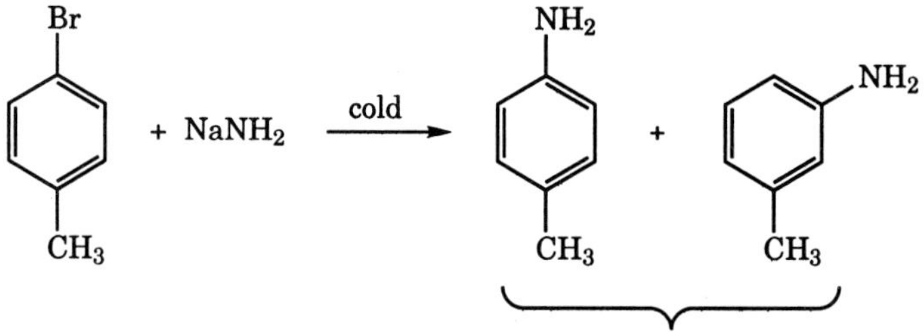

equal amounts!

FIGURE 14.34: Nucleophilic substitution yielding para and meta isomers.

The combination of para and meta isomers is unusual. Remember that when large amounts of two isomers are obtained in electrophilic aromatic substitution, the isomers are always ortho and para. The simultaneous formation of meta and para isomers is explained by the elimination addition mechanism of nucleophilic substitution in Figures 14.35 and 14.36.

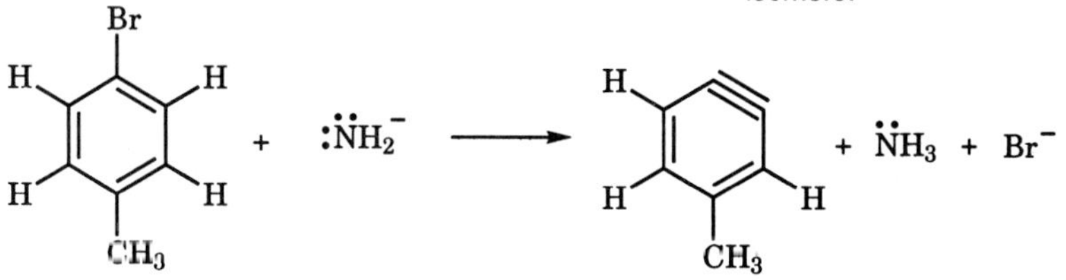

FIGURE 14.35: Formation of a benzyne intermediate via elimination.

The benzyne molecule has two alkyne carbon atoms using sp hybrid orbitals. These carbon atoms demand a bond angle of 180 degrees. But the bond angle in a flat six member ring is 120 degrees. This strain results in benzyne being a very unstable and highly reactive molecule. As such, benzyne reacts readily with the protonated nucleophile. Addition to the triple bond occurs with the nucleophile being equally likely to bond to either of the triply bonded carbons. Thus large amounts of both para and meta toluidine result.

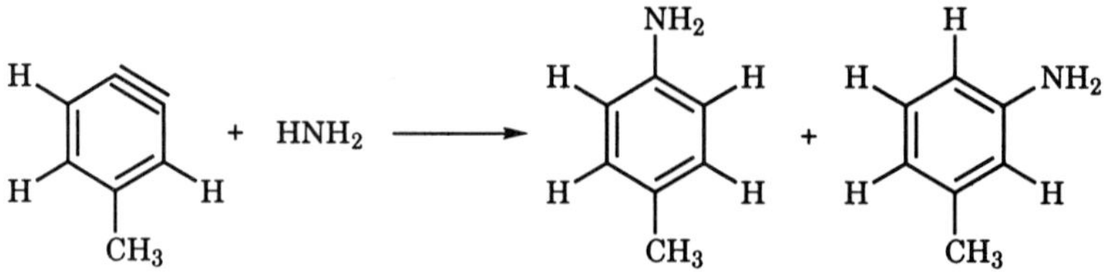

FIGURE 14.36: Formation of toluidine from benzyne.

Activators: in electrophilic aromatic substitution reactions, any substituent on the aromatic ring that increases the electron density of the ring.

Acylation: the substitution of hydrogen by an acyl group on an aromatic ring.

Alkylation: the substitution of hydrogen by an alkyl group on an aromatic ring.

Benzyne intermediate: a short lived species in which benzene contains a triple bond.

Carbanion: a negatively charged carbon anion.

Carbocation: a positively charged carbon ion.

Catalyst: a substance that increases the rate of a reaction, but is not consumed by the reaction.

Deactivators: in electrophilic aromatic substitution reactions, any substituent on the aromatic ring that decreases the electron density of the ring.

Electrophile: literal meaning, "electron loving"; any positively charged or electron-deficient species.

Electrophilic aromatic substitution: any of a variety of reactions in which an electrophile is substituted for hydrogen on an aromatic ring.

Friedel-Crafts Reaction: typically an alkylation or acylation of an aromatic ring using an electrophile generated from an organic halide and aluminum chloride.

Halogenation: the substitution of hydrogen by a halogen atom on an aromatic ring.

Intermediate: a chemical species formed then rapidly destroyed in a reaction mechanism.

Mechanism: a sequence of elementary steps by which reactants become products.

Nitration: the substitution of hydrogen by a nitro group on an aromatic ring.

Nucleophile: literal meaning, "nucleus loving"; any negatively charged or electron-rich species.

Nucleophilic aromatic substitution: any of a variety of reactions in which a nucleophile is substituted for another species on an aromatic ring.

Sulfonation: the substitution of hydrogen by a SO_3H group on an aromatic ring.

HOMEWORK Problems

1. Give the major products(s) formed when each of the following aromatic rings react with HNO_3/H_2SO_4.
   (a) ethylbenzene
   (b) ethyl benzoate
   (c) bromobenzene
   (d) ethoxybenzene
   (e) *para*-dichlorobenzene
   (f) *ortho*-dichlorobenzene

2. Starting with benzene, indicate the other reagents used and show the reactions used to prepare each of the following. (assume ortho and para isomers can be separated)
   (a) 2-chlorotoluene
   (b) 3-chlorobenzenesulfonic acid
   (c) 3-acetyltoluene
   (d) *ortho*-nitrotoluene
   (e) *meta*-nitrobenzene

3. Write reactions to show how nitronium ion could be formed in nitration reactions in which only concentrated nitric acid is present.

4. Write a mechanism to show how toluene reacts with A in the presence of phosphoric acid to form B.

A

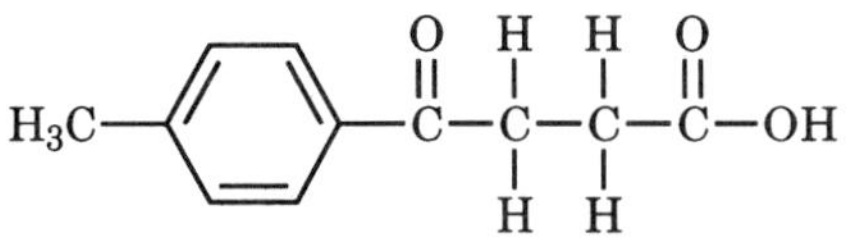

B

5. If chlorination of methoxybenzene were to give the monochloromethoxybenzene isomers on a random basis (*i.e.,* based only on a statistical probability without considering chemical properties) what would be the percentages of ortho, meta and para isomers?

6. Chlorination of $CF_3C_6H_5$ yields primarily the meta isomer. Explain.

7. What product would you predict upon monobromination (using Br_2 and $AlBr_3$) in each case.

a.

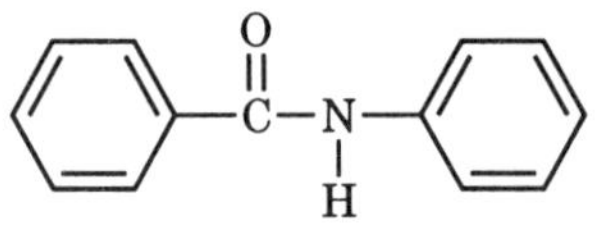

b.

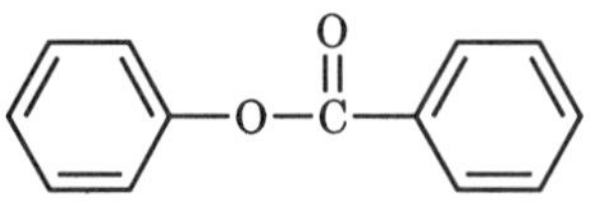

8. Draw the mechanism for the reaction of 4-methylbenzenesulfonic acid with steam to give toluene.

9. The reaction of ethyl bromide and benzene with aluminum bromide gives two products, A and B, which were separated by column chromatography. Elemental analysis gave $C_{10}H_{14}$ for both A and B. The CNMR spectrum of A showed 4 peaks and B showed 5 peaks. Give the structures of A and B and explain the CNMR spectra.

10. A method often used to label carbons on a ring utilizes C-14. When the chlorobenzene labeled with C-14 (as shown in A) was reacted with potassium amide, the aniline produced was analyzed for the position of the C-14. Approximately equal amounts of B and C were formed. Explain this result by drawing a mechanism.

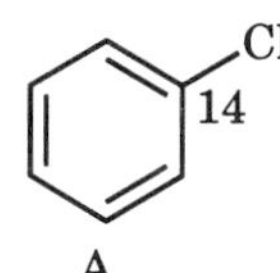

A

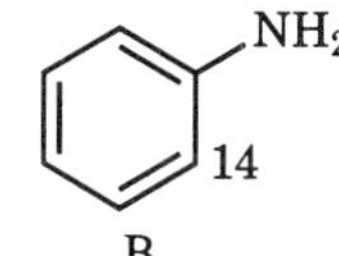

B

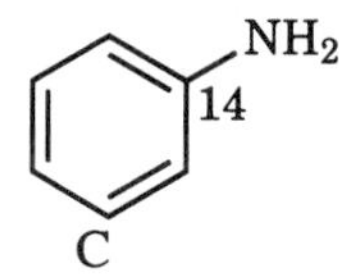

C

11. Write the mechanism for the reaction of methoxybenzene with acetyl chloride and aluminum chloride which produces 4-methoxy-acetylbenzene.

12. Write the mechanism for the reaction of methoxybenzene with acetic anhydride and phosphoric acid which produces 4-methoxy-acetylbenzene.

13. Draw all steps in the mechanism by which KCN reacts with 2,4-dinitrobromobenzene to an give excellent yield of 2,4-dinitrocyanobenzene.

14. Write the mechanism for the reaction of sodamide ($NaNH_2$) with *para*-chlorotoluene which produces both meta and para-amino-toluene.

CHAPTER FOURTEEN

OVERVIEW Problems

15. Iodobenzene can be made by reacting benzene with iodine in the presence of an oxidizing agent such as hydrogen peroxide.

(a) Write a balanced reaction showing the formation of I^+ from I_2 and H_2O_2 in acid solution.

(b) Show the rest of the mechanism by which iodobenzene is formed from I^+ and benzene.

16. In the text, the electrophile NO_2^+ was shown to form from nitric acid and sulfuric acid, which ultimately yielded aromatic nitro derivatives. The nitrosonium ion, NO^+, can similarly be formed by reacting nitrous acid (HONO) and a strong acid like HCl. A mixture of phenol, sodium nitrite, and excess HCl is known to yield primarily para-nitrosophenol (HOC_6H_4NO.)

(a) Write a balanced reaction showing the formation of nitrosonium ion from the nitrite ion (NO_2^-) and excess H^+.

(b) Show the mechanism by which the nitrosonium ion undergoes electrophilic substitution reaction with phenol.

17. Aromatic compounds are typically resistant to reduction reactions that are effective on typical alkenes.

(a) Draw the structure of Z-1-phenyl-2-butene.

(b) Predict the product of the reaction of Z-1-phenyl-2-butene with H_2 over a Pd catalyst.

18. Aromatic rings can be reduced using alkali metals in a process known as the Birch reduction. One example is benzene reacting with lithium to form 1,4-cyclohexadiene.

(a) Draw the structure of 1,4-cyclohexadiene.

(b) Write a balanced redox reaction between lithium and benzene (assume an acidic environment).

(c) In practice, one can not use lithium in an aqueous medium. Instead, a mixture of ethanol and liquid ammonia is used as the solvent, with ethanol serving as the proton donor. Write a better version of the reaction, starting with lithium and benzene in an ethanol solution.

19. The monochlorination of mesitylene (1,3,5-trimethylbenzene) could potentially result in chlorine substituting on the ring to form (A) 2-chloro-1,3,5-trimethylbenzene, or substituting on a methyl group to form (B) 5-chloromethyl-1,3-dimethylbenzene.

(a) Draw the structures of (A) and (B).

(b) Do you expect A or B to be the more common product, and why?

(c) How many carbon environments are present in A? in B?

(d) How would one distinguish between (A) and (B) based on proton NMR?

20. Toluene reacts with hot $KMnO_4$ to form benzoic acid. The benzoic acid formed from this reaction is used to make ethyl benzoate as the final product.

(a) Write the balanced reaction between toluene and $KMnO_4$.

(b) Write a balanced reaction using benzoic acid to form ethyl benoate.

(c) If 10.0 g of ethylbenzoate are to be produced, what is the minimum amount (in grams) of toluene one must have at the start? What minimum amount of $KMnO_4$ (in grams) must be used in the synthesis?

21. Two compounds (C and D) share the same molecular formula (C_8H_{10}). Vigorous oxidation of (C) results in benzoic acid, while (D) is oxidized to form ortho-phthalic acid. The carbon NMR for (C) shows 6 carbon environments, while (D) has only 4 carbon environments. The proton NMR spectrum of (C) results in signals at 7.3, 2.6, and 1.2 ppm. The proton NMR spectrum of (D) results in signals at 7.1 and 2.3 ppm.

(a) Draw the structures of (C) and (D).

(b) Indicate the expected integration ratio of the three peaks in the proton NMR spectrum of (C). What splitting does one expect for the peaks at 2.6 and 1.2 ppm?

(c) Indicate the expected integration ratio of the two peaks in the proton NMR spectrum of (D). What splitting does one expect for the peak at 2.3 ppm?

(d) As possible confirmation, what differences might one expect to find in the IR spectra of (C) and (D)?

22. In some respects, 1,3,5-trinitrobenzene (TNB) is a more effective explosive than the far more common 2-methyl-1,3,5-trinitrobenzene (TNT). Even though benzene is cheaper than toluene, TNT has a synthetic advantage over TNB.

(a) Explain the advantage of synthesizing TNT over TNB.

(b) Assuming that the products are N_2, CO_2 and H_2O, write the balanced reactions for the combustion of TNB and TNT.

(c) The combustion of TNT releases 3,434 kJ of energy per mole of TNT. Calculate the energy released per gram of TNT.

23. The following compounds have the indicated dipole moments: aniline (1.53 D); bromobenzene (1.70 D); para-bromoaniline (2.91 D).

(a) Draw the structures of the three compounds.

(b) Based upon your knowledge of activating and deactivating groups, explain the dipole moments. Your answer should include a clear indication of the direction of the dipole moment.

24. The following compounds have the indicated dipole moments: anisole (1.38 D); nitrobenzene (4.22 D); para-nitromethoxybenzene (5.26 D).

(a) Draw the structures of the three compounds.

(b) Based upon your knowledge of activating and deactivating groups, explain the dipole moments. Your answer should include a clear indication of the direction of the dipole moment.

CHAPTER fifteen

COMPLETE Chaos AND IDEAL BEHAVIOR

CHAPTER FIFTEEN

COMPLETE CHAOS and Ideal Behavior

15.1 THE MODEL OF A GAS

Below, we show models of the three phases of matter: solid, liquid and gas. In the next two chapters we examine the behavior of these phases of matter in more detail. The gas phase is treated first since it is the simplest phase to understand. The liquid and especially the solid phase are in many respects more orderly than the gas phase. Their order occurs as a result of liquid and solid atoms/molecules being generally much closer together than gas molecules. Note that the distances between molecules are not to scale in Figure 15.1. Gas molecules are typically much farther apart than indicated by this figure.

Attractive and repulsive forces that depend upon the distance between molecules complicate our understanding of liquids and solids. This is much less of a problem in the gas phase. Gas molecules move more or less independently of one another since the distance separating them is so great.

One model of the gas phase is pictured by "marbles in a box", with the marbles (molecules) in constant, random motion. Although the model assumes numerous gas molecules are present, the volume inside the "box" is considered completely empty space, *i.e.,* the "marbles" represent a negligible fraction of the total volume. Collisions between molecules and between a molecule and the container are elastic (no net loss of kinetic energy.) The molecules are assumed to move in straight lines, with no attraction or repulsion between molecules. And finally, the average kinetic energy of a group of gas molecules depends on the temperature; increasing the temperature increases the speed and thus the kinetic energy of the molecules. The theoretical implications of such a model, known as the kinetic molecular theory, were developed in the late 1800's through the work of Ludwig Boltzmann (1844-1906), James Clerk Maxwell (1831-1879), and Rudolph Clausius (1822-1888.) But the practical understanding of gases and their behavior began much earlier.

FIGURE 15.1:
The solid, liquid and gas phases.

15.2 EARLY UNDERSTANDING OF THE PROPERTIES OF GASES

The gas phase has been known since the time of the ancient Greeks. Empedocles (484-424 B.C.) demonstrated the material existence of air by using a clepsydra, a kind of water lifter (Figure 15.2.)

Normally, as in Figure 15.2(a), this hollow vessel was dipped into water, a finger was then placed over the narrow opening on top, and the tube lifted from the water. Removing the finger from the top opening allowed the water to slowly sprinkle out. But, as shown in Figure 15.2(b), by blocking the narrow opening with his finger before immersing the water lifter, Empedocles showed that "something" blocked the entrance of water into

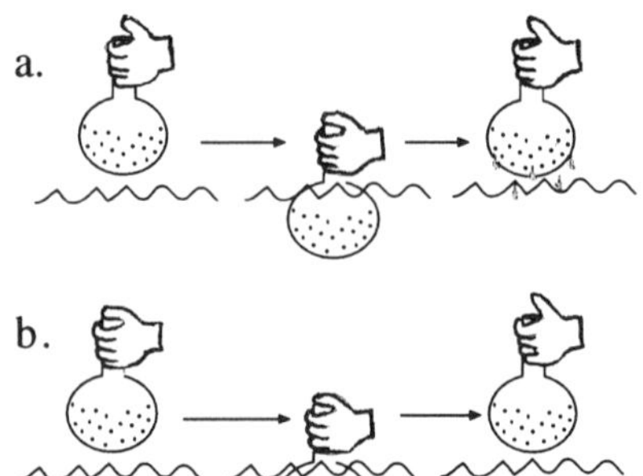

FIGURE 15.2:
Clepsydra.

the clepsydra. That "something" was air, an invisible but demonstrably material gas.

That gases occupied more space than liquids, and that gases had energy when heated was shown by Hero of Alexandra (1st century A.D.), whose "aeolipile" (wind-ball) is pictured in Figure 15.3:

FIGURE 15.3:
The aeolipile.

When water in the bottom of the ball was heated by fire, steam formed. As the steam rapidly exited the side arms, the jet-like thrust caused the ball to spin. Although the aeolipile was simply a novelty item of the day, it proved to be a significant example that aided later understanding not only of gas behavior, but of thermodynamics as well.

15.3 PRESSURE AND VOLUME

The first quantitative experiment on the properties of gases was conducted by Robert Boyle (1627-1691.) Boyle was fascinated by the compressibility of gases, and investigated air by squeezing and expanding it systematically. His report, "Touching the Spring of the Air" likened the behavior of air to that of a wound metal spring: elastic, with definite laws describing its behavior. His analogy is not surprising, given that his co-worker, Robert Hooke (1635-1703), also studied the physics of springs. Hooke showed that the restoring force exerted by a spring was proportional to its distance of displacement: i.e., the farther one compresses a spring, the greater its force pushing against the compression. The length of a spring is related to the force applied to it. The volume of a gas, Boyle determined, is also related to a force. More specifically, an increase in the force applied per unit area (or pressure) of the gas causes a decrease in the volume.

A device for measuring pressure had been invented earlier by Evangelista Torricelli (1608-1647.) A glass cylinder closed at one end was immersed in a liquid (commonly mercury.) The tube was then partially lifted from the pool of liquid in which it was immersed, taking care that the open end remained always under the liquid. The result was that a certain length of liquid remained supported in the column (Figure 15.4a). The length of the column of liquid was always the same: roughly 30 inches for mercury. Changing the diameter of the glass cylinder did not change the height of liquid supported (Figure 15.4b). Making the glass cylinder longer did not alter the length of the column of mercury (Figure 15.4c). If the opening of the liquid filled tube was closed, then the plug removed when the extra-long cylinder was upright but still submerged, the liquid level in the tube would fall to the same height seen in shorter tubes. The empty space above the mercury in the glass cylinder was essentially a vacuum.

FIGURE 15.4:
A mercury barometer.

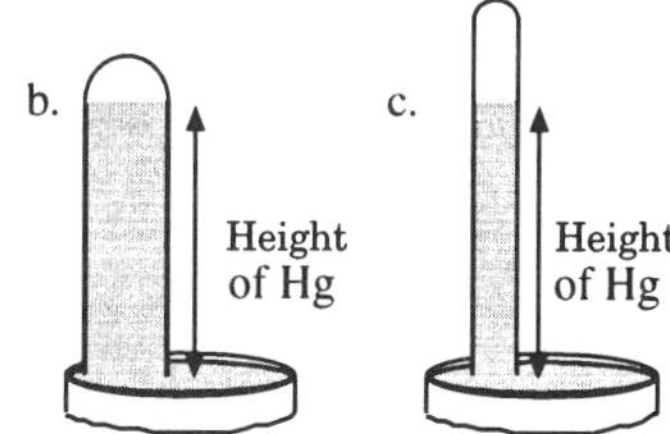

That the mass of the mercury was pushing down with a certain force was undeniable; there had to be a compensating force pushing back. This force was related to atmospheric pressure. Torricelli's device, called a **barometer,** was later examined by

Blaise Pascal (1623-1662), who discovered that varying the liquid used changed the height of the liquid supported by the atmosphere: 30 inches of mercury, 34 feet of water, 34.6 feet of red wine, and 40 feet of oil were supported by the atmosphere. The pressure units of **"torrs"** and **"Pascals"** are used in honor of these researchers. An exact relationship between the height of liquid supported and units of pressure is shown in Appendix G-1.

Figure 15.5 shows an experiment that measures the same variables that Boyle did in his classic study. The volume of a gas is varied, and the pressure the gas exerts at each volume is measured. It is assumed that during the course of the experiment, the temperature remains constant and the gas does not leak into or out of the system.

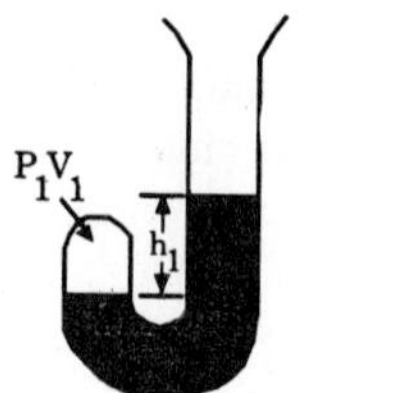

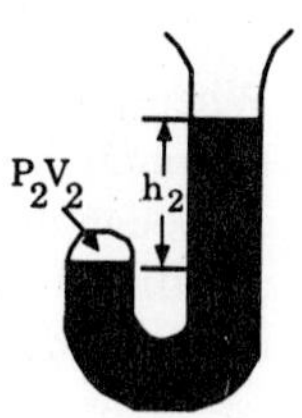

FIGURE 15.5: Boyle's Law shown by trapping gas in a J-tube.

The data from this experiment are ordered pairs, (P, V). A sample table of data is included. A graphical representation of the results is given in Figure 15.6

TABLE 15.1: RELATION BETWEEN P AND V

P/psi	V/cm^3
5	20.0
10	10.0
15	6.73
17	5.88
20	5.00
22	4.55
30	3.33
40	2.50

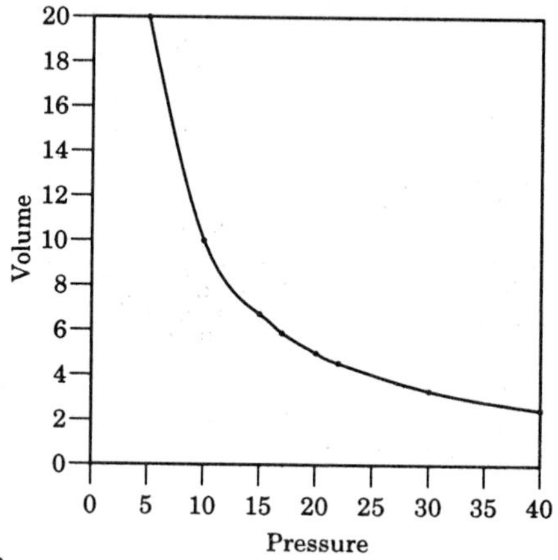

FIGURE 15.6: The volume of gas vs. its pressure.

Analysis of the data shows a simple relationship. The product of pressure times volume is shown to be a constant. This relationship is known as **Boyle's Law,** and is stated:

$$P \times V = \text{constant}$$

(with a fixed temperature and amount) EQ. 15.1

If the initial amount of gas examined, or the temperature of the experiment is varied, the value of the constant changes, but the relationship remains the same.

15.4 THEORETICAL INTERPRETATION OF BOYLE'S LAW

Boyle's Law was an empirical result. When the volume was halved, the pressure doubled. There was no theoretical justification for this relationship. That gas was composed of separate molecules was not even theorized until 1738 by Daniel Bernoulli (1700-1782.) With the emergence of kinetic molecular theory in the late 1800's, the question "Why?" could finally be answered.

Consider the gas sample shown in Figure 15.7. The pressure exerted by the gas is the result of the force of collisions between the gas molecules and the container. Other things being equal, the more collisions the gas molecule makes per unit time, the higher the pressure. Consider a box, with a single gas molecule in it, with a velocity in the x direction of v_x .

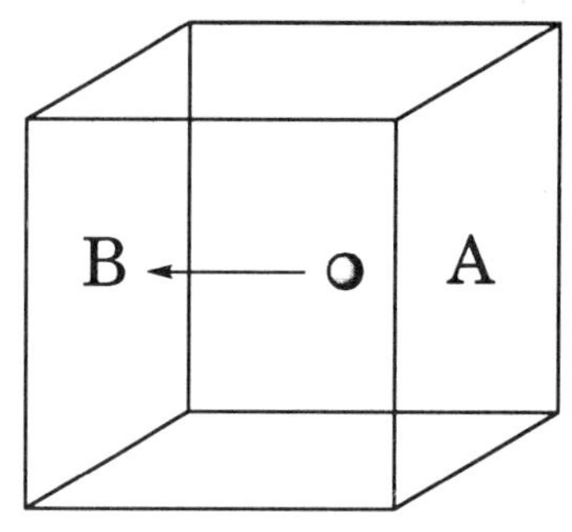

FIGURE 15.7: Box with one gas molecule.

The molecule moves with constant velocity v_x until it hits side B. Then it reverses itself and travels in the opposite direction with velocity $-v_x$ until it hits side A. The molecule then heads to side B again. The kinetic molecular theory shows that if the temperature remains constant, the average speed of this gas molecule remains constant. If the dimension of the box is decreased by a factor of two, the distance between side A and side B is halved. Since the speed is constant, half the distance can be traveled in half the time. Instead of hitting side B "n" times per minute, halving the distance means the molecule will hit side B "2n" times per minute. Decreasing the volume by half effectively doubles the number of collisions per second, and thus doubles the pressure. A more exact treatment of this is provided in Appendix G-2.

15.5 TEMPERATURE AND CHARLES'S LAW

Temperature is most commonly measured by a mercury in glass thermometer. A fixed amount of liquid mercury is enclosed in glass, the top part of which is a thin capillary along which marks are made referring to the corresponding temperature. As the thermometer is heated, the mercury expands. This drives the mercury thread up the capillary, and the level of mercury reaches a higher value on the scale. The following table shows the volume of a gram of mercury at various temperatures:

TABLE 15.2: VOLUME PER GRAM OF LIQUID MERCURY AT SELECTED TEMPERATURES

t/°C	volume/mass (ml/g)
0	.0735540
10	.0736877
20	.0738215
30	.0739552

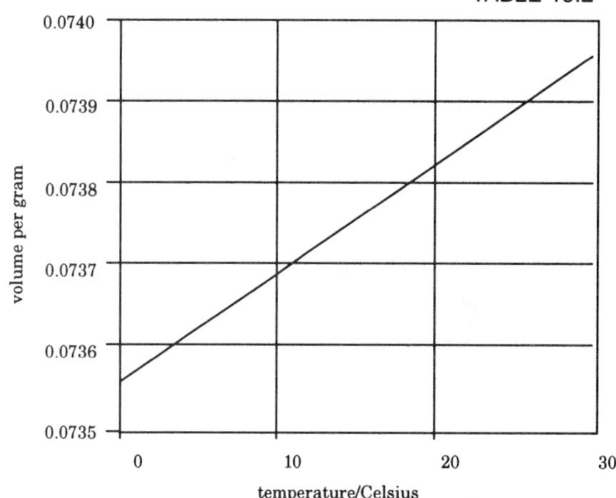

FIGURE 15.8: Volume of liquid mercury as a function of temperature.

The volume occupied by a gram of liquid mercury is seen to increase with temperature. This is true for most materials, including gases. We can measure similar data for the volume occupied by a gram of hydrogen gas. From Boyle's Law, we know that pressure will influence volume, so we will keep the pressure constant and measure the volume of hydrogen gas as a function of temperature.

TABLE 15.3: VOLUME PER GRAM OF HYDROGEN GAS AT SELECTED TEMPERATURES

T/°C	volume/mass (ml/g)
0	11.11843
10	11.52547
20	11.93252
30	12.33956

TABLE 15.3

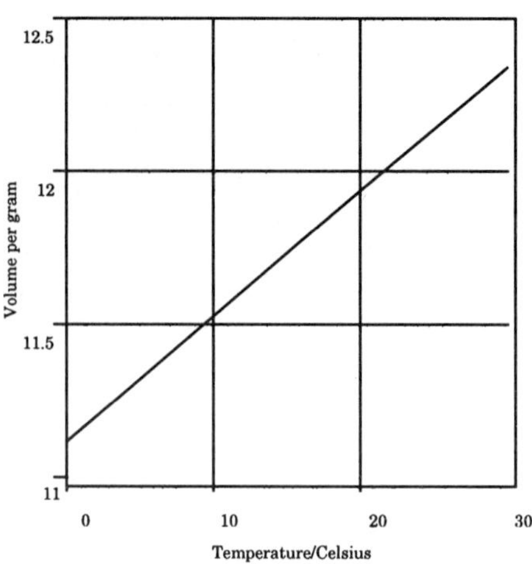

FIGURE 15.9: Volume of gaseous hydrogen as a function of temperature.

The graphs of the volumes of liquid mercury and gaseous hydrogen are both linear over the temperature range studied. They differ in several respects, however. The values for the volumes and the change in volumes for mercury are both much less than the corresponding values for hydrogen. Another difference is what happens when lower and lower temperatures are examined. Mercury will freeze at about -39°C, while hydrogen gas does not liquefy until about -253°C. As we shall soon see, extrapolating the linear behavior of gases such as hydrogen to lower and lower temperatures is quite revealing.

The trend to occupy more volume per mass with increasing temperature is common for nearly all materials. Note that this is the same as saying that the density of most materials decreases as temperature increases. In June of 1783, Joseph (1740-1810) and Etienne (1745-1799) Montgolfier were the first humans to have a controlled flight, rising in a hot air balloon. As the air within the balloon was heated, it became less dense than the surrounding air. The hot air balloon rises for the same reason that objects which are less dense than water float upwards when submerged in water: buoyancy.

Two months later, Jacques Alexander Cesar Charles (1746-1823) launched the first hydrogen balloon. He generated the hydrogen gas by reacting large amounts of iron and mineral acids. Hydrogen balloons became known as "Charliers" in his honor. The other notable achievement of J. A. C. Charles was his discovery of **Charles's Law.** By collecting data, Charles showed that the volume of a gas increases as temperature increases. Expressing this relationship in terms of an equation (or even a proportionality) was difficult, however. For example, one could not simply state:

$$\text{volume (in ml)} \propto \text{a temperature (in °C)}$$

because at temperatures below 0°C, the volume does not become negative. As noted by William Thomson (1824-1907), later to become Lord Kelvin, it would be convenient to have a temperature scale composed of only positive values. The 0 value on such a scale would be the lowest conceivable temperature, an "absolute" zero. The centigrade scale was based on the freezing and boiling point of water being 0° and 100° respectively. The absolute temperature scale keeps these same intervals, but moves the absolute zero to that temperature where the volume of gases (as shown in Fig 15.10) extrapolate to zero volume. This occurs at -273.15°C. The degrees in the absolute temperature scale are known as Kelvin, symbolized K, in his honor. This convention makes the freezing and boiling points of water 273.15 and 373.15 K respectively.

Charles's Law can now be written as:

$$V \propto T$$

since volume goes to zero as T goes to zero, and negative values are impossible for both terms. The assumptions built into Charles's Law are that there are no gas leaks, and that pressure is constant. Another way of writing this relation would be as:

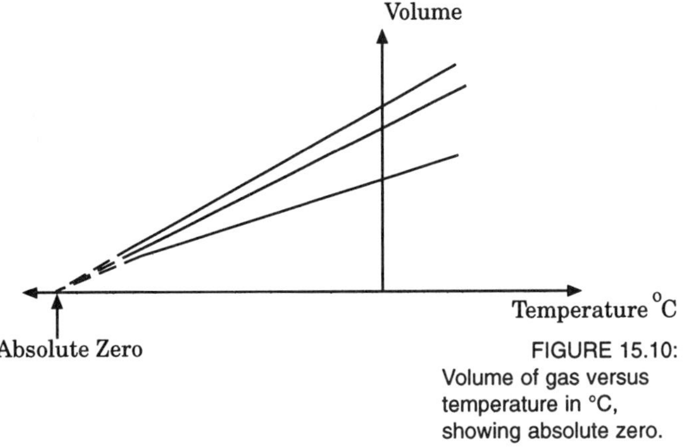

FIGURE 15.10: Volume of gas versus temperature in °C, showing absolute zero.

$$\frac{V}{T} = \text{constant} \quad \text{(amount, pressure of gas constant)} \qquad \text{EQ. 15.2}$$

Note that capital T by convention only refers to absolute temperature (i.e., in Kelvin.) That volume is proportional to temperature means that gases, as well as liquids, could be used as the material inside a thermometer. In fact, one type of highly sensitive thermometer does just that.

15.6 AMONTONS'S LAW

Guillaume Amontons (1663-1705) studied the relation between pressure and temperature. If we examine the results for 1.00 mole of hydrogen gas measured in a constant 1.00 liter vessel, the results are:

TABLE 15.4: PRESSURE OF HYDROGEN GAS AT SELECTED TEMPERATURES

t/°C	Pressure/ (atm)
0	22.41
10	23.23
20	24.06
30	24.88

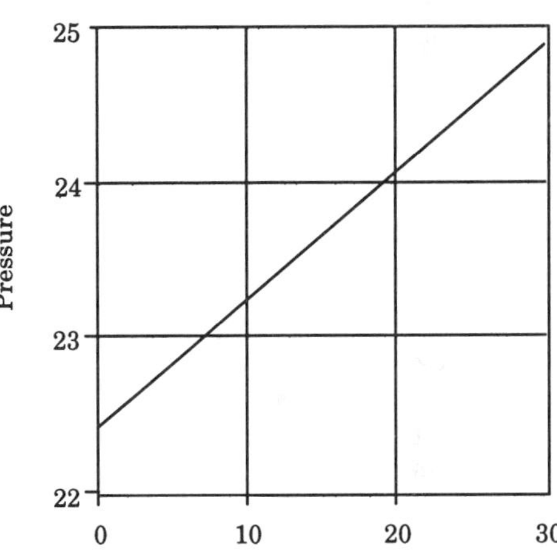

FIGURE 15.11: Pressure of hydrogen gas as a function of temperature.

The problems associated with a temperature scale that includes negative numbers again precludes a direct relationship between pressure and temperature in degrees Celsius. Instead, the plot is redone with T in Kelvin, including extra data at lower temperatures:

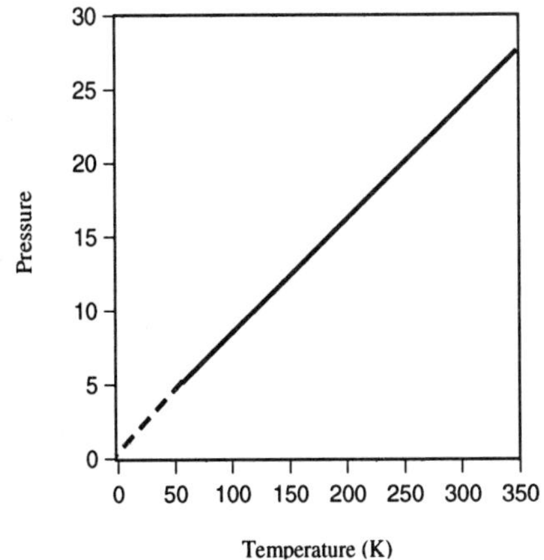

FIGURE 15.12: Pressure vs. absolute temperature.

Note that the intercept at 0 Kelvin is an extrapolation. Hydrogen gas no longer exists at that temperature. Other gases, provided the temperature is high compared to their boiling point and the pressure is reasonably low, show the same pattern observed by Amontons and Charles: linear P vs T and linear V vs T. **Amontons's Law** is shown below in the form of an equation.

$$\frac{P}{T} = \text{constant (amount, volume of gas constant)} \qquad \text{EQ. 15.3}$$

15.7 CHARLES'S LAW, AMONTONS'S LAW AND KINETIC MOLECULAR THEORY

If we return to our model of a gas, we note that the velocity at which the gas molecules move increases as temperature increases. If the sides of the container were rigid, the collisions against the walls would occur more frequently and with greater momentum. This generates a higher pressure. This is in accord with Amontons's Law.

If the walls of the container are flexible, such that the pressure inside the container must always equal the ambient pressure outside the container, something must happen to offset the faster motion of the gas molecules caused by an increase in temperature. What happens is the gas expands. By increasing the volume, the gas molecules (although traveling faster due to higher temperature) have farther to go before colliding with the walls and generating pressure. Thus an increase in temperature increases the volume.

You may have observed this if you have watched the initial inflation of a hot air balloon. The balloon starts out sideways, flat on the ground, with the walls all collapsed. A large burner is put into the opening. The gas molecules of the air, being heated, move more rapidly and push out the walls of the balloon. A more common occurrence happens when a partially filled 2 liter plastic pop bottle is removed from the refrigerator. If the cold pop bottle originally had an indentation, permitting it to sit out at room temperature warms the gas in the bottle, and often causes the bottle to "snap back" into its original shape.

15.8 AVOGADRO'S HYPOTHESIS

Besides Charles, another French chemist and balloon daredevil was Joseph Louis Gay-Lussac (1778-1850). His ascent in a hot air balloon in 1804 set a long-standing altitude record of 23,000 feet. In chemistry, Gay-Lussac is remembered for his law of combining volumes: Gases react at constant temperature and pressure such that their volume ratios can be given by small whole numbers. For example, at constant temperature and pressure, 1 liter of hydrogen reacts with 1 liter of chlorine to make two liters of hydrogen chloride. Note that volume is not necessarily conserved; 2 liters of hydrogen and 1 liter of oxygen combine to make 2 liters of water vapor.

Years later, Amadeo Avogadro explained Gay-Lussac's results by postulating that equal volumes of gases at the same temperature and pressure contained equal numbers of moles. **Avogadro's Hypothesis** can be expressed via:

$$V \propto n \text{ (amount); [provided P, T constant]}$$

This could also be written as:

$$\frac{V}{n} = \text{constant} \ \ [\text{assuming P, T constant}] \qquad \text{EQ. 15.4}$$

The ratio V/n is known as the molar volume.

15.9 COMBINING THE GAS LAWS

The various gas laws we have seen previously can be combined in various ways. One common way is the combined gas law, which states that:

$$\frac{P\,V}{T} = \text{constant} \ [\text{assumes n constant}] \qquad \text{EQ. 15.5}$$

In terms of relating one set of parameters to another for a fixed amount of gas, this means that:

$$\frac{P_1\,V_1}{T_1} = \frac{P_2\,V_2}{T_2} \qquad \text{EQ. 15.6}$$

In general the most encompassing and useful expression of ideal gas behavior is the ideal gas law, written as:

$$\frac{P\,V}{n\,T} = \text{constant} \qquad \text{EQ. 15.7}$$

This combination of Boyle's, Charles's, Amontons's, and Avagadro's Laws assumes nothing other than that the values used are not changing during the course of the experiment. The constant in the above equation is so common in science, it is given its own symbol and name: R, the gas constant. Depending upon the units used for pressure and volume, R can have different values. Note that n and T are always given in moles and Kelvin. The ideal gas law is also often written in the form:

$$PV = nRT \qquad \text{EQ. 15.8}$$

where the units can be, for example:

$$P\,(\text{atm})\ V\,(\text{liters}) = n\,(\text{moles})\ R\left(\frac{\text{lit atm}}{\text{mole K}}\right) T\,(\text{K}) \qquad \text{EQ. 15.9}$$

AUTHORS' NOTE:

Some students may have previously learned to solve gas law problems using a memorized value of 22.414 liters, or learned to memorize a series of equations involving ratios of initial and final parameters. The authors of this text strongly advise against such methods. One reason is that ratio multiplying either ignores or cancels units, which makes remembering the order of the ratio difficult for the student. It also permits (mathematically) silly things to happen, like using a ratio of temperatures in Celsius in a problem, without the chance to catch the problem by the units appearing in the calculation. Trying to memorize multiple equations, instead of one master equation, often leads to unnecessary mistakes, especially in test situations. Solving problems using the ideal gas law allows the student to follow the units carefully. As we shall soon see, the ideal gas law is also the basis for finding equations to model real gas behavior (instead of ideal gas behavior.) In short, we advise you to stick with
$P\,V = n\,R\,T$.

15.10 TYPES OF GAS LAW PROBLEMS

VARYING ONE PARAMETER TO CAUSE A CHANGE IN ANOTHER EXPERIMENTAL VARIABLE.

This type of problem usually specifies the initial values for two parameters and one final value for only one of those parameters. The final value of the other parameter is to be determined. The other variables in the ideal gas law equation are assumed constant.

PROBLEM: A sample of gas at 20.0°C and 5.00 atm pressure is heated to a temperature of 100.0°C. Assuming that the volume of the container is constant, the amount of gas is constant, and ideal behavior by the gas, calculate the new pressure.

This type of problem can be done several ways. If you start with the ideal gas equation, you first realize that there are two sets of experimental conditions: the initial and final. The constant that relates the two sets of variables is the gas constant R. We can express this via an equation:

$$R = \frac{P_i V_i}{n T_i} = \frac{P_f V_f}{n T_f}$$ EQ. 15.10

Note that we nearly always assume that n is constant, so it is simply termed n for both the final and initial cases. In this particular case, the initial volume equals the final volume, or Vi = Vf. Canceling n and V and rearranging the remaining terms yields:

$$\frac{P_i}{T_i} = \frac{P_f}{T_f}; \quad \text{or } P_f = \frac{T_f P_i}{T_i}$$ EQ. 15.11

Recalling that T must be in Kelvin (not Celsius) yields the final pressure via:

$$P_f = \left(\frac{100.0 + 273.15\text{K}}{20.0 + 273.15\text{K}}\right)(5.00 \text{ atm}) = 6.3\bar{6}4 \text{ atm} = \boxed{6.36 \text{ atm}}$$ EQ. 15.12

Problem: A sample of gas with an initial pressure of 752.0 torr occupies a volume of 1.50 liters. As shown in the figure at right, a valve is opened which adds an additional 3.00 liters of volume. Calculate the final pressure of the gas. We assume in this problem ideal behavior and constant temperature and amount of gas. Again, we start with the ideal gas law

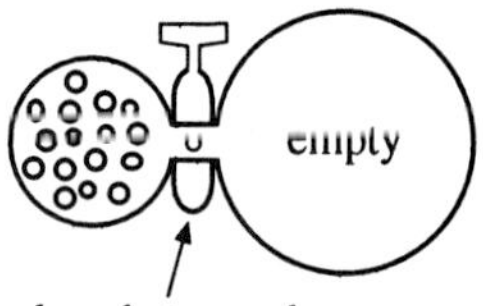

FIGURE 15.13: Two gas bulbs, one filled, one evacuated, connected by valve.

$$R = \frac{P_i V_i}{n T_i} = \frac{P_f V_f}{n T_f}$$

which under conditions of constant n and T becomes Boyle's law:

$$P_i V_i = P_f V_f$$ EQ. 15.13

Solving for the final pressure yields

$$P_f = \frac{P_i V_i}{V_f} = (752.0 \text{ torr})\left(\frac{1.50 \text{ lit}}{1.50 + 3.00 \text{ lit}}\right) = 25\bar{0}.6 \text{ torr} = \boxed{251 \text{ torr}}$$ EQ. 15.14

DETERMINING A SINGLE MISSING VARIABLE GIVEN INFORMATION ABOUT ALL OTHER VARIABLES IN THE GAS LAW.

In this type of problem, three of the four variables (n, P, V and T) are specified or sufficient information is given in order to determine their values. Then the value of the missing variable is determined.

PROBLEM: Calculate the volume of a balloon filled with 100.0 grams of helium at a pressure of 748.0 torrs and a temperature of 21.5°C. We have the temperature in Celsius, which must be converted to Kelvin. The pressure in torr is often converted to atmospheres in order to use a common value for R. The amount of gas in moles is determined from its mass via the molar mass. The volume is solved by:

$$V = \frac{nRT}{P} = \frac{(100.0\ g)\left(\frac{1\ mol}{4.0026\ g}\right)\left(0.0820578\frac{lit\ atm}{mol\ K}\right)(21.5 + 273.15K)}{(748.0\ torr)(1\ atm/760\ torr)} = 613.\bar{7}5\ lit = \boxed{613.8\ lit} \quad \text{EQ. 15.15}$$

Note that the temperature in Kelvin has four significant figures, even though the temperature in Celsius has three significant figures. Also note that the value of 760 torr equaling one atmosphere is a defined value, and essentially has an infinite number of significant figures.

DETERMINING OTHER PROPERTIES OF A GAS USING IDEAL GAS LAW PARAMETERS.

In this type of problem, several variables in the ideal gas law are specified, and some property of the gas is thereby determined.

PROBLEM: Calculate the density (in grams per liter) of helium at a pressure of 0.992 atm and a temperature of 310.0 K. At first it might seem that we do not have enough information. Density is the ratio of mass to volume, and we know neither from the information given. The ideal gas law, however, provides a ratio of moles to volume:

$$\frac{n}{V} = \frac{P}{RT} = \frac{0.992\ atm}{(.0820578\ lit\ atm\ mol^{-1}\ K^{-1})(310.0K)} = 0.038\bar{9}9\ \frac{mol}{lit} \quad \text{EQ. 15.16}$$

Combining this information with the literature value for helium's molar mass yields the density.

$$density = \left(0.038\bar{9}9\ \frac{mol}{lit}\right)\left(\frac{4.0026\ g}{1\ mol}\right) = 0.15\bar{6}1\ \frac{g}{lit} = \boxed{0.156\ \frac{g}{lit}} \quad \text{EQ. 15.17}$$

PROBLEM: Determine the molecular formula of an unknown gas 0.9014 grams of which occupies 495.8 ml at a temperature of 120.0°C and a pressure of 742.3 torr. The gas is 40.00% carbon, 6.71% hydrogen and the remainder is oxygen.

The empirical formula of the gas is determined by the methods used earlier. In Chapter 5, section 8, the listed % composition of carbon, hydrogen and oxygen was shown to result in an empirical formula of CH_2O. Remember that this means that the molecular formula is some multiple of this value: CH_2O (MW = 30.026), $C_2H_4O_2$ (MW = 60.053), $C_3H_6O_3$ (MW = 90.079), etc. The actual molecular identity is detemined from the molecular weight, which in this case can be determined from the gas law data. Solving the ideal gas law for moles yields:

$$n = \frac{PV}{RT} = \frac{(742.3\ \text{torr})\left(\frac{1\ \text{atm}}{760\ \text{torr}}\right)(495.8\ \text{ml})\left(\frac{1\ \text{lit}}{1000\ \text{ml}}\right)}{(.0820578\ \text{lit atm/mol K})(120.0 + 273.15\text{K})} = 0.0150\bar{1}0\ \text{moles}$$ EQ. 15.18

Molecular weight is determined from the mass of gas and the moles of gas:

$$\text{Mol Wt} = \frac{\text{mass}}{\text{mole}} = \frac{0.9014\ \text{g}}{0.0150\bar{1}0\ \text{mol}} = 60.0\bar{5}3\ \frac{\text{g}}{\text{mol}} = \boxed{60.05\ \frac{\text{g}}{\text{mol}}}$$ EQ. 15.19

Since the empirical formula CH_2O has a molar mass of about 30, the experimental molecular weight of 60 indicates that the molecular formula is twice the empirical formula: $C_2H_4O_2$.

DETERMINATIONS BASED UPON THE GAS LAW AND CHEMICAL STOICHIOMETRY.

In this type of problem, gas law data and balanced chemical equations are used in combination.

PROBLEM: Some commercial drain cleaners contain solid aluminum and sodium hydroxide. When added to water, the aqueous sodium hydroxide reacts with the aluminum to generate heat (to help melt the clog) and hydrogen gas (to help dislodge the clog). The balanced reaction is:

$$2\ \text{Al(s)} + 2\ \text{NaOH(aq)} + 6\ H_2O \rightarrow 2\ \text{NaAl(OH)}_4\text{(aq)} + 3\ H_2\text{(g)}$$ EQ. 15.20

What volume of hydrogen gas can form from a reaction of 5.00 g of aluminum with excess sodium hydroxide at 28.0°C and 1.000 atm?

As a first step, the mass of aluminum can be related to moles of hydrogen gas:

$$(5.00\ \text{g Al})\left(\frac{1\ \text{mol Al}}{26.982\ \text{g}}\right)\left(\frac{3\ \text{mol}\ H_2}{2\ \text{mol Al}}\right) = 0.27\bar{7}9\ \text{mol}\ H_2$$ EQ. 15.21

This number of moles can then be combined with the rest of the gas law data to determine the volume of hydrogen gas:

$$V = \frac{nRT}{P} = \frac{(0.27\bar{7}9\ \text{mol})\left(0.0820578\ \frac{\text{lit atm}}{\text{mol K}}\right)(28.0 + 273.15\text{K})}{1.000\ \text{atm}} = 6.8\bar{6}8\ \text{lit} = 6.87\ \text{lit}$$ EQ. 15.22

15.11 MIXTURES OF GASES

In the preceding problem, we assumed ideal gas behavior, that the hydrogen was insoluble in water, and that the only gas of concern was pure hydrogen. If in fact we collected the gas generated from this reaction, we would find not only hydrogen gas, but also water vapor. "Vapor" is a term commonly used to denote a gas phase species that has evaporated from a liquid. (Another way of describing vapor is as a gas that can be condensed to a liquid by increasing the pressure.) The amount of water vapor present in the gas phase will depend upon the temperature; the higher the temperature, the greater the amount of water vapor. Table 15.5 shows the relationship between temperature and the pressure of water in air when saturated with water vapor.

TABLE 15.5: SATURATION VAPOR PRESSURE OF WATER

TABLE 15.5

T/°C	P/torr
0.0	4.579
10.0	9.209
20.0	17.535
30.0	31.824
40.0	55.324
50.0	92.51
60.0	149.38
70.0	233.7
80.0	355.1
90.0	525.8
100.0	760.0

Figure 15.14 shows a method of collecting gas over water. As the water is displaced from the bottle, a volume of gas is collected. The gas bubbling up into the collection bottle is saturated with water vapor. The total pressure of the gas can be measured as explained earlier, as can the volume and the temperature. But what can be said concerning the pressure of the gas of interest, that generated via a chemical reaction? It is not simply the total measured pressure (e.g., as determined with a manometer.) This is because the total pressure depends upon the individual pressures of each of the gases in the mixture. This is known as **Dalton's Law of Partial Pressure**, and is formalized in the following definition:

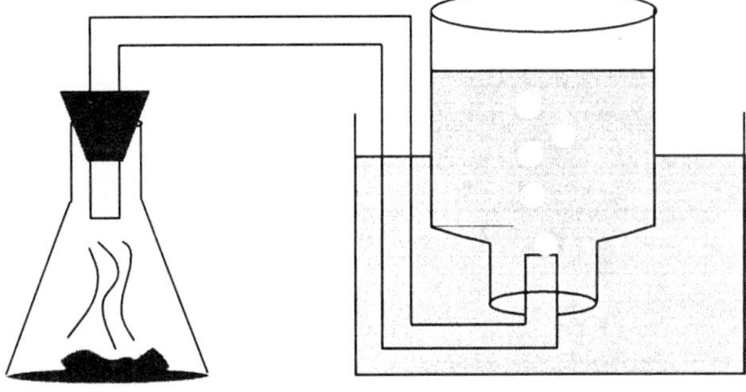

FIGURE 15.14: Collecting a gas over water.

$$P_{total} = P_A + P_B + P_C + \ldots. \qquad \text{EQ. 15.23}$$

where P_A represents the partial pressure of gas component A, P_B the partial pressure of gas component B, and so on.

We can show the validity of Dalton's Law of Partial Pressures if it is assumed that each of the gases in the mixture and the mixture as a whole behaves ideally. Then the pressures of A, B, C etc. depend on the moles of each via the equations:

$$P_A = \frac{n_A RT}{V}$$ EQ. 15.24

$$P_B = \frac{n_B RT}{V}$$ EQ. 15.25

$$P_C = \frac{n_C RT}{V} \text{ (e1}$$ EQ. 15.26

and that

$$P_{total} = \frac{n_{total} RT}{V} = \frac{(n_A + n_B + n_C + etc)}{V} = P_A + P_B + P_C + etc.$$ EQ. 15.27

PROBLEM: What would be the partial pressure of hydrogen gas collected over water at 20°C if the total measured pressure was 745.0 torr?

In this case, Dalton's Law of Partial Pressure can be written as:

$$Ptotal = P_{H_2} + P_{H_2O}$$ EQ. 15.28

from which the partial pressure of hydrogen can be solved:

$$P_{H_2} = Ptotal - P_{H_2O} = 745.0 - 17.535 \text{ torr} = 727.\bar{4}65 = \boxed{727.5 \text{ torr}}$$ EQ. 15.29

A common weather term is "relative humidity". Unless there is 100% relative humidity, the partial pressure of water in the air is less than the saturated water vapor pressures listed in Table 15.5. Relative humidity is the ratio of these two terms at a specified temperature:

$$\text{Relative Humidity} = \frac{\text{actual } P_{H_2O}}{\text{saturated } P_{H_2O}}$$ EQ. 15.30

15.12 REAL GASES VS. IDEAL GASES

So far in this chapter we have assumed ideal behavior. Careful measurements of real gases show behavior that approximates the ideal gas law at low pressures and high temperatures; but under other conditions, real gases show significant, measurable differences. We have also trusted in the assumptions upon which our model of a gas (and **kinetic molecular theory**) was based. A close examination of those assumptions prove that they can not be exactly true under all conditions.

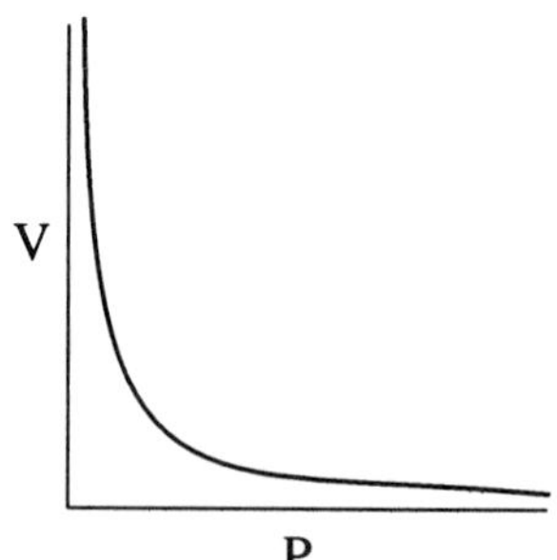

FIGURE 15.15: Plot of volume vs. pressure as pressure goes to infinity.

Earlier we assumed that the volume of the gas molecules themselves (the marbles) was completely negligible compared to the volume of the container of the gas (the box). We also assumed no repulsive forces between the gas molecules. Let us push these assumption to an extreme and see what happens. Consider the behavior of Boyle's law for an ideal gas and for a real gas. For an ideal gas, Boyle's law (PV = constant) is obeyed exactly. If the pressure is increased to infinity, for example, the volume of the gas would have to go to zero. This is shown graphically in Figure 15.15.

While the size of the marbles is decidedly small under atmospheric conditions, as the volume of the box decreases, the marbles themselves occupy a finite volume. The fraction of the volume inside the box occupied by the marbles themselves becomes larger and larger as the volume of the box decreases with increasing pressure. In the extreme case, molecules of gas at very high pressure push directly against one another (Figure 15.16.)

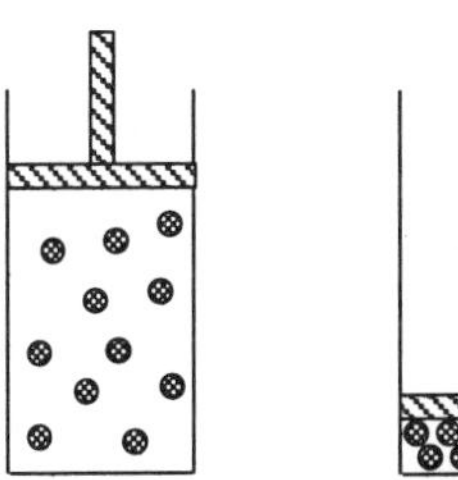

FIGURE 15.16: A model of gas as pressure increases.

Boyle's Law can not be obeyed in the limit of very high pressure. The volume of an ideal gas (solid line) approaches the pressure axis asymptotically (*i.e.*, approaching a zero volume). But for a real gas (dotted line), the corresponding curve approaches a finite limiting volume (close to the volume of the marbles themselves).

In our model of a real gas, the volume can not be driven to zero with increasing pressure because the marbles have a finite volume. Put another way, molecules of gas pushed too close together exhibit a repulsive force that resists a further decrease in the distance of separation between the molecules.

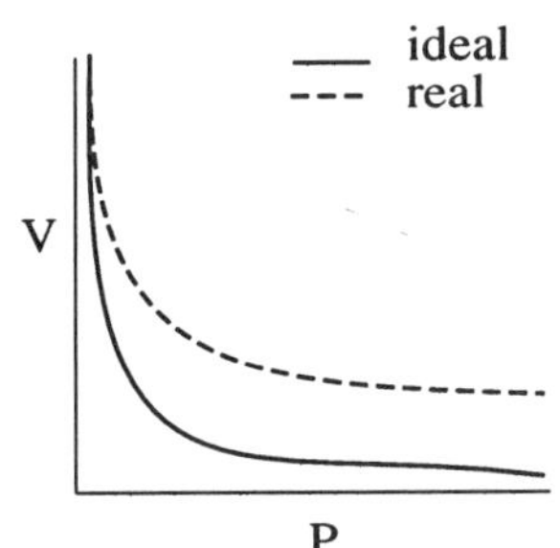

FIGURE 15.17: Ideal and real gas volume vs. pressure.

Another extreme leads to a different kind of behavior in many real gases. As the temperature is lowered, many gases will condense into the liquid phase. As we will see in the next chapter, there are attractive forces that are responsible for holding the molecules together in the liquid and solid phases. If real gas phase molecules in fact experienced no attractive forces, then simply lowering the temperature would never liquefy a gas. The fact that most gases can be liquefied in this fashion indicates that there are attractive forces present in the gas phase.

The behavior of real gases is commonly expressed in ways related to the ideal gas law. We will discuss two attempts to relate real gas behavior that, along with the ideal gas law, are known as equations of state.

The ideal gas law can be written:

$$\frac{PV}{nRT} = 1 \qquad \text{EQ. 15.31}$$

For a real gas, however, the ratio is denoted as the **compressibility factor Z.**

$$\frac{PV}{nRT} = Z \qquad \text{EQ. 15.32}$$

If the real gas behaves ideally, the value of Z will equal 1. Real gases can also have values of Z greater than or less than 1. One way to express the behavior of a real gas with greater accuracy than the ideal gas law, is to account for the value of Z. The value of Z will depend upon the identity of the real gas as well as the actual pressure and temperature. Figure 15.18 shows how values of Z typically vary. Note that all gases approach ideal behavior (Z = 1) when the pressure approaches zero, and that all gases deviate from ideal behavior when P becomes very large.

The compressibility factor equation has the advantage of simplicity in that it adds only one additional term to the ideal gas equation. This approach, however, is complicated by the fact that the value of Z changes when temperature or pressure are varied. The other common approach is a two-term correction to the

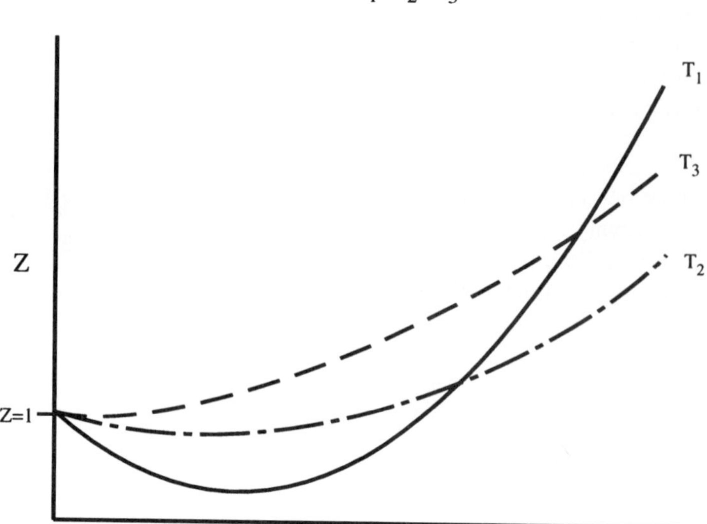

FIGURE 15.18: Z vs. P at varying T for real gases.

ideal gas law. The most common of these is the **van der Waals equation of state.**

The van der Waals equation corrects the real pressure and real volume to account for the attractive and repulsive forces exhibited by real gases. Consider the volume term in the ideal gas law. This is the ideal volume, i.e., the actual empty space within the box, not including the volume of the marbles themselves. The measured volume of the container (V) is greater than the actual "empty" or ideal volume (V'). This is expressed mathematically by:

$$V' = V - nb \quad \text{EQ. 15.33}$$

where n, as before, represents the number of moles of gas, and b represents the volume excluded per mole of gas (i.e., the size of the marbles themselves). The b term is the result of repulsive forces between molecules that resist compression, and is related to that volume difference between real and ideal gas behavior at high pressures seen earlier in Figure 15.17.

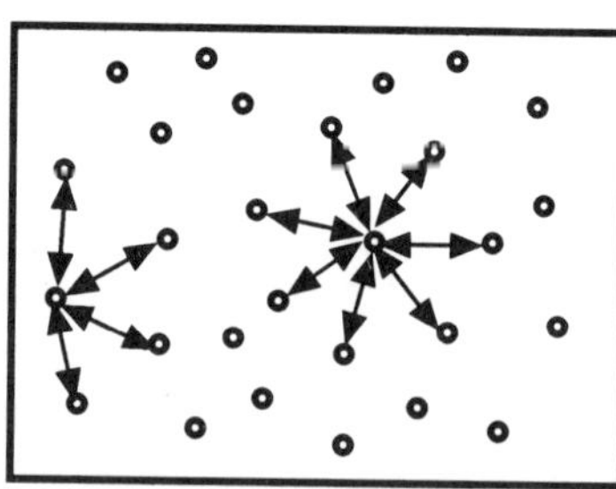

FIGURE 15.19: Model of gas with attractive forces.

The attractive forces influence the measured pressure (P). As shown in Figure 15.19, the attractive forces act asymmetrically on those molecules about to hit the wall.

By reducing the force of such collisions, the attractive forces act to reduce the measured pressure below that pressure (P') that would be observed in the ideal gas without forces of attraction. The ideal pressure (P') and the observed pressure (P) are related by the equation:

$$P' = P + \frac{a\,n^2}{V^2} \quad \text{EQ. 15.34}$$

The n and V terms represent the moles and volume of the gas; the a term is related to the force of attraction. The two coefficients in the van der Waals equation, a and b, are referred to as the attractive and repulsive terms respectively. The values of a and b depend upon the identity of the specific gas, but are independent of the temperature and pressure.

Substituting the corrected pressure (P') and corrected volume (V') into the ideal gas law yields van der Waals equation:

$$\left(P + \frac{a\,n^2}{V^2}\right)(V - n\,b) = n\,R\,T \qquad \text{EQ. 15.35}$$

PROBLEM: Calculate the pressure of 1.000 moles of xenon confined to a volume of 2.000 liters at a temperature of 25.0°C

(a) using the ideal gas law:

$$\text{Pideal} = \frac{nRT}{V} = \frac{(1.000\text{ mole})(.0820578\text{ lit atm/mol K})(298.\bar{1}5\text{K})}{(2.000\text{ lit})} = 12.2\bar{3}2\text{ atm} = \boxed{12.23\text{ atm}} \qquad \text{EQ. 15.36}$$

(b) using the van der Waals Equation:

$$P = \frac{n\,RT}{V - nb} - \frac{a\,n^2}{V^2};\ a = 4.250\text{ L}^2\text{ atm mol}^{-2};\ b = 0.5105\text{ L mol}^{-1} \qquad \text{EQ. 15.37}$$

$$P = \frac{(1.000\text{ mol})(.0820578\text{ lit atm/mol K})(298.\bar{1}5\text{K})}{(2.000\text{ lit}) - (1.000\text{ mol})(.05104\text{ Lit/mol})} - \frac{\left(4.250\,\frac{\text{Lit}^2\text{ atm}}{\text{mol}^2}\right)(1.000\text{ mol})^2}{(2.000\text{ lit})^2} =$$

$$12.5\bar{5}32\text{ atm} - 1.06\bar{2}5\text{ atm} = 11.4\bar{9}13\text{ atm} = \boxed{11.49\text{ atm}}$$

The two approaches differ by about 6% in their prediction of pressure. Note that the values of a and b are literature values that depended only on the identity of the gas: xenon. These same values of a and b could be used to calculate the properties of xenon under other experimental conditions.

Amontons's Law: a statement that the pressure of a gas is directly proportional to its temperature, provided that the amount and volume of the gas remain constant.

Avogadro's Hypothesis: the assumption that the number of moles of any gas is directly proportional to its volume, provided that the temperature and pressure are both constant.

Barometer: an instrument that measures the pressure of a gas, usually in terms of the height of a column of liquid mercury.

Boyle's Law: the statement that the volume of a gas is inversely proportional to its pressure, provided that the amount and temperature of the gas are both constant.

Charles's Law: the statement that the volume of a gas is directly proportional to its temperature, provided that the amount and pressure of the gas are both constant.

Compressibility factor (Z): the ratio of PV/nRT. It expresses how "non-ideal" a gas is, by the amount it differs from the ideal value of unity.

Dalton's Law of Partial Pressure: the total pressure of a mixture of gases equals the sum of the partial pressures of each component gas.

Kinetic molecular theory: a mathematical model of the behavior of the gas phase based upon some drastic assumptions, which include treating the gas as composed of volumeless point masses in constant, random, straight line motion until collisions (which are elastic) alter the path. The speed of the motion depends on the temperature of the gas.

Pascals: the metric unit of pressure, equivalent to kilograms per meter per second squared. 1 atmosphere equals 101325 Pascals.

Torrs: a non-metric unit of pressure. 760 millimeters of mercury at 0°C is called 760 torr, and equals one atmosphere.

van der Waals Equation of State: an extension of the ideal gas law, which accounts for both attractive and repulsive forces between the gas molecules by means of factors "a" and "b", unique for each real gas, and called its van der Waals constants.

HOMEWORK Problems

PROPERTIES OF GASES

1. In a few sentences, clearly distinguish how the gas phase differs from the liquid and solid phases.

2. Almost all substances (solids, liquids and gases) expand when heated. How is the thermal expansion of a typical gas different from that of a typical solid or liquid?

3. Pressure is an important property of gases, and is measured in various units. Convert a pressure of 30.0 inches of mercury into
   (a) torrs
   (b) atmospheres
   (c) Pascals

4. Convert a pressure of 250.0 kilopascals into
   (a) atmospheres
   (b) torrs
   (c) pounds per square inch

5. Part of the kinetic molecular theory of gases holds that gas molecules move fast, with a speed related to temperature. One measure of the speed of a gas is V_{rms}, the root mean square velocity. Given the equation below, calculate the velocity (in meters per second) of helium at 300.0 K. **HINT:** Helium has a molecular weight (MW) of 4.0026 x 10^{-3} kg $mole^{-1}$; use R in units of Joules $mole^{-1}$ K^{-1}.

$$V_{rms} = \sqrt{\frac{3RT}{MW}}$$

GAS LAW PROBLEMS

6. An ideal gas is held in a 250.0 ml bulb at a pressure of 742 torrs. By opening a valve, the gas can occupy both the original and an additional bulb with an added volume of 1.000 liters. What is the expected pressure of the expanded gas?

7. Suppose that the air in an automobile tire has an absolute pressure of 45.0 pounds per square inch when inflated on a cold day (0.0°C). What is the pressure (in psi) inside the tire when the temperature has risen to 30.0°C? (You may assume the tire does not leak, and that it has a constant volume.)

8. A weather balloon has a volume of 3.00 m^3 at the earth's surface, where the pressure inside and outside the balloon is 1.00 atmosphere and the temperature is 20.0°C. When the balloon has risen a certain height, the pressure inside and outside the balloon has fallen to 367 torr, and the temperature has fallen to -20.0°C. What volume should the balloon have at this height?

9. Liquid helium is used in cooling the superconducting magnet in FTNMR spectroscopy. A common model of this instrument has about 14.7 kg of helium gas/liquid mixture at about -270°C. A critical failure in the system when filling

the helium reserve could release all the helium into the instrument room. What would be the partial pressure of this amount of helium in a room of 100.0 m^3, assuming the final temperature in the room would be 0.0°C? (note: the instrument makers advise keeping the door open during this procedure.)

GASES: COMPOSITION AND STOICHIOMETRY CALCULATIONS

10. An unknown compound X is a hydrocarbon (*i.e.*, contains only carbon and hydrogen). Combustion of X yields water (MW = 18.01528) and carbon dioxide (MW = 44.0099). When 2.500 g of X were combusted in excess oxygen, the combustion products were swept onto a water absorbing column. This column gained 3.208 g, indicative of the amount of water formed. After all the water was removed, the excess oxygen was absorbed. The remaining gas was carbon dioxide, and it occupied 4.048 liters at a pressure of 750.0 torr and a temperature of 0.0°C. What is the empirical formula of X?

11. The compound arsenolite is known to contain only arsenic and oxygen. Analysis shows that the compound is 75.74% arsenic by mass.

(a) Calculate the empirical formula of arsenolite.

(b) When 4.030 grams of arsenolite vaporizes into the gas phase at 193°C, a vessel of 400.0 ml is filled to a pressure of 740.0 torr. Calculate the molecular weight of arsenolite.

12. Under certain conditions copper metal reacts with nitric acid to form nitric oxide gas (NO) according to the equation below. Assuming that only NO is captured, and that all the NO is captured, what mass of copper is required to form 20.0 liters of NO gas at 1.00 atm and 25.0°C?

$$3Cu(s) + 8\ H^+(aq)\ 2\ NO_3^-(aq) \rightarrow 2\ NO(g) + 4\ H_2O(l) + 3\ Cu^{2+}(aq)$$

13. Calculate the number of kilograms of iron (atomic mass = 55.847) that would be necessary to produce sufficient hydrogen gas (MW = 2.0158) to inflate the Bullwinkle Moose Balloon in the Macy's Parade on Thanksgiving. The balloon has a volume of 2.50 x 10^5 liters when filled with a pressure of 745 torr and a Thanksgiving Day temperature of 5.0°C. The reaction generating hydrogen is shown below. It is assumed that the acid is in excess.

$$2\ Fe(s) + 6\ H^+(aq) \rightarrow 2\ Fe^{3+}(aq) + 3\ H_2(g)$$

14. Air bags for automobiles are inflated during collision by the explosion of sodium azide, NaN_3 (MW = 65.0099). One possible equation for the decomposition is given below. What mass of sodium azide would be needed to inflate a 30.0 liter bag to a pressure of 1.50 atm at a temperature of 25.0°C?

$$2\ NaN_3(s) \rightarrow 2\ Na(s) + 3\ N_2(g)$$

OVERVIEW Problems

15. When 1.00 gallon (3.784 L) of liquid gasoline is burned, air supplies the oxygen. Assume that the air used is at 30.0°C and 752 torr, and that air is 20.0% oxygen by volume. You may also assume that the combustion process uses all the oxygen in air. Assume that gasoline has the formula of C_8H_{18}, and has a density of 0.703 g/ml.
    (a) Write the balanced combustion reaction for gasoline.
    (b) What volume of air (not oxygen) at 30.0°C and 752 torr must be input to completely combust 1.00 liquid gallon of gasoline?
    (c) What volume of exhaust gas at 200.0°C and 752 torr is produced? Note: the exhaust gas contains carbon dioxide, water vapor, and unreacted nitrogen from the air.

16. When calcium carbide (CaC_2, FW = 64.10) reacts with water, calcium hydroxide and acetylene gas are produced. When 100.0 g of an impure solid (containing calcium carbide and unreactive materials) reacts with excess water, the acetylene generated is dried and collected. At 25.0°C and 742.5 torr, 15.6 liters of acetylene are collected.
    (a) Write the balanced reaction forming acetylene.
    (b) Draw the Lewis dot structure of acetylene, and indicate the hybrid orbitals used by the carbon atom.
    (c) Calculate the moles of acetylene actually formed.
    (d) What is the % purity of the calcium carbide sample?

17. Sulfur dioxide gas reacts in water to form a water soluble covalent compound. When 1.00 liter of sulfur dioxide gas at 20.0°C and 740.0 torr is dissolved in water, 750.0 ml of solution is formed. This solution is then reacted with 2.00 molar aqueous sodium hydroxide to form sodium sulfite.
    (a) Write the balanced chemical reaction between sulfur dioxide and water.
    (b) What concentration (in moles per liter) of the product in (a) is formed?
    (c) Write the balanced reaction between the product of (a) and sodium hydroxide.
    (d) What volume of sodium hydroxide solution was required for this reaction?

18. A liquid is analyzed to contain 40.44%C, 7.92%H, 15.72%N, and 35.92%O by mass. When 2.50 g of the liquid are heated to 200.0°C in a 1.000 liter evacuated vessel, the liquid completely vaporizes, and a pressure of 828.0 torr results. When the liquid is analyzed by proton NMR, two signals result: a septuplet and a singlet, with a ratio of integrated areas of 1 to 6 respectively.
    (a) Calculate the empirical formula of the compound.
    (b) Calculate the IHD of the compound.
    (c) Calculate the molecular weight of the compound.
    (d) Calculate the molecular formula of the compound.
    (e) Draw the molecular structure of the compound.
    (f) Explain the proton NMR spectrum by labelling the proton environments in (e) in relation to your structure in (e)
    (g) How many peaks in the carbon NMR are expected based on your structure?

19. In the Marsh test for arsenic poisoning, the stomach contents (possibly containing aresenate ion, AsO_4^{3-}) are mixed with hydrochloric acid and granulated zinc. Zinc and HCl react to give hydrogen gas. The hydrogen gas formed in this reaction reduces arsenate ion to arsine, AsH_3, a gas. The arsine gas bubbles out of solution along with extra hydrogen gas. This gaseous effluent is brought through a heated tube, where the arsine decomposes to form hydrogen gas and elemental arsenic, which is deposited on the walls of the tube as a metallic mirror.

(a) Write the balanced reaction between zinc and HCl.
(b) If 5.00 g of zinc react with excess hydrochloric acid, what volume of hydrogen gas at 20.0°C and 745.0 torr will be produced?
(c) Write the balanced reaction between hydrogen gas and arsenate.
(d) Draw the Lewis dot diagram of arsine, and list its electronic geometry, molecular geometry, and the hybrid orbitals used by arsenic.
(e) What volume of pure arsine gas, measured at 20.0°C and 745.0 torr, is required in order to obtain a mirror of arsenic weighing 1.0 mg?

20. Fermentation tanks are large gas-tight vessels used to convert glucose ($C_6H_{12}O_6$, MW = g/mole) into ethanol and carbon dioxide. A 30.0% (by mass) aqueous glucose solution has a density of 1.1246 g/ml at 20.0ºC. When 500.0 liters of a 30.0% glucose solution is put into a fermentation tank, 1,500.0 liters of gas space remain. Assume that all the glucose is converted to ethanol and carbon dioxide at 20.0°C.

(a) Write the balanced fermentation reaction.
(b) Calculate the maximum number of moles of ethanol and carbon dioxide produced.
(c) Calculate the pressure of carbon dioxide that would result from this number of moles, assuming all of it was in the gas phase.

21. An alternate method of forming ethanol is the hydration of ethylene (ethene) under conditions of high pressure and high temperature (300.0°C). This gas phase reaction is conducted with an excess of water vapor in a vessel of 1.00 m^3. Initially, 6.00 kg of ethene and 23.0 kg of water are introduced into the vessel.

(a) Assuming ideal behavior, what are the partial pressures of ethene and water in the reaction vessel before any reaction occurs?
(b) The van der Waal's constants for water vapor are $a = 5.536\ L^2\ atm\ mol^{-1}$, $b = 0.03049\ L\ mol^{-1}$. What is the initial partial pressure of water vapor (before any reaction) assuming that it behaves as a van der Waal's gas, not as an ideal gas?
(c) Write the balanced gas phase reaction.
(d) Assuming ideal behavior and a complete reaction, what is the partial pressure of all components present at the end of reaction?

22. Potassium chlorate ($KClO_3$, FW = 122.55) is used as a source of oxygen in some chemical demonstrations. When heated, potassium chlorate decomposes to form oxygen gas and a non-volatile byproduct. Assume that one wishes to collect 1.00 mole of oxygen gas, and that the gaseous product is collected over

water at a temperature of 23.6°C and a total pressure of 738.5 torr. The saturation vapor pressure of water at 23.6°C is 21 845 torr.

(a) Write the balanced decomposition reaction of potassium chlorate.

(b) What mass of potassium chlorate must be decomposed?

(c) What mass of non-volatile byproduct is formed by the decomposition?

(d) What volume of gas is collected under these conditions?

23. Another means of producing oxygen is by the catalyzed autooxidation of hydrogen peroxide, H_2O_2. A 30.0% (by mass) aqueous solution of hydrogen peroxide has a density of 1.11 g/ml. At 26.2°C, the saturation vapor pressure of water is 25.509 torr. Assume that one wishes to obtain 100.0 g of oxygen gas, and that the oxygen is collected over water, at 26.2°C and 746.8 torr.

(a) Write the balanced autooxidation reaction of aqueous hydrogen peroxide.

(b) How many moles of hydrogen peroxide must be decomposed to obtain the
100.0 g of oxygen gas?

(c) What volume of 30.0% aqueous hydrogen peroxide must be used?

(d) What volume of gaseous product is collected under these conditions?

24. One form of solid animal fat is glycerol tristearate ($C_{57}H_{110}O_6$, MW = 891.51). The body oxidizes fat to form carbon dioxide gas and liquid water. Assume that air is 20.0% oxygen by volume, and that the air is at 25.0°C and 745.0 torr.

(a) Write the balanced redox reaction converting fat to CO_2 and H_2O.

(b) How many moles of fat are present in 1.00 pound (453.59g) of fat?

(c) What volume of air must be inhaled for the body to oxidize 1.00 pound of fat?

25. There is considerable concern over the concentration of radon gas in our homes. Although radon is a noble gas (and not chemically reactive) the radioactive ^{222}Rn isotope poses a problem. As a gas, it is readily inhaled into the lungs. Once there, it may decompose via alpha emission. A partial pressure of ^{222}Rn as low as 3.00×10^{-15} torr is a serious problem. The half-life of ^{222}Rn is 3.30×10^5 seconds. A common unit of radioactivity is the curie, corresponding to 3.7×10^{10} disintegrations per second. A picocurie is 10^{-12} curies.

(a) Write the balanced nuclear decay of ^{222}Rn yielding alpha radiation.

(b) Calculate the number of radon atoms per liter present at a pressure of 3.00×10^{-15} torr.

(c) The radioactive half life is the time for one-half of the atoms in a sample to disintegrate. From the half life of ^{222}Rn and the answer to part (b), calculate the number of disintegrations per second liter.

(d) Express the answer to part (c) in picocuries per liter. (As a check, the current EPA guidelines set a maximum exposure of 4 picocuries per liter.)

CHAPTER sixteen

LIQUIDS AND Solids

LIQUIDS AND Solids

16.1 THE EMPIRICAL EVIDENCE

In chapter 15 we learned about the behavior of gases. Our knowledge of the behavior of gases is sufficiently complete that we can make many quantitative predictions of the properties of gases under almost any conditions, especially when the temperature is high and the pressure is low. But, near the end of chapter 15, we learned that at low temperatures and/or high pressures, the ideal behavior that allows us easily to predict gas properties begins to break down.

Why is this? The reason advanced is that under these conditions the molecules of the gas are squeezed much closer together. This leads to a breakdown of the assumptions (that the gas molecules occupy no volume and the gas molecules exhibit no mutual forces of attraction or repulsion) upon which the Ideal Gas Law is based. It is easy to see that the molecules of a liquid or solid must be much closer together than the molecules of a gas if we look at densities of common materials at room temperature. Quick calculations using the Ideal Gas Law will show us that the densities of most common gases are approximately 1 gram per liter near room temperature and 1 atmosphere pressure. Under the same conditions of temperature and pressure, densities of common liquids and solids are approximately 1 gram per milliliter. This is a factor of 1000 in the density with which the molecules must be packed. While it is possible that the molecules in a condensed phase might still occupy negligible volume and exhibit negligible forces of attraction and/or repulsion, certainly it is much less likely for liquids and solids than was the case for gases.

Let us first look at the evidence that suggests that there must be significant forces operating between adjacent atoms or molecules in the liquid and solid phases that were not there in the gas phase. Consider first properties of the liquid state:

(1) Surface Tension: This is a property of the liquid state that makes penetration of the surface of the liquid difficult. Because of surface tension, certain insects are able to walk on top of water without sinking below the surface, droplets of liquid dispersed in air assume a spherical shape, water moves up a capillary tube, and a given quantity of liquid has a definite volume.

(2) Vapor pressure: This is the pressure exerted by a liquid at a given temperature. The vapor pressure of a liquid is dependent on the temperature, being higher at higher temperatures as shown in Figure 16.1. When the vapor pressure of the liquid equals the pressure of the surroundings, the liquid has reached its **boiling point**.

(3) Boiling point variations: Boiling points normally vary regularly within groups of chemically similar compounds, with larger molecules having higher boiling points. There are, however, a number of striking examples of variation from this behavior. Figure 16.2 presents the boiling points of the noble gases. Figure 16.3 plots the boiling points of the Group IV, V, VI, and VII chlorides as a function of the group number. The behavior shown by these graphs is typical of most compounds. However, Figure 16.4 is a similar plot for the hydrides of these same elements. Here the abnormal behavior shown by the group V, VI, and VII compounds provides some important insights into the nature of the forces between molecules in liquids.

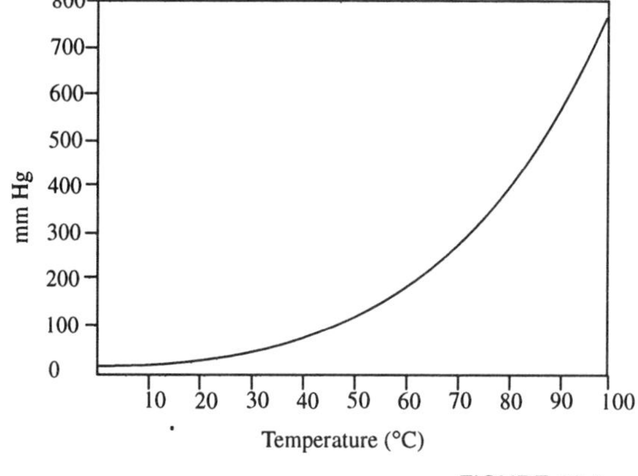

FIGURE 16.1: Vapor pressure of H_2O vs. temperature.

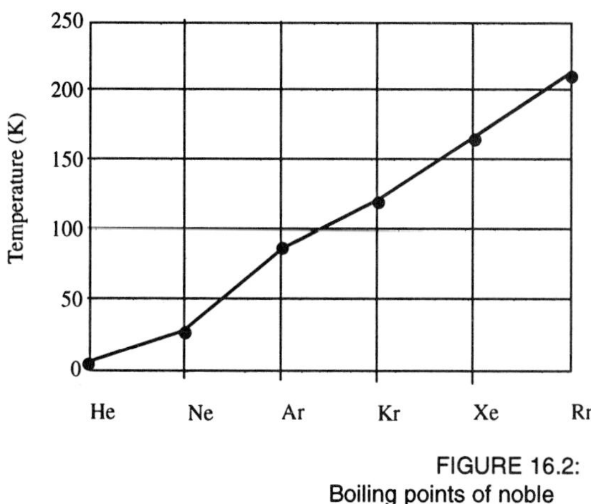

FIGURE 16.2: Boiling points of noble gases.

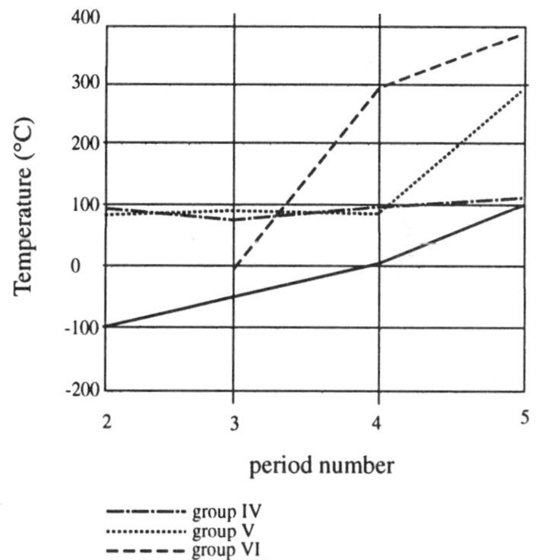

FIGURE 16.3: Boiling points of chlorides by period.

While Group IV hydrides show boiling points that smoothly increase with the period number, the hydrides of Groups V, VI and VII do not. Specifically, the boiling points of the second period hydrides of nitrogen, oxygen and fluorine are anomalously high. This point will be addressed later.

If we look at the properties of solids, there are two key features: (1) the melting and boiling points of covalent solids are much lower than those of ionic solids, and (2) the melting points of ionic solids are generally higher for those solids composed of multiply charged ions than they are for those solids composed of singly charged ions. For example: the melting and boiling points of oxygen difluoride, a covalent compound with a molecular weight of 54 g mol^{-1}, are -218°C and -145°C respectively, while the melting and boiling points of potassium fluoride, an ionic compound of almost the same formula weight (58 g mol^{-1}) are 858°C and 1505°C respectively. Comparing the properties of ionic solids reveals that their melting points depend upon the ionic charges: alkali metal chlorides (with +1 and -1 charges on the ions) have melting points which range from about 600°C–800°C while alkali metal sulfates (+1 and -2 charges on the ions)

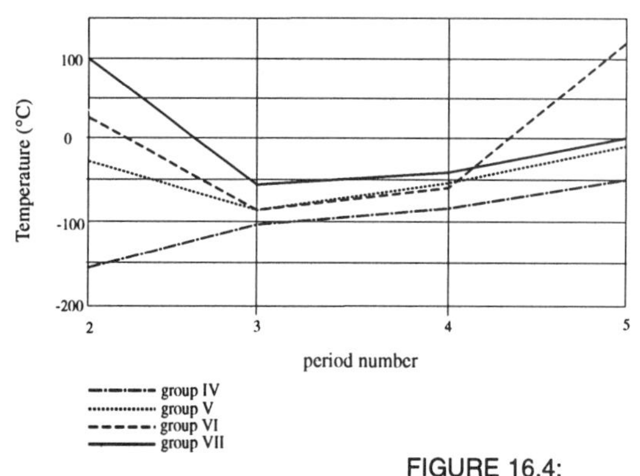

FIGURE 16.4: Boiling points of selected binary hydrides.

have melting points ranging from ca. 900°C–1100°C. The trend continues if we examine the melting points of the alkaline earth sulfates which, with the exception of beryllium sulfate that has considerable covalent character, range from ca. 1100°C–1600°C.

All of these things can be explained by an extension of the ideas advanced in the previous chapters. The strengths of the interaction between molecules govern how easy it is to separate molecules. If a great deal of force is required to push molecules apart, then an object lying on top of a liquid may not sink in the liquid because it is incapable of generating enough force to separate the molecules. Therefore an insect, even though it is more dense than water, is able to walk on top of water without sinking into it. Similarly, the boiling and melting points of substances will be determined by how difficult it is to break the attractions between adjacent molecules to break down the structure of the solid or to allow the molecule to break completely free of the liquid. Let us now look at the forces that exist between species in more detail.

16.2 THE FORCES

If we arrange the forces possible between species from the strongest possible forces to the weakest, the following list is obtained: Ion—Ion Forces, Dipole—Dipole Forces, Dipole—Induced Dipole Forces, and Induced Dipole—Induced Dipole Forces. Let us consider each of these separately starting with the ion—ion forces.

1. Ion—Ion Forces

These are the forces of attraction between ions of opposite charge. We have previously noted that the strength of these forces can be calculated using Coulomb's Law,

$$F = \frac{kq_1q_2}{r^2}$$ EQ. 16.1

where k is a proportionality constant, q_1 and q_2 are the charges on the ions, and r is the distance separating the centers of charge. While we are not concerned here with the exact values, it is easy to see that with the very small distances separating ionic centers in solids (ca. 10^{-8} cm), these forces can be quite large. They are so large that we normally consider these forces to be bonds and term them "ionic bonds". The strength of these forces vary depending upon the identity of the compound, and typically range between 400—1000 kJ mol^{-1}. Such forces are strong enough that, unless a great deal of energy is put into the compound as heat, the ions stay very closely and regularly spaced in the crystals. Thus these ionic compounds have very high melting points. We have seen, however, that the melting temperatures of these compounds depend upon the charges of the ions, with compounds made up of highly charged ions generally having higher melting temperatures than compounds of ions with lower charges, as expected from Coulomb's Law.

2. Dipole—Dipole Forces.

These are the strongest intermolecular forces, but dipole-dipole interactions are significant only for some covalent compounds. A "pure" covalent bond, i.e., one formed by sharing of an electron pair between two identical atoms, has no separation of charge. Hence a molecule such as chlorine, Cl_2, has no **dipole moment**. However, when the atoms involved in a bond have different electronegativities, the electrons are no longer shared equally between the partners to the bond. The extreme example of this is, of course, an ionic bond where one of the atoms involved gets an extra electron and the other loses an electron. If the difference in electronegativities is not large enough to support complete transfer of the electron, but the electrons in the bond are still not shared equally, the atoms acquire a partial instead of a complete charge. This partial charge separation is known as a dipole moment.

The dipole moment has a magnitude measured in units termed Debyes, a unit named to honor Peter **Debye** (1884–1966) who won the Nobel Prize in 1936 for work involving molecular structure. The magnitude of the dipole moment is calculated as $\mu = \delta d$ where μ is the magnitude of the dipole moment, δ is the magnitude of the charges and d is the distance separating the charges. The dipole moment also has a direction, usually indicated on drawings of molecules as an arrow with a plus sign on the end of the arrow opposite the point. It is, in mathematical terms, a vector quantity. Since a molecule can involve more than one bond, the dipole moment of a molecule is determined by the vector sum of all the dipole moments of the individual bonds. If the geometric arrangement of the bonds in a molecule is such that the bond dipole moments do not cancel, the molecule as a whole has a dipole moment.

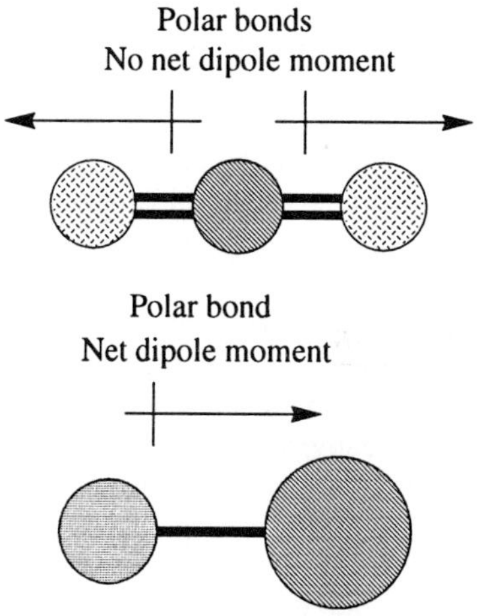

FIGURE 16.5: Polar covalent bonds and net dipole moments.

When molecules having dipole moments get close to one another, these dipole moments can interact in much the same way as ions, with the positive end of one dipole attracting the negative end of another dipole. Of course, since the "charges" are smaller in dipoles than in ions, the force of interaction is also smaller. Typically dipole–dipole forces are between 5 and 50 kJ mol^{-1}. These are strong enough to account for compounds being liquids and/or low–melting solids at room temperature, but are definitely not strong enough to make the molecules hold together so tightly that they are high melting solids.

A special case of dipole–dipole interaction is so common and so strong that it has come to have a special name: **Hydrogen Bonding** (H–bonding). This occurs between hydrogen (a very small, electropositive atom) and small, highly electronegative atoms (nitrogen, oxygen, fluorine, and, to some extent, chlorine). It is these dipole–dipole interactions, now called hydrogen–bonding interactions, that are responsible for the markedly higher boiling points of the hydrides of the second period elements as we saw in Figure 16.4. Those species with anomalously high boiling points (NH_3, H_2O, and HF) all engage in hydrogen bonding as shown in Figure 16.6.

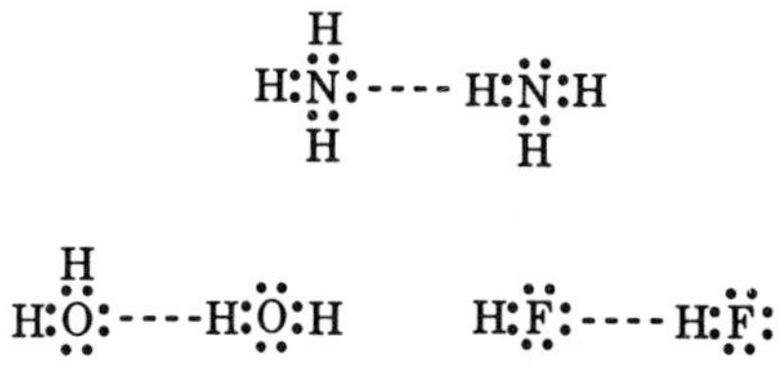

FIGURE 16.6: Hydrogen bonding in second period hydrides.

Hydrogen bonding plays a very important role in biochemistry as well, being an important factor in the secondary structure of proteins and the base–pairing behavior that is responsible for the actions of DNA and RNA.

adenine thymine guanine cytosine

adenine uracil

FIGURE 16.7: Hydrogen bonding and the pairing of bases in DNA and RNA.

3. Induced Dipole–Dipole Forces

Even molecules that do not have a permanent dipole moment may have a dipole induced in them by close approach to an ion or a molecule with a dipole moment. One might think of this as occurring by the attraction or repulsion of the symmetric electron cloud of a molecule by the charge on another.

This uneven distribution of electrons generates a dipole moment that disappears immediately when the perturbing species is removed. Such a dipole moment is known as an induced dipole moment. The magnitude of the these forces again varies greatly, and depends upon how strongly charged (the technically correct word is "polarized") is the perturbing species and how "loosely held" are the electrons of the molecule having the dipole moment induced. In general larger molecules have more loosely held electron clouds and, therefore, if all else is constant, the magnitude of the force increases with the molecular weight of the substance. These forces can vary between ca. 2–10 kJ mol^{-1}.

a) Before Dipole Interaction

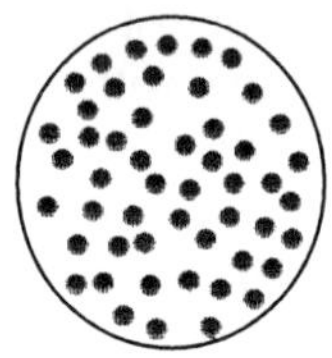

b) After Dipole Interaction

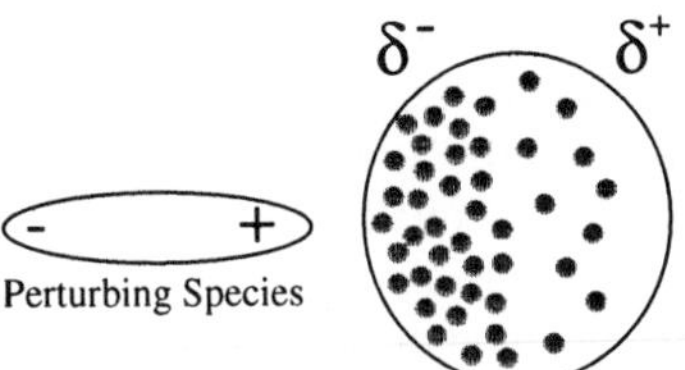

FIGURE 16.8: Dipole induced in a non-polar molecule.

4. Induced Dipole–Induced Dipole Forces

These forces, sometimes called London Dispersion Forces after the eminent German physicist Fritz London (1900–1954) who first explained them in 1928, are necessary to explain why non polar gases such as the noble gases liquefy. This explanation relies on the constant motion of the electrons surrounding an atom or molecule. Because of this random motion there are instances when the electron distribution is not totally symmetric. During these fleeting times the molecule has a defacto dipole moment, albeit only for a very short time and not always in the same direction. During these short times, this dipole moment in one molecule can induce a dipole moment in another molecule. The net result is a small intermolecular force between the molecules. In general, the magnitude of the force is greater for molecules whose electron clouds are not tightly held. This is more likely to be the case if the molecule is large than if the molecule is small. As a result intermolecular forces increase as the molecular weight of a species increases, all other things held constant. This explains the regular increase in boiling point of the noble gases with atomic weight as well as the increase in boiling point for the hydrides of the period 3–5 elements and the chlorides of the period 2–6 elements shown in Figures 16.2 to 16.4.

In Chapter Two, Table 2.1, we previously saw that as the number of carbons in a hydrocarbon increased, so did the boiling point. This may now be explained; as the length of the hydrocarbons increases, the polarizability increases, the dispersion forces become stronger, and thus the boiling point will increase. In Table 16.1, we extend our examination of the trend to see the effect of structural isomers on the boiling point.

TABLE 16.1: BOILING POINTS OF BRANCHED AND STRAIGHT CHAIN HYDROCARBONS

TABLE 16.1

FORMULA	NAME	b.p./°C
C_4H_{10}	n-butane	-0.5
	2-methylpropane	-11.7
C_5H_{12}	n-pentane	36.1
	2-methylbutane	27.8
	2,2-dimethylpropane	9.5
C_6H_{14}	n-hexane	68.9
	2-methylpentane	60.3
	2,2-dimethylbutane	49.7
C_7H_{16}	n-heptane	98.4
	2-methylhexane	90.0
	2,2-dimethylpentane	79.2

In general, we confirm the previously noted trend that lower molecular weight hydrocarbons have lower boiling points than the corresponding higher molecular weight hydrocarbons. But we also note that as branching of the hydrocarbon increases, the melting point decreases. This also is attributable to the London dispersion forces. As the hydrocarbon becomes more branched, its overall surface area is decreased. For example, 2,2-dimethylpropane is much more like a sphere (with minimum surface area) than is n-pentane. Branched molecules are smaller, more compact than linear molecules. The branched molecules have correspondingly lower polarizability, weaker dispersion forces, and thus lower boiling points.

16.3 THE LIQUID STATE

What, then, can we say about the liquid state? If we take a sample of liquid water and slowly heat it, the molecules begin moving, on average, faster as the temperature increases. As in the gas state, this motion is in random directions with molecules continually bouncing off each other and changing direction.

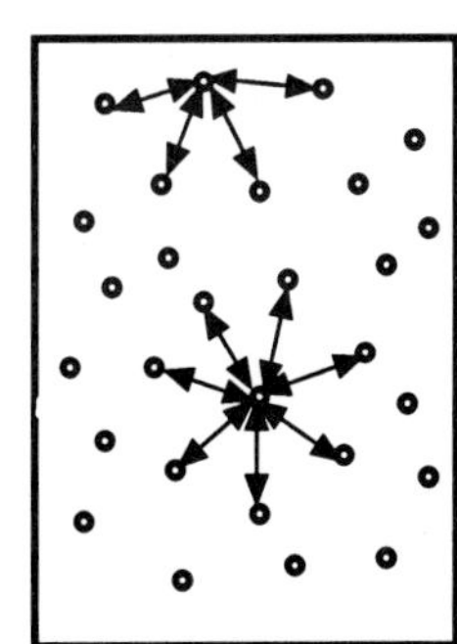

FIGURE 16.9: Internal forces in a liquid.

From the previous section, we can say that adjacent molecules in the liquid state experience one or more forces of attraction. A molecule in the center of the liquid experiences forces that are symmetrically distributed, since that molecule interacts with molecules all about it. There is, therefore, no net force pulling the molecule one direction or another except that provided by the kinetic motion of the molecule. This is shown in figure 16.9. This figure also shows a molecule at the surface of the liquid. The forces on a surface molecule are not distributed equally in all directions. The net result of summing these asymmetric forces is to pull a surface molecule back into the center of the liquid. It is these forces that are responsible for the phenomenon of surface tension to which we made reference earlier.

Despite all the forces of attraction acting on a molecule, we must not forget that all molecules in a sample of a liquid are in constant random motion. This means that there will be some molecules moving in a direction that is straight toward the surface of the liquid. If these molecules acquire sufficient kinetic energy as a result of their velocity, it is possible that they will be able to break free from the backward "pull" of the other molecules in the liquid and escape into the gas phase. The fraction of molecules that will have this amount of kinetic energy will vary with temperature and will be larger at higher temperatures than at lower temperatures.

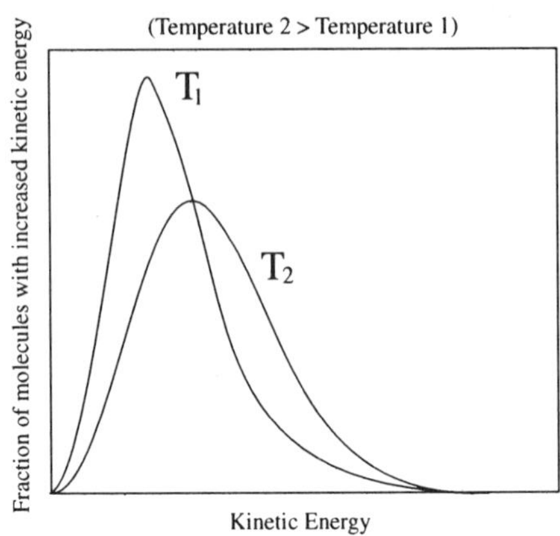

FIGURE 16.10: Fraction of molecules with a given KE vs. KE at temperature T_1 and a higher temperature T_2.

This means that the rate at which water molecules escape into the gas phase will be larger at higher temperatures than at lower temperatures. The result is that the concentration of water molecules in the gas phase will be higher at higher temperature. Recall from chapter 15 that the concentration of a species in the gas phase is directly related to the partial pressure of that species, hence the partial pressure of water in the gas above a sample of water will vary with the temperature, being higher at high temperatures and lower at low temperatures.

Consider a sample of liquid water at constant temperature pictured in Figure 16.11. When there is a low pressure of the gas above the liquid, evaporation readily occurs and molecules move predominantly from the liquid to the gas phase. As the concentration of water molecules in the gas phase increases, a larger number of them per unit time will travel toward the surface of the liquid and will be recaptured by the liquid.

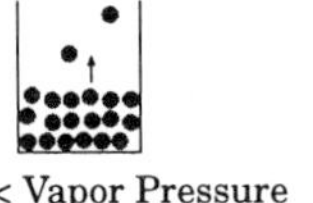

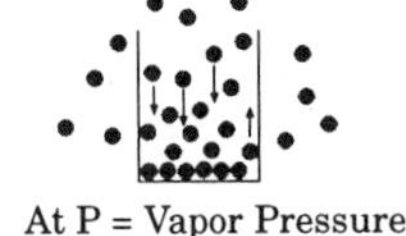

FIGURE 16.11: A volatile liquid evaporating, recondensing.

Thus, at some partial pressure of water the rate at which water molecules escape from the liquid and the rate at which water molecules re–enter the liquid will be equal. This partial pressure is termed the equilibrium vapor pressure (or, more commonly, just the vapor pressure) at this temperature. It is this that is plotted as a function of temperature in Figure 16.1. When the vapor pressure equals the ambient (surrounding) pressure, the liquid is said to boil. If the ambient pressure is exactly 1 Atm., the temperature is called the normal boiling point for the liquid. When the ambient pressure is reduced (e.g., at higher elevations) the boiling point is lowered. When the ambient pressure is raised (e.g., in a pressure cooker) the boiling point is raised.

16.4 THE SOLID STATE

As a liquid is cooled even further, there comes a point at which the molecules no longer have enough kinetic energy to break free from the forces of attraction operating on them from the surrounding molecules. At this point they just vibrate back and forth in position. Such a collection of molecules no longer can assume the shape of the container and is now considered to be a solid.

There are, however, many different types of solids with quite different properties. These solids can be classified by the long range structure of the solid as well as by the forces responsible for that long range structure.

1. Amorphous Solids

Amorphous solids have no long range structure. These materials are sometimes considered to be simply extremely viscous liquids because they have some properties in common with liquids. In these solids the molecules are arranged in random positions throughout the solid. The structure of these materials is, therefore, similar to that of a liquid in which the molecules do not have enough kinetic energy to overcome the weak dipole–dipole or London forces operating between the molecules and slide past each other. Amorphous solids are characterized by softening temperatures rather than true melting points. Glass, butter, and some organic polymers are a few of the materials that are amorphous solids.

2. Crystalline Solids

Crystalline solids, on the other hand are characterized by all that we expect of a solid. There is long–range order in the solid and the melting point is sharply defined for a pure solid. A crystal is most easily imagined as an elementary solid shape that is repeated in all three directions many times. The elementary shape is termed the unit cell. This unit cell is the smallest arrangement of the components of the crystal which, when repeated exactly in all three dimensions, can reproduce the crystal. The components of the crystal may be atoms, ions, or molecules. The forces holding the crystal together, and therefore the properties of the crystalline solid, vary greatly depending upon the components of the crystal. Let us look at the properties of each of these different types of solids before considering crystal structure itself.

a. Metals

In metallic crystals, the components of the unit cell are metal atoms arranged in a repeating pattern that is usually twelve–coordinate and based on one of the closest–packing crystal structures (see the discussion of crystal structure below). The forces holding the metal atoms in place are metallic bonds, described earlier in Chapter 11. Because these metallic bonds can be fairly strong, the melting points of most metals are rather high, although not as high as the melting points of some of the other crystal types in which the interatomic forces are even stronger. The absence of strong directional forces in metallic bonding leads to metallic properties of ductility and malleability in which the metal atoms simply slide past one another to change position in the crystal.

b. Molecular Crystals

Many different types of molecules crystallize in molecular crystals and many different crystal types are represented among molecular solids. Since the forces between molecules can vary widely in strength, from the very weak London forces to the rather strong dipole–dipole forces between molecules such as water, the melting points of molecular crystals vary greatly from well below 0°C for molecules with weak London or dipole–dipole forces, to 300–400°C for other molecules with stronger dipole–dipole forces. The weak forces between crystal components make these crystals rather soft and easily broken. The exact crystal types represented by molecular solids vary depending upon the ratio of dimensions in the x, y, and z directions for the molecules. Typical of materials in this category are ice (Figure 16.12), dry ice (solid CO_2), sucrose, elemental iodine, elemental sulfur, etc.

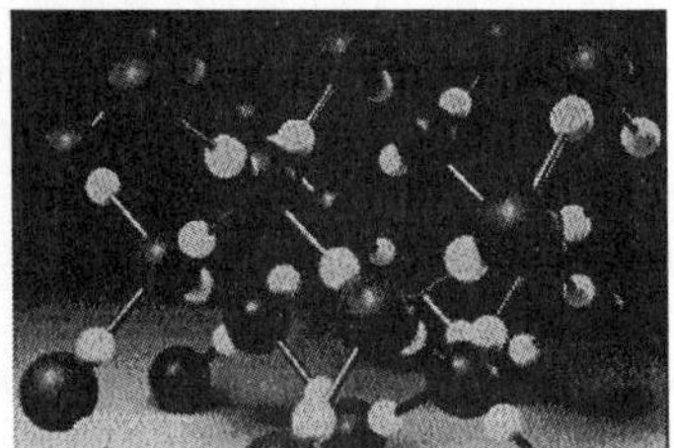

FIGURE 16.12: Ice, a molecular crystal, with hydrogen bonding as the primary intermolecular force.

c. Molecular Network Solids

These solids consist of atoms in a three dimensional matrix determined by the covalent bonding requirements of the particular atom. The forces holding the atoms together in the solid are covalent bonding forces. These crystals are, in essence, gigantic molecules in which every atom in the crystal belongs to the molecule. These forces are, therefore, quite strong forces that give the crystals very high melting points and make them quite hard and difficult to break apart. When they do break, they tend to shatter into smaller pieces, each of which is another crystal. Typical crystals in this category include diamond (Figure 16.13), graphite, SiC, and SiO_2 (quartz).

FIGURE 16.13: Diamond, a molecular network solid, with carbon atoms on the lattice sites held together by covalent bonds.

d. Ionic Solids

These crystals contain intermixed cations and anions arranged in some type of crystal lattice, as in Figure 16.14. Because the forces holding the components of the crystal in place are strong ion–ion forces, these solids have universally high melting points and require a great deal of force to break. When they do break, the crystals tend to shatter into small pieces that retain the symmetry of the original crystal.

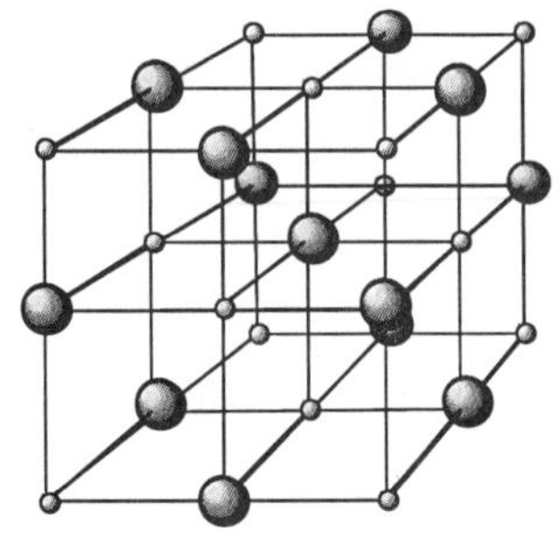

FIGURE 16.14: Model of NaCl, an ionic solid, with ions on lattice sites held together by coulombic forces.

3. Crystal Structure

A substance is crystalline if its long–range structure is very ordered. This order takes the form of a repeated short–range pattern known as the unit cell. This unit cell, when translated and reproduced in 3 dimensions produces the overall crystal as shown in Fig. 16.15.

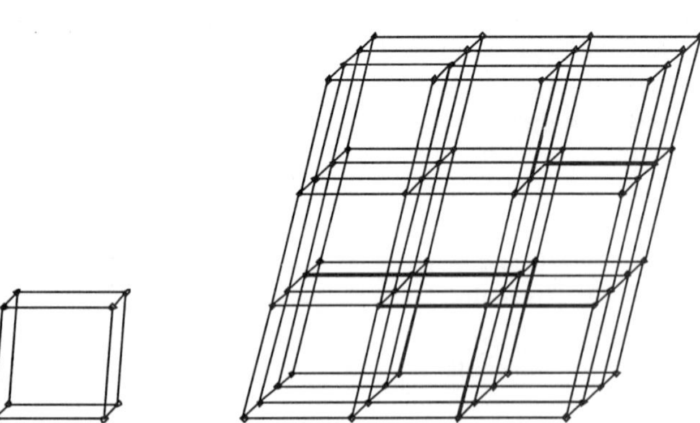

FIGURE 16.15: Crystal formation from a unit cell.

cubic
a=b=c; $\alpha=\beta=\gamma=90°$

tetrahedral
a=b≠c; $\alpha=\beta=\gamma=90°$

orthohombic
a≠b≠c; $\alpha=\beta=\gamma=90°$

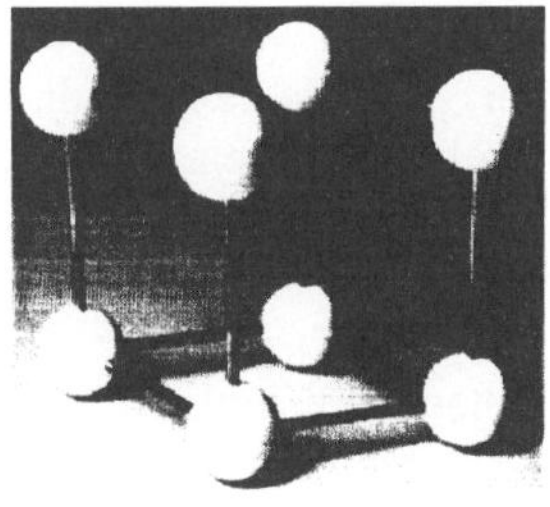

hexagonal
a=b≠c; $\alpha=\beta=90°;\gamma=120°$

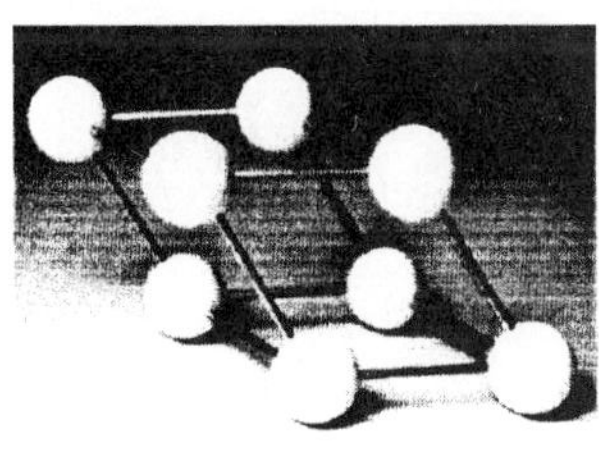

rhombohedral
a=b=c; $120°>\alpha=\beta=\gamma\neq90°$

monoclinic
a≠b≠c; $\alpha=\gamma=90°$

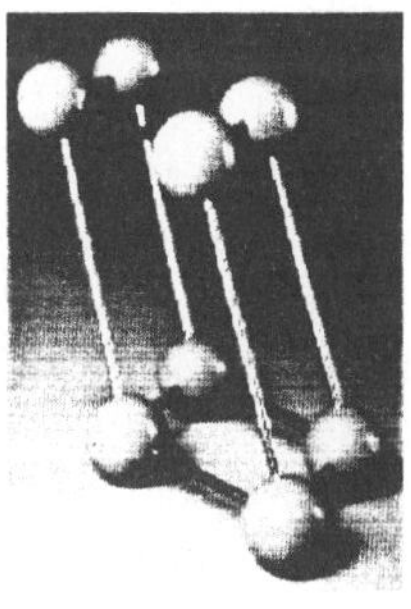

triclinic
a≠b≠c; $\alpha\neq\beta\neq\gamma\neq90°$

FIGURE 16.16:
Unit cell types.

In Figure 16.15 a unit cell and the resulting crystal formed by its replication are shown for one particular type of unit cell. There are seven different unit cell types , and several sub–types, found in nature. These are pictured in Figure 16.16.

Of these, probably the most important is the cubic crystal unit cell. This is the one we will spend our time analyzing. Most of what is discussed about the cubic unit cell is also true of the other unit cells with appropriate changes for differences in angles and side lengths.

A. Evidence for Crystal Structure

Most of the evidence for the structure of crystalline solids comes from x–ray diffraction studies. In Chapter 9, we saw that light exhibits constructive and destructive interference patterns (Figure 9.10.) When x–rays are diffracted from the different planes of atoms within a crystal, they form an interference pattern that produce spots on a photographic plate only when constructive interference of the x–rays occurs. While the exact details of an x–ray crystallographic study of a crystal are very complicated, all the points at which a spot appears on the photographic plate are described by the Bragg equation:

$$n\lambda = 2d \sin\theta \qquad \text{EQ. 16.2}$$

where λ is the wavelength of the x–ray, d is the distance separating the crystal planes from which reflection comes, and θ is the angle of incidence and reflection. X–rays are the radiation of choice because their wavelength is approximately the same as the crystal dimension, d. By careful study of x–ray diffraction data, scientists can determine the entire structure of the crystal, including the angles and dimensions of the unit cell.

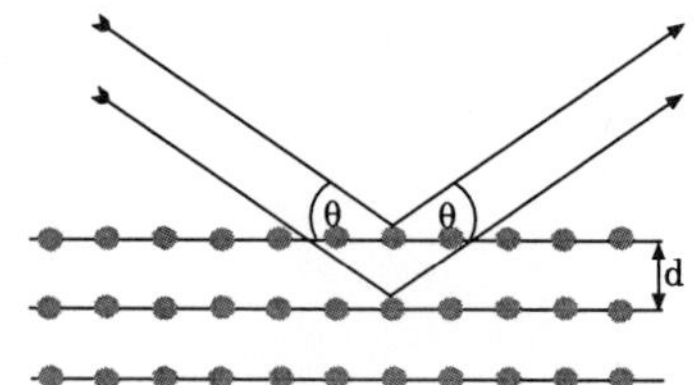

FIGURE 16.17: Diffraction of X-rays and the determination of crystal structures.

Although examples of crystal structures often use elements or inorganic salts, it is worth noting that much advanced x-ray crystallographic work has been done on more complicated organic molecular structures. It is now fairly common for the structures of proteins to be determined by x-ray crystallography. Even the original determination of the structure of DNA was done by a combination of the results of x-ray crystal structures and simple model building.

B. Simple Cubic Structures

FIGURE 16.18: Cubic unit cell.

The simplest type of unit cell is the cubic unit cell in which there is an atom at each corner of a cube. (Fig. 16.18.) Such a structure is known as the simple cubic structure. While there are examples of simple cubic structure in nature, the simple cubic structure is an inefficient packing arrangement of atoms and is not favored. Note that in simple cubic packing, each atom has six nearest neighbors, four in an imaginary plane, one above that plane, and one below that plane. This means that each atom will interact most strongly with only six other atoms.

FIGURE 16.19: Body centered cubic.

In a small modification of the simple cubic structure, termed body centered cubic (BCC), another atom is positioned at the exact center of the cube. The number of nearest neighbor interactions in this structure rises to eight (Fig 16.19.)

This increase in the number of nearest neighbors helps improve the overall energy of interaction, but is not the best that can be done. Two other structures exist which are "closest packed" structures. One of these closest packed structures is a cubic structure known as cubic closest packed or face centered cubic (FCC).

C. Closest Packing

Closest packing is best understood if one takes a box and fills it with a monolayer of marbles. If all the marbles are the same size, the packing arrangement that will be favored is one in which each marble touches six other marbles. When layers of such closest packed atoms are placed on top of one another, there are two ways in which this can be accomplished (Fig 16.20 a and b.) In both of these ways, the second layer of atoms is placed in such a way that the atoms lie in hollows in the lower layer structure. When the third layer is placed, it, too, goes into the hollows in the second layer. However, there are two different sets of hollows that it can go on. In one arrangement, the atoms in the third layer are directly over the atoms of the first layer. Such an arrangement is described as ababab... and leads to a geometry termed hexagonal closest packing (Fig 16.20 a). In the other arrangement, the third layer lies in another set of hollows that is not directly above the first layer atoms. In this arrangement the repetition of the first layer does not occur until the fourth layer

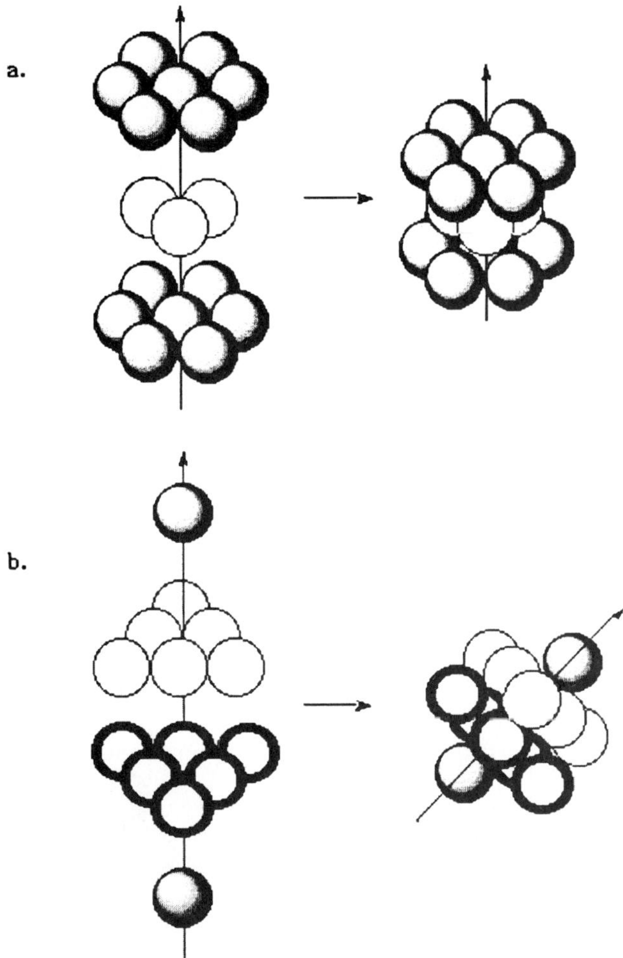

FIGURE 16.20: Closest packed structures: a) hexagonal and b) cubic.

is added. Such an arrangement is termed abcabcabc ... and leads to another cubic crystal structure known as cubic closest packing.

Both types of closest packing lead to each atom having 12 nearest neighbors, six in its plane, three in the plane above, and three in the plane below. This is the largest number of nearest neighbor interactions that are possible in nature when spheres are packed. As a consequence, nature uses closest packed structures in most instances.

D. Real Ionic Structures

When dealing with crystals of molecules or atoms, the components of the crystal are either atoms or molecules and the crystal structure is simply the arrangement of those atoms or molecules. The crystal structures of ionic materials are somewhat different since, in these crystals, two different species, the cation and the anion, must be accommodated. This is made even more difficult in that, in many compounds, there may not be equal numbers of cations and anions because the charges of the two may differ.

In a closest packed arrangement there are two types of holes available. One type, termed octahedral holes, are arranged such that there are six such holes octahedrally surrounding each crystal position. Each of these holes is shared by six of the crystal positions, hence the ratio of octahedral holes to crystal positions is 1:1. A second group of holes are arranged in tetrahedral positions. These holes are smaller than octahedral holes but they are twice as numerous. Hence, the ratio of tetrahedral holes to crystal positions is 2:1.

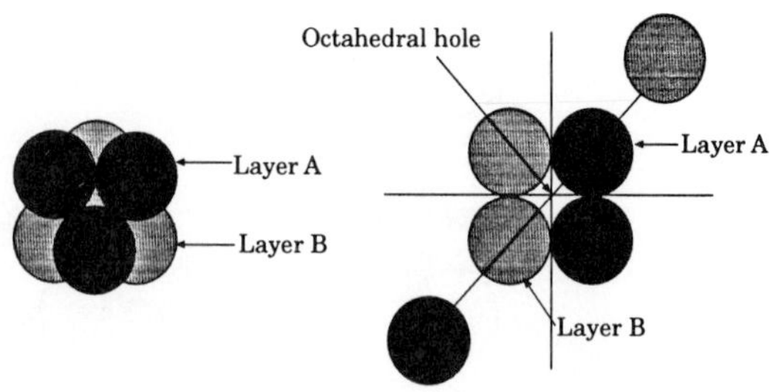

FIGURE 16.21(a): Octahedral hole.

The structures of ionic materials can often be considered by thinking of them as arrangements of one of the ions in a closest packed structure with the counter ions occupying some or all the holes that exist in the structure. Since anions are usually larger than cations, the anions are usually the ones that are arranged in the closest packed structure. This approach to the structure of ionic crystals works well until the sizes of the ions become such that the smaller of the two will not fit easily into the remaining holes or until the stoichiometry of the compound cannot be accommodated by the number of holes available. Thus, the sodium chloride structure may be considered to be a cubic closest packed array of chloride ions with sodium ions occupying each of the octahedral holes. This crystal structure is one of the most common structures for 1:1 salts as it leads to the best cation–anion contact possible for a wide range of cation:anion radius ratios.

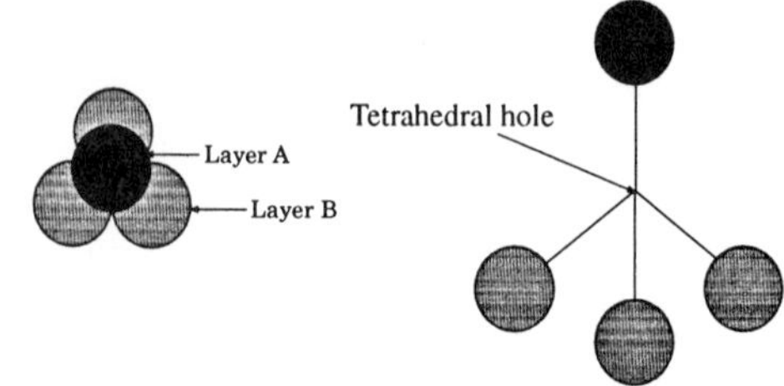

FIGURE 16.21(b): Tetrahedral hole.

FIGURE 16.22: NaCl, with Cl^- on lattice sites of a face centered cubic unit cell, Na^+ in octahedral holes.

As the cation becomes larger and no longer fits into the octahedral holes, the cesium chloride structure is used. In this structure the chloride ions are arranged in a simple cubic lattice with the cesium ion occupying the center of the cube. While this structure provides each cesium ion with 8 chloride ions surrounding it (vs. 6 for the sodium chloride structure), the sodium chloride structure provides a number of other interactions with chloride ions just slightly farther away. These extra interactions provide the stabilization required for the structure with the smaller number of nearest neighbors to be the more stable structure. Thus, the sodium chloride structure is preferred if the radius ratios will permit.

Other salts, with different stoichiometries, crystallize in similar fashions. For instance, cadmium chloride, $CdCl_2$, features cubic closest packed chloride ions with cadmium ions occupying half of the octahedral holes. Aluminum oxide, Al_2O_3, has oxide ions in a hexagonal closest packed structure with aluminum ions occupying 2/3 of the octahedral holes. Finally, sodium sulfide, Na_2S, has sulfide ions in a cubic closest packed structure with sodium ions occupying all the tetrahedral holes.

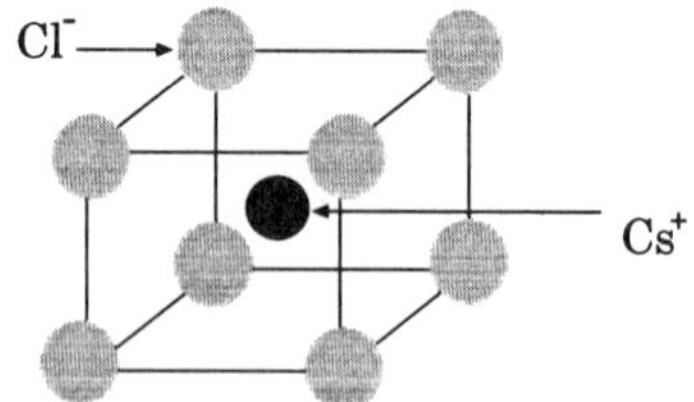

FIGURE 16.23: Unit cell for CsCl, with Cl^- on lattice sites of a simple cubic cell.

E. Structure/Density Relationships

There is a direct relationship between the unit cell structure and the density of a material. Indeed, measurements of the density of materials may rule out certain structures. For instance, let us calculate the expected density of solid neon from the following data: neon crystallizes in a face centered cubic structure, and the atomic radius of the neon atom is 71 pm.

Since an overall crystal is built by replicating the unit cell in 3 dimensions, we must first determine how many atoms there are per unit cell. While, at first glance, this may appear quite simple, we must remember that any particular atom may "belong" to more than one unit cell. There are four places an atom may be placed within a cubic unit cell: (1) the exact center of the cell, (2) in the middle of one of the edges of the unit cell, (3) in the middle of one of the faces of the unit cell, and (4) at the corner of the unit cell.

FIGURE 16.24 (a): Exploded view of face-centered cubic cell.

Atoms at each of these sites contribute different amounts to the unit cell. An atom in the center of the unit cell is not shared with any other unit cell (Fig 16.25 a). Therefore it contributes 1 to the count of atoms in the unit cell. An atom placed in the middle of an edge belongs to four different unit cells and, therefore, contributes 1/4 to the unit cell count (Fig 16.25 b). Since there are a total of 12 edges on a cubic unit cell, atoms in the center of each edge will contribute a total of 3 atoms (12/4) to the unit cell count. Atoms in the middle of a face belong to two different unit cells and contribute 1/2 to each unit cell (Fig 16.25 c). There are a total of six faces; this leads to a total contribution of 3 (6/2) to the atom count for the unit cell. Finally, an atom at the corner of the unit cell belongs to eight different unit cells and counts 1/8 toward the atom count (Fig 16.25 d). Since a unit cell has eight corners, the corner atoms contribute a total of 1 (8/8) to the count.

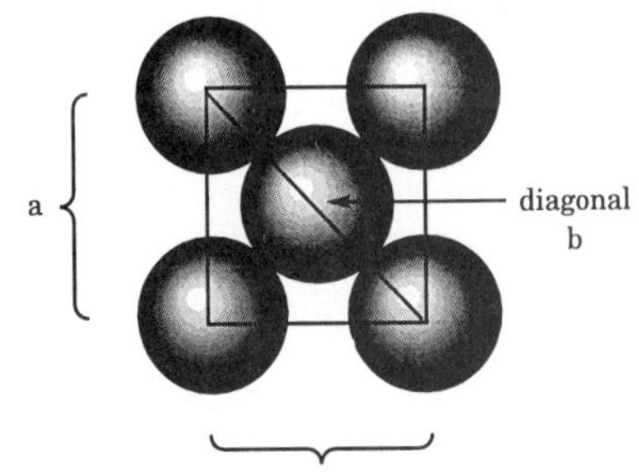

FIGURE 16.24 (b): Atoms touching in one face of face centered cubic cell

A face centered cubic unit cell has atoms at each corner and one in the center of each face. As a result the unit cell contains 4 atoms (8/8 + 6/2) = 4. If we assume that the neon atoms touch across the face diagonal, the length of the face diagonal must be 4 times the radius of the neon atom or 284 pm. From a consideration of the geometry of right triangles: $a^2 + a^2 = b^2$ where a is the length of the side and b is the length of the diagonal. We can easily calculate the length of the side of the cubic unit cell as:

$$2a^2 = b^2 \qquad \text{EQ. 16.3a}$$

$$a = \sqrt{\frac{(284)^2}{2}} \qquad \text{EQ. 16.3b}$$

or $200.\overline{8}$ pm. The volume of the unit cell can then be determined to be:

$$\left(200.\overline{8}\ \text{pm}\right)^3 = 8.09\overline{6} \times 10^6\ \text{pm}^3 \qquad \text{EQ. 16.4}$$

Converting this to the more familiar cm^3 we obtain:

$$\left(8.09\overline{6} \times 10^6\ \text{pm}^3\right)\left(\frac{1\ \text{m}}{10^{12}\ \text{pm}}\right)^3\left(\frac{100\ \text{cm}}{1\ \text{m}}\right)^3 = 8.09\overline{6} \times 10^{-24}\ \text{cm}^3/\text{unit cell} \qquad \text{EQ. 16.5}$$

The weight, in grams, of one unit cell would be:

$$\left(\frac{4\ \text{atoms}}{1\ \text{unit cell}}\right)\left(\frac{1\ \text{mol}}{6.023 \times 10^{23}\ \text{atoms}}\right)\left(\frac{20.18\ \text{g}}{1\ \text{mol}}\right) = 1.340 \times 10^{-22}\ \text{g}/\text{unit cell} \qquad \text{EQ. 16.6}$$

This leads to an expected density for solid neon of:

$$\frac{1.340 \times 10^{-22}\ \frac{\text{g}}{\text{unit cell}}}{8.096 \times 10^{-24}\ \frac{\text{cm}^3}{\text{unit cell}}} = 16.55\ \frac{\text{g}}{\text{cm}^3}$$

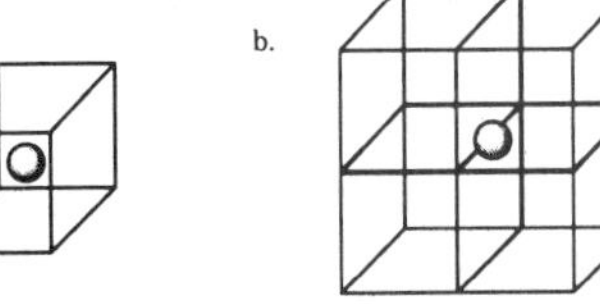

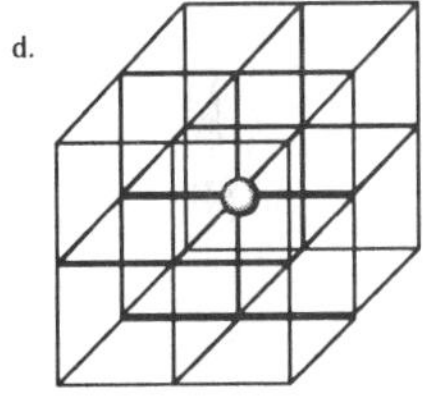

FIGURE 16.25: Fraction of atom/ion belonging to a unit cell for various positions in the cell.

KEY WORDS & Concepts

Amorphous solid: a solid with no long–range structure.

Boiling point: that temperature at which the vapor pressure of a liquid is equal to the pressure of the surroundings.

Closest packing: the most efficient way of packing identical spherical objects: each sphere has 12 nearest neighbors.

Coulombic forces: forces existing between charged particles.

Crystalline solid: a solid characterized by a 3–dimensional regular repeating arrangement of sub–units.

Debye: an English chemist who contributed significantly to the understanding of molecular structure.

Dipole moment: a partial separation of charge within a molecule resulting from the unequal distribution of electrons.

Dispersion forces: intermolecular forces which exist because of temporary dipole moments formed in molecules as a result of the motion of the electrons around the molecule.

Hydrogen bonding: an especially strong dipole–dipole interaction between hydrogen atoms and small, electronegative atoms on other molecules.

London forces: see dispersion forces – named for the German physicist Fritz London.

Network solids: solids in which the sub–units are held together by actual covalent bonds making the entire crystal one molecule.

Surface tension: a property of a bulk liquid which makes the surface of the liquid behave as if it were a "skin."

Unit cell: the repeating unit in a crystalline solid.

Vapor pressure: the partial pressure exerted by the vapor of a substance in contact with the liquid phase of the substance.

X–rays: electromagnetic radiation with wavelengths of ca. 1Å.

HOMEWORK Problems

1. Pick the species you expect to have the higher boiling point and give the reason:

(a) Ar and He
(b) F_2 and ClF
(c) CO_2 and NO_2
(d) CH_3OCH_3 and CH_3CH_2OH
(e) $CH_3CH_2CH_2CH_2CH_3$ and $CH_3CH_2CH_2CH_2CH_2CH_2CH_3$

2. Pick the species that you would expect to have the higher melting point and give the reason:

(a) NaCl and KCl
(b) NaCl and MgO
(c) Na_2O and H_2O
(d) CO_2 and NO_2
(e) NCl_3 and NH_3
(f) NCl_3 and NI_3

3. Acetone and ethanol are both common laboratory solvents. Explain how, even though initially dry (i.e., free of water) both solvents are usually found to contain water after storage in the lab for some time. HINT: show the structures of acetone, ethanol, and water, and indicate the direction of the dipole of each.

4. Liquid carbon monoxide has a surface tension of 9.8 dyne/cm, while liquid carbon dioxide has a surface tension of 1.16 dyne/cm. Draw Lewis dot diagrams of both species and explain the difference in surface tension.

5. Oxygen molecule has a 0 dipole moment, while ozone has a dipole moment of 0.53 debyes. Draw Lewis dot diagrams of both and explain the difference in dipole moment.

6. Pure sulfuric acid is very viscous (e.g., 25.4 centipoise, compared to about 1 centipoise for water). Explain, based on the molecular structure of sulfuric acid.

7. There are two geometric isomers of 1,2-dichloroethene. One achieves a vapor pressure of 400 torr at 41.0°C; the other isomer reaches this vapor pressure at 30.8°C. Draw and name the two isomers. Explain which is likely to be more volatile.

8. Acetic acid (CH_3COOH) is known to dimerize in both the liquid and gas state, forming $(CH_3COOH)_2$. The dimer is held together by hydrogen bonding. Draw a reasonable structure of the dimer of acetic acid.

9. Iodine monochloride (ICl) boils at 97.4°C. Bromine (Br_2) boils at 58.78°C. Explain.

10. Indicate which interparticle force is most important in the interaction between:

(a) Ethanol dissolved in water?
(b) Iodine dissolved in carbon tetrachloride?
(c) Sodium chloride dissolved in water?
(d) Formaldehyde dissolved in acetone?
(e) n-pentane and 2-methylbutane?

11. Cesium chloride crystallizes in a simple cubic structure in which there is a cesium ion at the corner of each unit cell and a chloride ion in the center of the cell. The structure could alternatively be considered as having a chloride ion at each corner with a cesium ion in the center. The radius of a cesium ion is 181 pm; the radius of a chloride ion is 167 pm. Calculate the length of the side of the CsCl unit cell. (Note: the edge length may be equal to twice the radius of the Cs^+ ion, twice the radius of the Cl^- ion, or it may be determined by the length of the body diagonal which may be determined by assuming that the cesium and chloride ions touch. Since cubic unit cells share corner atoms, the length of the edge of a cubic unit cell is measured between the centers of the atoms at the corners of the unit cell.)

12. Calculate the volume of a unit cell of sodium metal. Sodium metal crystallizes in a body–centered cubic structure with a sodium atom at each corner of the cube and a sodium atom in the center of the body of the cube. Assume that sodium has an atomic radius of 186 pm.

13. A body centered cubic structure is considered to have two atoms per unit cell. (Each of the eight corner atoms belongs to 8 different unit cells and, therefore each contributes 1/8 of an atom, or one total atom, to the cell. The center atom belongs entirely to the cell.) Calculate the density of sodium metal using the information in problem 12 assuming that the sodium atoms touch across the body diagonal.

14. Nickel crystallizes in a face–centered cubic structure in which there is one atom at each corner of the cube and one additional atom in the middle of each face of the cube. Each unit cell of a face centered cubic structure contains 4 atoms. (Each of the 8 corner atoms counts 1/8 and each of the 6 face atoms contributes 1/2). The atomic radius of the nickel atom is 124 pm. Calculate the expected density of nickel. Assume that the nickel atoms touch across the face diagonal.

15. The density of iron metal is 7.86 g/cm3, and the unit cell has a side with a length of 287 pm. Are these data most consistent with a face centered cubic structure, a body centered cubic structure, or a simple cubic structure (atoms at the corners of the cube only, 1 atom per unit cell)?

16. Gold (Au, 196.97) uses a closest cubic packing structure and has a density of 19.3 g cm^{-3}. Calculate the atomic radius of a gold atom.

17. Potassium fluoride has the same crystal structure as sodium chloride. The unit cell of KF has an edge length of 0.539 nm. Calculate the density of KF.

OVERVIEW Problems

18. Ammonia and NF_3 have dipole moments of 1.47 and 0.235 D respectively.
   (a) Draw Lewis dot diagrams of each.
   (b) Draw a three dimensional representation of both molecules; show the direction of the dipole moment with an arrow with appropriate orientation.
   (c) Explain the difference in dipole moments based on molecular structure and electronegativity differences.

19. Methyl acetate and propionic acid both have a molecular formula of $C_3H_6O_2$, but have boiling points of 57.8 and 141°C respectively. Draw structures of each and use them to explain the difference in boiling points.

20. Three compounds with the formula $C_6H_{15}N$ have boiling points of 89, 110, and 130°C. The three names (in no particular order) are *n*-hexylamine, *n*-propyl-1-propanamine, and N,N-diethylethanamine. Draw the structure of the three named compounds, match each with its boiling point, and explain the order of boiling points.

21. Two liquid compounds, A and B, have the same composition: 64.81% carbon, 13.60% hydrogen, and 21.59% oxygen. Both show the same molecular ion peak in the mass spectrum at 74. The surface tension at 20.0°C for A and B are 17.01 and 24.6 dyne/cm respectively. The carbon NMR shows absorbances at 15 and 66 ppm for A, and absorbances at 14, 19, 35, and 62 ppm for B. The IR spectra show a broad absorbance between 3600 and 3100 cm^{-1} for B, but no absorbance above 3100 cm^{-1} for A.
   (a) What is the molecular formula of A and B?
   (b) What is the IHD for A and B?
   (c) Draw the structural formula of A and B.
   (d) Explain the significance of the IR spectra in relation to the structures for A and B.
   (e) Explain the significance of the NMR spectra in relation to the structures for A and B.
   (f) Explain the difference in surface tension between A and B.

22. Two liquid compounds, C and D, have the same composition: 52.14% carbon, 13.13% hydrogen, and 34.74% oxygen. Both show the same molecular ion peak in the mass spectrum at 46. A vapor pressure of 100.0 torr is reached by C at 34.9°C, and by D at -62.7°C. The carbon NMR shows absorbances two absorbances for C, and one absorbance for D. The IR spectra show a broad absorbance between 3600 and 3100 cm^{-1} for C, but no absorbance above 3100 cm^{-1} for D.
   (a) What is the molecular formula of C and D?
   (b) What is the IHD for C and D?
   (c) Draw the structural formula of C and D.
   (d) Explain the significance of the IR spectra in relation to the structures for C and D.
   (e) Explain the significance of the NMR spectra in relation to the structures for C and D.
   (f) Explain the difference in vapor pressure between C and D.

23. Two liquid compounds, E and F, have the same composition: 62.04% carbon, 10.41% hydrogen, and 27.55% oxygen. Both show the same molecular ion peak in the mass spectrum at 58. At 30.0°C, E and F have viscosities of 0.295 and 1.07 centipoise respectively. The carbon NMR shows two absorbances for E, and three absorbances for F. The IR spectrum for E shows a strong absorbance at 1710 cm^{-1}, but no absorbance above 3100 cm^{-1}. The IR spectrum for F shows a broad absorbance between 3600 and 3100 cm^{-1} and a peak around 1640 cm^{-1}.

(a) What is the molecular formula of E and E?

(b) What is the IHD for E and F?

(c) Draw the structural formula of E and F.

(d) Explain the significance of the IR spectra in relation to the structures for E and F.

(e) Explain the significance of the NMR spectra in relation to the structures for E and F.

(f) Explain the difference in viscosity between E and F.

APPENDIX  THE SI SYSTEM

The SI (or metric) System has the following base units:

PHYSICAL QUANTITY	NAME OF UNIT	SYMBOL
Length	meter	m
Mass	kilogram	kg
Time	second	s
Temperature	Kelvin	K
Electric Current	ampere	amp
Luminous Intensity	candela	cd
Amount of substance	mole	mol

The SI (or metric) System includes the following derived units:

PHYSICAL QUANTITY	NAME OF UNIT	SYMBOL
Energy	joule	J (kg m^2 s^{-2})
Frequency	hertz	Hz (cycles s^{-1})
Force	newton	N (kg m s^{-2})
Pressure	pascal	Pa (kg m^{-1} s^{-2})
Power	watt	W (J s^{-1})
Electric Charge	coulomb	C (amp s)
Electric Potential	volt	V (J C^{-1})

The following prefixes and symbols are used in modifying SI units:

PREFIX/SYMBOL	MULTIPLIER	PREFIX/SYMBOL	MULTIPLIER
exa(E)	10^{18}	atto(a)	10^{-18}
peta(P)	10^{15}	femto(f)	10^{-15}
tera(T)	10^{12}	pico(p)	10^{-12}
giga(G)	10^{9}	nano(n)	10^{-9}
mega(M)	10^{6}	micro(m)	10^{-6}
kilo(k)	10^{3}	milli(m)	10^{-3}
hecto(h)	10^{2}	centi(c)	10^{-2}
deka(d)	10^{1}	deci(d)	10^{-1}

APPENDIX  OTHER COMMON UNITS

The following are units in common use, and their relation to corresponding SI units:

PHYSICAL QUANTITY	COMMON UNIT	VALUE IN SI UNITS
Length	angstrom	10^{-10} m
	inch	0.0254 m
Mass	pound (avdp)	0.45359 kg
	amu	1.6606×10^{-24} g
Temperature	°C	(K - 273.15)
Energy	calorie	4.184 J
	electron volt	1.6022×10^{-19} J
Pressure	atm	101,325 Pa
	bar	10^5 Pa
	torr	133.32 Pa
Volume	liter	10^{-3} m^3
Concentration	Molarity	mol L^{-1}; mol dm^{-3}
Density		g cm^{-3}

APPENDIX  SELECTED FUNDAMENTAL CONSTANTS

Avogadro's Number	6.022045×10^{23} mol^{-1}
Electron Charge	1.602189×10^{-19} C
Electron Mass	9.109534×10^{-28} g
Proton Mass	1.672648×10^{-24} g
Neutron Mass	1.674954×10^{-24} g
Speed of Light	2.997925×10^{8} m s^{-1}
Planck's Constant	6.626176×10^{-34} Js
Gas Constant	8.31441 J mol^{-1} K^{-1}
	0.0820578 L atm mol^{-1} K^{-1}
	1.98717 cal mol^{-1} K^{-1}

APPENDIX  THE CHEMICAL ELEMENTS

ATOMIC NUMBER	NAME	MASS	ELECTRON CONFIGURATION
1	hydrogen	1.00794	$1s^1$
2	helium	4.002602	$1s^2$ = [He]
3	lithium	6.941	[He] $2s^1$
4	beryllium	9.012182	[He] $2s^2$
5	boron	10.811	[He] $2s^2\ 2p^1$
6	carbon	12.011	[He] $2s^2\ 2p^2$
7	nitrogen	14.00674	[He] $2s^2\ 2p^3$
8	oxygen	15.9994	[He] $2s^2\ 2p^4$
9	fluorine	18.998403	[He] $2s^2\ 2p^5$
10	neon	20.1792	[He] $2s^2\ 2p^6$ = [Ne]
11	sodium	22.989768	[Ne] $3s^1$
12	magnesium	24.3050	[Ne] $3s^2$
13	aluminum	26.981539	[Ne] $3s^2 3p^1$
14	silicon	28.0855	[Ne] $3s^2 3p^2$
15	phosphorus	30.973762	[Ne] $3s^2 3p^3$
16	sulfur	32.066	[Ne] $3s^2 3p^4$
17	chlorine	35.4527	[Ne] $3s^2 3p^5$
18	argon	39.948	[Ne] $3s^2 3p^6$ = [Ar]
19	potassium	39.0983	[Ar] $4s^1$
20	calcium	40.078	[Ar] $4s^2$
21	scandium	44.955910	[Ar] $4s^2\ 3d^1$
22	titanium	47.88	[Ar] $4s^2\ 3d^2$
23	vanadium	50.9415	[Ar] $4s^2\ 3d^3$
24	chromium	51.9961	[Ar] $4s^1\ 3d^5$ *
25	manganese	54.93805	[Ar] $4s^2\ 3d^5$
26	iron	55.847	[Ar] $4s^2\ 3d^6$
27	cobalt	58.93320	[Ar] $4s^2\ 3d^7$
28	nickel	58.6934	[Ar] $4s^2\ 3d^8$
29	copper	63.546	[Ar] $4s^1\ 3d^{10}$ *
30	zinc	65.39	[Ar] $4s^2\ 3d^{10}$
31	gallium	69.723	[Ar] $4s^2\ 3d^{10} 4p^1$
32	germanium	72.61	[Ar] $4s^2\ 3d^{10} 4p^2$
33	arsenic	74.92159	[Ar] $4s^2\ 3d^{10} 4p^3$
34	selenium	78.96	[Ar] $4s^2\ 3d^{10} 4p^4$
35	bromine	79.904	[Ar] $4s^2\ 3d^{10} 4p^5$
36	krypton	83.80	[Ar] $4s^2\ 3d^{10} 4p^6$ = [Kr]
37	rubidium	85.4678	[Kr] $5s^1$
38	strontium	87.62	[Kr] $5s^2$
39	yttrium	88.90585	[Kr] $5s^2\ 4d^1$
40	zirconium	91.224	[Kr] $5s^2\ 4d^2$
41	niobium	92.90638	[Kr] $5s^1\ 4d^4$ *
42	molybdenum	95.94	[Kr] $5s^1\ 4d^5$ *
43	technetium	(98)**	[Kr] $5s^2\ 4d^5$

44	ruthenium	101.07	[Kr] $5s^1$ $4d^7$ *
45	rhodium	102.90550	[Kr] $5s^1$ $4d^8$ *
46	palladium	106.42	[Kr] $5s^0$ $4d^{10}$ *
47	silver	107.8682	[Kr] $5s^1$ $4d^{10}$ *
48	cadmium	112.411	[Kr] $5s^2$ $4d^{10}$
49	indium	114.82	[Kr] $5s^2$ $4d^{10}$ $5p^1$
50	tin	118.710	[Kr] $5s^2$ $4d^{10}$ $5p^2$
51	antimony	121.757	[Kr] $5s^2$ $4d^{10}$ $5p^3$
52	tellurium	127.60	[Kr] $5s^2$ $4d^{10}$ $5p^4$
53	iodine	126.90447	[Kr] $5s^2$ $4d^{10}$ $5p^5$
54	xenon	131.29	[Kr] $5s^2$ $4d^{10}$ $5p^6$ = [Xe]
55	cesium	132.9054	[Xe] $6s^1$
56	barium	137.327	[Xe] $6s^2$
57	lanthanum	138.9055	[Xe] $6s^2$ $5d^1$
58	cerium	140.115	[Xe] $6s^2$ $5d^1$ $4f^1$
59	praseodymium	140.90765	[Xe] $6s^2$ $5d^0$ $4f^3$ *
60	neodymium	144.24	[Xe] $6s^2$ $5d^0$ $4f^4$ *
61	promethium	(145)**	[Xe] $6s^2$ $5d^0$ $4f^5$ *
62	samarium	150.36	[Xe] $6s^2$ $5d^0$ $4f^6$ *
63	europium	151.965	[Xe] $6s^2$ $5d^0$ $4f^7$ *
64	gadolinium	157.25	[Xe] $6s^2$ $5d^1$ $4f^7$
65	terbium	158.92534	[Xe] $6s^2$ $5d^0$ $4f^9$ *
66	dysprosium	162.50	[Xe] $6s^2$ $5d^0$ $4f^{10}$ *
67	holmium	164.93032	[Xe] $6s^2$ $5d^0$ $4f^{11}$ *
68	erbium	167.26	[Xe] $6s^2$ $5d^0$ $4f^{12}$ *
69	thulium	168.93421	[Xe] $6s^2$ $5d^0$ $4f^{13}$
70	ytterbium	173.04	[Xe] $6s^2$ $5d^0$ $4f^{14}$ *
71	lutetium	174.967	[Xe] $6s^2$ $5d^1$ $4f^{14}$
72	hafnium	178.49	[Xe] $6s^2$ $5d^2$ $4f^{14}$
73	tantalum	180.9479	[Xe] $6s^2$ $5d^3$ $4f^{14}$
74	tungsten	183.85	[Xe] $6s^2$ $5d^4$ $4f^{14}$
75	rhenium	186.207	[Xe] $6s^2$ $5d^5$ $4f^{14}$
76	osmium	190.2	[Xe] $6s^2$ $5d^6$ $4f^{14}$
77	iridium	192.22	[Xe] $6s^2$ $5d^7$ $4f^{14}$
78	platinum	195.08	[Xe] $6s^1$ $5d^9$ $4f^{14}$ *
79	gold	196.96654	[Xe] $6s^1$ $5d^{10}$ $4f^{14}$
80	mercury	200.59	[Xe] $6s^2$ $5d^{10}$ $4f^{14}$
81	thallium	204.3833	[Xe] $6s^2$ $5d^{10}$ $4f^{14}$ $6p^1$
82	lead	207.2	[Xe] $6s^2$ $5d^{10}$ $4f^{14}$ $6p^2$
83	bismuth	208.98037	[Xe] $6s^2$ $5d^{10}$ $4f^{14}$ $6p^3$
84	polonium	(209)**	[Xe] $6s^2$ $5d^{10}$ $4f^{14}$ $6p^4$
85	astatine	(210)**	[Xe] $6s^2$ $5d^{10}$ $4f^{14}$ $6p^5$
86	radon	(222)**	[Xe] $6s^2$ $5d^{10}$ $4f^{14}$ $6p^6$ = [Rn]
87	francium	223.0197***	[Rn] $7s^1$
88	radium	226.0254	[Rn] $7s^2$
89	actinium	227.0278***	[Rn] $7s^2$ $6d^1$
90	thorium	232.0381	[Rn] $7s^2$ $6d^2$ $5f^0$ *
91	protactinium	231.03588	[Rn] $7s^2$ $6d^1$ $5f^2$
92	uranium	238.0289	[Rn] $7s^2$ $6d^1$ $5f^3$

93	neptunium	237.0482***	[Rn] $7s^2$ $6d^1$ $5f^4$
94	plutonium	(240)**	[Rn] $7s^2$ $6d^0$ $5f^6$ *
95	americium	243.0614***	[Rn] $7s^2$ $6d^0$ $5f^7$ *
96	curium	(247)**	[Rn] $7s^2$ $6d^1$ $5f^7$
97	berkelium	(248)**	[Rn] $7s^2$ $6d^0$ $5f^9$ *
98	californium	(250)**	[Rn] $7s^2$ $6d^0$ $5f^{10}$ *
99	einsteinium	252.083***	[Rn] $7s^2$ $6d^0$ $5f^{11}$ *
100	fermium	257.0951***	[Rn] $7s^2$ $6d^0$ $5f^{12}$
101	mendelevium	(257)**	[Rn] $7s^2$ $6d^0$ $5f^{13}$
102	nobelium	259.1009***	[Rn] $7s^2$ $6d^0$ $5f^{14}$
103	lawrencium	292.11	[Rn] $7s^2$ $6d^0$ $5f^{14}$ $7p^1$ *
104	rutherfordium	261.11***	[Rn] $7s^2$ $6d^2$ $5f^{14}$ *
105	dubnium	262.114***	[Rn] $7s^2$ $6d^3$ $5f^{14}$ *
106	seaborgium	263.118***	[Rn] $7s^2$ $6d^4$ $5f^{14}$ *
107	bohrium	262.12***	[Rn] $7s^2$ $6d^5$ $5f^{14}$ *
108	hassium		[Rn] $7s^2$ $6d^6$ $5f^{14}$ *
109	meitnerium		[Rn] $7s^2$ $6d^7$ $5f^{14}$ *

n.b. elements 104-109 named according to IUPAC's 1997 proposal

* exceptional electron configuration

** average mass of the stable isotopes of radioactive element

*** exact mass of the most stable isotope of radioactive element

APPENDIX E COMMON IONS

CATIONS

COMMON ELEMENTAL CATIONS WITH A +1 OXIDATION STATE:

H^+, Li^+, Na^+, K^+, Rb^+, Cs^+, Cu^+ (cuprous), Ag^+, Tl^+, Hg_2^{2+}(mercurous)

COMMON ELEMENTAL CATIONS IN A +2 OXIDATION STATE:

Cr^{2+}, Fe^{2+}(ferrous), Co^{2+}, Ni^{2+}, Cu^{2+}(cupric), Zn^{2+}, Cd^{2+}, Sn^{2+}(stannous), Hg^{2+} (mercuric), Pb^{2+}

OTHER COMMON ELEMENTAL CATIONS:

Al^{3+}, Cr^{3+}, Fe^{3+}(ferric), Sn^{4+}(stannic), Bi^{3+}

(Note: When an element forms two common cations, the lower oxidation state species has the "-ous" ending, while the higher oxidation state species has the "-ic" ending; e.g., stannous, stannic. Elemental cations may also be refered to by the name of the element, followed by the parenthetical charge of the cation, expressed in Roman numerals; e.g., Iron (II), Iron (III).

ANIONS

COMMON ELEMENTAL ANIONS:

H^-	hydride
F^-	fluoride
Br^-	bromide
I^-	iodide
O^{2-}	oxide
S^{2-}	sulfide
Se^{2-}	selenide
Te^{2-}	telluride
N^{3-}	nitride
P^{3-}	phosphide
C^{4-}	carbide

(Note: elemental anions are characterized by the ending "-ide". The only common exceptions are OH^- (hydroxide) and CN^- (cyanide).

Common Binary Oxo Anions:

CO_3^{2-}	carbonate		
$C_2O_4^{2-}$	oxalate		
NO_3^-	nitrate	NO_2^-	nitrite
PO_4^{3-}	phosphate	PO_3^{3-}	phosphite
AsO_4^{3-}	arsenate	AsO_3^{3-}	arsenite
SO_4^{2-}	sulfate	SO_3^{2-}	sulfite
SeO_4^{2-}	selenate	SeO_3^{2-}	selenite
ClO_4^-	perchlorate	ClO_3^-	chlorate
ClO_2^-	chlorite	ClO^-	hypochlorite
BrO_3^-	bromate		
IO_4^-	periodate		
CrO_4^{2-}	chromate	$Cr_2O_7^{2-}$	dichromate
MnO_4^-	permanganate		

(Note: When an element forms two oxo anions, the one in which the element has the higher oxidation state has the "-ate" ending; the one in which the element has the lower oxidation state has the "-ite" ending. When more than two common oxidation state occur in binary oxo anions, the highest oxidation state species is designated by a prefix "per-" in addition to the suffix "-ate". The lowest oxidation state species is designated by a prefix "hypo-" in addition to the suffix "-ite".)

Miscellaneous Common Ions

NH_4^+	ammonium ion	
HCO_3^{2-}	hydrogen carbonate	bicarbonate
HSO_4^-	hydrogen sulfate	bisulfate
HSO_3^-	hydrogen sulfite	bisulfite
HPO_4^{2-}	mono-hydrogen phosphate	
$H_2PO_4^-$	di-hydrogen phophate	

APPENDIX F.1 THOMSON'S CHARGE TO MASS DETERMINATION FOR THE ELECTRON.

A charged particle moving perpendicular to a magnetic field experiences a centripetal force that is perpendicular both to the velocity vector and magnetic field. (This is sometimes known as the right hand rule, and is illustrated in the figure below.)

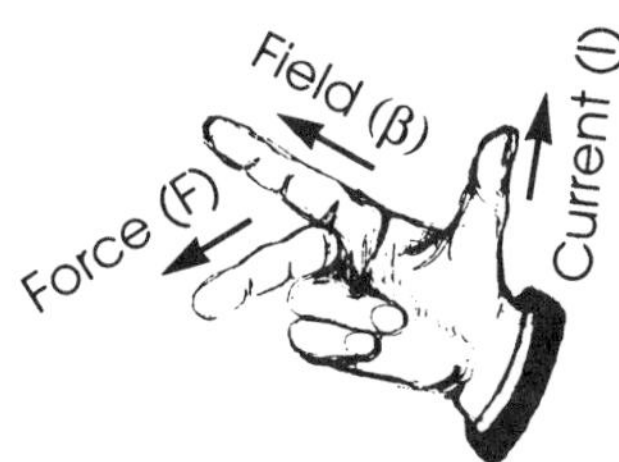

The result is that a charged particle moving perpendicular to a magnetic field will travel in a circle. When a particle with charge q is moving with velocity v perpendicular to a magnetic field of strength β, the force experienced is given by:

$$F = q\,v\beta \sin\theta$$

Where θ is the angle between the velocity vector and the magnetic field. In this case, θ equals 90°, and Sin θ is one. The force F has units of newtons when q is in coulombs, v is in meters per second, and β is in Tesla (note, the mks units are used, specifically <u>not</u> using the common unit of Gauss for magnetic field strength.)

The force exerted on the charged particle equals the centripetal force (that responsible for keeping the charged particle in a circular orbit.)

$$\text{Centripetal force} = \frac{m\,v^2}{r}$$

By setting the two equations for force equal, the charge to mass ratio can be determined:

$$\frac{q}{m} = \frac{v}{\beta\, r}$$

Thomson experimentally controlled the magnetic field strength β. That value, along with the electric field strength, allowed Thomson to independently determine the velocity of the electron (which he determined to be about 2.58 x 10^8 meters per second, or about 86% the speed of light.) Finally, Thomson measured r, the radius of curvature of the "cathode rays" in the magnetic field. Knowing all the terms on the right hand side allowed him to determine that:

$$\frac{q}{m} = 1.759 \times 10^{8}\text{coulomb gram}^{-1}$$

APPENDIX F.2 BOHR'S THEORY OF THE HYDROGEN ATOM - PREDICTION OF r.

Bohr proposed stable orbits for electrons based upon an offsetting of the electrostatic force of attraction and the centripetal force of the moving electron. The electron has mass m, charge e, velocity v, and is moving in a circular path of radius r in space with a permitivity of ε_o. Equating the two forces in terms of these variables yields:

$$\frac{e^2}{4\,\pi\varepsilon_0 r^2} = \frac{m\,v^2}{r}$$

Rearranging and solving for the velocity of the electron yields:

$$v^2 = \frac{e^2}{4\,\pi\varepsilon_0 r\,m}$$

In his theory, Bohr allowed only those orbits which had angular momentum quantized in units of $h/2\pi$. This corresponds to the equation:

$$mvr = n\left(\frac{h}{2\,\pi}\right)$$

where n is any positive integer. Solving this equation for v yields:

$$v = \frac{n\,h}{2\,\pi m\,r}, \text{ or } v^2 = \frac{n^2h^2}{4\pi^2m^2r^2}$$

Setting the two equations for v^2 equal yields:

$$\frac{e^2}{4\,\pi\varepsilon_0 r\,m} = \frac{n^2h^2}{4\pi^2m^2r^2}$$

From which the radius r can be solved as:

$$r = \frac{n^2h^2\varepsilon_0}{\pi me^2}$$

All the terms in this equation are known constants with the exception of n. If n is set to 1, the lowest energy, smallest orbit, the resulting calculation yields the value of r:

$$r = \frac{1^2\left(6.6262 \times 10^{-34}\text{J sec}\right)^2\left(8.8542 \times 10^{-12}\text{ F m}^{-1}\right)}{\left(3.141\right)\left(9.1095 \times 10^{-31}\text{ kg}\right)\left(1.602 \times 10^{-19}\text{ Coulomb}\right)^2} = 5.2918 \times 10^{-11}\text{m}$$

APPENDIX F.3 BOHR'S THEORY OF THE HYDROGEN ATOM - ENERGY OF ORBITALS.

An electron in one of the stable orbits predicted by Bohr would have a total energy equal to the sum of its potential and kinetic energy. For the n-th orbit, the radius is r_n and the velocity of the electron is v_n. The potential energy, from Coulombic attraction, is:

$$\text{Potential Energy} = \frac{-e^2}{4\pi\varepsilon_0 r_n}$$

The kinetic energy can be determined from the previous equation

$$\frac{e^2}{4\,\pi\varepsilon_0 r^2} = \frac{m\,v^2}{r}$$

which rearranges to:

$$\text{Kinetic Energy} = \frac{1}{2}\,m\,(v_n)^2 = \frac{e^2}{8\pi\varepsilon_0 r_n}$$

The total energy is given by:

$$\text{Total Energy} = \frac{e^2}{8\pi\varepsilon_0 r_n} - \frac{e^2}{4\pi\varepsilon_0 r_n} = \frac{-e^2}{8\pi\varepsilon_0 r_n}$$

Substituting the previously determined expression for r,

$$r_n = \frac{n^2 h^2 \varepsilon_o}{\pi m e^2}$$

yields an expression for the total energy of an electron in the n-th orbital:

$$E = \frac{-1}{n^2}\left(\frac{me^4}{8(\varepsilon_o)^2 h^2}\right) = \frac{-2.18 \times 10^{-18}\,\text{J}}{n^2}$$

Note that when an electron moves from an initial orbit of n_i to a final orbit of n_f, the change in energy of the electron is:

$$\Delta E = -2.18 \times 10^{-18}\,\text{J}\left(\frac{1}{(n_f)^2} - \frac{1}{(n_i)^2}\right)$$

This last equation is one form of the Rydberg equation. Note that the change in energy of the electron corresponds to the energy of a photon associated with that change. If $n_f > n_i$, ΔE is a positive value (the photon is absorbed). If $n_f < n_i$, ΔE is a negative value (the photon is emitted).

APPENDIX F.4 DE BROGLIE'S THEORY OF ELECTRON WAVELENGTH

Bohr's quantization of angular momentum of an electron in a Bohr orbit is expressed by the equation:

$$mvr = \frac{nh}{2\pi}$$

Bohr was unable to explain why this should be, other than it resulted in the correct final answer. De Broglie explained this result by combining Einstein's general theory of relativity, $E = mc^2$, with Planck's quantum theory, $E = h\nu$:

$$E = mc^2 = h\,\nu$$

For an electron with velocity v approximating c, and a frequency $\nu = c/\lambda$, the prior equation can be written:

$$m\,v^2 = h\nu = \frac{hv}{\lambda};\ \lambda = \frac{h}{m\,v}$$

A particle with mass m and velocity v has a characteristic de Broglie wavelength λ. For λ to be measurable, given the small magnitude of Planck's constant h, the mass must be very small (e.g., that of an electron.)

In order for a particle in a circular orbit with radius r to exhibit constructive interference, the circumference of the circle must equal an integral number of wavelengths. This is expressed by the equation:

$$2\pi r = n\lambda$$

Solving for r and substituting the de Broglie wavelength yields:

$$r = \frac{n\lambda}{2\pi} = \frac{n\left(\frac{h}{m\,v}\right)}{2\pi} = \frac{nh}{2\pi m\,v}$$

Combining de Broglie's expression for the radius with the definition of angular momentum (mvr) justifies Bohr's original postulate:

$$mvr = \frac{nh}{2\pi}$$

APPENDIX G.1 PRESSURE AND HEIGHT OF LIQUID IN A BAROMETER

Pressure is a force per unit area. In English units, force is given in pounds, and area in square inches. This results in a pressure unit of pounds per square inch, or psi. Tire pressure gauges frequently use this unit. The metric alternate, ***Pascals***, uses force in Newtons (kilogram meters seconds^{-2}) and area in square meters. The combined units of the Pascal are thus:

$$\text{Pressure} = \frac{\text{force}}{\text{area}} = \frac{\text{kg m s}^{-2}}{\text{m}^2} = \frac{\text{kg}}{\text{m s}^2} = \text{Pascal}$$

A liquid has a density ρ with units of kilograms per cubic meter. If a column of length h is supported by pressure P, the relationship among these terms is given by the equation:

$$P = \rho\, g\, h = \left(\frac{\text{kg}}{\text{m}^3}\right)\left(\frac{\text{m}}{\text{s}^2}\right)(\text{m}) = \frac{\text{kg}}{\text{ms}^2} = \text{Pascals}$$

where g is the gravity constant, 9.806 meters seconds^{-2}.

One standard atmosphere of pressure is about 14.7 pounds per square inch. In terms of height of liquid, this corresponds to 760 mm of mercury (when the mercury is at 0°C.) The density of mercury at this temperature is about 13.596 grams per ml, which is 13596 kg m^{-3}. The pressure in Pascals corresponding to 760 mm of mercury is about:

$$P(1\text{ atm}) = \left(13596\,\frac{\text{kg}}{\text{m}^3}\right)\left(9.806\,\frac{\text{m}}{\text{s}^2}\right)(0.7600\text{ m}) = 101\bar{3}25\frac{\text{kg}}{\text{ms}^2}$$

APPENDIX G.2 DERIVATION OF BOYLE'S LAW FROM KINETIC MOLECULAR THEORY

In Chapter 15.4, a gas particle was placed in a box. Let the box have dimensions a, b and c in the x, y and z directions as shown below.

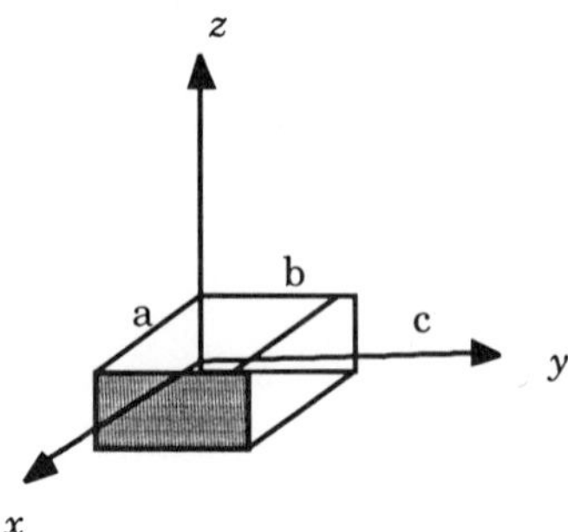

A particle with mass m was intially travelling with velocity v_x before colliding with wall B, shown as shaded in the diagram above. After collision with the wall, its velocity was $-v_x$ (since it was headed in the opposite direction.) Let Δt be the time interval between two collisions with wall B. Let the dimension of the box between sides A and B be represented by a. Then the distance the particle travels between two successive collisions with wall B is 2a. Since speed is distance travelled per time interval, we can write:

$$v_x = \frac{\text{distance traveled}}{\text{time interval}} = \frac{2a}{\Delta t}; \quad \text{or } \Delta t = \frac{2a}{v_x}$$

The force of the collision of the particle with side B is given by mass times acceleration; acceleration is the change in velocity with respect to time. Thus the force is related to momentum, the product of mass times velocity:

$$\text{Force} = m\,(\text{acceleration}) = m\left(\frac{dv_x}{dt}\right) = \frac{d(mv_x)}{dt} = \frac{\text{change in momentum}}{\text{change in time}}$$

Before collision the particle's momentum was mv_x. After collision the particle's momentum was $-mv_x$. Thus the change in momentum during collision was
$mv_x - (-\,mv_x) = 2\,mv_x$. Thus the force of the collision is given by:

$$\text{Force} = \frac{d(mv_x)}{dt} = \frac{2\,mv_x}{\Delta t}$$

Substituting in the earlier expression for dt yields:

$$F = \frac{2\,mv_x}{(2a/v_x)} = \frac{m\,(v_x)^2}{a}$$

This force is applied to wall B, which has dimensions of b x c. Since pressure is force per area, we can write:

$$P = \frac{\text{force}}{\text{area}} = \frac{m\left(v_x\right)^2/a}{b\,c} = \frac{m\left(v_x\right)^2}{a\,b\,c} = \frac{m\left(v_x\right)^2}{\text{Volume}}$$

We simplified the explanation by discussing the velocity only in terms of its vector component in the x direction, v_x. This can be clarified by recognizing that motion along each axis is on average equal, and is related to the overall velocity by:

$$v^2 = \left(v_x\right)^2 + \left(v_y\right)^2 + \left(v_z\right)^2; \text{ or } \left(v_x\right)^2 = \frac{1}{3}v^2$$

Substitution into the pressure equation yields:

$$P = \frac{m\left(v_x\right)^2}{\text{Volume}} = \frac{\left(\frac{1}{3}\right) m\left(v\right)^2}{\text{Volume}}$$

Finally, we note that the box contains not 1 molecule of gas, but n′ such molecules. It can be shown that the pressure of n′ particles is simply n′ times the pressure for one particle. Thus, the final relationship is:

$$P = \frac{\left(\frac{1}{3}\right) n' m\left(v\right)^2}{\text{Volume}}; \quad \text{or, } P\,V = \text{constant (Boyle's Law)}$$

Pressure times volume is constant provided no gas leaks (n′ constant) and that temperature, which determines the average velocity, is constant.

APPENDIX H

Answers to Selected End of Chapter Problems

1.1 (a) ^{18}O has 8 protons, 10 neutrons, 8 electrons
(b) ^{40}Ar has 18 protons, 22 neutrons, 18 electrons
(c) $^{29}Al^{3+}$ has 13 protons, 16 neutrons, 10 electrons
(d) $^{32}S^{2-}$ has 16 protons, 16 neutrons, 18 electrons.

		#protons	#neutrons	#electrons
1.4.	(a) ^{132}Xe	54	78	54
	(b) ^{52}Cr	24	28	24
	(c) $^{114}Cd^{2+}$	48	66	46
	(d) $^{127}I^-$	53	74	54

1.5 (a) ^{31}P
(b) $^{127}I^-$
(c) $^{56}Fe^{3+}$
(d) $^{240}U^{2+}$

1.9 (a) $MgCl_2$ is ionic (metal and non-metal)
(b) P_4O_{10} is covalent (non-metal and non-metal)
(c) SF_6 is covalent (non-metal and non-metal)
(d) $CsNO_3$ is ionic (metal and non-metal)

1.10. (a) N_2O_5 covalent, since both N and O are non-metallic;
(b) $NiCl_2$ ionic, since Ni is metallic and Cl is non-metallic;
(c) XeF_4 covalent, since both Xe and F are non-metallic;
(d) $Sr(IO_3)_2$ ionic, since Sr is metallic, and I,O are non-metallic.

1.13 (a) Cs < Cu < C
(b) Fr < P < F
(c) Ni < N < Ne

1.15. Ranking in terms of increasing electronegativity:
(a) Be < C < O
(b) Ba < Au < Pb
(c) Po < Te < Se

1.16 (a) oxygen: $1s^22s^22p^4$
(b) aluminum: $1s^22s^22p^63s^23p^1$
(c) calcium: $1s^22s^22p^63s^23p^64s^2$

1.17 (a) oxygen has 6 valence electrons
(b) aluminum has 3 valence electrons
(c) calcium has 2 valence electrons.

1.20. (a) Cl: [Ne] $3s^2$ $3p^5$;
(b) Ca: [Ar] $4s^2$;
(c) Sb: [Kr] $5s^2$ $4d^{10}$ $5p^3$.

1.30. (a) :Ö:S::Ö: or :Ö—S=Ö

$N = 3(8) = 24$
$A = 3(6) = 18$
$S = 6$

(b) H:Äs:H (with H below) or H—Äs—H (with H below)

$N = 8 + 3(2) = 14$
$A = 5 + 3(1) = 8$
$S = 6$

(c) :F̤:S̤:F̤: or :F—S—F:

$N = 3(8) = 24$
$A = 6 + 2(7) = 20$
$S = 4$

(d) H₂C::CH₂ or H₂C=CH₂

$N = 2(8) + 4(2) = 24$
$A = 2(4) + 4(1) = 12$
$S = 12$

2.5 35.1% by mass methanol; 64.9% by mass ethyl bromide

2.6 47.3% by volume methanol

2.7 4.57 m^3 of ore; 1.50×10^4 kg waste

2.10 (a) $SO_3 + H_2O = H_2SO_4$ (strong acid)

(b) $SO_2 + H_2O = H_2SO_3$ (weak acid)

(c) $Cl_2O_7 + H_2O = 2\ HClO_4$ (strong acid)

(d) $P_4O_{10} + 6\ H_2O = 4\ H_3PO_4$ (weak acid)

2.17 (a) lead acetate
(b) cobalt (II) carbonate
(c) ammonium phosphate
(d) ferric hydroxide, or iron (III) hydroxide
(e) nickel (II) chloride

2.18 (a) soluble, since all acetates are soluble
(b) insoluble, since carbonates insoluble
(c) soluble (ammonium solubility has precedence over phosphate insolubility)
(d) insoluble, since hydroxides are insoluble
(e) soluble, since chlorides are soluble

3.2 aldehyde, aromatic ring

3.4 alcohol

3.6 amine, aromatic ring

3.10 2 alcohols, ester, amine, aromatic ring

3.14 (a) IHD = 1; $CH_3CH_2CH{=}CH_2$ (alkene)
(b) IHD = 1; $CH_3CH_2CH_2CHO$ (aldehyde)
(c) IHD = 0; $CH_3CH_2CH_2CH_2CH_2Cl$ (aklyl chloride)
(d) IHD = 4; $C_6H_5\,CH_3$ (toluene, an aromatic compound)

3.17 (d) IHD = 5; carbonyl at 1750 and broad OH from 3500 to 2500 cm^{-1} indicates carboxylic acid. High value of IHD suggests benzene ring. Compound is likely benzoic acid, $C_6H_5\,COOH$.

3.18 (a) no, it could not be a carboxylic acid; that would require an IHD of at least 1.
(b) no, it could not be an ether, since there is clearly an OH stretch.
(c) yes, it could be an alcohol.
(d) no, it could not be an aldehyde; that would require an IHD of at least 1.

3.25 (a) 603 g of ethyl ether
(b) 9367 g of water

4.4 (a) 4 carbon environments
(b) 5 carbon environments
(c) 7 carbon environments

4.14 (a) 2 proton environments
(b) 2 proton environments
(c) 5 proton environments

4.22 (a) 3-methyl-1-pentanol
(c) 2-methyl-4-heptanal
(e) pentanoic acid
(g) pentamide
(i) 2-chlorobutanal

4.24 (a) 1-isopropyl-4-methylbenzene
(c) 3-heptanone

4.25 (a) IHD = 1;
(b) aldehyde or ketone

4.26 $(CH_3CH_2)_2C{=}O$; 3-pentanone

5.1 (a) $2\,Li(s) + Br_2(l) \rightarrow 2\,LiBr(s)$
(b) $CH_3CH{=}CHCH_3 + H_2 \rightarrow CH_3CH_2CH_2CH_3$
(c) $MgO(s) + CO_2(g) \rightarrow MgCO_3(s)$
(d) $C_6H_{10} + Br_2 \rightarrow C_6H_{10}Br_2$ (1,2-dibromocyclohexane)

5.2 (a) $BaCO_3(s) \rightarrow BaO(s) + CO_2(g)$
(c) $Ba(OH)_2(s) \rightarrow BaO(s) + H_2O(g)$
(e) $KCl(l)$ + electricity $\rightarrow K(l) + 0.5\,Cl_2(g)$

5.3 (a) $3\,BaCl_2(aq) + 2K_3PO_4(aq) \rightarrow Ba_3(PO_4)_2(s) + 6\,KCl(aq)$
$3\,Ba^{2+}(aq) + 2\,PO_4^{3-}(aq) \rightarrow Ba_3(PO_4)_2(s)$

(d) $CH_3CH_2CBr(CH_3)_2 + H_2O \rightarrow CH_3CH_2C(OH)(CH_3)_2 + HBr$

(f) $Ba(OH)_2(s) + H_2SO_4(aq) \rightarrow BaSO_4(s) + 2\ H_2O(l)$
$Ba(OH)_2(s) + H^+(aq) + HSO_4^-(aq) \rightarrow BaSO_4(s) + 2\ H_2O(l)$

or

$Ba(OH)_2(s) + 2H^+(aq) + SO_4^{2-}(aq) \rightarrow BaSO_4(s) + 2\ H_2O(l)$

(h) $FePO_4(s) + 3\ HCl(aq) \rightarrow FeCl_3(aq) + H_3PO_4(aq)$
$FePO_4(s) + 3\ H^+(aq) \rightarrow Fe^{3+}(aq) + H_3PO_4(aq)$

5.5 (a) 1.259 moles Na
(b) $2.750\text{x}10^{-5}$ moles U

5.6 (a) $7.582\text{x}10^{23}$ atoms of Na
(b) $1.656\text{x}10^{19}$ atoms of U

5.7 (a) 352.07 g/mole
(b) 342.30 g/mole

5.9 Empirical formula CH_2; Molecular Formula C_4H_8

5.12 (a) shift right
(b) shift right
(c) shift left
(d) shift right
(e) shift left
(d) shift right

5.15 *n*-pentane has 3 carbon environments; 2-methylbutane has 4 carbon environments; 2,2-dimethylpropane has 2 carbon environments.

5.16 major product = 2-bromobutane

5.18 (a) IHD = 0
(b) alcohol

5.20 (a) IHD = 2
(b) alkene

6.1 (a) Na = +1; Br = +5; O = -2;
(c) S = +4; O = -2;
(e) K = +1; Cr = +6; O = -2;
(h) I = - 1/3

6.2 (a) pentanoic acid
(c) cyclohexanol
(e) 2,3-dihydroxy-4-methylpentane

6.4 $C_5H_{12}O$

6.5 (b) $FeCO_3(s) + 2\ HCl(aq) \rightarrow FeCl_2(aq) + H_2O(l) + CO_2(g)$
$FeCO_3(s) + 2\ H^+(aq) \rightarrow Fe^{2+}(aq) + H_2O(l) + CO_2(g)$

(d) $C_8H_{18}(l) + 12.5\ O_2(g) \rightarrow 8\ CO_2(g) + 9\ H_2O(l)$

6.6 (b) $7(2e^- + S_2O_8^{2-} \rightarrow 2\ SO_4^{2-})$
$2(2\ H_2O + NH_3 \rightarrow NO_2 + 7\ H^+ + 7\ e^-)$
NET: $7\ S_2O_8^{2-} + 4\ H_2O + 2\ NH_3 \rightarrow 14\ SO_4^{2-} + 2\ NO_2 + 14\ H^+$

(d) $3(Cu \rightarrow Cu^{2+} + 2\ e^-)$
$2(3\ e^- + 3\ H^+ + HNO_3 \rightarrow 3\ Cu^{2+} + 2\ NO + 4\ H_2O)$
NET: $3\ Cu + 6\ H^+ + 2\ HNO_3 \rightarrow 3\ Cu^{2+} + 2\ NO + 4\ H_2O$

6.7 (a) $4(5\ e^- + 8\ H^+ + MnO_4^- \rightarrow Mn^{2+} + 4\ H_2O)$
$5(H_2O + CH_3CH_2CH_2OH \rightarrow CH_3CH_2COOH + 4\ H^+ + 4e^-)$
NET:
$12H^+ + 5CH_3CH_2CH_2OH + 4MnO_4^- \rightarrow 11H_2O + 4Mn^{2+} + 5CH_3CH_2COOH$

(d) $3H_2O + NaBH_4 \rightarrow H_3BO_3 + Na^+ + 7\ H^+ + 8\ e^-$
$4(2e^- + 2H^+ + CH_3CH_2COCH_3 \rightarrow CH_3CH_2CHOHCH_3)$
NET:
$H^+ + 3H_2O + NaBH_4 + 4CH_3CH_2COCH_3 \rightarrow 4CH_3CH_2CHOHCH_3 + H_3BO_3 + Na^+$

6.10 4.91×10^6 g of SO_2 produced; 9.21×10^6g $CaSO_3$ produced.

6.13 33.8 g C_9H_{12} formed

6.14 76.6% yield

6.16 (a) C=O stretch of the aldehyde
(b) O-H stretch of an alcohol
(c) reduced
(d) reducing agent

6.19 (a) oxidized
(b) ketone
(c) a carbonyl carbon
(d) $CH_3CH_2C{=}OCH_3$

6.24 (a) C_7H_8O
(b) C_7H_8O

6.25 (a) $C_6H_5CH_2OH$ (benzyl alcohol)
(b) 3500-3200 cm^{-1} is O-H stretch
3150-3000 cm^{-1} is aromatic C-H stretch
3000-2800 cm^{-1} is aliphatic C-H stretch
(c) singlet at 7.4 due to aromatic ring protons
singlet at 4.7 due to CH_2 protons
singlet at 2.5 due to OH proton

7.1 (a) N = 32; A = 26; S = 6;
(b) N = 16; A = 10; S = 6;

7.2 (a) tetrahedral
(b) linear

7.3 (a) trigonal pyramidal
(b) linear

7.7 Boron in BCl_3 has trigonal planar electronic geometry. This yields trigonal planar molecular geometry, and a symmetric arrangement of bonds. This causes a non-polar molecular. Phosphorous in PCl_3 has tetrahedral electronic geometry. This yields a trigonal pyramidal molecular geometry. This asymmetrical arrangement causes a polar molecule.

7.16 (a) 1,2-dibromoethane

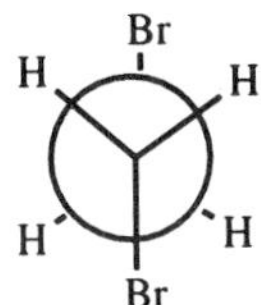

7.17 (a) 1,2-dibromoethane

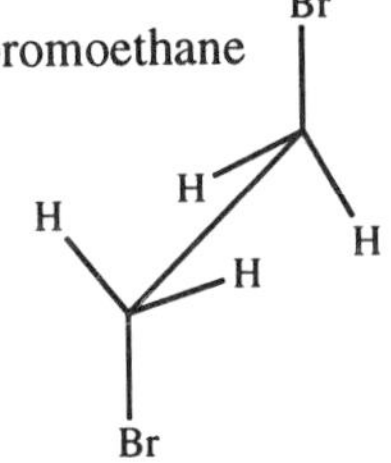

7.18 (a) $3\ Cl_2 + 6\ OH^- = 4\ Cl^- + 2\ ClO_3^- + 3\ H_2O$
(c) about 109°

7.23 (a) $N_2H_4 + HNO_2 = HN_3 + 2\ H_2O$
(b) N is in −1/3 oxidation state
(c) $H—\ddot{N}—N\equiv N:$ or $H—\ddot{N}=N=\ddot{N}:$

7.26 (a) $6\ HF + Sb_2O_3 = 2\ SbF_3 + 3\ H_2O$
(c) tetrahedral electronic structure around Sb means a trigonal pyramidal molecular geometry. Thus SbF_3 is expected to be polar and water soluble.

8.1 (a) F < Cl < Br < I
(c) $CO_2H < CO_2CH_3 < NH_2 < OH$

8.2 (a) *E*
(c) *Z*

8.3 (a) (*Z*) – 2 – hexene
(c) (*Z*) – 2-methyl-3-ethyl-2-hexene-1-ol
(e) (*Z*) – cyclodecene

8.7 (a) (*S*)
(c) (*S*)
(e) (*R*)

8.8 (a) (*S*)-3-bromohexane
(c) (1*R*, 2*R*) -2-methylcyclopropanol

8.10 (a) structural isomers
(c) geometric isomers

8.13 +14.5°

8.16 58.2% enantiomeric excess

8.18 -113°

8.21 (a) CH_2O
(b) $C_3H_6O_3$
(c) carboxylic acid
(d) $CH_3CH(OH)COOH$
C#1 is singlet; C#2 is doublet; C#3 is quartet
(e) yes; carbon #2 has 4 different substituents.

9.1 (a) ${}^{9}_{4}Be + {}^{1}_{1}H \rightarrow {}^{6}_{3}Li + {}^{4}_{2}He$

(b) ${}^{32}_{16}S + {}^{1}_{0}n \rightarrow {}^{1}_{1}H + {}^{32}_{15}P$

(c) ${}^{43}_{20}Ca + {}^{4}_{2}He \rightarrow {}^{46}_{21}Sc + {}^{1}_{1}H$

9.2 (b) ${}^{7}_{3}Li \left({}^{1}_{1}p,\ {}^{1}_{0}n\right) {}^{7}_{4}Be$

(c) ${}^{27}_{13}Al \left({}^{1}_{0}n,\ {}^{4}_{2}a\right) {}^{24}_{11}Na$

9.3 (a) 416 m
(b) 4.77 x 10-28 J

9.5 $5.08 \times 10^{14}\ s^{-1}$

9.8 277 nm;

9.9 $- 3.7729 \times 10^{-19}$ J (light given off)

9.10 (a) $E = 2.1193 \times 10^{-18}$ J
(b) wavelength = 93.72 nm
(c) frequency = $3.198 \times 10^{15}\ s^{-1}$

9.13 7.3×10^{-12} m
9.15 299 nm

9.18 (a) 5.882×10^{-6} m
(b) $5.097 \times 10^{13}\ s^{-1}$
(c) 3.377×10^{-20} J

9.20 (a) Reaction: $H_2 + Br_2 = 2\ HBr$
(b) 623.0 nm

10.5 Sb

10.7 IV B transition metals (Ti, Zr, Hf) are similar to C, Si;
All form oxides with formula MO_2; all form halides with formula MX_4.

10.8 N < B < Li < Na

10.10 As < Se < S < O < F

10.11 B^{3+}

10.14 Hg: [Xe] $6s^2\ 4f^{14}\ 5d^{10}$; Hg^{+1}: [Xe] $6s^1\ 4f^{14}\ 5d^{10}$;
note: like atomic H, Hg^{+1} has singly occupied s orbital which it can use to form a diatomic species.

10.16 Ag: [Kr] $5s^1\ 4d^{10}$; Ag^+: [Kr] $5s^0\ 4d^{10}$;
like the alkali metals, silver is easily oxidized to a stable +1 state.

10.20 (a) Ce: [Xe] $6s^2\ 5d^1\ 4f^1$;
(b) Ce^{3+}: [Xe] $6s^0\ 5d^0\ 4f^1$; thus Ce^{3+} is relatively stable.
(c) $2\ Ce^{4+} + CH_2O_2 \rightarrow CO_2 + 2\ H^+ + 2\ Ce^{3+}$
(d) $2Ce^{4+} + 2H_2O + C_4H_6O_2 \rightarrow CH_2O_2 + C_3H_6O_2 + 2H^+ + 2Ce^{3+}$

10.22 (a) covalent
(b) covalent
(c) VIA
(d) QO, QO_2, QO_3;
(e) 128 g/mole (i.e., Te)

11.2 (a) metallic
(b) ionic
(c) covalent

11.3 conducts as solid: (a)
conducts as liquid: (a), (b), (e)

11.4 BN has higher melting point

11.5 41.8 kJ consumed

11.7 (a) :N:::N: steric number = 2;

11.8 (a) sp hybrid orbitals

11.9 (a) linear molecular geometry

11.15 $CH_3C(=O)OCH(CH_3)_2$
carbonyl oxygen is sp^2 hybridized; C-O-C oxygen is sp^3 hybridized
4 unique carbon environments

11.17 (a) Na_2O
(b) ionic
(c) Na = +1; O = -2;

11.20 (a) SCl_2O_2
(b) covalent

11.23 (a) +5
(b) $N_2O_5 + H_2O \rightarrow 2\ HNO_3$
(d) sp^2

12.5 Any balanced reaction in which the species listed first donates a proton is an example of it acting as a Bronsted Lowry acid.

(a) $H_2SO_4 + H_2O \rightarrow HSO_4^- + H_3O^+$
(b) $NH_4^+ + H_2O \rightarrow NH_3 + H_3O^+$

(c) $HC_2H_3O_2 + H_2O \rightarrow C_2H_3O_2^- + H_3O^+$

(d) $OH^- + H_2O \rightarrow O^{2-} + H_3O^+$

12.6 Any balanced reaction in which the species listed first accepts a proton is an example of it acting as a Bronsted Lowry base.

(a) $HSO_4^- + H_2O \rightarrow H_2SO_4 + OH^-$

(b) $NH_3 + H_2O \rightarrow NH_4^+ + OH^-$

(c) $C_2H_3O_2^- + H_2O \rightarrow HC_2H_3O_2 + OH^-$

(d) $OH^- + HCl \rightarrow H_2O + Cl^-$

12.8

	Compound listed	Conjugate Base Formula	Name
(a)	$HClO_4$	ClO_4^-	perchlorate
(b)	HSO_4^-	SO_4^{2-}	sulfate
(c)	$HC_2H_3O_2$	$C_2H_3O_2^-$	acetate
(d)	NH_3	NH_2^-	amide

12.9

	Conjugate Acid / Base	Conjugate Base / Acid
(a)	HNO_3 / NO_3^-	H_2O / H_3O^+
(b)	HNO_3 / NO_3^-	NH_3 / NH_4^+
(c)	H_3O^+ / H_2O	$C_2H_5O_2^-$ / $HC_2H_3O_2$
(d)	H_3O^+ / H_2O	NH_2^- / NH_3

12.10 (a) $H_2Te > H_2Se > H_2S > H_2O$
(b) $HClO_4 > HClO_3 > HClO_2 > HClO$
(c) $HBr > H_2S > NH_3$
(d) $CF_3COOH > CHF_2COOH > CH_2FCOOH > CH_3COOH$

12.11 (a) Lewis base
(b) Lewis acid
(c) Fe^{3+} Lewis acid (empty 4s, 3d orbitals)
(d) Lewis base

12.12

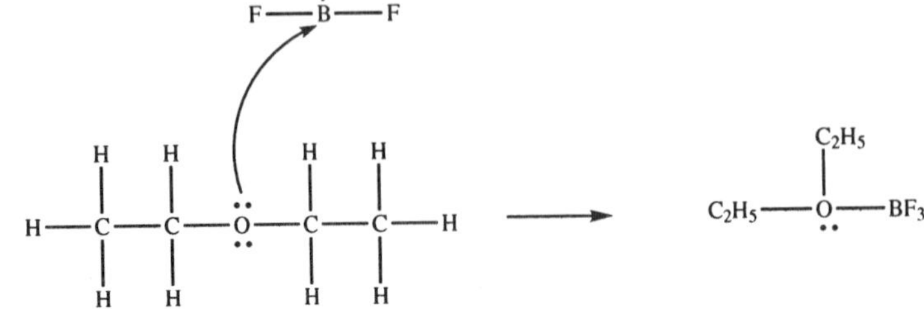

12.14 (a) $Cl_2O_7 + H_2O \rightarrow 2\ HClO_4$ (a strong acid)
(b) 0.656 M H^+;
(c) since $[HClO] \rightarrow 0.656M$, $[H^+] < 0.656$ M;

12.16 (a) $(CH_3CO)_2O + H_2O \rightarrow 2\ CH_3COOH$
(b) 1.96 M
(c) 7.84 g

12.22 (b) 4 proton environments
(c) 4 carbon environments
(d) $H_3C_6H_5O_7 + 3\ NaHCO_3 \rightarrow 3\ H_2O + 3\ CO_2 + 3\ Na^+ + C_6H_5O_7^{3-}$
(e) 3.27 g

13.1 (a)  :O::S:O: ⟷ :O:S::O:

13.2 H—O—C≡N: ⟷ H—O—C=N: ⟷ H—O=C=N

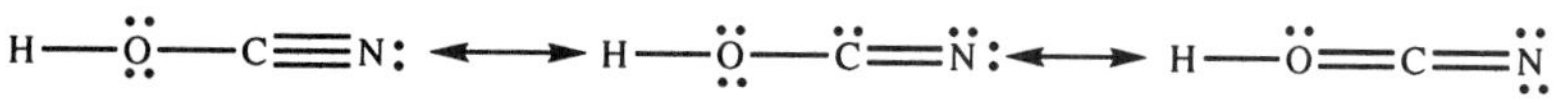

C = 0	C = +1	C = 0
H = 0	H = 0	H = 0
O = 0	O = 0	O = +1
N = 0	N = −1	N = −1

13.3 (a) 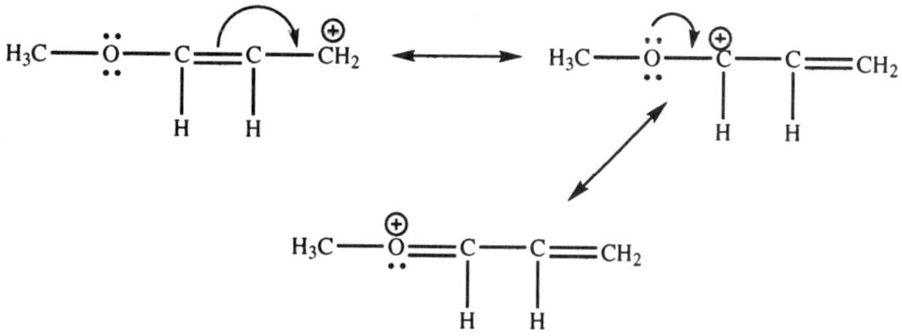 H_3C—O—CH=CH—$CH_2^{\oplus}$ ⟷ H_3C—O—$CH^{\oplus}$—CH=CH_2 ⟷ H_3C—$O^{\oplus}$=CH—CH=CH_2

(b)

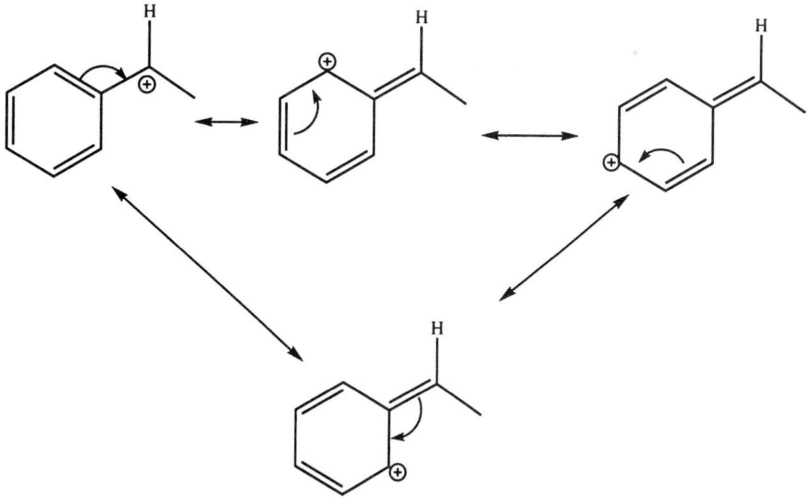

(c)

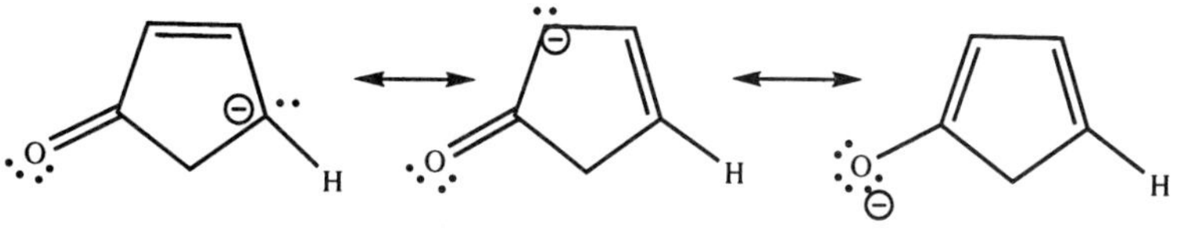

13.4 (a) isopropylbenzene
(c) *m*-ethylbenzene (3-ethyltoluene)

13.5 (a) CH3 / OH (e) Cl / Cl

13.9 (a) 1,4-dibromobenzene yields only 1,4-dibromo-2-nitrobenzene

13.10 (a) 4-methylbenzoic acid;
(b) 5 unique carbon environments
(c) the singlet at 9.8 ppm is the carboxylic acid proton;
the peaks at 7.0 and 7.8 are the 4 ring protons;
the singlet at 3.9 are the methyl protons.

13.12 (a) C_4H_5
(b) C_8H_{10}
(c) 1,4-dimethylbenzene (*p*-xylene)
(d) strong IR absorbance at 800 cm^{-1} shows para substitution
(e) 6 methyl protons are a singlet at 2.3 ppm
4 ring protons resonate at 7.0 ppm.

14.1 (a) 1-ethyl-2-nitrobenzene; 1-ethyl-4-nitrobenzene

14.2 (a)

+ CH_3Cl $\xrightarrow{AlCl_3}$ CH3 + HCl

CH3 + Cl_2 $\xrightarrow{AlCl_3}$ CH3 / Cl + HCl

(plus para isomer)

14.5 ortho: 40%; meta: 40%; para: 20%

14.7 (a) O / C—N / H / Br (or ortho isomer)

14.9 A is 1,4-diethylbenzene; 4 carbon environments

14.13

:C≡N:

Br / N=O: / :O: / NO2 → Br / CN / N—O: / :O: / NO2 → CN / N=O: / :O: / NO2

14.16 (a) $NO_2^- + 2\,H^+ = H_2O + NO^+$

(b)

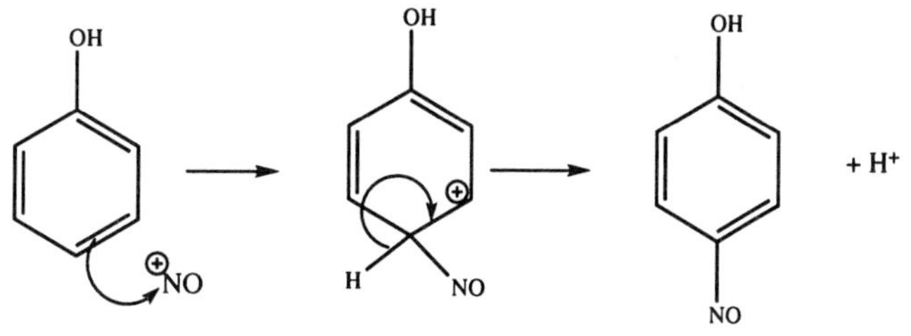

OH → OH → OH + H^+

NO H NO NO

plus other resonance forms

14.20 (a) $18\,H^+ + 6\,MnO_4^- + 5\,C_7H_8 \rightarrow 14\,H_2O + 5\,C_7H_6O_2 + 6\,Mn^{2+}$

(b) $C_6H_5COOH + HOCH_2CH_3 \rightarrow C_6H_5COOCH_2CH_3 + H_2O$

(c) 6.14 g toluene; 12.63 g $KMnO_4$

15.3 1.00 atm; 762 torr; 14.7 atm;

15.7 49.9 psi

15.9 0.823 atm

15.12 77.9 g Cu

15.14 79.7 g NaN_3

15.16 (a) $CaC_2 + 2\,H_2O \rightarrow Ca(OH)_2 + C_2H_2$
(b) H:C:::C:H; C is sp hybridized
(c) 0.623 moles
(d) 39.9%

16.1 (a) Ar; larger atom has greater polarizability
(b) ClF; polar molecule allows dipole-dipole interaction

16.2 (a) NaCl; smaller ion has greater coulombic attraction
(b) MgO; more highly charged ions have greater attraction

16.4 CO_2 is a symmetric molecule, CO is not. CO_2 has a zero dipole moment, CO does not. CO_2 would have a lower surface tension than CO.

16.7 trans-1,2-dichloroethene would have a lower dipole moment than cis-1,2-dichloroethene. The trans isomer would be more volatile.

16.12 volume = 7.92×10^{-14} cm^3

16.13 0.964 g/cm^3

16.16 radius = 1.44×10^{-8} cm

16.21 (a) $C_4H_{10}O$
(b) IHD = 0
(c) A = $CH_3CH_2OCH_2CH_3$
B = $CH_3CH_2CH_2CH_2$

INDEX

IA	IIA	IIIB	IVB	VB	VIB	VIIB	VIIIB			IB	IIB	IIIA	IVA	VA	VIA	VIIA	VIIIA
1 H 1.00794																	2 He 4.00260
3 Li 6.941	4 Be 9.01218											5 B 10.81	6 C 12.011	7 N 14.0067	8 O 15.9994	9 F 18.998403	10 Ne 20.1797
11 Na 22.98977	12 Mg 24.305											13 Al 26.98154	14 Si 28.0855	15 P 30.97376	16 S 32.066	17 Cl 35.453	18 Ar 39.948
19 K 39.0983	20 Ca 40.078	21 Sc 44.9559	22 Ti 47.88	23 V 50.9415	24 Cr 51.996	25 Mn 54.9380	26 Fe 55.847	27 Co 58.9332	28 Ni 58.69	29 Cu 63.546	30 Zn 65.39	31 Ga 69.72	32 Ge 72.61	33 As 74.9216	34 Se 78.96	35 Br 79.904	36 Kr 83.80
37 Rb 85.4678	38 Sr 87.62	39 Y 88.9059	40 Zr 91.224	41 Nb 92.9064	42 Mo 95.94	43 Tc (98)	44 Ru 101.07	45 Rh 102.9055	46 Pd 106.42	47 Ag 107.8682	48 Cd 112.41	49 In 114.82	50 Sn 118.710	51 Sb 121.757	52 Te 127.60	53 I 126.9045	54 Xe 131.29
55 Cs 132.9054	56 Ba 137.33	57 La 138.9055	72 Hf 178.49	73 Ta 180.9479	74 W 183.85	75 Re 186.207	76 Os 190.2	77 Ir 192.22	78 Pt 195.08	79 Au 196.9665	80 Hg 200.59	81 Tl 204.383	82 Pb 207.2	83 Bi 208.9804	84 Po (209)	85 At (210)	86 Rn (222)
87 Fr (223)	88 Ra 226.0254	89 Ac 227.0278	104 Rf (261)	105 Db (262)	[106] Sg (263)	[107] Bh (262)	[108] Hs (265)	[109] Mt (266)									

58 Ce 140.12	59 Pr 140.9077	60 Nd 144.24	61 Pm (145)	62 Sm 150.36	63 Eu 151.96	64 Gd 157.25	65 Tb 158.9254	66 Dy 162.50	67 Ho 164.9304	68 Er 167.26	69 Tm 168.9342	70 Yb 173.04	71 Lu 174.967
90 Th 232.0381	91 Pa 231.0359	92 U 238.0289	93 Np 237.048	94 Pu (244)	95 Am (243)	96 Cm (247)	97 Bk (247)	98 Cf (251)	99 Es (252)	100 Fm (257)	101 Md (258)	102 No (259)	103 Lr (260)